T0328950

Sustainable food production and ethics

Sustainable food production and ethics

Preprints of the 7th Congress of the European Society for Agricultural and Food Ethics

EurSAFE 2007
Vienna, Austria
September 13 - 15, 2007

edited by:
Werner Zollitsch
Christoph Winckler
Susanne Waiblinger
Alexander Haslberger

Wageningen Academic
Publishers

All rights reserved.
Nothing from this publication may be
reproduced, stored in a computerised system
or published in any form or in any manner,
including electronic, mechanical, reprographic
or photographic, without prior written
permission from the publisher, Wageningen
Academic Publishers, P.O. Box 220, NL-6700
AE Wageningen, The Netherlands.
www.WageningenAcademic.com

The individual contributions in this
publication and any liabilities arising from
them remain the responsibility of the authors.

The publisher is not responsible for possible
damages, which could be a result of content
derived from this publication.

ISBN 978-90-8686-046-3

First published, 2007

Wageningen Academic Publishers
The Netherlands, 2007

EurSafe 2007 Committees

Organizing Committee
Christoph Winckler (Chair)
Michaela Bürtlmair
Alexander Haslberger
Knut Niebuhr
Susanne Waiblinger
Werner Zollitsch

Scientific Committee
Werner Zollitsch (Chair, Austria)
Frans Brom (The Netherlands)
Alexander Haslberger (Austria)
Matthias Kaiser (Norway)
Jos Metz (The Netherlands)
Susanne Padel (United Kingdom)
Peter Sandoe (Denmark)
Susanne Waiblinger (Austria)
Christoph Winckler (Austria)

The EurSafe 2007 conference is organized by:

- Department of Sustainable Agricultural Systems,
  University of Natural Resources and Applied Life Sciences (BOKU) Vienna

- Institute for Animal Husbandry and Animal Welfare,
  University of Veterinary Medicine Vienna

- Faculty of Life Sciences, University of Vienna

Faculty of Life Sciences, University of Vienna

In co-operation with:

- Austrian Scientists for Environmental Protection

Austrian Scientists for Environmental Protection

The organizers gratefully acknowledge financial support by:
- The Austrian Federal Ministry for Health, Family and Youth
- The Austrian Federal Ministry for Science and Research
- The City of Vienna

Wageningen Academic Publishers

Table of contents

Preface: Sustainable food production and ethics   17
Werner Zollitsch and Christoph Winckler

Keynote papers

On sustainability, dogmas, and (new) historical roots for environmental ethics   21
Ben A. Minteer

How do the ethical values of organic agriculture relate to standards and to current practice?   26
Susanne Padel

Animal welfare and intensive animal production: are they compatible?   31
David Fraser

Animal welfare in intensive and sustainable animal production systems   37
Vonne Lund

Ethics and action: a relational perspective on food trends and consumer concerns   43
Unni Kjærnes

Coexistence? What kind of agriculture do we want?   49
Louise W.M. Luttikholt

Coexistence and ethics: NIMBY-arguments reconsidered   53
Matthias Kaiser

Vertical gene flow in the context of risk/safety assessment and co-existence   57
A. De Schrijver, Y. Devos and M. Sneyers

Part 1 - Theoretical, conceptual and foundational issues: concepts and approaches

Ethical bases of sustainability   63
Paul B. Thompson

Building a sustainable future for animal agriculture: an environmental virtue ethic of care
approach within the philosophy of technology   69
Raymond Anthony

Emergence and auto-organisation: revising our concepts of growth, development and
evolution toward a science of sustainability   75
Sylvie Pouteau

Values behind biodiversity: ends in themselves or knowledge-based attitudes   81
Arne Sveinson Haugen

Part 2 - Theoretical, conceptual and foundational issues: assessment and models

A structuring pathway to tackling ethical problems   89
Michael Zichy

Standing on the shoulders of a giant: the promise of multi-criteria mapping as a decision-
support framework in food ethics   95
Volkert Beekman, Erik de Bakker and Ronald de Graaff

Sustainability concept in agricultural scientific papers   101
Matias Pasquali

Part 3 - Theoretical, conceptual and foundational issues: bringing ethics into practice

Practice-oriented ethics   109
S. Aerts and D. Lips

How do stakes and interests shape the discursive strategies for framing (multi-)causality?   115
Laura Maxim and Jeroen P. van der Sluijs

Depoliticizing technological decisions?   121
Bernice Bovenkerk

Implicit normativity in scientific advice - a case study of nutrition advice to the general public   127
Anna Paldam Folker, Hanne Andersen and Peter Sandøe

Trustworthiness: the concrete task to take vague moral ideals seriously   132
Franck L.B. Meijboom

Part 4 - Diversity, resilience, global trade

Using past climate variability to understand how food systems are resilient to future climate
change   139
Evan D.G. Fraser, Mette Termansen, Ning Sun, Dabo Guan, Kuishang Feng and Yang Yu

Integration of resilience into sustainability model for analysis of adaptive capacities of regions
to climate change: the EASEY model   145
R. Paulesich, K. Bohländer and A.G. Haslberger

Sustainability, corporate social responsibility and food markets: the role of cooperatives   151
Valeria Sodano

A method for construction and evaluation of scenarios for sustainable animal production,
with an application on future Swedish dairy farming   157
Stefan Gunnarsson and Ulf Sonesson

Protecting local diversity in scenarios of modern food biotechnology, globalised trade and intellectual property rights 163
A.G. Haslberger, A.H. Gesche, M. Proyer, R. Paulesich and S. Gressler

Does free trade in agriculture promote 'one planet farming'? 169
Tom MacMillan and Neva Frecheville

Constructing sustainable regional food networks: a grounded perspective 174
Dirk Roep and Johannes S.C. Wiskerke

Ethical use of Andean tomato germplasm 180
Bart Gremmen

A legal examination of the requirements relating to developing countries' exports of organic products to the European Union 184
Morten Broberg

Part 5 - Sustainability and animal welfare: animal welfare and basic value

Vegan agriculture: animal-friendly and sustainable 193
Tatjana Visak

Understanding farmers' values 198
Carolien de Lauwere, Sabine de Rooij and Jan Douwe van der Ploeg

Consumers versus producers: a different view on farm animal welfare? 204
Filiep Vanhonacker, Els Van Poucke, Frank Tuyttens and Wim Verbeke

Market segmentation based on perceived importance and evaluation of farm animal welfare 210
Filiep Vanhonacker, Els Van Poucke, Frank Tuyttens and Wim Verbeke

Part 6 - Sustainability and animal welfare: different practices

Stacking pigs: Dutch pig tower debates and the changing nature of ethical livestock production 219
Clemens Driessen

Towards sustainable livestock production systems 223
John J. McGlone and Mhairi Sutherland

Economic, ecological and societal performance of organic versus conventional egg production in the Netherlands 229
Eddie A.M. Bokkers, Sophia van der Ploeg, Erwin Mollenhorst and Imke J.M. de Boer

Considering the farmer-animal relationship in the development of sustainable husbandry systems for cattle production 233
X. Boivin, S. Waiblinger, A. Brulé, N. L'hotellier, F. Phocas and G. Coleman

From the backyard, through the farm, to the laboratory - changes in human attitudes to the pig 239
Reinhard Huber, I. Anna S. Olsson, Mickey Gjerris and Peter Sandøe

Logistics at transport to slaughter: food and environment-optimised animal transport 244
Sofia Wiberg, Anne Algers, Bo Algers, Ulrika Franzén, Magnus Lindencrona, Olof Moen, Sofia Ohnell and Jonas Waidringer

The ethics of automatic milking systems and grazing in dairy cattle 249
Leonie F.M. Heutinck and Clemens Driessen

Part 7 - Sustainability and animal welfare: implementation and legislation

Conflicting areas in the ethical debate on animal health and welfare 257
Albert Sundrum

The concept of sustainable agriculture necessitates a duty of stewardship in contemporary animal welfare legislation 263
Ian A. Robertson

The Australian Animal Welfare Strategy: sustaining food production drivers within broader societal agendas 269
H.R. Yeatman

Sheep welfare in the welfare state: ethical aspects of the conventionalisation of Norwegian organic production 273
Marianne Kulø and Lill M. Vramo

Animal welfare in assurance schemes: benchmarking for progress 279
S. Aerts

Best Practice in animal production: a grassroots evaluation framework 285
S. Aerts and D. Lips

Part 8 - Sustainability and animal welfare: workshop 'Science and public expectations'

Can animal science meet the expectations in the 'animal welfare' debate? 293
Karel (K.H.) de Greef and Bram (A.P.) Bos

Empirical facts in farm-animal welfare discourses 300
Herwig Grimm

Part 9 - Ethics of organic farming

On the ethical dimension of organic agriculture 309
Tatjana Kochetkova

Organic farming: for the sake of nature?   316
Karsten Klint Jensen

Mutilations in organic animal husbandry: ethical dilemmas?   321
Susanne Waiblinger, Christoph Menke and Knut Niebuhr

The development of organic aquaculture as a sustainable food production system: exploring
issues and challenges   328
S. Tomkins and K.M. Millar

Part 10 - Sustainable disease control

Keeping backyard animals as a way of life   337
Nina E. Cohen, Elsbeth N. Stassen and Frans W.A. Brom

Animal disease policy as a moral question with respect to risks of harm   342
Franck L.B. Meijboom, Nina Cohen and Frans W.A. Brom

Avian influenza and media coverage: a qualitative study in four European countries   347
Asterios Tsioumanis

Using participatory methods to explore the social and ethical issues raised by bioscience
research programmes: the case of animal genomics research   354
K.M. Millar, C. Gamborg and P. Sandøe

System approach to improve animal health   360
Albert Sundrum, Klaas Dietze and Christina Werner

Will nano-enabled diagnostics make animal disease control more sustainable?   365
Johan Evers, Stef Aerts and Johan De Tavernier

Is it possible to make risk-reduction strategies socially sustainable?   371
Sara Korzen-Bohr and Jesper Lassen

New understanding of epigenetics and consequences for environmental health and sustainability 376
Veronika Sagl, Roman Thaler, Astrid H. Gesche and Alexander G. Haslberger

Local knowledge and ethnoveterinary medicine of farmers in Eastern Tyrol about wild plant
species: a potential basis for disease control according to EC Council Reg. 2092/91 in organic
farming   382
Brigitte Vogl-Lukasser, Christian R. Vogl, Susanne Grasser and Martina Bizaj

Part 11 - Sustainable food production and consumption

Sustainable value chain analysis   389
David Simons

Sustainable green marketing of SMEs' ecoproducts   395
Maarit Pallari

The role of women in promoting ethical consumerism: an historical perspective   401
Laura Terragni

Waste not want not - the ethics of food waste   407
André Gazsó and Sabine Greßler

Ruminant feeding in sustainable animal agriculture   412
Wilhelm Knaus

What is 'regional food'? The process of developing criteria for regional food in Eastern Tyrol,
Austria   416
Julia Kaliwoda, Heidrun Leitner and Christian Vogl

Part 12 - GMO: GM plants

Contested GMO's: how questions of global justice and basic structures matter in the debate
on GM plants   425
Kristian Høyer Toft

Benefits at risk: does the motivation of farmers counteract potential benefits of GM plants?   431
Jesper Lassen and Peter Sandøe

Typology of ethical judgements on transgenic plants   436
Catherine Baudoin

Marketing GM Roundup Ready rapeseed in Norway? Report from a value workshop   442
Ellen-Marie Forsberg

GM-food: how to communicate with the public?   450
Ursula Hunger, Brigitte Gschmeidler and Elisabeth Waigmann

Part 13 - GMO: transgenic animals

Ethical concerns related to cloning of animals for agricultural purposes   455
Mickey Gjerris and Peter Sandøe

Animal welfare aspects of creating transgenic farm animals   461
Kristin Hagen

Cloning for meat or medicine?   466
C. Gamborg, M. Gjerris, J. Gunning, M. Hartlev and P. Sandøe

Assessing people's attitudes towards animal use and genetic modification using a web-based
interactive survey   472
Catherine A. Schuppli and Daniel M. Weary

Part 14 - Nature conservation and ethics

Environmental values and their impact on sustainability and nature conservation 477
Diana Ehrenwerth and Alexander G. Haslberger

Aspects of ethics and animal protection in animal husbandry under semi-natural conditions,
in animals returned to the wild, and in wild animals 483
K.M. Scheibe

The ethics of catch and release in angling: human recreation, resource conservation and
animal welfare 489
Cecilie M. Mejdell and Vonne Lund

Part 15 - Other contributions

Why farmers need professional autonomy 495
F.R. Stafleu and F.L.B. Meijboom

Reflections from cultural history: the story of the Bohemian landscape in Romania - a
sustainable past? 500
Pavel Klvac and Zbynek Ulcak

Landscape, land use and soul: ecopsychology: mending a troubled relationship 506
Diana Voigt, Thomas Lindenthal and Andreas Spornberger

Organic farmers' experiments and innovations: a debate 512
Susanne Kummer, Racheli Ninio, Friedrich Leitgeb and Christian R. Vogl

Detecting pathogens using real-time PCR: a contribution to monitoring animal health and
food safety 515
*Sophie Kronsteiner, Konrad J. Domig, Silvia Pfalz, Philipp Nagel, Werner Zollitsch and
Wolfgang Kneifel*

Analysis of diet-induced changes in intestinal microbiota: microbial balance as indicator for
gut health 518
*Silvia Pfalz, Konrad J. Domig, Sophie Kronsteiner, Philipp Nagel, Werner Zollitsch and
Wolfgang Kneifel*

Analysis of food associated bacterial diversity with molecular methods comparing lettuce
from organic and conventional agriculture and cheese from pasteurized *versus* unpasteurized
milk 521
Jutta Zwielehner, Michael Handschur, Norbert Zeichen, Selen Irez and Alexander G. Haslberger

Non-physical interaction between humans and plants and its impact on plant development 526
F. Leitgeb, C. Arvay, K. Dolschak, A. Spornberger and K. Jezik

A tailor-made molecular biological strategy for monitoring the gut microbiota of pigs
reflecting animal health   532
*Agnes Petersson, Konrad J. Domig, Elisabeth Moser, Karl Schedle, Wilhelm Windisch and
Wolfgang Kneifel*

Evaluating novel protein sources for organic laying hens   537
Heleen A. Van de Weerd and Sue H. Gordon

Authors index   543

Keyword index   547

Preface: Sustainable food production and ethics

The concept of sustainability has come to occupy a prominent place in discussions of the future of food production. At the same time, consumers show increasing interest in foods produced in a sustainable way. However, the concept of sustainability has come to encompass a growing number of concerns and this has given rise to conceptual ambiguity and dilemmas.

In its earliest usage, the notion of sustainability was connected with strategies to maintain renewable resources for harvest and consumption in perpetuity, i.e. sustained yield. The term sustainability was further extended, beyond an anthropocentric concern for human livelihood, to cover the preservation of species and ecosystems. The integration of fair intra-generational distribution and benefit sharing between the developed and developing countries constituted another important extension of the original concept of sustainability

As the discussions of sustainability have become more politicised, more and more issues have been connected with the concept. Different dimensions of sustainability have been defined (e.g. ecological, economical and social), and different instruments to measure sustainability have been developed (such as ecological footprinting and environmental accounting). In the agricultural sector matters as diverse as the provision of a good working environment, the profitability of the individual farm, and improved animal welfare are regarded as aspects of sustainable production. In addition to this heterogeneity, in discussions about sustainability in food production the focus may significantly differ between continents, countries and regions.

The premise underlying the main theme of this conference is that the principal concepts of sustainability are inexplicably linked to the future of food production. Continuous ethical reflection is needed to clarify and evaluate underlying concerns, as well as to facilitate a transparent discussion of socio-political strategies to meet the challenges of sustainable food production.

The contributions within this volume cover a wide field, thus providing an image of the diversity of concepts and approaches of sustainability and the relevance of ethical considerations to the overall theme. This includes among others theoretical issues and concepts of sustainability in food production, animal welfare considerations and dilemmas, organic agriculture, and case studies of food supply chains which integrate different stakeholder interests.

Contributions have been peer reviewed prior to being accepted for the conference. The editorial board is grateful to the many colleagues who provided their expertise in the review process. Because of the tight schedule and other limitations not all contributions are included herein. We hope that this book will provide an overview over the many fascinating aspects covered during the conference and will help to stimulate fruitful discussions about issues which are most important for the future of food production and agriculture.

Werner Zollitsch and Christoph Winckler

Keynote papers

On sustainability, dogmas, and (new) historical roots for environmental ethics

Ben A. Minteer
School of Life Sciences, Arizona State University, Tempe, AZ 85287-4501, USA, ben.minteer@asu.edu

Most formulations of the sustainability idea emphasize (often fairly explicitly) duties to future generations; i.e., negative duties to avoid harming future persons through present consumption and development, or positive duties to maintain a level of economic well-being, perpetuate just social institutions, and so on. In other words, the sustainability discussion has been largely *anthropocentric* in nature, focused primarily on human interests and values. This is certainly the case with food production, as sustainable agriculture is all about maintaining the productive capacity of the land by following a more ecological and adaptive path than that taken by industrial agriculture.

This very same ethical humanism driving the sustainability idea has, I believe, marginalized it in the very field that would seem to have much to say about sustainable agriculture and the broader transition to a sustainable society: environmental ethics. Although a few environmental ethicists (e.g., Thompson, 1995; Norton, 2005) have stepped into the breach, contributing valuable studies of the ethical foundations and philosophical implications of agricultural practices and sustainability theory, we have not seen nearly as many treatments of these themes as we should have.

Why is this so? It is not because the questions are undeserving of philosophical attention, or that they lack sufficient practical and social urgency. Rather, I believe this neglect of sustainability concerns in agriculture (and other areas of resource production) by environmental ethicists has more to do with the ideological foundations of the field and its selective reading of its own intellectual history. Environmental ethics has long been in the grip of a set of philosophical dogmas that have shaped its development, directing the questions it asks, the landscapes it cares about, and the narratives it tells to support this work.

The first of these dogmas is the embrace of a nonanthropocentric ethical system. Most writers in the field have historically rejected anthropocentric ethical theories terminating in human interests in favor of biocentric (life-centered) or ecocentric (ecologically-centered) moral projects that grant some sort of moral standing on nonhuman nature (see, e.g., Rolston, 1988, 1994; Callicott, 1989, 1999; Westra, 1994; Katz, 1997). The second, related dogma is a tendency to think in terms of sharp and absolute moral dualisms, especially in the area of environmental value theory - and then to read these dualisms back into the historical tradition. Philosophically, this has supported the erroneous belief that one must hold *either* a nonanthropocentric position recognizing the intrinsic value of nature *or* an anthropocentric view in which nature can 'only' possess instrumental value (e.g. Stone, 1987; Varner, 1991; Wenz, 1993).

The third dogma in environmental ethics is often seen as a logical entailment of the first two (though it need not be): A traditional preference for the pristine and the wild over other geographies and conservation targets, including rural and working landscapes, sustainable agriculture, urban land use, and so on. Although the wilderness myth has been under intellectual assault in environmental studies since the early 1990s (see, e.g., Callicott and Nelson, 1998), it continues to have a powerful influence on the policy goals and moral sensibilities of environmental philosophers (e.g., Rolston, 1991; Westra, 2001). To the degree that agricultural and other productive landscapes are seen as diminished by human influence, the wilderness values of many nonanthropocentrists can thus also carry anti-agrarian sentiments.

We can see all three dogmas - nonanthropocentrism, ethical dualism, and an anti-agrarian attitude - at play in what is probably one of the most cited and anthologized papers in the environmental studies literature, Lynn White, Jr's 1967 essay, 'The Historical Roots of our Ecological Crisis' (White, 1967). White's essay is famous for its reading of the anti-environmental implications of the Judeo-Christian tradition, in particular the creation story recounted in Genesis I. Specifically, White argued that Christian teachings placed humans in a cosmologically privileged position whereby 'Man' was fundamentally separate from and superior to the rest of creation. According to White, Genesis conveyed the view that 'no item in the physical creation had any purpose save to serve man's purposes' (White, 1967: 1205). Indeed, Christianity was, in his opinion, 'the most anthropocentric religion the world has ever seen' (White, 1967: 1205). This was not meant to be construed as high praise.

Interestingly, White singled out agricultural practices in 'Historical Roots' for what he argued were their alienating and ecologically destructive consequences. He discussed in particular the introduction of the heavy plow, which in the late 7th century replaced the older and less productive scratch-plow employed by European peasants. This new plow, White argued, 'attacked the land' with great 'violence.' As a result (and in a fit of bold technological determinism), White asserted that 'Man's relation to the soil was profoundly changed. Formerly man had been part of nature; now he was the exploiter of nature. Man and nature are [now] two things, and man is master' (White, 1967: 1205). While the steel plow may have been the proximate cause of this rupture, White argued that the cultural and religious context of the creation of this technology - the Judeo-Christian belief system - was ultimately to blame for this metaphysical and moral rift between humans and nature.

White's essay has cast a long shadow over a generation of writers in environmental ethics. In fact, I'm sure many in the field - possibly most - would agree with J. Baird Callicott, who observed that 'Historical Roots' is the 'seminal paper in environmental ethics', and that following its publication in 1967, 'The agenda for a future environmental philosophy thus was set' (Callicott, 1995: 31). White's condemnation of anthropocentrism, as I've said, became one of the foundational commitments of environmental ethics in its first decade (e.g., Routley, 1973, Rolston, 1975, Regan, 1981). And his negative attitude toward agriculture generally would also echo throughout the work of many subsequent environmental philosophers, who do not seem very comfortable with the transformation and modification of nature for human purposes, even well-motivated efforts such as environmental restoration (e.g., Elliot, 1997; Katz, 1997).

The confluence of doctrinaire nonanthropocentrism, dualistic thinking, and a wilderness bias explains why the field has not devoted much energy to the kinds of questions being asked, for example, at this conference. As a normative subject, sustainable agriculture is apparently deemed as too anthropocentric in scope; it reflects a 'fallen' landscape in which humans have alienated themselves from the mysterious and complex workings of the natural order. Pushing back just as hard against such a view are the familiar utilitarian, deontological, and other humanist arguments, positions that do not (with a few exceptions) confer moral considerability upon plants, animals, or the land as a whole. Thus we have the warring tribes of environmental ethics.

Here is where I believe a return to the roots of the American environmental tradition - and sustainability theory - may be of some help, though I would propose a different path through this history than that taken by most in the field. In the standard potted history of environmental ethics, writers usually return to the first American Conservation Movement of the early 20th Century, the era of John Muir, Teddy Roosevelt, and Gifford Pinchot. In these narratives, Muir is positioned as an early advocate of biocentrism and the wild, a staunch moralist and preservationist squaring off against the utilitarian 'wise use' policies of Pinchot (first head of the U.S. Forest Service). Muir would become for many a legendary biocentric thinker and wilderness defender in the tradition, while Pinchot assumed the role of environmental scoundrel - a judgment that has only begun to soften in recent years (see Miller, 2001). As the story goes, also Leopold would pick up the Muir legacy in the 1930s and 1940s, investing the biocentric worldview with a strong dose of new community ecology and managerial insight. The rise of biocentrism is in these narratives depicted in almost teleological terms: It is as if Muir began a metaphysical and moral process that required Leopold to 'single handedly' bring to fruition by midcentury.

This account, with little variation, has become a fixture in the intellectual history of environmental ethics (see, e.g., Katz, 1995). As a 'usable past' for environmental philosophers, it has provided a powerful foundation for justifying the contemporary nonanthropocentric project as well as important continuity with the earlier American naturalist and conservation traditions. And it has provided a kind of moral indictment for the foundations of sustainability theory, in as much as Pinchot's narrow materialist model of 'sustained yield' resource management is seen in hindsight as irredeemably anthropocentric and ecological destructive; a rotten plank in the foundation.

The difficulty with this account, though, is twofold. First, it misreads the work of key figures in the tradition (like Leopold; see Norton, 1988; Minteer, 2006) because it has insisted on pressing them into an ideological mold that exists partly to bolster contemporary dogmas in environmental ethics and conservation advocacy. Second, it is incomplete, missing the work of thinkers of the era who do not fit so neatly into the rigid narratives constructed by environmental ethicists.

Consider, for example, a name that almost never appears in environmental ethicists' discussion of the field's historical roots: Liberty Hyde Bailey. Despite the fact that he authored a series of important and environmental ethics-relevant books in the early years of the 20th Century and was a key figure in the U.S. Progressive-Era Country Life and Nature Study Movements, Bailey is all but invisible in environmental ethics and discussions of sustainability theory today. Yet I think perhaps more than anyone - and certainly more than his fellow Progressive conservationists - Bailey illustrated how the sustainability vision was firmly embedded in good agricultural and conservation practice.

Conservation, Bailey believed, started on the farm. It was an attempt to recognize the intergenerational responsibilities of good husbandry, which in practice translated as socially responsible and ecologically constrained forms of farming and resource development (Bailey, 1915: 178). Bailey sought to steer the mainstream conservation agenda toward the care of the soil and the culture of rural life as well as the forests and wildlife, and he brought a much-needed civic vision to the movement by highlighting the moral ties connecting the farmer to the community, both within and across generations. 'The man who plunders the soil is in very truth a robber', Bailey wrote, 'for he takes that which is not his own, and we withholds bread from the mouths of generations yet to be born. No man really owns his acres: society allows the use of them for his lifetime, but the fee comes back to society in the end' (Bailey, 1915: 188-189).

Bailey's writing in this vein therefore does not follow the thin utilitarian line usually attributed to his generation of conservationists; i.e., the era of Pinchot, Roosevelt and their allies. The orthodox view of classic American conservation philosophy was heavily shaped by an earlier generation of historical accounts, such as Samuel Hays's classic text, *Conservation and the Gospel of Efficiency* (1959). According to Hays, mainstream conservation was at its core an efficiency-obsessed, scientific movement, with little regard for the public interest or nonmaterial values. The narrow technocratic reading of the first conservation movement has, I believe, has done much to shape the intellectual histories provided by environmental ethicists. Prominent philosophers such as Callicott, for example, frequently describe the Progressive conservation movement as dominated by philosophy of 'resourcism', a strongly anthropocentric outlook in which nature was accorded only an ethically flimsy form of material use value (Callicott, 2003).

While Bailey's conservationism did contain utilitarian elements typical of the era (e.g., much talk of 'mastery', 'control', 'products', and the like), it was more philosophically multi-dimensional than this. Bailey also argued for the need to conserve the aesthetic and communal values of the land, and was motivated by a strong sense of duty to future generations. His conservationism, in other words, may be read as an early sustainability ethic, one that displayed a quasi-deontological form in its emphasis on the various duties attaching to the farming life. As such, it was a normative project that departed from the prevailing utilitarianism evident in the sustained yield models of resource management taking hold on the nation's forests, ranges, and waters during the first decades of the 20th Century.

But Bailey's environmental thought burrowed even deeper than this broad humanism. In addition to its emphasis on biophysical restraint, civic duty, aesthetic values, and future generations, Bailey's conservation - or, more properly as this point, environmental - ethic was shaped by his belief in the human duty to practice a benign and environmentally responsible dominion over a valuable earth (Bailey, 1919). His was a far more nuanced and careful reading of the spiritual obligation to nature than the despotic interpretation later advanced by Lynn White. In fact, Bailey crossed over into biocentric territory in his 1915 book, *The Holy Earth*. There, Bailey articulated a more sweeping, holistic environmental ethic, one that appeared to extend ethical consideration to the entire globe: 'The earth is good in itself, and its products are good in themselves. The earth sustains all things. It satisfies. It matters not whether this satisfaction is the result of adaptation in the process of evolution; the fact remains that the creation is good' (Bailey, 1919: 8).

Bailey's environmental ethic in *The Holy Earth* was shaped by his underlying commitments to evolutionary naturalism and a religious conception of a divine creation, resulting in a biocentric worldview that elevated nonhuman nature and de-privileged humans in the natural order. This, combined with the more traditional anthropocentric values stemming from his agrarian convictions and his concern with the social and moral conditions of country life, positions Bailey as a pluralist in the jargon of environmental ethics: he openly embraced *both* anthropocentric and biocentric commitments, and did not feel compelled to choose between them.

In light of this reading, I believe Bailey falls into a distinct tradition in American environmental thought, a lineage I have elsewhere called 'third-way' environmentalism (Minteer, 2006). Although I do not have space to document it fully here, this third way is marked by several features, including an ethically pluralistic approach toward environmental values that defies the dualisms (anthropocentric-nonanthropocentric, conservation-preservation, nature-culture, etc) thrown down by environmental ethicists and historians over the years. The thinkers in this tradition are also geographically ecumenical, writing approvingly about the countryside, the city, and the region as well as the wilderness. Finally, third-way writers in environmental ethics view environmental values not as freestanding expressions of 'nature philosophy', but as normative commitments thoroughly wrapped up with American civic life, including such concerns as community identity, social regeneration, and the public interest. The third way remains a largely obscured tradition in environmental ethics, as its primary thinkers - i.e., Bailey, Leopold, Benton MacKaye, Lewis Mumford, and others - are either neglected by environmental philosophers (Bailey, MacKaye, Mumford), or shoehorned into ideological categories that do not capture the wider humanist and political dimensions of their work.

I think Bailey's project has something vital to teach us today about the relationship between culture and human productive work (notably agriculture) and the environment - and the possibility of accommodating human will and activity within a pluralistic notion of conservation stewardship. When we look at his body of work - especially the series of environmentally-themed books he published in the first two decades of the 20th Century - we see the headwaters of not only third-way environmentalism, but also sustainable agriculture. Indeed, Bailey offers us the hope that a more adaptive and ecologically constrained agriculture is possible, and that this path is not only the prudent one to take, but reflects a moral obligation on the part of the farmer to the land and his fellow citizens - both now and in the future. By articulating biocentric commitments within a broad stewardship ethic, Bailey provides us with a different way of thinking about the interlocking normative dimensions of sustainability theory, and in his work we may find as well the roots of an alternative, and forgotten, third-way environmental ethics.

References

Bailey, L.H. (1915; orig. 1911). The Country-Life Movement in the United States. The Macmillan Company, New York, USA.

Bailey, L.H. 1919 (orig. 1915). The Holy Earth. Charles Scribner's Sons, New York, USA.

Callicott, J.B. (1989). In Defense of the Land Ethic. State University of New York Press, Albany, NY, USA.

Callicott, J.B. (1995). Environmental Philosophy IS Environmental Activism: The Most Radical and Effective Kind. In: D.E. Marietta and L. Embree (eds.) Environmental Philosophy and Environmental Activism. Rowman and Littlefield, Lanham, MD, USA, pp. 19-35.

Callicott, J.B. (1999). Beyond the Land Ethic. State University of New York Press, Albany, NY, USA.

Callicott, J.B. (2003). The Implication of the 'Shifting Paradigm' in Ecology for Paradigm Shifts in the Philosophy of Conservation. In B. A. Minteer and R. E. Manning (eds.) Reconstructing Conservation: Finding Common Ground. Island Press, Washington, D.C., USA, pp. 239-261.

Callicott, J.B. and Nelson, M.P. (eds.; 1998). The Great New Wilderness Debate. University of Georgia Press, Athens, Georgia, USA.

Elliot, R. (1997). Faking Nature: The Ethics of Environmental Restoration. Routledge, London, UK.

Hays, S.P. (1959). Conservation and the Gospel of Efficiency: The Progressive Conservation Movement, 1890-1920. Harvard University Press, Cambridge, MA, USA.

Katz, E. (1995). The Traditional Ethics of Natural Resources Management. In: R.L. Knight and S.F. Bates (eds.), A New Century for Natural Resources Management, Island Press, Washington, D.C., USA, pp. 101-116.

Katz, E. (1997). Nature as Subject: Human Obligation and Natural Community. Rowman and Littlefield, Lanham, MD, USA.

Miller, C. (2001). Gifford Pinchot and the Making of Modern Environmentalism. Island Press/Shearwater, Washington, D.C., USA.

Minteer, B.A. (2006). The Landscape of Reform: Civic Pragmatism and Environmental Thought in America. MIT Press, Cambridge, MA, USA.

Norton, B.G. (1988). The Constancy of Leopold's Land Ethic. Conservation Biology 2: 93-102.

Norton, B.G. (2005). Sustainability: A Philosophy of Adaptive Ecosystem Management. University of Chicago Press, Chicago, USA.

Regan, T. (1981). The Nature and Possibility of an Environmental Ethic. Environmental Ethics 3: 19-34.

Rolston, H. III. (1975). Is There an Ecological Ethic? Ethics 85: 93-109.

Rolston, H. III. (1988). Environmental Ethics. Temple University Press, Philadelphia, USA.

Rolston, H. III. (1991). The Wilderness Idea Reaffirmed. Environmental Professional 13: 370-377.

Rolston, H. III. (1994). Conserving Natural Value. Columbia University Press, New York, USA.

Routley, R. (1973). Is There a Need for a New, an Environmental Ethic? Proceedings, 15th World Congress of Philosophy 1: 205-210.

Stone, C. (1987). Earth and Other Ethics: The Case for Moral Pluralism. Harper and Row, (New York, USA.

Thompson, P.B. (1995). The Spirit of the Soil: Agriculture and Environmental Ethics. Routledge, New York, USA.

Varner, G.E. (1991). No Holism without Pluralism. Environmental Ethics 13: 175-79.

Wenz, P.S. (1993). Minimal, Moderate, and Extreme Moral Pluralism. Environmental Ethics 15: 61-74.

Westra, L. (1994). An Environmental Proposal for Ethics: The Principle of Integrity. Rowman and Littlefield, Lanham, MD, USA.

White Jr., L. (1967). The Historical roots of our Ecologic Crisis. Science 155: 1203-1207.

How do the ethical values of organic agriculture relate to standards and to current practice?

Susanne Padel
Institute of Rural Sciences, University of Wales AberystwythLlanbadarn Campus, Aberystwyth, SY23 3 AL, United Kingdom

Abstract

The concern that in a growing market organic farming will be more intensive and industrialised has led to renewed interest in organic principles. This paper examines what organic values are covered by standards. It uses the ethical value basis of organic agriculture as described in the four IFOAM Principles of Organic Farming of Health, Ecology, Fairness and Care which encompass the values of sustainability, naturalness and of systems approach. It further explores whether these values are reflected in current organic farming practice and what challenges arise from them in relation to the future development of organic farming.

Keywords: organic farming, standards, ethical value

Introduction

The growing market for organic food involving large companies and global trade has led to concerns about a lack of respect for the core values and principles of organic farming. For example, Guthman (2004) was concerned about the increasing involvement of agri-business creating a lighter version of 'organic' vegetable growing in California by influencing both the rule setting (standards) and the agronomic practice. Such 'conventional' organic farming would be conducted in a more intensive, industrialised fashion and no longer function effectively as more sustainable alternative (Reed, 2005).

The main purpose of organic farming standards and certification is to provide a guarantee about organic production practices. The first standards were based on practices that producers were undertaking, but concerns of consumers or the general public have led to changing the standard in certain areas. For example, awareness of pesticide residues in breast milk led to a restriction of the conventional feed ingredients, and the awareness of the suffering of animals in intensive systems resulted in minimum requirements for outdoor access and space (Padel *et al.*, 2004). Standards therefore represent a compromise between the values of different actors like consumers and producers. Nevertheless, consumers may associate a broader range of values with organic farming that are not part of the standards and producers may practise organic farming in a way that goes beyond what the standards require. The ongoing discussion about 'conventionalisation' of organic farming may well originate from differences between the value expectations and what values are explicated in organic standards.

In this paper, I will therefore examine what organic values are covered by standards. In order to do that it is necessary to identify what ethical values are considered to be at the core of the organic idea. I will further evaluate how this is reflected in current practice and what challenges arise in relation to the future development of organic farming.

What are the ethical values of organic agriculture?

Since 2000, there have been a number of publications aiming to identify the core value base and the principles of organic farming that guide practice (for example DARCOF, 2000; Vogt, 2000). This is comparable to deontological ethics, in which certain principles are formulated to assure respect for a range of fundamental values. Such ethical principles (Box 1) can function both as a source of inspiration and as setting boundaries to certain activities (Padel *et al.*, 2007).

Of particular importance for the identification of values is the formulation of agreed 'Principles of Organic Agriculture' initiated by the International Federation of Organic Agriculture Movements (IFOAM, 2005). The IFOAM Principles were based on a process of stakeholder consultation and a democratic decision amongst the word-wide members of IFOAM (Luttikholt, 2007). In the preamble, these four principles of organic agriculture are clearly identified as ethical principles and as a vision to improve agriculture in a global context (see Box 1). The four principles together act as a whole and each principle also contains a set of explanations in which a range of value elements are referred to. Even if they do not necessarily use the same terms, they also refer to three integrative values that are frequently mentioned in the literature, namely sustainability, naturalness and systems thinking (see Figure 1). The core value basis of organic agriculture can therefore be described by referring to these four IFOAM principles of Health, Ecology, Fairness and Care. Identifying the value elements of the Principles is useful for further analysis and especially for the comparison with standards (Padel *et al.*, 2007).

Box 1. The IFOAM Principles of Organic Farming (IFOAM, 2005).

Principle of health
Organic Agriculture should sustain and enhance the health of soil, plant, animal and human as one and indivisible.

Principle of ecology
Organic Agriculture should be based on living ecological systems and cycles, work with them, emulate them and help sustain them.

Principle of fairness
Organic Agriculture should build on relationships that ensure fairness with regard to the common environment and life opportunities.

Principle of care
Organic Agriculture should be managed in a precautionary and responsible manner to protect the health and well-being of current and future generations and the environment.

Which core values are covered by standards?

Many organic standards (including the current EU regulation 2092/91) do not clearly state the value base on which they are based and there is widespread concern that core organic values are not well represented. The production rules focus on values that are easy to codify and audit through the inspection and certification process, such as what inputs are permitted or excluded (Lockie *et al.*, 2006; van der Grijp, 2006). Values more difficult to operationalise are not translated into rules. This includes agro-ecological systems values such as bio-diversity and nutrient recycling expressed in the Principle of Ecology. Lockie *et al.* (2006) comment also on the paucity of social considerations in most organic standards, again because of difficulties in developing auditing mechanism that refer to them.

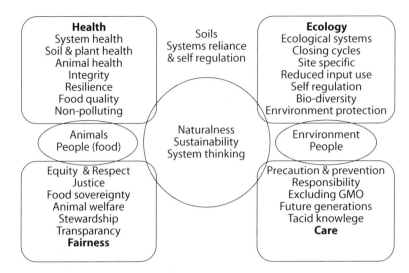

Figure 1. Value elements in the four IFOAM Principles of Organic Agriculture.

The fact that some core values are not part of the standards does, however, not mean they are less important to organic stakeholders, as was confirmed by a comparison of core values and principles with the literature (Padel *et al.*, 2007) and in focus group research about the values of organic stakeholders (Padel *et al.*, 2005).

Which values are implemented in practice?

Producers follow minimal standards but many also adopt practices that go beyond. Much research has focused on categorising organic producers according to their value system, by identifying clusters of *'pragmatic'* and *'committed'* or *'ideological'* organic farmers (for example Darnhofer *et al.*, 2005). Studies often use a distinction between individualistic or financial and altruistic values, producers that have more altruistic values are believed to be more organic (Meeusen *et al.*, 2003). However, because of the well known gap between attitudes and behaviour the fact that producers have a more 'organic' value system does not necessarily mean that their practices are also more 'organic'. This type of research also ignores the learning process leading to changes in attitudes that producers undergo during conversion.

For two reasons it is quite difficult to carry out a well founded assessment of the implementation of core organic farming values in practice: (1) The amount of statistical and representative survey data in relation to organic farming remains limited and (2) It would be necessary to define suitable indicators for the implementation of each of the core values.

To (1): Detailed audits of farm practices of representative samples of organic farms would need to be carried out, but the availability of statistical data remains limited (Rippin *et al.*, 2006). As part of the Organic Revision Project we carried out a comparison of core organic values with current practice in relation to intensification (Padel *et al.*, 2007). Intensification is characterized by higher use of production factors, in particular external inputs and resources. Organic standards regulate what inputs can be used through the positive lists (annexes) and have restricted the use of many non-organic inputs, as illustrated by the stepwise reduction of non-organic feeds introduced in the European Regulation in 2005. They have, however, been less consistent in monitoring or restricting the overall use of external inputs. For example, the amount of organic feed and the overall use of permitted fertilisers are not always limited. Certain farm types can rely to a large extent on external inputs even if these have to come from organic sources, in particular in arable, horticulture and pig and poultry production. Padel *et al.* (2007) concluded that some practice of organic farms that are currently permitted under the standards appears to contradict some of the values expressed in the Principles of Organic Farming, such as recycling of nutrients, the systemic approach of self-reference and self regulation, bio-diversity and environmental protection.

To (2): Implementing core values in practice is easier said than done. This is illustrated by the example of the IFOAM Principle of Fairness that states: 'Organic Agriculture should build on relationships that ensure fairness with regard to the common environment and life opportunities'. The explanations to the principle further state that 'those involved in organic agriculture should conduct human relationships in a manner that ensures fairness at all levels and to all parties - farmers, workers, processors, distributors, traders and consumers' (IFOAM, 2005).

The most prominent standard in this area is those of the fair trade movement which has disadvantaged producers in developing countries as its main target group (www.fairtrade.net). Fair trade standards, however, do not consider the fairness of whole trade chain in a way that the IFOAM principle suggests. Nevertheless, they can provide some inspiration and guidance as to how aspects of the fairness principle could be implemented in organic standards. They also recognise the need for development, for example by the way in which they distinguish between minimum requirements that producers and their organizations must fulfill in order to be certified, and progress requirements that foster continuous improvement in relation to sustainable, social, economic and environmental impact.

To increase the range of values implemented in the standards it would be necessary to develop suitable indicators for more value, monitor performance in relation to them, and develop clear pass/fail criteria. This is very similar to developing tools for sustainability assessment in relation to multiple environmental, social, and economic objectives. Developing such practical tools that help farmers monitor their achievements in relation to a broad range of sustainability objectives and indicators would be a first step to raise awareness. Such tools need to approach sustainability in a very practical way and consider the relationship between various objectives, in particular between non-financial and financial outcomes of a farm. Apart from policy or market incentives, the ability to encourage farmers to strive for greater sustainability would also depend being able to predict the impacts of one particular action on the range of other sustainability objectives.

What are the challenges arising from the ethical principles of organic farming?

The very reason for talking about the core values of organic farming in the sense of deontological ethics is so that they are respected in practice. IFOAM has carried out an important first step in formulating four ethical principles of organic farming. These principles gain further legitimacy through the consultation process and democratic acceptance by the membership of IFOAM.

However, formulating principles alone will not guarantee that core values are respected by organic operators. I have shown that current organic standards and certification systems only implement a proportion of the core values. Other values represent a greater challenge for implementation, as illustrated by the example of the fairness principle. Both the private (producer organisation, certification bodies and companies) and the public sector need to gain more experience how more of these core values can be reflected in the setting of standards and in the certification procedures and it is important to consider in what other ways the awareness about the organic principles among all operators can be increased. Overall, three values that are part of the Organic Farming Principles appear particularly important in this context:

1. Transparency: There should be complete openness in relation to which values are covered and not covered by standards and certification and which values express aspirations for further development.
2. Participation: A process of participative and deliberative democracy allowing representation of relevant stakeholders should be adopted in implementing core values in standards.
3. Respect: There is a need for respect among the discussion partners, which includes respect for arguments and for emotions and sensitivity for specific contexts. Developing a common of the understanding and relating the theory (the value) to the practice should also be an important part of this ethical dialogue (Röcklinsberg, 2006).

Acknowledgements

Funding from the EU Commission for the Organic Revision Project (Contract No. FP6-502397) is gratefully acknowledged (www.organic-revision.org). The views expressed are those of the author, not of the Commission. I would also like to thank all my colleagues from the Organic Revision Project for their support, in particular Helena Röcklingsberg, Henk Verhoog and Otto Schmid.

References

DARCOF (2000). Principles of Organic Farming. Danish Research Centre for Organic Farming, Tjele, Denmark.

Darnhofer, I., Schneeberger, W. and Freyer, B (2005). Converting or not converting to organic farming in Austria: Farmer types and their rationale. Agriculture and Human Values 22: 39-52.

Guthman, J. (2004). The trouble with 'Organic Lite' in California: a rejoinder to the 'conventionalisation' debate. Sociologia Ruralis 44: 301-316.

IFOAM (2005). Principles of Organic Agriculture Bonn, International Federation of Organic Agriculture Movements.

Lockie, S., Lyons, K., Lawrence, G. and Halpin, D. (2006). Going Organic. Mobilizing Networks for Environmentally Responsible Food Production. Wallingford, CABI Publishing.

Luttikholt, L.W.M. (2007). Principles of organic agriculture as formulated by the International Federation of Organic Agriculture Movements. NJAS Wageningen Journal of Life Sciences 54: 347-360.

Meeusen, M.J.G., Beekman, V., de Graaff, R.P.M. and Kroon, S.M.A.(2003). Biologische waarden in tweevoud. Waarden als determinanten van communicatie en samenwerking in biologische voedselketens. Den Haag, Landbouw Economisch Instituut (LEI).

Padel, S., Röcklinsberg, H., Verhoog, H., Schmid, O., de Wit, J., Alrøe, H.F. and Kjeldsen, C. (2007). Balancing and integrating basic values in the development of organic regulations and standards: proposal for a procedure using case studies of conflicting areas. Aberystwyth, Tjele, University of Wales, Aberyswyth and DARCOF.

Padel, S., Schmid, O. and Lund, V. (2004). Organic livestock standards. Animal Health and Welfare in Organic Agriculture. M. Vaarst, S. Roderick, V. Lund and W. Lockeretz. Wellingford, CAB International, p. 57-72.

Reed, M. (2005). The socio-geographies of organic farming and 'conventionalisation': an examination through the academic road trip. XXI Congress of the ESRS,, Keszthely, Hungary.

Rippin, M., Vitulano, S., Zanoli, R and Lampkin, N. (2006). Synthesis and final recommendations on the development of a European Information System for Organic Markets Eisfom - A European Information Systems for Organic Markets - QLK5-2002-02400. Bonn, Ancona and Aberystwyth, Zentrale Markt- und Preisberichtstelle, Polytechnic University of Marche, University of Wales, Aberystwyth, p. 78.

Röcklinsberg, H. (2006). Consent and Consensus in Policies Related to Food - Five Core Values. Journal of Agricultural and Environmental Ethics 19: 285-299.

van der Grijp, N. (2006). Private regulatory approaches and the challenge of pesticide use. Ethics and the politics of food. Proceedings of Eursafe Congress 2006. Oslo, 85.

Vogt, G. (2000). Entstehung und Entwicklung des oekologischen Landbaus im deutschsprachigen Raum. Bad Duerkheim, Stiftung Oekologie und Landbau.

Animal welfare and intensive animal production: are they compatible?[1]

David Fraser
Animal Welfare Program, University of British Columbia, Canada

To answer the question posed in the title, we first need to ask two preliminary questions: 'What constitutes a good life for animals?' and 'What exactly do we mean by intensive animal production?'. We will then be in a better position to ask whether good animal welfare is compatible with intensive production.

A good life for animals

The first question - what constitutes a good life for animals? - is partly a scientific question. We can, for example, use science to identify what environments animals prefer, whether animals are more likely to become sick or injured in one environment than another, and whether they experience frustration if the environment does not allow them to behave in certain ways. These types of factual information play important roles in shaping our conclusions about whether animals can have a good life in a given environment.

Underlying the science, however, are certain value-based presuppositions. Scientists (often tacitly) use value-based judgements when they decide which variables to study in assessing animal welfare, and how much relative importance to attach to such factors as freedom from injury, freedom from frustration, and freedom to perform natural behaviour. These decisions are value-based judgements in the sense that they invoke beliefs about what is more important or less important for animals to have good lives. The beliefs and values that underlie these decisions are complex, but for the purposes of this paper I want to describe two contrasting sets of values that we see clearly in the contemporary debate about animal welfare.

The values can perhaps best be understood by looking back roughly two centuries to the time of the Industrial Revolution. During the Industrial Revolution, the so-called 'factory system' became the predominant way of producing textiles and other goods throughout much of Europe. Thousands of factories were erected, and they proved so efficient that traditional handicraft production of textiles disappeared almost completely. Workers moved from villages and rural areas into large cities; and instead of working at looms in their homes, people operated machines in the factories. It was a profound social change, and it touched off an intense debate over whether the new industrial system was good or bad for the quality of human life.

On one side of the debate were fierce critics of the factory system who insisted that industrial manufacturing led to miserable and unwholesome lives. Critics claimed that the cities where workers were forced to live were cramped, unhealthy, and robbed people of contact with nature. The machines themselves were unsafe and often caused injuries. Moreover, working at machines was seen as an unnatural form of labour that caused extreme fatigue and even physical deformities from the abnormal strain it placed on the body. Perhaps even worse, critics claimed, by doing repetitive work with machines, the workers themselves became like machines and lost touch with their human nature.

---

[1] Many of the sources and data presented in this paper are given in detail in Fraser (2005). Some of the ideas are from a forth-coming book, *Understanding Animal Welfare*, to be published by Blackwell Science.

But the factory system also had staunch defenders. One was Adam Smith who opened his book *The Wealth of Nations* by noting that the quality of life in a nation depends on the goods that are available to supply the citizens with what they need and desire; hence, increasing the productivity of the work force should improve the lives of a nation's people. Changing from a less productive to a more productive system might involve some adjustment, but ultimately it would allow people to live better lives.

But were the workers themselves not being sacrificed for this progress and prosperity? Not at all, insisted the industrialists. Instead of imposing unnatural stress, automation relieved workers of much of the drudgery that manual handicrafts required, and it represented an intermediate step toward a time when automation would make human labour unnecessary. Moreover (industry supporters claimed) it is in the factory owner's own interests to ensure that workers are healthy and happy because maximum productivity would not otherwise be achieved. In fact, the high productivity of factories was proposed as proof that the factories were well suited to human workers.

Clearly, these contrasting claims reflected self-interest, both by the industrialists and by those who wanted to preserve a more traditional life-style, but the claims also reflected two very different world-views. The world-view of the critics involved a set of values that we see extending from the rural poetry of Virgil to the Romantic artists and writers of the 1800s who were among the chief critics of industrial society. According to this world-view, a good life must be a natural life that avoids the artificiality created by technology and urban living. Reacting against the Age of Reason, this world-view emphasized emotion ahead of rationality, and it looked back to a Golden Age when people had lived in harmony with nature.

The world-view of the industrialists was more a product of the Enlightenment. It viewed nature not as an ideal state that we should strive to return to, but as an imperfect state to be controlled and improved. Instead of looking back to a Golden Age of harmony with nature, it looked forward to a Golden Age when the rational application of science, technology and commerce would make a better life possible; and the factory system with its greater productivity represented a form of progress toward that better world.

The conflict that we see today between the critics and the promoters of intensive animal production has strong parallels with the debates during the Industrial Revolution, even to the extent that critics commonly adopt the terms 'factory' and 'machine' to make their points. For example, animal rights advocate Anna Sequoia (1990) described modern farming as a world where, 'animals have become the immobilized machine parts of great automated assembly lines in darkened factories ..... Absolutely no consideration is given to the comfort of these animals, except to keep them alive in large enough numbers to make the animal factories profitable.'

In contrast, defenders of modern animal production commonly speak of exerting control over environmental variables for the benefit of animals and their health. For example, a brochure promoting the U.S. swine industry claims that many grower-finisher pigs, 'are kept indoors in buildings where the pork producer can control the temperature, humidity and other environmental factors. These buildings are well lit and clean, so that the producer can monitor and promote the health of the pigs. Sows may live in individual stalls because it helps give them their own space and also allows the producer to feed and observe each sow individually to meet her needs.' (National Pork Producers Council, 1997).

In these and many other passages we see rhetorical descriptions designed to win public support either for or against modern animal production, but if we look beyond the rhetoric, I think we also find two very different world-views at work.

For the promoters of confinement systems, including many animal producers and the veterinarians and animal scientists who support them, a good life for animals must, first and foremost, be a healthy life marked by good functioning of the body. Promoters of this view see nature not as an ideal state for animals, but as a sub-optimal state where animals are challenged by parasites, infectious diseases, harsh weather and predation. From this viewpoint, the application of science and technology, through vaccines, scientifically formulated diets, well engineered environmental controls and so on, can provide animals with a better life than they would have in a state of nature; and one proof of this better life is the high level of productivity that these systems commonly achieve.

For the critics of intensive production, a good life for animals must, first and foremost, be a natural life. To achieve good animal welfare (the critics claim) we need to approximate natural living conditions through such means as free-range and pasture systems where animals are exposed to nature, and where they are free to carry out their full range of natural behaviour.

Thus, when we ask 'What is animal welfare?' or 'What constitutes a good life for animals?', we receive different answers from different people. The differences were nicely captured in an interview-based study of animal producers and consumers in the Netherlands. The study found that the producers saw animal welfare as 'mainly about health'; they believed that the welfare of their animals was good; and they were skeptical of seemingly natural rearing systems which they characterized as 'chickens eating each other' and 'little piglets, dying from cold'. The consumers, in contrast, perceived animals on modern farms as having 'a short and miserable life, with a lack of space, fresh air, and light'; and they differed from producers in considering that animal welfare requires 'freedom to move and freedom to fulfill natural desires' (te Velde *et al.,* 2002: 208-210).

The disagreement in such cases does not necessarily involve disagreements about facts. Both sides may agree, for example, about the incidence of mastitis, the neonatal mortality rate and the level of ammonia in the air on a given farm. And science can provide valuable guidance, for example by determining what level of ammonia is actively avoided by animals, and what level reduces their growth rate. But disagreements still arise because people attach different levels of importance to the various factors that contribute to animal welfare.

Intensive animal production

What exactly do we mean by intensive animal production? Again, a little history helps set the stage. During the past 50 years we have seen two changes in animal production that are often combined under the heading of 'intensification'. One is the concentration of animal production on fewer and fewer units, with some countries reporting that the number of farms producing animals is now only 10% of the number that existed in the mid 1900s. The second is a move toward indoor, confinement systems, especially for non-ruminant animals. Thus there is intensification in fewer hands, and intensification in limited space.

These two aspects of intensification were no doubt due to a variety of factors. Some factors were demographic: in particular, the expansion of industrial economies meant that agriculture had trouble retaining the work force needed to operate the older methods of animal production, and confinement systems provided a means of reducing labour requirements. Others were policy-related: many governments saw larger, specialized units as a way to improve food security and farm incomes. Other factors were technological: for example, new medications allowed large numbers of animals to be raised successfully in small spaces.

But I would hypothesize that the intensification of animal production occurred partly in response to a radical change in economics driven by two major technological developments of the 20th century. One development was the road system which allowed animals to be transported from almost any farm to distant slaughter and processing plants. The second was refrigeration and other methods of preserving perishable products. These allowed animal products to be transported and sold in other regions, other countries, or other continents. Thus, a single slaughter or processing plant could source animals from a large geographic area and sell the resulting products throughout the world. The result was a massive consolidation in the slaughter and processing industries, and farmers therefore found themselves competing, not with a handful of other local farmers selling to a local butcher or dairy, but competing with thousands of other producers selling to a small number of powerful buyers.

Given this change in the slaughter and processing industries, we might predict that profit to the farmer would fall significantly. We need historical data to test this prediction, but data from the U.S. swine industry provide a striking example. The data show that the profit from farrow-to-finish swine production in the United States averaged about US$21 per animal sold in the late 1970s when the data collection began, and then declined to about US$7 per animal in the 1980s, and to about US$4 in the 1990s, with years of loss mixed with years of modest profit.[2]

These declining profits must have had a major impact on the production sector. One effect will have been a major increase in farm size. Whereas a farm with 100 sows could have generated a good family income in the 1970s, roughly 1000 sows would have been needed in the 1990s to give the same purchasing power. Low profits would also have forced producers to change their production methods so as to reduce costs and losses; and the shift to confinement was a way to reduce the cost of labour, and to achieve higher levels of hygiene and disease-prevention that would reduce losses due to disease and death. Thus, low profits must have been a powerful stimulus for the change both to larger units and to confinement facilities.

But in addition to these effects, low profits would also have had important effects on the details of how animals are kept in confinement systems. With adequate profit per animal, producers could afford to provide animals with space and bedding at levels that would promote comfort even if they were not strictly cost-effective; at low profit levels, these amenities would need to be severely constrained. With adequate profit per animal, producers could spend time caring for individuals, attending births and treating the sick; with lower profits, staff time per animal needed to be minimized. Thus, the economic climate that led to larger units and confinement housing, must also have led to cost-cutting in important aspects of animal housing and care.

My purpose here is to make a distinction between intensification - in the sense of larger units and confinement housing - and the cost-cutting measures that commonly accompanied it. With this distinction in mind, I think we are in a better position to answer our original question.

Animal welfare and intensive animal production

Is good animal welfare compatible with intensive animal production? We need to give one answer regarding intensification itself and another for some of the associated cost-cutting measures.

Whether animals have better lives on large, specialized farms or small mixed farms, and whether they have better lives in indoor housing or on free-range systems - these are issues where we are likely to see profound disagreement, partly because of different value assumptions. For many intensive animal

---

[2] Data kindly provided by Dr. John Lawrence of Iowa State University and reported in Fraser (2005).

producers, along with many animal scientists and veterinarians who support them, large units that have access to specialized veterinary services, an active program of vaccination and disease prevention, scientifically formulated diets, and well designed ventilation, represent an improvement in animal welfare over what might be found on a small, mixed farm with animals kept partly outdoors. However, for many consumers, along with many organic producers and animal advocates, large-scale confinement systems are inherently bad for animal welfare because they do not allow animals to live in a 'natural' environment or carry out their natural behaviour.

But when we focus on the cost-cutting measures that commonly accompanied intensification, we are more likely to find agreement. For example, all informed participants acknowledge that crowding can be a significant animal welfare problem: with too little space animals cannot rest comfortably or gain adequate access to food and water, as reflected (in extreme cases) by reduced survival and growth. Another problem is poor air quality: air-borne dust, ammonia and other harmful gasses are aversive to animals at certain levels and detrimental to health at others. A third factor is staff time and animal handling skill: studies show that inconsistent or negative handling by humans is a significant cause of fear and reduced functioning in many animal species. The list continues, and it includes effective prevention and management of disease, adequate temperature control, the comfort afforded by lying and standing surfaces, and management of pain. In all these areas, cost-cutting - in space, ventilation rate, veterinary costs, bedding and so on - is a major contributor to animal welfare problems.

If intensification was partly a response to a need to reduce costs, then it may seem merely academic to distinguish between the move to larger units and confinement housing, versus the cost-cutting measures that commonly accompanied these changes. By making the distinction, however, I believe we are in a better position to move the discussion beyond mere contradictory rhetoric, and into areas where consensus can be achieved.

Moving forward

In some (mainly European) countries, animal welfare has become a sufficiently potent political issue that governments have imposed animal welfare standards on a sometimes unwilling industry. In much of the world, however, animal welfare comes lower on the political agenda, and change is likely to be achieved only if the different players - farmers, consumers, retailers and others - share a common understanding of the problems and work toward widely supported goals.

The debate over large versus small farms, and indoor versus outdoor systems, has created acrimony between intensive producers and their critics, but it has not led to a consensus on how to resolve animal welfare problems. But if we focus instead on the cost-cutting measures that unquestionably reduce animal welfare, then we may be able to achieve enough consensus to form the basis of effective change.

Given that many problems arise from cost-cutting that is made necessary by low profits, a large part of the solution will need to be economic. Specifically, producers will need to be protected from the market pressures that force them to cut back on space, bedding, ventilation, staff time, salary levels and other factors that play a key role in animal welfare. Examples of such remedies include: (1) product-differentiation programmes that provide premium prices for products produced according to specific standards, (2) government programmes to help producers adjust to animal welfare standards, perhaps modeled after monetary incentives to encourage conversion to organic methods, (3) purchasing agreements whereby corporate customers (chain restaurants, retail chains) agree to pay higher prices in return for guarantees of animal welfare standards, and (4) supply-management or other guaranteed-price programmes that ensure that prices paid to producers reflect the cost of producing animal products in a manner that conforms to agreed animal welfare standards.

Such programs, however, are not likely to be effective if they ignore the values of important participant groups. An intensive system that achieves a high level of health and hygiene but prevents virtually all natural behaviour is not likely to be widely accepted as promoting good welfare. A non-intensive, outdoor system that has high levels of parasitism and neo-natal mortality may be superficially appealing to the public, but ultimately may well bring such production methods into disrepute. To be effective, reform measures will need to respect the values inherent in the different world-views that shape people's beliefs about animal welfare.

References

Fraser, D. (2005). Animal welfare and the intensification of animal production: an alternative interpretation. FAO Readings in Ethics Number 2. Food and Agriculture Organization of the United Nations, Rome.

National Pork Producers Council (NPPC) (1997). How hogs are raised today. NPPC, Des Moines, USA.

Sequoia, A. (1990). 67 Ways to Save the Animals. HarperCollins, New York.

te Velde, H., Aarts, N. and van Woerkum, C. (2002). Dealing with ambivalence: farmers' and consumers' perceptions of animal welfare in livestock breeding. Journal of Agricultural and Environmental Ethics 15: 203-219.

Animal welfare in intensive and sustainable animal production systems

Vonne Lund
National Veterinary Institute, P.O. Box 8156 Dep., N-0033 Oslo Norway, vonne.lund@vetinst.no

Abstract

A crucial factor for animal welfare appears to be the aim to get maximum production from the animals in the production system in relation to invested resources. As a result many modern production systems are designed so that the animal lacks adaptations to meet challenges in the system or it has adaptations that lack a function. Whether or not an animal is considered to have good welfare depends partly on the definition of welfare. If measured in terms of the animal's biological functioning, welfare in intensive production systems may not be severely compromised. However, there are reasons to measure welfare in terms of 'natural living', thus, the animal should be able to feed according to its physiology and live in a physical and social environment that can give functional feedback to its behaviour. With this definition, and if animal integrity is to be respected, production systems are animal welfare friendly only in as much as they can offer environments that correspond to animal adaptations. Animal welfare should also be viewed in a sustainability perspective. Organic farming is an attempt to realize a sustainable agro-ecosystem. It addresses animal welfare on several levels: regarding how an animal's quality of life is to be defined, regarding its moral status and also how animal adaptations and role in the agro-ecosystem can guide the design of welfare friendly husbandry in practice. The challenge for both intensive and alternative farming systems is to make animal welfare economically possible in a global market.

Keywords: animal welfare, intensive production, sustainable agriculture, organic farming

Introduction

The intensification of agriculture and its effects on farm animals and their welfare have been a matter of discussion for nearly half a century. Animal farming systems emerging after the Second World War have been heavily criticised and accused of having sinister effects on livestock welfare. The alarm bell was rung by Ruth Harrison in 1964 through her book 'Animal Machines: The New Factory Farming Industry', which raised extensive public debate. Harrison's book lead to the establishment of the Brambell committee, which was given the task by the British government to enquire into the welfare of intensively farmed animals (Brambell Committee, 1965). A modified version of the recommendations introduced by the committee (the Five Freedoms) is still widely used. Based on the biology and behavioural needs of the animals, the report concluded that a rearing system is acceptable only if the innate behaviour of the animal is not unreasonably violated.

However, the term 'intensive production' can be interpreted in several ways. Discussing whether good animal welfare is possible in intensive production systems, the first question must be: What is meant by 'intensive'? A look in the dictionary doesn't quite help. It reveals that the term refers to 'aiming to achieve maximum production within a limited area' (www.askoxford.com) or 'constituting or relating to a method designed to increase productivity by the expenditure of more capital and labour rather than by increase in scope' (www.m-w.com). However, a capital intensive system usually relies on technical aids, non-renewable energy and concentration of animals in high numbers in small spaces to minimize input of human resources and costs per unit of production, and may have other welfare implications than management of intensive systems offering an abundance of human resources. Fraser (2005) points out that intensification has brought about changes in production methods as well as made production concentrate on fewer units with a consequent increase in farm size. Commonly the term 'intensive production' is used synonymous with 'industrial' or 'factory' farming, or CAFOs, confined animal feeding operations. When it comes to welfare, a crucial factor appears to be the aim to get maximum production from the animals in the production system in relation to invested resources.

Intensive production is nothing new, although the scale has changed over time. One example is the former Scandinavian practice of keeping a maximum number of livestock over the winter, near the upper limit of winter fodder production during normal years. This gave farmers benefits such as a higher share of common grazing areas. However, in years with poor summer harvest or late spring, winter feed was insufficient and the animals sometimes had to be carried out into the spring pastures since malnutrition and starvation had made them too weak to walk (Cserhalmi, 2004). This exemplifies a kind of intensive production on farm level in a self-contained farming system. When local markets for agricultural products emerged in Europe it opened for intensification on a much larger scale. This took place in the Netherlands in the 17th century while for example in Sweden only 350-400 years later with the emergence of an established working class with (some) purchasing power. The technical and economical development after 1945 opened the global market for agricultural animal products, with further intensification of production systems as a logical consequence. This was accompanied by a cultural change, so that livestock farming changed from being a life-style of the practicioners to an enterprise - regardless of whether the company was family owned or not. This change was highlighted in the Swedish educational systems as agricultural colleges in the early 1970's changed the discipline name from 'animal husbandry' to 'animalian production'. The mindset for factory farming was established (which doesn't mean that it was adopted by even a majority of Swedish farmers).

Problems in 'intensive production'

So what's the problem with intensive farming? From an ethical perspective, the problem can be (at least) two-fold:

- One deals with the animal's quality of life in intensive systems, that is, whether the animal has good welfare or not.
- The other concern relates to whether the animal is not considered to have a value of its own or not, and if it is only as a means to fulfil human interests, usually to achieve a goal of profitability. Not ascribing inherent worth or dignity to the animal is in itself problematic to many people, regardless of its actual welfare status.

A third aspect, which is linked to animal welfare, relates to the establishment of sustainable agricultural systems and protection of ecosystems.

Farm animals in intensive systems are often kept under conditions that induce restrictions on their life quality. The following are often mentioned as characteristics of intensive production systems: confinement; crowding; large groups and unnatural social groupings; barren and polluted environments; 'therapeutic manipulations' to increase production or maintain health under poor conditions; inappropriate diets; etc. (For a further discussion regarding the welfare effects see, for example, the homepage of Compassion in World Farming, www.ciwf.org). Many of these welfare challenges occur either because the animal has an adaptation that no longer has a function in modern production systems or because the animal lacks adaptations to meet challenges of the system (Figure 1; Fraser *et al.,* 1997). Figure 1 can be understood in an evolutionary perspective: as the species has evolved in interplay with a particular biotope, it has developed physiological and behavioural adaptations to the challenges of that environment. When removed from its natural (original) environment the animal's ability to adapt may be challenged. Our farm animal species are with few exceptions 'generalists' characterized by their great adaptability. Still, the bigger the discrepancy between the two circles, the bigger the welfare problems the

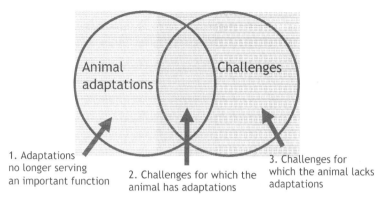

Figure 1. Many welfare challenges in contemporary farming occur either because the animal has an adaptation that no longer can find a function in modern rearing systems or because the animal lacks adaptations to such systems (after Fraser et al., 1997).

animal has to cope with. The general solution to welfare problems is thus to make the circles (re)unite: either through making the animal fit the environment, or through changing the environment so that it doesn't offer insurmountable challenges. The first option may involve different forms of therapeutic manipulations such as mutilations (e.g. tail docking or beak trimming) or chemical treatments such as low-dosage antibiotics, anti-stress medication or 'happiness pills'. Genetic change is another option, and quite a few see the breeding of 'unnatural' animals (such as blind hens) as a handy way of dealing with welfare problems (e.g. Rollin, 1995). However, manipulation of the animal has ethical implications, for example if it is considered to have integrity, as well as food safety aspects. The other possibility, then, is to adapt the environment to the animal.

The environment consists of physical as well as social components. The stockperson also plays an important role and can be considered an important part of the animal's social environment. The task is to create an environment to which the animal can adapt, that will allow it to express its 'natural' behaviour. The concept of 'natural' is considered difficult by both ethologists and philosophers, since often it is not only difficult to identify what is 'natural', but even so, the 'natural' is not always desirable or inherently good. However, a biological approach to the concept has been suggested that can make it a useful tool when discussing animal welfare. Natural behaviour is then defined as the behaviour which the animal is strongly motivated to perform, and when performed it will generate a functional feed back that will lower motivation and make the animal cease performing the behaviour. Motivation is the internal state of an animal that causes a particular, immediate behavioural response. 'Being able to perform a natural behaviour' then refers to being able to carry out behaviours for which the animal is motivated (Algers and Lund, 2007). The functional feedback may not necessarily be elicited by elements in the animal's 'natural' environment, but an environment based on ethological insights may serve the same purpose. One example where this has been applied is the work by ethologists David Wood-Gush and Alex Stolba. They designed experiments to study the natural behaviour of domesticated pigs in a semi-wild enclosure to determine which behaviours were most important to the animals and what functions they served (Stolba, 1984). Based on these studies they could suggest what elements might be changed in the production environment to improve pig welfare. However, contact with nature, as in free-range systems, may add certain favourable qualities to the life of an animal.

An important question is of course how much the genetic disposition in current breeds of production animals differs from that of their wild ancestors. Could it for example be that 'the cage is a natural environment for an unnatural hen'? This seems not to be the case. Although a changed selection pressure has induced significant differences, for example in fur colours and growth rates, there is little evidence that domestication has resulted in the loss of behaviours from the species repertoire or that the basic structure of the motor patterns for such behaviours has been changed. In nearly all cases, behavioural differences between wild and domestic stocks are quantitative in character and best explained by differences in response thresholds (Andersson, 2000).

Definitions of animal welfare and their implications

Before discussing the welfare effects of intensive systems it is also necessary to state what the notion of animal welfare encompasses. Philosophers as well as scientists have tried to elaborate on the concept, and the debate has resulted in three kinds of definitions, based on what is considered the most important for the welfare of the animal (Duncan and Fraser, 1997).

- The biological functioning approach. Here good quality of life or welfare is met when the animal's biological systems are functioning in a normal or satisfactory manner.
- The subjective experience approach. Here animal feelings, such as suffering or pleasure, should count when welfare status is evaluated.
- The natural living approach. Here an animal's welfare depends on the possibility of expressing its natural behaviour and living a natural life according to its genetically encoded nature.

In reality these positions overlap, but the classification is useful since it indicates how to prioritize when different aspects of welfare come into conflict. Depending on the definition used one may reach different conclusions regarding the welfare of animals in intensive systems. Which definition to prefer is basically an ethical choice, since animal welfare not only is about factual issues but also related to one's values and beliefs about what is good to have in life.

Using biological functioning as criteria, animal welfare in industrialized systems may not be compromised since physiological needs may get fulfilled in regards to nutrient requirements and climatic conditions, or performance records may be high (milk yield, growth rate, etc.). Modern farming systems are (at least to some extent) designed with the animal's biological functioning in mind. Focusing on the animal's subjective experience of its situation has the disadvantage of being very difficult to accurately measure. Also, the approach may favour 'simple' solutions involving altering the animal to make it more insensitive to bad rearing conditions. The third alternative, the natural living approach, imposes bigger welfare challenges on intensive production. Natural living implies that the animal should be able to perform its species-specific behaviour, be fed according to its physiology and live in a physical and social environment that can give functional feedback to its behaviour. Thus, it requires that the animal is given a certain amount of freedom, in terms of space, freedom of choice and control over its environment. It may favour the idea of animal integrity, implying that the two circles in Figure 1 ought to be united through adaptation of the environment rather than through manipulation of the animal.

There are arguments favouring the natural living approach. It seems to correspond to many people's gut feelings (Te Velde *et al.,* 2002), and the value of moral intuition is important in ethical consideration. Also, it has been shown that subjective judgements of animal expression are possible to use as a reliable measure of animal welfare (Wemelsfelder, 2005). More importantly, this is a precautionary approach: It is more likely for a species to possess adaptations to challenges in environments similar to its original biotope. It can of course be argued that the subjective experience approach also appeals to deeply held intuitions. However, to a certain extent negative experiences can be viewed as an important part of the functional feedback system connecting individual behaviour and the surrounding world - which is not to say that animal suffering should not be avoided (Lund, 2002). But although 'a natural life' does not guarantee absence of frustration and discomfort, natural systems in general (and free-range systems in particular) will offer a wide range of stimuli, and animals in a stimulating environment where they can perform a natural behaviour generally have better welfare than animals in barren environments and crowded conditions. Challenges in free-range systems are generally caused by the 'dilemma of freedom', that is, linked to the difficulties of exerting control over the individual animal as well as over its environment in systems that allow animals freedom to exercise species-specific characteristics.

Animal welfare in sustainable systems

If animal welfare is defined in terms of natural living, intensive systems are animal welfare friendly only in as much as they can offer environments that correspond to animal adaptations. Good management can alleviate problems caused by poor physical environment but it cannot compensate for it. Interestingly, farm size is not necessarily the main obstacle for the (short-term) welfare of the individual animal, as long as group sizes and management intensity within the unit are kept reasonable. (Some research indicates that the biggest welfare problems actually occur in intermediate sized groups; Keeling *et al.*, 2003). However, there are valid arguments against very big units, for example that disease dynamics change with population size, making health status more fragile in a bigger population. Consequences of failures are also much greater in large systems. Another important argument deals with the environmental impact of large units and its connection to animal welfare. Thus, animal welfare should not be considered as an isolated phenomenon. Farm animals are part of the agro-ecosystem and the surrounding ecosystem, and as such their welfare is in the long term dependent on the sustainability and viability of these systems (as is human welfare).

Ideally, sustainable agricultural systems should function as mimics of natural ecosystems, creating agro-ecosystems based on local resources, renewable energy and recirculation of nutrients where farm animals are an integrated part of the system (Lund *et al.,* 2004). Organic farming is an attempt to realize such sustainable systems. It lends itself to think of farm animals in another way, where they become partners in the agro-ecosystem and members of the mixed agrarian community, rather than production units. As such they have a value of their own. They should not be subjected to therapeutic manipulations, but the environment should be adapted to their coping capability. Hence, in organic farming, animal welfare is understood in terms of natural living. Large scale farming operations are acceptable only as long as they function in harmony with surrounding ecosystems. The aim to allow free-ranging and grazing puts even bigger restrictions on farm size and intensity. Animal welfare directly links to sustainability since use of substances that may negatively affect the ecosystem, such as hormones or antibiotics, must be kept to a minimum. Instead preventive health care, of which good welfare is a vital part because it promotes low stress levels and thereby health, becomes mandatory. In a short-term perspective this is a more costly production but in a longer perspective it is a necessary way to go if sustainability is to be achieved.

As changing market opportunities throughout history have favoured intensification, it is important to create economical frameworks now, that make sustainable and animal welfare friendly agriculture economically possible. The organic labelling scheme as well as the subsidy schemes applied within the EU are at least partly successful attempts to realize this. Education of consumers as well as farmers is also important in this context. However, it is also necessary to achieve a general acceptance for animal welfare restrictions in international trade agreements. The OIE initiative towards establishing global minimum standards for animal welfare must be seen as a very positive step in this direction. The challenge for both intensive and alternative farming systems is how to make animal welfare concerns to work in a global market.

Conclusions

Whether or not animal welfare is judged acceptable depends in part on the definition given to the animal welfare concept. If measured in terms of the animal's biological functioning, welfare in intensive production systems may not be severely compromised. However, there are reasons to measure welfare in terms of 'natural living', which implies that the animal should be able to feed according to its physiology and live in a physical and social environment that can give functional feedback to its behaviour. These requirements are difficult to meet for many current intensive systems, although it is not impossible in principle, if insights in animal behaviour are applied when designing the system. However, animal welfare ought also to be viewed in a long-term perspective, taking system sustainability into account. Organic farming is an attempt to realize ethically, ecologically and economically sustainable farming systems. The approach allows it to address the animal welfare issue on several levels: Regarding how to define an animal's quality of life, farm animals' moral status but also how animal adaptations and role in the agro-ecosystem can guide the design of welfare friendly husbandry in practice. However, a major challenge for intensive as well as alternative farming systems is to make good animal welfare economically possible in a global market.

References

Algers, B. and Lund, V. (2007). A biological approach to the concept of natural behaviour. Paper presented at the IX Annual Swedish Symposium on Biomedicine, Ethics and Society, Searching for the Animal in Animal Ethics. 11-12 June, Sandhamn, Sweden.

Andersson, M. (2000). Domestication effects on behaviour: foraging, parent-offspring interactions and antipredation in pigs and fowl. Acta Universitatis Agriculturae Sueciae. Veterinaria, 86. Swedish Univ. of Agricultural Sciences, Uppsala. PhD Thesis.

Cserhalmi, N. (2004) Djuromsorg och djurmisshandel 1860-1925: synen på lantbrukets djur och djurplågeri i övergången mellan bonde- och industrisamhälle. Gidlund, Hedemora, Sweden, 368 p.

Duncan, I.J.H. and Fraser, D. (1997). Understanding animal welfare. In: Appleby, M.C. and B.O. Hughes (eds.) Animal welfare. CAB International. Wallingford, pp. 19-31.

Fraser, D. (2005). Animal welfare and the intensification of animal production. An alternative interpretation. FAO Readings in Ethics 2. FAO, Rome, 32 pp.

Fraser, D., Weary, D.M., Pajor, E.A. and Milligan, B.N. (1997). A scientific concept of animal welfare that reflects ethical concerns. Animal Welfare 6: 187-205.

Keeling, L.J., Newberry, R.C. and Esteves, I. (2003). Production related traits of layers reared in different sized flocks: The concept of problematic 'intermediate' group sizes. Poultry Science 82: 1-4.

Lund, V. (2002). Ethics and animal welfare in organic animal husbandry - an interdisciplinary approach. Acta Universitatis Agriculturae Sueciae, Veterinaria 137. Dept. of Animal Environment and Health, Swedish University of Agricultural Sciences, Skara.

Lund, V., Anthony, R. and Röcklinsberg, H. (2004). The ethical contract as a tool in organic animal husbandry. Journal of Agricultural and Environmental Ethics 17:23-49.

Rollin, B. (1995). The Frankenstein Syndrome. Ethical and Social Issues in the Genetic Engineering of Animals. Cambridge Studies in Philosophy and Public Policy, Cambridge, 241 p.

Stolba, A. (1984). The identification of behavioural key features and their incorporation into a housing design for pigs. Annales de Recherche Veterinaire 15: 287-298.

Te Velde, H., Aarts, N. and van Workum, C. (2002). Dealing with ambivalence: Farmers' and consumers' perceptions of animal welfare in livestock breeding. Journal of Agricultural and Environmental Ethics 15: 203-219.

Wemelsfelder, F., Hunter, E.A., Mendl, M.T. and Lawrence, A.B. (2001). Assessing the 'whole animal': a Free-Choice Profiling approach. Animal Behaviour 62: 209-220.

Ethics and action: a relational perspective on food trends and consumer concerns

Unni Kjærnes
The National Institute for Consumer Research, P.O. Box 4682 Nydalen, 0405 Oslo, Norway,
unni.kjarnes@sifo.no

Abstract

Consumption as 'choice' dominates many contemporary discourses on food consumption, often understood as decision-making at the point of purchase. But ideas concentrating on individual choice are problematic when trying to understand how social and ethical issues are dealt with (or not) in everyday food habits. I argue in this paper that practices are formed within interrelations between households and other societal institutions. Such interrelations are characterised by their practical arrangements as well as issues of power, trust and responsibility. The question of ethics without action is better understood within such frames, seeing mobilisation as well as inertia and disinterest as emerging within specific constellations and contexts. Disregard of such conditions means that promoting good causes may end up being moralising rather than mobilising.

Keywords: food, consumption, political consumerism, ethics, practice, trust

Introduction

The last years' discourses on food have had increasing references to 'the consumer' (Reisch, 2004). In several fields of food policy, consumers are attributed agency through their choices, for a number of societal issues: health, food quality, animal welfare, and environmental sustainability. The notion of 'choice' seems to epitomise individual autonomy and agency as well as responsibility. New demands towards norms and standards in food provisioning are coming from people as consumers. However, powerful demands are also coming from other societal actors that consumers should take on more responsibility. Yet, a repeated observation is that while people may be interested, engaged and concerned about a wide range of food issues, this is often not reflected in what they do (Vermeir and Verbeke, 2006). Standard answers are that this is either a matter of failing convictions or due to market related barriers, such as lack of information or high prices. But what if people think that they are doing the right thing, that others should take responsibility, or that specific purchases are not an efficient form of action? Such questions direct attention towards social processes and conditions; like the distribution of responsibility for key food issues, the role of trust in social interrelations, and power to change conditions in the food supply chain. It is also evident from several studies of environmental and ethical issues that there are vast variations in consumption practices - across countries and between social groups. I will argue that a social and relational conception of food consumption represents a viable approach to study how social and ethical concerns are reflected in what people do.

Approaches to food consumption

In neoclassical economic theory consumption 'is the sole end and purpose of all production' and the system of production is seen as responding as a servant to the needs and wishes of consumers (Fine and Leopold, 1993). These sovereign consumers are seen as rational choosers, driven by an individual utilitarian orientation, seeking to maximize personal benefits at the lowest possible cost. This conceptualisation implies a tendency to see individual consumers as independent and autonomous actors and their choices of consumer goods as driven solely by their individual needs and demands. While economic theory takes needs and demand for granted, giving them little attention, consumer research has focussed much more on preferences, i.e. on how consumer needs and demands are constituted. The widely used 'Theory of Planned Behaviour' assumes that choices are reflected and deliberate, behaviour being seen as the result of behavioural intention (Ajzen, 1991). This is determined by the individuals' attitudes and the subjective norm. Norms and values are a matter of individual priority, not socially structured. Extra variables are added to the model in order to strengthen its predictive power (e.g. habits, knowledge, perceived control) (Frewer *et al.,* 1996). Socio-demographic variables may differentiate results, presenting systematic differences in attitudes, social norms and intentions according to age, gender and social class. Further, perceptions of risk and danger, credibility of and trust in public authorities and other actors in the food chain may be included too (de Jonge *et al.,* 2007). If we look at sustainable consumption, large parts of the literature refer to individualistic and cognitive notions of consumption and the consumer (Uusitalo, 2005).

Many contemporary theories of consumption suggest that the consumer is far from being a champion of individualistic forward-looking choices, based on deliberate calculation of self interest. There are predictable societal patterns of behaviour related to food provisioning and consumption, emerging from social structures, norms and conventions, and formed by the particular contexts and situations within which consumption takes place, like the family, work, and the marketplace (Warde, 2005). Food represents an intersection between public arenas and the private sphere, the collective and the individual. Meal structure and cuisine affect how people procure food, but the character of various forms of supply as well as governance structures will also have significant influence on people's expectations and actions.

Eating is something that everybody does - usually several times every day in a highly routinised manner. It is just something that we 'normally do'. The normalisation implies predictability and taken-for-grantedness as well as strong normative regulation. Still, practices that involve eating are also very diverse, like the practices of making and consuming family meals, of maintaining health, strength and functionality as part of doing other things - work or leisure activities, as well as socialising with others, of pausing and resting, of celebrating, etc. (Gronow, 2004). Unlike eating, food procurement and preparation may or may not stand out as particular and significant parts of practices in which the individual is involved (because somebody else can take care of it). This introduces an important issue of division of labour and responsibility. Focus is therefore shifted from individual perceptions of particular foods to the logic of situations where food is purchased, cooked, served, and eaten. The attention is on how activity generates wants, rather than *vice versa*. People may and will often challenge or alter practices through their actions, but they will do so with reference to the established practices and the structures which form these practices. Social differentiation and inequality are not a matter of randomly choosing a lifestyle, but a structurally contingent disposition influenced by economic resources, upbringing, social networks, and cultural codes.

Eating is a strongly normative issue. Zwart (2000) distinguishes between three aspects; the logic of more or less (dietetics), the binary logic of either/or and the idea of moral contamination, and the social dimension, referring to issues such as biodiversity, the extinction of species, and other global moral issues. To this third direction, I think that it is crucial also to add justice and social welfare - in provisioning (Coff, 2005; Korthals, 2001) as well as in consumption (Miller, 1998). Studies of food in everyday life show that there are no necessary oppositions between for example concern for own health and concern for the environment (Halkier, 2001). Organic food is, to most people, good for health, taste, the environment, as well as the local community. Taking good care of farm animals will not only benefit the animals, as such a legitimate cause; it will also benefit society - and the consumer (in terms of reduced health risks and better taste) (Kjærnes and Lavik, 2007). Eating food means that the external and the internal are combined.

Daily practice implies a balancing of concerns about what is good and bad, right and wrong. This normativity is rarely a matter of explicit priorities and decisions, but closely associated with normality. Rather than referring to ultimate goals and ideals, the 'normal' is about what is good enough and the appropriate thing to do in a given situation. The appropriate thing to do is, in turn, based on socialisation, what others do, and what we did the last time. The drive for normality implies a high degree of stability. But changes do occur and the normality is gradually adjusted. What we see now is that new aspects of 'appropriate' eating have come to the fore, considering not only new technologies and organisation of food provisioning, but also, perhaps even more, nutrition and responsibility for own health and wellbeing. Giddens (1991) has suggested that we see a process of opening-up and increasing reflexivity. But reconsidering a lot of issues all the time is both time consuming, bordering to the impossible, and uncomfortable. I will instead suggest that we see waves of opening-up and closing down of issues and aspects of eating, with moments of norms turning explicit and contested, then (perhaps new ones) again gradually becoming taken for granted. These waves may be characteristically modern, but they are, nevertheless, emphasising the continued importance of norms and normality.

A relational perspective on food consumption

Purchases form the most direct connection between commercial systems of provisioning and what we do outside the market sphere. We can only purchase what is available in the shops. The purchase situation and the selection of items on offer are strongly influenced by the character of the distribution and retailing system, whether it is a supermarket, a small shop, or a farmers' market - which, in turn, is influenced by the structure and character of production and distribution. Instead of analysing markets and politics as a matter of an external 'context' the analysis should focus on institutionalised relationships between societal actors. Interrelations influence on and are part of established patterns of consumption. Patterns of buying, cooking and eating vary in time and space, shaped by, and in turn shaping, the specific context formed by the food provisioning system. We need to consider relationships between consumers and the provisioning system, as well as interrelations with the state and public authorities and, more indirectly, interrelations between the market and the state. Together they form what can be called 'triangular affaires', characterised by particular divisions of responsibilities, relations of power, as well as forms and levels of trust (Kjærnes *et al.,* 2007). The triangular affaires are built up by characteristics of the poles as well as the complex direct and indirect relationships between them. Each pole of the triad has its own internal organisation. Consumers of food are organised in households in ways that vary across countries. Provisioning systems, including the way retail shops sell food to consumers, differ greatly. And state regulatory and political systems are strongly contrasted between countries.

A simple example is organic food, where such relationships are highly decisive for how organic production, consumption and regulation are organised and framed. Organic food is sold mainly in 'reform shops' in Germany and in the major supermarket chains in the UK, influencing not only the practicality of shopping, but also the products on offer and the social relations involved. Concerns over food are quite similar in Denmark and Norway, but the consumption of organic food is highly diverging, due to different supplies and politics.

These examples are associated with different conditions of trust. Two dimensions are important when discussing trust and distrust in relation to food consumption. *First*, trust in food is relational and conditional (Mishler and Rose, 2001). In most cases we depend on others to provide us with food. Modern food provisioning is based on advanced division of labour, complex organisations and technologies. We cannot control these food institutions. So we need to trust. Trust refers to general trust in institutionalised procedures as well as expectations and experiences regarding specific actors and organisations. *Second*, trust is part of the taken-for granted normality that characterise our everyday food practices, making our world (reasonably) predictable (Misztal, 1995). Trust develops within specific social contexts, referring to collectively formed norms and expectations. Trust is formed with reference to the 'normal' - and normality is strengthened with the help of trust. Normality may also be formed on the basis of general distrust. But expecting little of others and distrusting most situations means we loose out on flexibility and the need of direct control increases.

Studies of trust in food over the last couple of decades indicate that trust is highly variable and it may ebb and flow (Kjærnes *et al.*, 2007). Repeated scandals that are poorly handled, corruption, deceit, disregard of consumer demands are reflected in scepticism and distrust, expressed as search for trustworthy alternatives, in public protest or, more severe, in powerlessness and resignation. These are aspects that evolve in the market as well as in the public domain. Visibility is an important part of the story. Visibility does not necessarily indicate distrust, but may be part of a process of re-establishing, perhaps even reinforcing trust. Considering the contemporary public attention towards trust issues, many are assuming that the modernisation of food provisioning is associated with distrust, linked to science and expertise, the growing complexity of provisioning systems, and new types of social relations and dependencies, like in the 'risk society' thesis. But empirical studies indicate that efforts improving predictability and organisational legitimacy may work. There are many examples of institutional modernisations that are successful in terms of enhancing trust. This is perhaps surprising, considering the immense imbalances of power and resources as well as differences in ultimate goals between the economic sphere and the sphere of the household. People are well aware of the imbalances of power, but trust that powerful actors actually do take their interests and demands into consideration. Distrust is much more widespread in situations where trustworthiness and power do not correspond. We (need to) trust powerful actors to ensure that our food is handled in ways that we find acceptable in terms of health and environmental hazards, social conditions, quality, etc.

Consumer responsibility?

But what about our own role as consumers? Producing and preparing food is increasingly moved away from the home and into technologically and organisationally advanced systems of provisioning and distribution. Consumer responsibility must be discussed with this in mind. Taking responsibility is a matter of power, power to set the standards and the resources to implement them. This leads me on to classical debates about relations between states and markets and between society and the individual. The literature on new forms of governance blurs these distinctions, outlining new forms of interrelations between state and market, but also a new role of individuals (Dean, 1999; Jordana and Levi-Faur, 2004). Nutrition is at the moment a highly contested field in that respect (Coveney, 2000). Conflicts over food safety seem on the other hand to have become more settled in terms of the division of responsibility, allocating more responsibility to market actors, but hardly more to individual consumers. But many ethical issues appear to be much more open in terms of the division of responsibility, such as the welfare of farm animals or fair conditions in trade with food. The allocation of responsibility can be contested as well as practically difficult. For example, allocating responsibility for animal welfare to consumers can be problematic if many consider that animal welfare should not be subject to market competition, if people are not made aware of concrete reasons for them to act, if they think that their actions will not make a difference, or if the supply is not there. At the moment, few market actors see sufficient economic advantages in using this to differentiate products (except eggs).

Does this mean that people in their capacity as consumers are not able or willing to take on new responsibilities? Considerable experience indicates that this would be a wrong conclusion. Not only actions characterised as ethical and political consumerism give support to that. Even the gradual transformation of food consumption practices to accommodate new concerns and demands point in the same direction. But the solutions and the framing may be different from what is wished or asked for by institutional actors. Quite a few people do consider animal welfare while shopping, but they do so as part of broader concepts of high quality or ethically good food, sometimes referring also to different ideas of what good welfare is, like extensive farming or a special regional or national origin (Kjærnes and Lavik, 2007). Such ideas are, in turn, influenced by their experiences in the market and in public discourse. These are very different in France, compared to the UK, Norway or Hungary.

During a food crisis, especially if safety is involved, buyers often change their actions abruptly. And it can make a difference. Somebody has to 'take responsibility' and leave their assignment, new routines may be reinstated, etc. In these cases mass media play a central role, often supported by non-governmental organisations and experts. Sometimes scares are driven out of proportions, but, as already indicated; openness is part of the process. What we have seen in Europe over the last decade is a re-regulation of food safety, brought about by, among other things, consumer actions. And this re-regulation appears to have improved trust in food safety. Even in less dramatic cases, consumer activism rarely emerges in isolation, but is commonly supported by public debates and collective mobilisation. Though appearing as individual choices, it is part of a social process: 'Consumer individual action and collective activism, although arguably in conflict with market liberal theory, is here the ultimate ethical enforcer.' (Newholm, 2000).

The politics of consumer choice

It is easy to sympathise with the claims that consumer concerns must be taken seriously and that proper channels for feedback and influence should be ensured. Such channels are often poorly developed in Europe today. However, my concern here has been that demands of consumer responsibility are often coming from other societal actors. Such demands may sometimes be in accordance with the expectations of many modern consumers. Korthals (2001) has argued that caring and ethical concerns in shopping challenge the distinctions between the consumer role and the citizen role. 'Political consumerism' as a form of political activism points in the same direction (Micheletti, 2003). In my opinion, this is not only a matter of value orientations or public debates. Values and debates as well as practice are embedded in particular social institutions; in everyday life, markets and politics. The ways in which consumption is institutionalised - the daily routines, the directions and priorities of food consumption, as well as the responsibility, power and resources of 'the consumer' - are not static preconditions. It is the institutionalisation of social interrelations that form consumers' actions and the ways such actions may influence the provisioning system. Assuming agency without considering concrete arrangements, power structures, and trust relations may end up being moralising rather than mobilising. There are numerous examples.

References

Ajzen, I. (1991). The theory of planned behaviour. Organisational Behaviour and Human Decision Processes 50: 179-211.

Coff, C. (2005). Smag for etik. På sporet efter fødevareetikken.Museum Tusculanums Forlag, Københavns Universitet, Copenhagen.

Coveney, J. (2000). Food, Morals and Meanings: The Pleasure and Anxiety of Eating. Routledge, London.

de Jonge, J., van Kleef, E., Frewer, L. and Renn, O. (2007). Perception of risk, benefit and trust associated with consumer food choice. In: L. Frewer and H. C. Van Trijp (eds.) Understanding Consumers of Food Products. Woodhead Publishing, Cambridge, pp. 12-150.

Dean, M. (1999). Governmentality. Power and Rule in Modern Society. Sage, London.

Fine, B. and Leopold, E. (1993). The World of Consumption. London, New York: Routledge.

Frewer, L.J., Howard, C., Hedderley, D. and Shepherd, R. (1996). What determines trust in information about food-related risks? Underlying psychological constructs. Risk analysis 16: 473-85.

Giddens, A. (1991). Modernity and self-identity. Self and society in the late modern age. Polity Press, Cambridge.

Gronow, J. (2004). Standards of taste and varieties of goodness: the (un)predictability of modern consumption. In: M. Harvey, A. McMeekin and A. Warde (eds.) Qualities of Food. Manchester University Press, Manchester, pp. 38-60.

Halkier, B. (2001). Risk and food: environmental concerns and consumer practices, International Journal of Food Science and Technology 36: 801-12.

Jordana, J. and Levi-Faur, D. (2004). The politics of regulation in the age of governance. In: J. Jordana and D. Levi-Faur (eds.) The Politics of Regulation. Institutions and Regulatory Reforms for the Age of Governance. Edward Elgar, Cheltenham, pp. 1-30.

Kjærnes, U., Harvey, M. and Warde, A. (2007). Trust in Food. A Comparative and Institutional Analysis. Palgrave Macmillan, London.

Kjærnes, U. and Lavik, R. (2007). Farm Animal Welfare and Food Consumption Practices: Results from Surveys in Seven Countries. In: U. Kjærnes, M. Miele and J. Roex (eds.) Attitudes of Consumers, Retailers and Producers to Farm Animal Welfare. Welfare Quality Reports No.2. Cardiff University, Cardiff.

Korthals, M. (2001). Taking consumers seriously: Two concepts of consumer sovereignty, Journal of Agricultural and Environmental Ethics 14: 201-15.

Micheletti, M. (2003). Political Virtue and Shopping. Individuals, Consumerism, and Collective Action. Palgrave Macmillan, New Yoirk and Houndmills.

Miller, D. (1998). The Dialectics of Shopping. The University of Chicago Press, Chicago and London.

Mishler, W. and Rose, R. (2001). What are the origins of political trust? Testing institutional and cultural theories in post-communist societies. Comparative Political Studies 34: 30-62.

Misztal, B. (1995). Trust in modern societies: the search for the bases of social order. Polity Press, Cambridge.

Newholm, T. (2000). Consumer exit, voice, and loyalty: indicative, legitimation, and regulatory role in agricultural and food ethics, Journal of Agricultural and Environmental Ethics 12: 153-64.

Reisch, L.A. (2004). Principles and visions of a new consumer policy. Journal of Consumer Policy 27:1-27.

Uusitalo, L. (2005).Consumers as citizens - Three approaches to collective consumer problems. In: K.G. Grünert and J. Thøgersen (eds.) Consumers, Policy and the Environment. A Tribute to Folke Ölander. Springer, New York, pp.127-150.

Vermeir, I. and Verbeke, W. (2006). Sustainable food consumption: exploring the consumer 'attitude-behavioural intention' gap. Journal of Agricultural and Environmental Ethics 19: 169-94.

Warde, A. (2005). Consumption and theories of practice. Journal of Consumer Culture 5: 131-53.

Zwart, H. (2000). A short history of food ethics, Journal of Agricultural and Environmental Ethics 12: 113-26.

Coexistence? What kind of agriculture do we want?

Louise W.M. Luttikholt
International Federation of Organic Agriculture Movements (IFOAM), Charles-de-Gaulle str. 5, 53113 Bonn, Germany

Abstract

The so called coexistence debate on Genetically Modified Organisms (GMO's) and conventional and organic agriculture focuses on arbitrary details like threshold levels and meters of distances. It forces those who do NOT want to accept GMO's, into a debate which is not theirs. There tends to be not enough room to explain the values of Organic Agriculture, the very ethical basis organic farmers wish to base their work on; values that can be explained to and shared with consumers and citizens. The debate on coexistence draws the attention away from a necessary debate that European citizens should have on 'what kind of agriculture do we want?'

Organic agriculture is value based

Organic agriculture is a holistic production management system, which enhances agro-ecosystem health, utilizing both traditional and scientific knowledge. Organic agricultural systems rely on ecosystem management rather than external agricultural inputs. Organic agriculture is based on the Principles of Organic Agriculture, as articulated by the International Federation of Organic Agriculture Movements (IFOAM) through a worldwide participatory stakeholder process. The process aimed to bridge the values from the pioneers of organic agriculture to the present time of globalization and to extended growth of the organic sector. As a result the Principles of Health, Ecology, Fairness and Care are now worldwide considered as the basis from which organic agriculture grows and develops.

The notion of 'ecological justice' as presented by Alrøe *et al.* (2006) overlaps several of the four principles of organic agriculture. Ecological justice is about a fair distribution of all environments - good and bad, including externalities - over all living creatures on the planet. With this it recognizes that the environment has value beyond the human nature in all its abundance and diversity. This notion defines the place of human beings as interdependent of ecological systems. It broadens the Principle of Fairness to all living beings. And exactly this attitude is one of the basics in organic farming, where man and nature are considered to be an integrated whole. The wording in the IFOAM principles of organic agriculture is such that it points to the duties and responsibilities of the practitioners, from farmers to consumers.

The principles confirm IFOAM's earlier formulated position on Genetic Engineering and Genetically Modified Organisms as well as the position based on it, from IFOAM's EU Regional Group on coexistence between GM and non GM crops.

Threshold levels do not comply with a process approach

Based on the system approach, control within the organic sector is process based and not product based. A threshold level for contamination of what ever unwanted residue or product does not make much sense and it will not protect against contamination. Therefore the organic sector cannot 'accept' a threshold level of whatever kind. A threshold for contamination at zero will not always be feasible and will put the burden (for testing etc) at the organic sector. The fact that there are threshold levels in food regulations, like for labelling of GMO's, does not mean that the organic sector accepts routine contamination up to an arbitrarily defined threshold; thresholds tend to become a 'new zero'. Anti contamination measures should aim at 0.1 (lowest measurable) contamination, regardless the rules for labelling. Otherwise there will be no co-existence at all; everything will be more or less (containing) GMO. This implies that 'adventitious and unavoidable' as formulated in the EU directive needs to be clearly articulated and improved. The organic sector favours clear liability rules, so that the 'polluter pays principle' count.

The pro-GM sector promotes separate thresholds for the organic and conventional agricultural sector, isolating the organic sector as the only sector that really cares about GMO free. However, this would legalise the contamination of all the (less strict) conventional farmers and thereby about 95% of agriculture. However, all farmers have the same right to be protected against GMO contamination and consumers have a right to eat organic and conventional GMO free food. As organic farmers live in the same surroundings as conventional farmers, they have no extra means to avoid contamination. Separate isolation distances would make it impossible for conventional farmers (once they are surrounded with GM- growers) to convert to organic.

How well contamination can be avoided will depend on the coexistence rules and the amount of GM crops grown in and outside Europe.

What kind of agriculture do current and future European citizens want?

This brings the debate to the core question that seems to get avoided when talking about coexistence: what kind of agriculture do the European citizens want? This question, that is concrete, inviting and understandable for citizens is of much more interest than detailed technical discussions on threshold levels and distances that are hard to follow and are often left to 'experts'.

References

NJAS volume 54/4 (2007). Values in Organic Agriculture.

Alrøe, H.F., Byrne, J. and Glover, L. (2006). Organic agriculture and ecological justice: ethics and practice. In: N. Halberg, H.F. Alrøe, M.T. Knudsen and E.S. Kristensen (eds.) Global Development of Organic Agriculture: Challenges and Prospects. CAB International, Wallingford, pp75 - 112.

The International Commission on the Future of Food and Agriculture (2006). Manifesto on the future of food.

Padel, S. (2005). Focus groups of value concepts of organic producers and other stakeholders. Report No D21 Organic Revision - Research to support revision of the EU Regulation on organic agriculture. University of Wales, Aberystwyth, 118 pp.

Padel, S., Röcklinsberg, H., Verhoog, H., Alrøe, H.F., De Wit, J., Kjeldsen, C. and Schmid, O. (2007). Balancing and integrating basic values in the development of organic regulations and standards: proposal for a procedure using case studies of conflicting areas. Report No D23 Organic Revision - Research to support revision of the EU Regulation on organic agriculture. Danish Research Centre for Organic Farming (DARCOF), Foulum, 118 pp.

www.ifoam.org

Annex - Principles of Organic Agriculture

Preamble

These Principles are the roots from which organic agriculture grows and develops. They express the contribution that organic agriculture can make to the world, and a vision to improve all agriculture in a global context.

Agriculture is one of humankind's most basic activities because all people need to nourish themselves daily. History, culture and community values are embedded in agriculture. The Principles apply to agriculture in the broadest sense, including the way people tend soils, water, plants and animals in order to produce, prepare and distribute food and other goods. They concern the way people interact with living landscapes, relate to one another and shape the legacy of future generations.

The Principles of Organic Agriculture serve to inspire the organic movement in its full diversity. They guide IFOAM's development of positions, programs and standards. Furthermore, they are presented with a vision of their world-wide adoption.

Organic agriculture is based on:
- the principle of health;
- the principle of ecology;
- the principle of fairness;
- the principle of care.

Each principle is articulated through a statement followed by an explanation. The principles are to be used as a whole. They are composed as ethical principles to inspire action.

Principle of health

Organic Agriculture should sustain and enhance the health of soil, plant, animal, human and planet as one and indivisible.

This principle points out that the health of individuals and communities cannot be separated from the health of ecosystems - healthy soils produce healthy crops that foster the health of animals and people.

Health is the wholeness and integrity of living systems. It is not simply the absence of illness, but the maintenance of physical, mental, social and ecological well-being. Immunity, resilience and regeneration are key characteristics of health.

The role of organic agriculture, whether in farming, processing, distribution, or consumption, is to sustain and enhance the health of ecosystems and organisms from the smallest in the soil to human beings. In particular, organic agriculture is intended to produce high quality, nutritious food that contributes to preventive health care and well-being. In view of this it should avoid the use of fertilizers, pesticides, animal drugs and food additives that may have adverse health effects.

Principle of ecology

Organic Agriculture should be based on living ecological systems and cycles, work with them, emulate them and help sustain them.

This principle roots organic agriculture within living ecological systems. It states that production is to be based on ecological processes, and recycling. Nourishment and well-being are achieved through the ecology of the specific production environment. For example, in the case of crops this is the living soil; for animals it is the farm ecosystem; for fish and marine organisms, the aquatic environment.

Organic farming, pastoral and wild harvest systems should fit the cycles and ecological balances in nature. These cycles are universal but their operation is site-specific. Organic management must be adapted to local conditions, ecology, culture and scale. Inputs should be reduced by reuse, recycling and efficient management of materials and energy in order to maintain and improve environmental quality and conserve resources.

Organic agriculture should attain ecological balance through the design of farming systems, establishment of habitats and maintenance of genetic and agricultural diversity. Those who produce, process, trade, or consume organic products should protect and benefit the common environment including landscapes, climate, habitats, biodiversity, air and water.

Principle of fairness

Organic Agriculture should build on relationships that ensure fairness with regard to the common environment and life opportunities

Fairness is characterized by equity, respect, justice and stewardship of the shared world, both among people and in their relations to other living beings.

This principle emphasizes that those involved in organic agriculture should conduct human relationships in a manner that ensures fairness at all levels and to all parties - farmers, workers, processors, distributors, traders and consumers. Organic agriculture should provide everyone involved with a good quality of life, and contribute to food sovereignty and reduction of poverty. It aims to produce a sufficient supply of good quality food and other products.

This principle insists that animals should be provided with the conditions and opportunities of life that accord with their physiology, natural behaviour and well-being.

Natural and environmental resources that are used for production and consumption should be managed in a way that is socially and ecologically just and should be held in trust for future generations. Fairness requires systems of production, distribution and trade that are open and equitable and account for real environmental and social costs.

Principle of care

Organic Agriculture should be managed in a precautionary and responsible manner to protect the health and well-being of current and future generations and the environment.

Organic agriculture is a living and dynamic system that responds to internal and external demands and conditions. Practitioners of organic agriculture can enhance efficiency and increase productivity, but this should not be at the risk of jeopardizing health and well-being. Consequently, new technologies need to be assessed and existing methods reviewed. Given the incomplete understanding of ecosystems and agriculture, care must be taken.

This principle states that precaution and responsibility are the key concerns in management, development and technology choices in organic agriculture. Science is necessary to ensure that organic agriculture is healthy, safe and ecologically sound. However, scientific knowledge alone is not sufficient. Practical experience, accumulated wisdom and traditional and indigenous knowledge offer valid solutions, tested by time. Organic agriculture should prevent significant risks by adopting appropriate technologies and rejecting unpredictable ones, such as genetic engineering. Decisions should reflect the values and needs of all who might be affected, through transparent and participatory processes.

Approved by the IFOAM General Assembly 2005.

Coexistence and ethics: NIMBY-arguments reconsidered

Matthias Kaiser
The National Committee for Research Ethics in Science and Technology (NENT), P.O. Box 522 Sentrum, Prinsensgate 18, N 0105 Oslo, Norway, matthias.kaiser@etikkom.no

Abstract

This paper has a focus on ethical issues involved in the debate about coexistence between conventional, organic and genetically modified crop farming. Some part of the opposition to coexistence is portrayed as a NIMBY reaction. Since NIMBY arguments are typically defective in terms of ethics, several modifications of the argument are considered. It is concluded that the right of regions to adopt voluntary schemes is defendable on ethical grounds, but do not make national or European coexistence schemes superfluous, in particular concerning good liability regulations.

Keywords: coexistence, GM crops ethics, NIMBY, NIABY

'Not-in-my-back-yard' (hereafter: NIMBY) is usually considered the typical response of individuals who on the one hand may benefit, albeit perhaps indirectly, from a certain development and thus support it in general terms, but who on the other hand do not want to have that development in their immediate vicinity out of fear of negative effects of some kind on them. Thus, the NIMBY response is also seen as anti-social, and perhaps even morally defective to the extent that one expects others to bear burdens that one is not prepared to bear oneself. Typical examples are people's reactions to plans to build incinerators or airports in their neighbourhood. Everybody produces waste that needs to be destroyed and deposited, but having an incinerator around reduces the values of properties and is rumoured to have polluting effects on the close environment.

In some sense, the discussions about coexistence between conventional, organic and genetically modified crops (hereafter: GM crops) resemble NIMBY-discussions. I shall clarify this claim below. In stating this, it is assumed that the more general food safety issues and health and environmental effects issues of GM crops have first been resolved positively, i.e. the products are found to be safe. The question then arises whether the GM crops can be planted in the vicinity of other crops, pasture or nature without affecting near-by plants. The generally accepted answer is no. Unless specific measures are put into place, the adventitious presence of one crop in another crop cannot be prevented. It may arise for a variety of reasons, like due to seed impurities, cross pollination between neighbouring fields, on-farm storage etc. On markets that differentiate between gm-products and gm-free products, this will cause disturbances and lead to economic losses. In order to avoid this, coexistence is proposed as a management scheme.

The present discussions, as I perceive them, are assumedly not so much whether coexistence in principle is desirable (- in principle it is -), but rather (1) whether coexistence with GM crops is practically possible at all, (2) whether subsidiarity implies regional autonomy to opt out of coexistence schemes in favour of gm-free areas and (3) whether stricter levels of adventitious presence of GM crops in other crops than those proposed by EU standards can be imposed. I understand that e.g. the European Commission endorses (1) and opposes (2) and (3), while NGOs (Declaration, 2005) are at least sceptical to (1), but endorse (2) and (3).

In the remainder of the paper I shall leave out all technical discussions about the possibility of coexistence. Coexistence between gm-free seeds of various kinds has been practised for a long time, and there is evidence that some such scheme may also work for GM crops. Certainly, specific proposed schemes of coexistence may be defective and not working in practice, for instance when minimal distances between fields are too small. But finding out what the optimal distances are is different from arguing that no such distance can be found. The latter seems obviously wrong. Therefore I assume the principle possibility of effective coexistence (cf. EC, 2006; GMO Free, 2005).

Points (2) and (3) on the other hand are more problematic. In spite of all the rhetoric used by industry and the EC (ISF, 2004, p.3; CEC, 2006, p.3), these points are not merely about economical aspects. They are also about ethics. It is stated in several documents that the main considerations are to find management schemes that satisfy the freedom of choice for both producers and consumers. This is an appeal to the ethical principle of autonomy. Furthermore, several elements of coexistence, as e.g. the inclusion of liability schemes, refer to the ethical principle of justice and fairness. The fair balancing of interests and preferences of one group against those of another group, regardless whether the groups are large or small, is at the core of the discussions about coexistence. I will therefore maintain, against the stated claims of the EU or industrial producers (seed), that the discussion is not (merely) about economical considerations, it is primarily about ethics.

I believe that the resistance by conventional and organic farmers against GM crops in their neighbourhood resembles the NIMBY phenomenon to some extent. First, the main conflict line is between two groups of colliding interests, those that promote a new development versus those that defend the status quo without that development. Second, the conflict is most intense before the actual development takes place since the most pressing issues are not about how the development is to be integrated into the existing region, but whether any efforts should be made to integrate the development at all. Third, the debate is focused on protecting the (perceived) value of the affected region, i.e. the specific value of 'my-back-yard', rather than on the appropriateness of the chosen site for the new development. Fourth, in terms of democratic decision-making, the debate is about the right of affected parties to have a say, or even the last word, on decisions that determine what goes on in their own neighbourhood, even when matters of the common good, e.g. national economy, are cited as reason for the development. These features are, I believe, common to the NIMBY responses to airport or incinerator developments, but are also underlying major parts of the debate on coexistence between conventional, or organic farming versus GM crop farming.

I stated already at the beginning that NIMBY responses are typically viewed as lacking in argumentative force. In terms of ethics they are seen as violating the old principle of not doing to others what one does not want others to do to oneself. Translated to our case, the principle would say that one should not expect others to tolerate risks that one is not prepared to tolerate oneself. Thus if the argument against coexistence is merely along the lines of: 'gm crops can be OK for some or even large groups of society, but let others carry the risks of production and not us', then the argument is clearly defective in terms of ethics.

However, it is not clear that this is actually what critics of coexistence would maintain. Actually it may be that critics often do not want to bring forward NIMBY responses, but rather tend to argue for NIABY responses, i.e. a 'not-in-anybody's-back-yard' argument. This would mean that some critics are moved by the conviction that GM crops carry with them environmental or other risks of such a magnitude that nobody or no region should be expected to have them produced in their neighbourhood. Arguing along these lines would not subject the critic to the objection that some more or less selfish interests are motivating their negative response to coexistence.

There is a difficulty with this line of reasoning. The difficulty is that state authorities are normally tasked to evaluate whether a given GM crop is of such a kind that a NIABY response is indicated. Any release and marketing of a GM product (crop) in Europe is required by law to be screened for possible health and / or environmental risks. If those risks are found to be significant, then the product is not allowed.

Only if the risks are found to be insignificant, can a GM crop be produced or marketed. In other words, state authorities evaluate whether there is a basis for a NIABY argument or not. It is only on the basis of assuming that this risk assessment has been carried out and concluded in favour of a GM crop, that the problem of coexistence arises at all. Now, if the opposition to coexistence is portrayed as a NIABY response, then the real issue is no longer about the region, but essentially about the role of the EU or state policy concerning GM crops in general. It might indeed be tempting for groups that oppose GM crops in general to simply move the debate from the state or EU level - a level where the general issue has more or less been decided against them - to the level of regions and continue the same debate there. But then the debate is a totally different one. It is a debate about civil disobedience. One fails to make the general case of showing by scientific evidence that a GM crop constitutes a significant risk *per se*, and thus moves the debate to a lower level without any further evidence. The claim is then, assumedly, that the regional constituency should not be bound by the assessment provided by state or EU authorities, but should determine for themselves what they find reasonable. Clearly, if the issue would be about e.g. driving with seat belts, many people would readily agree that such a claim is highly problematic and undermining state authority without good reason. I therefore conclude that opposition to regional coexistence cannot be based on NIABY arguments, unless one wants to make a general case for civil disobedience.

The question is then whether there can be more to the NIMBY responses than what we have outlined so far. If a NIMBY response is to have ethical legitimacy, then one needs to point out an ethically relevant property of one region that separates it from other regions. Ethically relevant properties are properties that potentially affect the scope or nature of normative statements. In the case of coexistence one would need to look for properties that would be weighty enough to justify an exemption or modification of national law.

There are indeed a number of examples where national law is modified or exempted in order to accommodate ethically relevant features of certain regions. For instance, while many countries accept only one official language, some regions may still be required to have street signs in two languages when linguistic minorities live in that region. Or another example: when driving a car on the island of Rügen in Northern Germany one is required to use headlights also during daytime, while this is not a requirement in the rest of Germany. The reason for this is the specific combination of shifts in the intensity of light resulting from the Northern location and the dense vegetation along the roads on that island.

The problem with this kind of reasoning is that many of those properties that would constitute appropriate candidates for providing justified exemptions to the policy of allowing the production of GM crops would also exclude any other production of crops. For instance, while a certain area of protected nature would exclude the production of GM crops within that area, it would also exclude the production of other crops within that same area. Where we want undisturbed nature, it makes no difference whether conventional farming or farming of GM crops disturbs the picture. Thus, one would need to find a regional property that makes an ethical difference between the production of GM crops on the one hand, and the production of conventional or organic crops on the other. I admit that I have doubts that such properties can be found.

So far I have considered whether there are justified NIMBY responses to coexistence that would be enforcible by law, i.e. whether a region can opt for a policy of GM free farm production that would provide a legally or ethically acceptable reason to prohibit individuals to engage in that production. I very much doubt that this is the case. However, the argument so far does not seem to exclude voluntary schemes adopted in a region.

It is quite common that regional producers of agricultural products form cooperatives and adopt voluntary schemes of agreed production standards that classify for labelling the products with regional quality labels. This is for instance quite common in wine production in Europe. Individual producers benefit from coming under the umbrella of such a regional label even if the standards are restrictive in terms of permissible forms of production. Certain intensive uses of pesticides are e.g. often not allowed, even if they are in principle allowed by national law. I lack the necessary insight to be definitive about this, but my impression is that these regionally adopted schemes would not be legally enforcible against individual producers who opt out of such a scheme. However, they would lose the market advantage of selling their products with the regional quality scheme. As far as I know, the EC has under different circumstances defended these schemes as acceptable and protected trade marks.

I believe a similar reasoning could be applied also in the case of GM crops. If a majority in a region is of the opinion that their products would sell better if that region markets its products as GM free products, and if the cooperative of producers sets down appropriate standards for production, then this is in itself a strong incentive for any producer to adopt that scheme. Such a scheme could surely not guarantee that no GM production ever will take place within that region since individuals could still opt out. Similarly, it will not make common standards of coexistence superfluous. But as a voluntary scheme the cooperative could impose stricter standards of coexistence, e.g. in regard to minimum distances between fields of similar crops, as those foreseen in national law or EU policy. Not covered by such voluntary regional schemes are the liability aspects of coexistence. These remain the domain of national law.

From an ethical point of view such voluntary schemes could be described as conditional NIMBY arguments. In effect they would say: 'not-in-our-back-yard-as-long-as-the-developer-wants-to-partake-in-benefits-of-the-cooperative'. As such they appeal to the autonomy of producers to adopt stricter quality standards than those required by law. I do not see any ethical problems with such a strategy[3]. What then remains of the ethical challenges of coexistence is basically related to the task of working out good and just liability schemes. This is in any case the task of national law. I understand that some European countries, like e.g. the Netherlands, have already demonstrated that agreement on these issues can be achieved. It is the task of national governments to ensure that similar arrangements can be made that satisfy the various interests.

I conclude that coexistence schemes seem a promising way forward, but that voluntary regional quality schemes that exclude GM crops is a viable alternative for regions where a majority of producers opts for maintaining conventional or organic farming. Those who find GM crops problematic in general should direct their attention to the provision of scientific arguments that demonstrate unacceptable risks.

References

CEC (2006). Communication from the Commission to the Council and the European Parliament - Report on the implementation of national measures on the coexistence of genetically modified crops with conventional and organic farming. {SEC (2006) 313} Brussels 9.3.2006.

Declaration (2005). Safeguarding sustainable European agriculture. Friends of the Earth Europe and Assembly of European Regions, Brussels 17 May 2005.

EC (2006). GMO Coexistence Research in European Agriculture, European Commission, Luxembourg: Office for Official Publications of the European Communities.

GMO Free (2005). Network's technical proposals on coexistence between gmo, and conventional and organic agricultures. GMO FREE EUROPEAN NETWORK, Rennes, 30 November 2005.

ISF (2004). Coexistence of Genetically Modified, Conventional and Organic Crop Production. International Seed Federation (ISF), Berlin, May 2004.

---

[3] In fact, it reminds me of the old slogan of the hippie-movement: 'Imagine it is war, but nobody takes part in it!'.

Vertical gene flow in the context of risk/safety assessment and co-existence

A. De Schrijver[1], Y. Devos[2] and M. Sneyers[1]
[1]Division of Biosafety and Biotechnology, Scientific Institute of Public Health, Juliette Wytsmanstraat 14, 1050 Brussels, Belgium, ads@sbb.ihe.be
[2]Department of Plant Production, Faculty of Bioscience Engineering, University of Ghent, Coupure Links 653, 9000 Ghent, Belgium, Yann.Devos@UGent.be

Abstract

In the pre-market phase, vertical gene flow is framed as a biosafety issue in which ecological consequences of the escape of transgenes are assessed, whilst it is framed as a socio-economic issue of co-existing cropping systems in the post-market phase. While no adverse effects have been identified on the environment due to vertical gene flow from the transgenic maize event MON810, its cultivation is expected to impact on-farm management. To reduce the adventitious GM-input due to pollen flow in non-GM maize produce, isolation distances between GM and non-GM fields are imposed. Numerous studies show that isolation distances ranging between 10 and 50 m would be sufficient to keep the adventitious GMO content resulting from cross-fertilisation below the 0.9% labelling threshold in maize.

Keywords: biosafety, co-existence, GM maize, pollen flow

Introduction

In the European Union (EU), the release into the environment of genetically modified (GM) crops continues to raise various concerns about their potential environmental and agronomic impacts, especially those associated with the escape of transgenes. Transgenes can be transferred beyond their intended destinations through vertical gene flow by transmitting pollen to sexually compatible plants. For (GM) crops, cross-compatible species can either be cultivars of the same species or wild/weedy relatives. Important steps in vertical gene flow are the spread of the (trans)genes by pollen and the formation of hybrids in case of successful fertilisation. A further potential step is the stabilisation of the (trans)genes into the population, a process that is termed introgression. Vertical gene flow is a natural process and well-known in certified seed production, but has received renewed attention in the context of environmental risk assessment of GM crops and in the context of co-existence between GM and non-GM crops.

In the EU, currently only varieties of the event MON810 are cultivated. MON810 is a 'Bt' maize - expressing a *Bacillus thuringiensis* (*Bt*) toxin gene - that confers resistance to the European and Mediterranean corn borer. In this paper, MON810 is taken as case study to illustrate vertical gene flow in the context of risk/safety assessment and co-existence. Isolation distances proposed by scientists and authorities to limit cross-fertilisation between GM and non-GM maize fields are described.

Vertical gene flow in the context of risk/safety assessment

Before a GM crop can be placed on the market and grown by European farmers, it needs an authorisation for marketing. This authorisation can only be given if it is shown that the GM crop imposes negligible risks for the environment and human health. One of the aspects that is looked at during the risk/safety assessment of GM crops, is vertical gene flow and its ecological consequences. In the context of risk/safety assessment, vertical gene flow can be specified as the flow of transgenes from the GM crop to other cultivars, weeds and wild relatives. The escape of transgenes is particularly considered, as the fitness of cross-compatible recipient plants may be altered by the acquisition of a transgene. Theoretically, and depending on which transgene is involved, altered fitness may exacerbate a weed problem, or enable wild/weedy relatives to go extinct or to expand and invade new habitats.

The questions posed in the context of risk assessment are the following: does vertical gene flow to cross-compatible relatives occur in the environment? And consequently, if it occurs will this pose any safety concerns? The probability of successful pollination depends on a great number of factors, including the level of pollen production of the (GM) plant, the rate of dispersion of viable donor pollen, the out-crossing rate, the spatial distance between pollen donor and recipient, flowering synchrony and the sexual compatibility between two parents. Safety concerns related to cross-fertilisation between a GM and non-GM crop will vary with species and traits. For example, a concern of pollen flow from a herbicide-resistant crop is the establishment of herbicide-resistant volunteers in subsequent crops. Herbicide-resistant volunteers may complicate weed control, as these plants can no longer be controlled by the herbicide to which they are resistant. Crosses between the GM crop and its wild/weedy relatives will only be considered as relevant if after pollination, viable and fertile hybrids are produced. Only then the transgene can remain in the wild/weedy population and pose a safety concern. It is thought that hybrids providing resistance to environmental stress - e.g. insects, pathogens, and drought - will more likely provide a selective advantage than genes that alter nutrient composition or confer male sterility.

Maize is a cross-pollinated crop, producing large amounts of viable pollen and relying on wind for the dispersal of its pollen. Maize has, however, no wild relatives in the EU. Therefore, vertical gene flow from GM maize to wild relatives does not pose any safety concerns for the environment. Pollen flow from GM maize to a conventional maize field may occasionally result in flowering volunteers in Mediterranean countries (Messeguer *et al.*, 2006). In such cases, MON810 volunteers can be controlled in the same way as non-GM volunteers, thereby not expected to pose a new ecological concern.

Vertical gene flow in the context of co-existence

In the context of co-existence, gene flow is not taken into account to cover risk/safety issues, but to avoid that the threshold of 0.9% for the adventitious presence of GM material in non-GM produce will be exceeded, as this will trigger the labelling of the product as containing genetically modified organisms (COM, 2003). In this respect, minimising the impact of pollen flow from the GM to the non-GM crop is a major objective, not forgetting that also other GM-inputs, such as uncontrolled GM volunteers and weed/wild relatives containing the transgene, may render it difficult to comply with the established labelling threshold.

As for risk assessment, the probability of cross-fertilisation - often re-termed in the context of co-existence as pollen contamination - is considered. While in risk/safety assessment the likelihood of occurrence of vertical gene flow and its potential consequences are evaluated, in co-existence studies focus is put on cross-fertilisation rates at various distances from the source. For several crops, the vast majority of cross-fertilisations have been shown to occur over short distances. Pollen concentrations, the viability of pollen and consequently successful cross-fertilisation levels tend to decline rapidly with distance from the source following a leptokurtic pattern with a tail containing long distance dispersal events. For maize, it is estimated that individual tassels produce 4.5 million pollen grains (for review on maize characteristics see Devos *et al.*, 2005). Maize pollen generally remains viable 1 to 2 hr after dehiscence, but may remain viable up to 24 hr. As maize pollen grains are one of the heaviest and largest among the wind-dispersed pollen, their travelling distance is limited. In general, cross-fertilisation levels rapidly decrease over short distances of up to 50 m, but low levels can be detected at several hundred meters (Figure 1). As there are no wild relatives of maize in the EU and as maize seeds or seedlings do

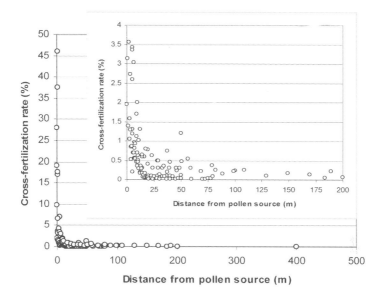

Figure 1. Cross-fertilisation levels in relation to the distance from the maize pollen source upper plot represents a magnification of the original graph (Reprinted with permission from Sanvido, O., Widmer, F., Winzeler, M., Streit, B., Szerencsits, E. and Bigler, F. (in press). Definition and feasibility of isolation distances for transgenic maize cultivations. Transgenic Research, DOI: 10.1007/s11248-007-9103-1.).

not survive winter cold in most EU countries, the major potential biological source of mixing is pollen flow from GM maize (e.g. MON810) fields. Cross-fertilisation rates detected in recipient fields as a result of pollen flow from donor (Bt) maize fields are generally higher in the borders and decrease towards the centre of the field (Messeguer *et al.*, 2006).

Because cross-fertilisation levels rapidly decrease with increasing distance from the pollen source, spatially isolating GM from non-GM fields is a valuable strategy to reduce the extent of out-crossing. A lot of studies have been done, including field experiments and modelling at the landscape level, to define isolation distances (and/or pollen barriers) that should be used to limit cross-fertilisations between a GM maize field and its neighbouring non-GM maize fields (for review see Devos *et al.*, 2005). During these studies, various factors have been taken into account because cross-fertilisation rates generally show a considerable amount of variation between years, sites, experimental designs and agricultural practices. More recent empirical cross-fertilisation studies mimicked worst-case commercial on-farm situations or were performed in real agricultural settings of co-existence (Messeguer *et al.*, 2006). The general conclusion of these studies is that in most actual situations, isolation distances ranging between 10 and 50 m would be sufficient to keep the adventitious GMO content resulting from cross-fertilisation below the 0.9% labelling threshold. However, it is important to bear in mind that this is under the pre-condition that GM presence in non-GM seeds remains low and proper machinery cleaning is performed.

Conclusions

Vertical gene flow through pollen from the transgenic maize event MON810 has no potential ecological impacts, but will affect on-farm management of GM maize grower if neighbouring farmers do not wish to label their maize produce as GMO. According to the European Commission, co-existence measures to limit gene flow should reflect the best available scientific evidence on the probability and sources of admixture between GM and non-GM crops (COM, 2003). The fact that co-existence is addressed at member states level provides an opportunity to develop regionally co-existence arrangements adapted to the local farming practices, landscape and meteorological conditions. Regional and national authorities competent for co-existence are currently discussing or implementing isolation distances ranging between 15 and 800 m, frequently favouring an isolation distance of 200 m (COM, 2006). In most cases, however, isolation distances ranging between 10 and 50 m - depending on e.g. field characteristics, crop types, differences in flowering times and meteorological conditions - would be sufficient to keep the adventitious GMO content resulting from cross-fertilisations below the 0.9% labelling threshold in the total yield of the maize field.

References

COM (2003). Commission recommendation of July 2003 on guidelines for the development of national strategies and best practices to ensure the co-existence of genetically modified crops with conventional and organic farming.

COM (2006). Report on the implementation of national measures on the coexistence of genetically modified crops with conventional and organic farming.

Devos, Y., Reheul, D. and De Schrijver, A. (2005). The co-existence between transgenic and non-transgenic maize in the European Union: a focus on pollen flow and cross-fertilization. Environmental Biosafety Research 4: 71-87.

Messeguer, J., Peñas, G., Ballester, J., Bas, M., Serra, J., Salvia, J., Palaudelmàs, M. and Melé, E. (2006). Pollen-mediated gene flow in maize in real situations of coexistence. Plant Biotechnology Journal 4: 633-645.

Part 1 - Theoretical, conceptual and foundational issues: concepts and approaches

Ethical bases of sustainability

Paul B. Thompson
Department of Philosophy, 503 South Kedzie Hall, Michigan State University, East Lansing, MI 48824, USA

Abstract

The literature on sustainability offers many different ways to integrate ethical norms (understood here as encompassing the full range of normative values) regarding the conceptualization of sustainability. This paper reviews several leading theoretical approaches to the achievement of sustainability within livestock production systems, and examines how ethics either are or might be analyzed as playing a role in each approach. In particular, I will emphasize approaches to sustainability that utilize 'indicators' of sustainability as a way to resolve value conflicts over environmental goals. One point of view sees each indicator representing alternative ethical norms, then utilizes empirical research to measure the relative effectiveness of alternative production systems in meeting achieving progress with respect to indicators. An alternative point of view sees indicators as themselves reflecting a consensus among parties holding different values, a consensus that should be reached before empirical work begins. In the end, it is important for livestock researchers and environmental policy analysts to continue debating the relationship between ethics and sustainability, if there is to be any hope of attaining any measure of coherence among alternative approaches.

Keywords: social movements, animal welfare, resource sufficiency, ecological integrity

Introduction

What is agricultural ethics? One answer is that we should understand agricultural ethics primarily as a debate about the social goals that any given configuration of production and distribution systems for food and fiber should fulfill. This approach differs from other areas of applied ethics and especially medical ethics or research ethics, where 'ethics' is understood primarily as an evaluation of individual conduct. In agriculture, individual agents may be guided by norms such as preserving a family lifestyle or seeking profits, but what matters ethically is whether the combination of these motivations along with technology and public policy produces an agriculture that meets our goals as a society (Thompson, 1986). Among the social goals of agriculture we can list the need to provide adequate supplies of food and fiber commodities, a safe and wholesome food supply, a becoming rural landscape, and a fair distribution of benefits and burdens. More controversial goals might include the preservation of autochthonous rural communities, the promotion of national identity and the encouragement of moral virtue. Agricultural ethics is, on this view, an articulation of these respective goals, and a deliberative assessment of the trade-offs between and amongst competing goals as applied to alternative ways of configuring the agrifood supply chain (Zimdahl, 2005).

Given this understanding of agricultural ethics, sustainability is arguably one goal among others that should be assessed and debated from in a deliberative fashion. Developing a normative conceptualization of sustainability has at least two components. One is the specification of what sustainability is and why it has normative content. The second is deliberation on how sustainability should be weighed against other social goals (productivity, efficiency or distributive justice among them) that can be appealed to in evaluating the food system from an ethical perspective (Thompson, 1995). In previous publications I have documented two general strategies for addressing these tasks, one which specifies sustainability in terms of *resource sufficiency*. This conceptualization has been particularly influential in debates over sustainable development. The alternative is to focus on the elements of the agrifood system that relate to its regeneration, or to its ability to reproduce itself over a series of cycles. This notion of sustainability, which can be called *functional integrity,* looks to models for sustainable yield that have been developed for fisheries management for its primary content, and that can be readily adapted to a variety of renewable resources including soil fertility and surface water management. It can also be extended to a number of socio-economic dimensions, including agricultural credit and conservation or tax policy (Thompson, 1997; Thompson and Nardone, 1999).

As noted already, both of these approaches leave much of the modeling of sustainability to agricultural and ecological science, including social science, and sees ethics as playing two roles. First, it will be necessary to make value judgments in order to specify how much is enough, in the case of resource sufficiency, or how the system is to be modeled, in the case of functional integrity. Is it 'enough' to be able to produce for 50 years, or should it be 100? Who is included? Should we presume that substitution for nonrenewable resources will go smoothly as they are depleted? Should the model of the food system be at farm, local, national or global scale? Should it presume a certain economic standard of living for farmers? For food consumers? Second, once sustainability is specified and measured, we will be presented with trade-offs that will need to be evaluated ethically. Do we accept some system vulnerability (that is some unsustainable elements) in order to promote humane treatment of animals, or cheaper food production? Do we accept some decline in our short term standard of living in order to help producers in less developed nations adopt more sustainable methods? These are just examples of the types of ethical question that might be posed in connection to sustainability. The relationship between ethics and sustainability can be summarized as follows: ethics is an evaluative and deliberative form of social inquiry; sustainability is one of several values or goals that this inquiry should address in connection with agriculture.

The alternative way to conceptualize sustainability almost reverses the relationship to ethics. Here, sustainability is seen as a social movement that incorporates agricultural ethics as one of its elements. On this view, agriculture and food system activities have been under assault from pressures associated with capitalism, the growth of global corporations and the development of industrial production technologies designed to increase return on capital investment. Resistance movements have struggled against this monolith, seeking thematic ways to consolidate their activity and recruit allies. While this orientation to sustainability is sometimes associated with socialist or at least strongly anti-capitalist political activism, it is not necessarily so. It is possible to see sustainability as committed to the creation of a social consensus on economic development and resource use that is capable of accommodating a fairly broad swath of the political spectrum. Nevertheless, sustainability is a 'banner' or rallying point for people concerned with health, personal autonomy, social justice and environmental protection. 'Ethics', is just one of the concerns that are included under the banner of sustainability (Allen and Sachs, 1993).

This way of understanding the relationship between sustainability and ethics does not provide much specificity about what is meant by ethics, but it is rare for ethics to be understood as a deliberative discussion about social goals. Instead, ethics is more typically conceptualized in terms of personal conduct and individual decision making. In the book *Food for the Future,* ethics is portrayed in an essay by philosopher Tom Regan as a personal commitment to vegetarianism. That is, people allied in the movement for sustainability should be morally committed to vegetarianism (Regan, 1993). This emphasis on accepting personal ethical responsibility in dietary decision making has been reiterated in recent books that promote sustainability, though most authors do not call for vegetarianism. Instead, ethics is associated with the consumption of products that promote fair trade, organic production, small-scale diversified farms, and humanely produced animal products, and with opposition to 'factory farming' and GMOs. Here, it is not so much that there is an ethical basis for sustainability, as the title of my paper suggests. Rather, sustainability is the basis for ethics.

Ethics and sustainability: debating the alternatives

What can be said in favour and against each of these alternatives? I will start with some advantages associated with the view that sustainability is a social movement that incorporates ethics, and then consider the alternative, that sustainability is a norm. This will not be a complete inventory of considerations that bear on this question, and I cannot be regarded as an unbiased evaluator of the debate, in any case. Nevertheless, I do think that the social movement conceptualization of sustainability has a number of points in its favour.

First, there is nothing wrong with encouraging people to take personal responsibility for their dietary choices, most of the moral advice the sustainability advocates offer is advice that I would endorse myself. I see nothing wrong with telling people to pay some attention to the way that their food and fiber products were produced, as well as evaluating characteristics of the product itself, and I think that moving away from a commodity based food system is mostly a good thing. Telling people that this supports sustainability may be a little misleading, and there are a few things on the list of ethical products that I do not think are particularly ethical at all, but these are not points I would press.

Second, along with promoting these personal behaviours, many advocates of the sustainability social movement associate sustainability with participatory decision making processes that encourage open-ended deliberative consideration of key options and choices. This is particularly evident in books by Norton (2005) and by De Jongh (1999), as well as with theorists who have seen the social movement toward sustainability as a symptom of reflexive modernization. Since my understanding of ethics relies heavily on the participatory discourse model developed by Jürgen Habermas (1990) I see this as a good thing. I am, again, reluctant to associate this too strongly with the idea of sustainability. When considering participatory models for environmental decision making, I am reminded of Oscar Wilde's observation that the trouble with socialism is that it takes too many evenings. I am not really sure how sustainable these participatory models are, but this is, again, not a point I would care to press.

Finally, I need to acknowledge a substantial weakness in my own approach, which is that most of the people who have broad agreement with me about how we should understand sustainability do not tend to see this notion as requiring any input from ethics. I have no trouble locating technically brilliant articles on the way that agricultural systems can be made more regenerative or that scientists can model agricultural ecosystems in ways that would reveal vulnerabilities in their integrity, but these articles tend to presume that this is a wholly technical exercise, that the authors are in full possession of all the values they need to conceptualize relevant systems, and that these results can simply be handed off to political decision makers who will choose based on the optimization of consumer preference satisfaction. I am, in short, fighting battles on two fronts. My potential allies in the battle against a social movement conceptualization of sustainability are dead set against carving out any role for ethics.

So what are the weaknesses in the social movement conceptualization? I will confine myself to three. First, I am concerned that the substantive understanding of sustainability as functional integrity in the social and biological reproduction of the agrifood system becomes wholly lost in the shuffle. I see no reason why a configuration of the supply chain that meets goals such as social equity for producers, humane conditions for animals and becoming landscapes will necessarily operate within the parameters needed to replenish soil, water and biodiversity, or even rural communities. For this reason, I think it is important that these values associated with the functional integrity of the agrifood system get singled out and acknowledged with a specific term. The word many people have suggested is 'sustainability', but if sustainability is just the name for a social movement aimed at resisting globalization and corporate control, then the commitment to functional integrity is substantially diluted, if it is not lost altogether.

Second, in conceptualizing 'ethics' as appropriate personal conduct in a resistance movement, this model implies that any of the reasons that might be offered in support of the industrial food system are 'unethical'. Thus, reducing the cost of food, increasing the availability of food, and decreasing the amount of land needed for food production all become irrelevant to the evaluation of an agricultural system. Worse than that, if the people who participate in a system that accomplishes these goals earn profits by doing so, they are disqualified from participating in any conversation about sustainability because their motives are wholly suspect. Of necessity, social movements must create a sense of common identity, a sense of the 'we'. Often this is achieved by demonizing a common enemy. The word 'ethics' seems to be playing both roles in the social movement for sustainability. Anyone who is not with us in the sense of endorsing our ethics is against us, and while I noted above that it is possible to accommodate a broad swath of the political spectrum in a consensus oriented view of sustainability social movement, it is also possible to frame ethical commitments in a manner that valorizes some while demonizing others. This aspect of sustainability is especially evident in terms of opposition to biotechnology. Biotechnology is not be *evaluated* ethically. Instead, one's stance on biotechnology is a critical test of whether one is in the movement or outside of it. I believe that this aspect of the sustainability movement is debilitating its ability for critical reflection and deliberative assessment of its own goals.

Finally, the very idea that ethics can be subsumed under the umbrella of social movement theory indicates the cynicism implicit in some quarters of contemporary social science. It is reflected in methodological commitments that interpret all talk about what *should* be the case in terms of subjective individual preferences that are incapable of rational evaluation, on the one hand, or in terms of strategically motivated attempts to control the behaviour of others, on the other. Whether deployed in service to the politics of the right or the left, these commitments in contemporary social science dictate that the analysis of sustainability cannot be conceived as a form of inquiry into what society should do. It is not as if social scientists conceive themselves to be immune from opinions on what society should do. It is simply that they must treat everyone's speech, including their own, as just another preference, just another rhetorical move within a social movement. Those who advocate an ethics debate or social inquiry into the goals of agriculture, such as I do, cannot be understood until we find out what secret preference or interest group they are trying to promote. As above, you are in the movement or you are against it, and social inquiry into norms is viewed as an impossibility.

Sustainability indicators and social movements

These abstract arguments can, I hope, be made more concrete by examining some indicators that have been suggested for promoting sustainability in livestock production. Indicators are measurable quantities that, while falling short of an exhaustive characterization for the phenomenon under investigation, but that can nevertheless be taken to be a sign of relative progress toward it in a given domain. Indicators have long been utilized in a variety of social policy contexts including development studies, where 'development' has been recognized to be a complex and multi-faceted phenomenon. Well before the 1987 Bruntland report called for 'sustainable development', development theorists had begun to develop means to integrate diverse development indicators into an index that would streamline evaluation and simplify decision making for development projects and policy planning. It is therefore not surprising that the use of indicators would be proposed for sustainable development.

Indicators that have been suggested for animal production systems include economic measures of farm viability such as productivity or net farmworker income, ecological measures such as eutrophication or acidification, and animal welfare measures such as mortality or air quality (Cornelissen *et al.*, 2001). These three groups of indicators provide a nice contrasting class for examining the difference between 'sustainability as a norm' and 'sustainability as a social movement'. If the word 'sustainability' is taken to refer to the broad social movement for change in agricultural production systems, then sustainability indicators should reflect the goals and values that participants in this social movement wish to promote. This will be a largely empirical question that can be answered through various kinds of social science research and political processes. The relevant question is whether the indicators chosen for farm viability, ecological impact and animal welfare (a) are in fact endorsed by individuals involved in the social movement; and (b) whether these reflect a sufficiently broad representation of the values that members of the social movement see as relevant to animal production systems.

Programs designed to achieve more sustainable animal production systems do, in fact, undertake social science research and organize public forums designed to ascertain just the answer to these questions. Projects that I have been able to identify generally tend to define the social movement toward sustainability as something that has been broadly endorsed by the public at large. Thus, they tend to utilize social science and public engagement techniques that involve representative samples of public opinion and open forums for discussion of indicators. This may be a very reasonable approach in the Netherlands, where recent events may be said to have created a national consensus on sustainability. But in the United States, at least, sustainability continues to be a highly contested concept. If there is a social movement for sustainability in the U.S., and I believe that there is, it would nonetheless need to be conceptualized as one that enrolls something much less than a majority of the United States public. For agriculture, at least, any movement for sustainability is one that is being actively opposed by many mainstream farm producers and producer organizations. As such, measures of opinion that reflect the U.S. population as a whole might not adequately reflect the values and goals that hold the social movement for sustainable agriculture together in a U.S. context.

In conclusion, when one sees sustainability as a social movement, the question of what indicators will represent movement values depends heavily on the local context. However, to the extent that specific indicators arise from or are endorsed by research or by a participatory process, they are indeed indicators of sustainability. Researchers may then develop indices or scenarios that reflect trade-offs amongst these indicators, which may in turn either be turned over to decision makers or fed back into an iterative process of public engagement. In either case, these indices or scenarios can reasonably be said to be informing public decision making. There is not really a clear role for ethics in this process except in so far as ethics has something to say about any process for involving members of the public or a specific group in decision making. There is no ethical content specifically associated with sustainability, in other words. Researchers must be ethical in reporting data and recruiting participation; members of the public can be expected to reflect their own ethical values in participating in the exercise. Followed faithfully, such an exercise can be expected to reflect prevailing opinion of those deemed to be included in the social movement for sustainability to a significant degree.

References

Allen, P. and Sachs, C. (1993). Sustainable Agriculture in the United States: Engagements, Silences and Possibilities for Transformation. In: P. Allen (ed.) Food for the Future. Wiley and Sons, New York: pp. 139-168.

Cornelissen, A.M.G., van den Berg, J., Koops, W.J., Grossman, M. and Udo, H.M.J. (2001). Assessment of the contribution of sustainability indicators to sustainable development: a novel approach using fuzzy set theory, Agriculture, Ecosystems and Environment 86: 173-185.

de Jongh, P. and Captain, S. (1999). Our Common Journey A Pioneering Approach to Cooperative Environmental Management. Zed Books, London.

Habermas, J. (1990). Moral Consciousness and Communicative Action. In: C. Lenhardt and S. Weber (eds.) The MIT Press, Cambridge, MA.

Norton, B. (2005). Sustainability. University of Chicago Press, Chicago.

Regan, T. (1993).Vegetarianism and Sustainable Agriculture: The Contributions of Moral Philosophy. In: P. Allen (ed.) Food for the Future, Wiley and Sons, New York, pp. 103-122.

Thompson, P.B. (1986). The Social Goals of Agriculture. Agriculture and Human Values 3: 32-42 .

Thompson, P.B. (1995). The Spirit of the Soil: Agriculture and Environmental Ethics. Routledge, London and New York.

Thompson, P.B. (1997). Sustainability as a Norm. Technology in Culture and Concept 2: 75-94.

Thompson, P.B. and Nardone, A. (1999). Sustainable Livestock Production: Methodological and Ethical Challenges. Livestock Production Science 61: 111-119.

Zimdahl, R. (2005). Extending Ethics. Journal of Extension 43: www.joe.org/joe/2005october/comm1.shtml

Building a sustainable future for animal agriculture: an environmental virtue ethic of care approach within the philosophy of technology

Raymond Anthony
Department of Philosophy, University of Alaska Anchorage, 3211 Providence Drive, Anchorage, AK 99508, USA, ranthon1@uaa.alaska.edu

Abstract

Our present philosophy of technology, which fails to consider the central ethical question, 'How should we live?' and which promotes an insular human-centricism, is inimical to sustainability, understood in terms of long-term health, security and vitality of economic, environmental and social systems. Within this paradigm, not only are farmed animals conceived as 'absent referents' and commodities but our ability to care for and respect these animals and the land as part of our broader community is impaired significantly. I begin by arguing that an ethically robust philosophy of technology must include both an account of how we ought to interact with agriculture and food, and an account of the character dispositions we ought to have regarding the land, farmed animals and those in the food system. I contend that the question related to the proper relationship between the two can be found by discovering the appropriate role of virtue in food production and consumption. I argue that we should accord an environmental virtue ethic of care (EVEC) a central place within a reformed philosophy of technology. Next, I discuss the merits of EVEC as a foundation for our philosophy of technology vis a vis animal agriculture. EVEC enjoins us to value the agricultural spaces and processes and its constituents for their own sake and for the sake of becoming better persons. I end by outlining particular virtues associated with the ethic of care and consider some likely practical implications as it relates to farming animals.

Keywords: ethic of care, animal ethics, philosophy of technology, environmental virtue theory

Introduction

Much of the contemporary discussion in farmed animal welfare ethics has centred either:
- On philosophical arguments concerning the moral status of animals, i.e., do animals possess the requisite moral status making properties that would admit them into the moral community (Singer, 1990; Regan, 1983 and 1990; Midgley, 1983, Nussbaum, 2004), or
- On the science of animal welfare, where the focus has been on how better to address the needs of farmed animals through a better understanding of their capacities, whether or not the adaptations they possess meet the demands of the production system and whether they are free from physiological, psychological and physical harm or disease (Fraser, 2001; Fraser and Weary, 2004).

While the various initiatives mentioned above have caught our collective imagination in terms of concern for animals, the move forward has been piecemeal and fragmented. What is often sidestepped is the fundamental ontological problem concerning the asymmetrical relationship that we have to animals (and the nonhuman world fort that matter), and how technology mediates this. Here, the underlying question that is missed is one that concerns 'how should we live?'. That is how should we live rightly with respect to the land and its constituents, understood in terms of ecological, economic and social sustainability. To address this shortcoming and in order to address the plight of animals on a more global scale and in a more sustainable manner, I want to suggest that we tackle the ontological question. This will involve revisiting our philosophy of technology.

This paper is divided into two parts:

In Part 1, I consider the ontological problem involving the conceptualizing of farmed animals as mere resources as a result of our current philosophy of technology. Appreciating the good of animals in their own right is obfuscated by the fact that we are blinded by our own self importance often distorted by innovation and technological systems that allow for things to be available on demand or 'on call'. I suggest that we challenge our existing philosophy of technology as a necessary first step in coming to terms with our 'being in the world'.

In Part 2, I suggest a virtues-motivated technology as a poignant way to 'turn' our philosophy of technology. Here, I delineate what an environmental virtue ethic of care (EVEC) within the philosophy of technology might look like vis a vis animals. I will sketch briefly what amounts to four elements that comprise the ethic of care approach as a possibility for reform of our philosophy of technology.

Part 1: revisiting and reforming our existing philosophy of technology

The desire for cheap food has come with moral and ecological costs. The quest for cheap food as the basis of our agricultural policy for the last half-century or so has not only raised health and environmental concerns, but also concerns related to the human-farmed animal relationship. The availability of 'cheap food' has (allegedly) allowed us to disburden ourselves from the need to spend much of our time and income on procuring food. This aspect of 'the promise of technology' has thus made it possible for us to maintain a certain quality of life. However, the technological or production systems that facilitate this availability have promoted a uni-dimensional valuing of food and our relationship to agriculture (Thompson, 2001). The way we farm has been under scrutiny for its failure to respect and strengthen social systems and moral relationships, minimize our collective ecological footprint and strengthen local economies.

In the case of farmed animals, our desire for cheap food has relegated them to the status of commodities or units of production. They have become 'absent referents' (Adams, 2000) for many of us, especially those of us who do not interact with farmed animals except at in their final form under the knife or fork. Animals are conceptualized and experienced as facsimiles of actual 'subjects-of-a-life.' When conceived and experienced as units of production and less as beings with moral status or a good of their own, farmed animals are easily forced into situations for which they lack the requisite adaptations in order to meet the demands of industry. That is, since 'they do not matter' in their own right, they are readily forced to fit into our plans and projects. Production practices that handle billions of these 'absent referents' annually continue to 'squeeze round pegs into square holes' (Rollin, 1995). While we enjoy the benefits of disburdenment, farmed animals suffer a disproportionate amount of the costs as a result of our values, desire and our obliging and expedient innovation.

Our treatment of farmed animals this way is a function of the fact that humans and animals occupy qualitatively different positions. Philosophical discussions in animal ethics typically sidestep this ontological problem regarding the asymmetrical relationship between ourselves and the nonhuman world. It is either taken for granted or more so, taken for granted that the ontological problem has been addressed adequately, simply by highlighting its existence. Thus, there is little critique of what continues to motivate this asymmetry and why among the general population, we do not take seriously enough the plight of resource animals.

Without going into too much detail here, the asymmetrical relationship that promotes a mentality of 'resource, here ready for us', is a reflection of our philosophy of technology. In the line of Heidegger (1977) and Borgmann (1992 and 1984), technology as such can be viewed as a form of life, with governing patterns that shape values, behaviour and ways of seeing the world. The deeper question that often is sidestepped by the standard critique regarding our treatment of animals is one that deals with how we should live, i.e., a question of Being in the world. It is one that deals with the moral position we occupy in the world, how we take up the 'other', and the kind of relationship we ought to have with the 'other'. In the case of resource animals, a genuine turn in our attitudes toward and treatment of them hinges on the prospects for a 'turn' or reformation of our philosophy of technology.

According to Heidegger, a new reconceptualizing and recontextualizing of Being is as a 'releasement toward our true nature', arguably as sympathetic or other-regarding beings not simply concerned with ourselves; who are not ruled by a technological system or technical devices which make us dependent upon them and which blinds us to the nature of things as they are in themselves and our proper relationship to them. A concerted effort to discuss this is still wanting among those interested in agricultural issues let alone those concerning animals. It is ultimately a question of not only an ethic of right action but that of right conduct too, of how we should live.

In the present context where our lives centre around consumption and commodities, we are seduced by 'the promise of technology' (Borgmann, 1984). In the case of animals, we view them as something that we can dispose of as we please. Animals continue to be viewed as beneath us and serve in the role of mere utility. Technology as a form of life here threatens our role as revealers of being so that we reveal Being only as 'standing-reserve' or resource (Heidegger, 1977). As mentioned above, our embrace of the industrial paradigm in the case of animals contextualizes them as having value in their final form, as protein or units of production. In light of this, we seek to command resource animals and pay little heed to what they are like in themselves, and their quality of life in our charge as a result of our technological systems. We have institutionally decided not to consider them in their own right as something for which we need skills, attention and other virtues in order to promote their welfare. As contended by Heidegger, things relegated to the realm of 'resource' are there merely 'on call', for our disposal. Our control of them occasions a kind of irreverence. Thus, we readily employ the controlling power of technology in such a way that situates these mere units of production, these 'square pegs into round holes'.

In Heideggerian terms, and in taking up the nonhuman world as 'available', we do not encounter ourselves in our essence. Questions related to the shape of our lives as regards food and agriculture is long overdue. As Heidegger points out, the danger of not understanding our relationship to technology (as form of life) is that we continue on as mere silhouettes of ourselves and continue to be dislocated from being in the world when we exalt ourselves over the nonhuman world and proclaim our self-importance. While often not conscious, the philosophy of technology that we adopt that perpetuates this kind of lifestyle and which conceals the nature of things and other possibilities of living more harmoniously and sustainably with others ultimately will be to our own detriment.

The prior question of being in the world is an important piece in tackling the kind of relationship we ought to have with agriculture and farmed animals; it is one that concerns the shape of our lives. The ontological challenge then is to cultivate and make room for genuine forms of Being in the world within a technological setting. Thus, the basic challenge to this asymmetry is to reinstitute a sort of counter weight that helps us resist the tendency towards domination and our own self-importance. So, if technology is to reform, my suggestion is that we should work towards cultivating certain virtues of character that serve to address the question of being in the world, especially how to be good at being human vis a vis resource animals in this case. The consideration of virtues teaches us to challenge the current ontology of technology and to find firm ground for a new way of Being with technology and the other.

Part II: an environmental virtue ethic of care within the philosophy of technology

The standard ways of critiquing this form of radical human-centricism fails (what I have termed elsewhere as first wave philosophical animal ethics) because it does not encourage moral imagination nor creativity in dealing with issues related to how our technology reinforces the conception of animals as necessarily inferior to us, nor does it address the antecedent or root concern of how we should live in relation to the nonhuman world, i.e., a question of 'Being.' What seems to be lacking in these standard forms is a steady focus on living sustainably with others, i.e., 'how to' care questions, especially for those in our charge. The hang up has been on 'whether' type arguments that deals with the responsibility to do so.

In helping us turn our attitudes towards being in the world and reforming our relationship to technology as a form of life that governs our patterns of behaviour, challenge then is to develop a more global environmental or agricultural ethics that not only contains a commitment to the welfare of farmed animals but which follows from what it is to be good at being human. It is a turn to virtue ethics that deals with the question of the shape of our lives that arguably will help promote these aims.

Environmental virtue ethics as has been discussed lately (Newton, 2003; Sandler and Cafaro, 2005) is a promising counterbalance to our existing philosophy of technology, since it advocates lives that are intentionally lived according to rational principles and empathy. Briefly, it focuses on developing personal as well as community based ethical standards that are open to critique and constant evaluation. More importantly, there is a tendency to be open to changing our orientation in order that we lead more authentic lives. Some central tenets of this ethic involves coming to terms with reason and emotion, pursuing the good through recognition of responsibility as a dynamic enterprise and taking ownership for choices that we make especially in the face of relationships that involve vulnerable or dependent others. Key to all of this is the view that the self is bound to others both personally and through the various institutions that formalize and facilitate life. We are thus, necessarily embedded within communities, both human and mixed human-animal-natural world ones.

While much has been taken for granted here for sake of brevity, virtue ethics within the environmental framework involves a desire to see not only human communities flourish, but nonhuman ones as well. Thus, the framework of virtue ethics can constitute a 'turn' in our philosophy of technology. It holds promise for addressing Heidegger's question concerning being (how should we live?) by recontexualizing our relationship with technology and with the other in fruitful ways. Considering virtue ethics also allows us to address Borgmann's concern that we have been seduced by the promise of technology somewhat heedlessly by turning attention on what else matters to us.

Back to animals and technology: how should we proceed?

Central to a turn in our philosophy of technology vis a vis resource animals is a genuine envisaging of the human-animal relationship that corrects the distorted view that animals are mere resources. Here, the virtue of care within the framework of an environmental virtue ethics is proposed as a counter weight between ourselves and farmed animals. However, we should be aware that the following is likely to persist and our ethic of care must consider these factors in order to be effective:

- We can expect that animal use will continue for the foreseeable future and may expand with technological innovation. Hence, any animal ethic will be mostly for nought if it is not developed within a philosophy of technology.
- Technical devices and technological systems can either mediate positively or negatively our relationships with the other and blind or reveal the true nature of things. In order that we do not

continue to experience the animals we farm in ways not limited to their end state as products, a more phenomenological understanding of animals should be primary.

- We should avoid privileging technology as good in its own right in order to forestall the tendency to be seduced by it and thus relegating other things to the status of 'relative good', dependent upon how they serve technology as good in its own right (Borgmann, 1984). Farmed animals become only a relative good, and continue to be degraded to the status of resource insofar as it serves our desire for consumption in one way or another.

Specifying the virtue of care

Environmental (and in this case agricultural) virtues are proper dispositions or character traits for human beings to have regarding their interactions and relationships with agriculture, farmed animals and food fort hat matter. The virtuous person in this case is disposed to respond to farmed animals in an excellent way. In the case of animals in our charge, we should consider the merits of an environmental virtue ethic of care. In the line of Simone Weil and other ethicists of care, to care adequately as virtue is a quality of the morally good person or society. What is central here is how best can we perform our caring responsibilities to those with whom we are situated. Thus, the ethic of care necessarily involves an engagement with the concrete, the local and the particular. Within this tradition, there are four key elements that provide a good starting point for developing a framework of care (an environmental virtue ethic of care if you will (EVEC)) that can serve as a counter weight to our existing philosophy of technology. They include (adapted from Joan Tronto, 1993 and Simone Weil's work on 'attentive love' in Little, 1988):

- Attentiveness: which involves being cognizant of what's going on in food production and paying heed to the plight of animals and how we contribute to their welfare (and to the capacity of those who care for them).
- Responsibility: which involves the recognition that there is a need to perform certain care functions as a result of the kinds of relationships we have formed with others.
- Competence: which involves discharging one's caring responsibilities in ways that actually bring about good welfare for the one's cared for.
- Responsiveness: which involves recognition of the dependency and vulnerability of those in our charge, and being alert to the possibilities of negligence, abuse or incompetence

Good care requires that the four elements work in an integrated fashion. Integration is important here so that we are not complacent and are able to see and address conflicts as they arise.

Practical implications

The ethics of care approach can serve as a counterweight to our relegating of animals as mere resources. It brings us one step closer in addressing more seriously some very important questions related to 'how we should be in the world' vis a vis farmed animals in this case, and the nature of the human animal relationship on the farm (Anthony, 2003). These questions include (Sandoe *et al.*, 2003):

- What is a good animal life?
- What is the baseline standard for morally acceptable animal welfare?
- What farming purposes are legitimate?
- What kinds of compromises are acceptable in a less-than-perfect world?

As we consider reforming our ethics as it applies to the context of farmed animals we should at least ensure that husbandry conditions, i.e., our technological system meet the following. That we:

- Work towards developing management practices and technologies that enhance the biological functioning of animals as one way to mitigate against unnecessary suffering and so that animals are in husbandry situations that meet their adaptations and capabilities.
- Consider farmed animals on thieir own terms (as much as possible). That is, to ensure that they feel well (for example, psychologically, socially and physiologically) and that their species specific tendencies that is consistent with positive welfare is promoted.
- More globally, those of us that are connected to the food system (and there are many of us), as consumer bare a responsibility to educate ourselves well about the plight of animals and the way in which we contribute to their care. Incumbent upon us is the goal of becoming more ethically savvy consumers. To do any less is to continue to demonstrate a kind of passiveness that is associated with a 'we are what we eat' framework which this essay is trying to regurgitate in favour of a more ethically digestible one where, motivated by the virtues of the care ethic outlined above 'we eat what we are'.

References

Adams, C. (2000). The sexual politics of meat: a feminist-vegetarian critical theory tenth anniversary addition. Continuum Publishers Co., NY.

Anthony, R. (2003). The ethical implications of the human-animal bond on the farm. Animal Welfare 12: 505-512.

Borgmann, A. (1992). Crossing the post modern divide. University of Chicago Press, Chicago.

Borgmann, A. (1984). Technology and the character of contemporary life: a philosophical inquiry. University of Chicago Press, Chicago.

Fraser, D. (2001). Farm animal production: changing agriculture in a changing culture. Journal of Applied Animal Welfare Science 4: 175-190.

Fraser, D. and Weary, D. (2004). Quality of life for farm animals: linking science, ethics, and animal welfare. In: G.J. Benson and B.E. Rollin (eds.) The well-being of farm animals: challenges and solutions. Blackwell Publishing, NY, 39-60.

Heidegger, M. (1977). The Question Concerning Technology and Other Essays, trans. William Lovitt. Harper and Row, New York.

Little, P. (1988). Simone Weil: waiting on truth. St. Martin's Press, NY.

Midgley, M. (1983). Animals and Why they matter. University of Georgia Press, Athens.

Newton, L. (2003). Ethics and sustainability: sustainable development and the moral life. Prentice Hall, NJ.

Nussbaum, M. (2004). Beyond 'compassion and humanity': justice for nonhuman animals. In: C. Sunstein and M. Nussbaum (eds.) Animal rights: current debates and new directions. Oxford University Press, NY, 299-320.

Regan, T. (1991). Defending animal rights. University of Illinois Press, Urbana, USA.

Regan, T. (1983). The case for animal rights. University of California Press, Berkeley.

Rollin, B. (1995). Farm animal welfare: social, bioethical, and research issues. Iowa State University Press, Ames, IA.

Sandler, R. and Cafaro, P. (eds.) (2005). Environmental Virtue Ethics. Rowman and Littlefield Publishers, UK.

Sandoe, P., Christiansen, S.B., and Appleby, M.C. (2003). Farm animal welfare: the interaction of ethical questions and animal welfare science. Animal Welfare 12:469-478.

Singer, P. (1990). Animal liberation revised edition. Avon Books, NY.

Thompson, P.B. (2001). Reshaping conventional agriculture: a North American perspective. Journal of Agricultural and Environmental Ethics 14: 217-229.

Tronto, J. (1993). Moral boundaries. Routledge/Taylor and Francis Books, Inc., NY.

Emergence and auto-organisation: revising our concepts of growth, development and evolution toward a science of sustainability

Sylvie Pouteau
Ethos INRA, UR Biologie Cellulaire, INRA, RD10, F78026 Versailles, France, Sylvie.Pouteau@versailles.inra.fr

Abstract

Our actual understanding of biodiversity and sustainability, hence our capacity to address and reverse adverse effects, is still limited. I would like to argue that this insufficiency is not (only) quantitative (learning through cumulative steps) but rather qualitative (raising adequate cognitive faculties). The common interpretation of biodiversity in terms of 'genetic resources' is based on a mechanistic scientific liberal worldview. But machines are not self-sustained as they rely on external principles for their coming into being and functioning. In fact, organisms have internal formative principles that are sufficient to both specify their kind and allow new emerging forms to arise from them. Metamorphosis (shaping oneself through becoming another) is an inherent property of living processes. Without variation, there would simply be no growth, no development, no evolution. Metamorphosis involves non linear changes resulting in emergence and auto-organisation. New theoretical frameworks do exist to analyse such processes. Yet, non linearity is still a challenge for our common sense. To think emergence, auto-organisation and individuality in life sciences we need to extend our cognitive faculties. Cognition is a value system and here the notion of paradigm shift takes all its importance as it links science and ethics in a new vision of the common good. I will discuss how changing our value system can impact on knowledge and call for a revision of the current theory of development and evolution. There is not only a need for new concepts but novel cognitive approaches must also be developed to address sustainability values.

Keywords: auto-organisation, biodiversity, cognitive values, emergence, genetic resources

Introduction

It is now widely recognised that biodiversity, meaning biological diversity for short, is a value under threat and that it should be protected. It is also clear that our human activity contributes to a large extent to its erosion. This concerns both the wild and the domesticate, i.e. the part of diversity that humans have created through domestication, acclimatisation and breeding and that constitutes a major source for our food supply. Obviously, conservation in national monuments and seed libraries can only offer a partial answer to the problems. The main challenge is to ensure that agricultural and technological practices are sustainable, i.e. able to maintain diversity 'in process' through continuous enhancement and dynamic management.

So what has gone wrong? Have we been merely foolish and vain in overlooking the consequences of our human activities ('we should have known') or was the 'state of the art' insufficient ('we simply did not know')? The answer to these questions - moral responsibility vs cognitive mastery - depends on the way we view and value biodiversity and the type of knowledge we consider appropriate to address it. Here, I would like to argue that sustainability is an intrinsic property of living systems. For this reason, raising adequate understanding of the nature of life processes is not only a cognitive issue but also a moral one. My approach will consist in examining the implicit values incorporated in current concepts and interpretations used in biosciences. I will discuss some conceptual and ethical implications with respect to a science of sustainability.

The liberal scientific value system: revolving around machines

Every world view relies on a specific value system that conveys meaning in our relationship to the world. This value system is a religion, taken here in the widest, etymological sense, i.e. a mediation between principles anchored in a representation of the world and our activities. Principles cannot be demonstrated. They are 'given' so to speak, or revealed through intuitive perception (Pouteau, 2004). Within a given principle framework, deliberative rationality provides specific internal coherence for a global cognition system. Knowledge proceeds not only through cumulative steps but also through qualitative changes in our representations of the world. Every value system allows to explore and describe a specific angle of the wider reality. Part of this reality is hidden - a precondition for any cognitive endeavour: to uncover the hidden is the purpose of science. Yet, in different value systems the ultimate source of moral intuitions and cause of phenomena - usually figured by god(s) and/or divinities - are thought to remain beyond our reach, making these systems to some extent compatible with each other.

The liberal scientific value system, however, establishes a unique metaphysical institution. First, its specific principles of method - in particular: objectification, empiricism and reductionism - lock the process of mediation itself (both moral and cognitive) by disqualifying intuitive perception and inner evaluation (Pouteau, 2004). Second, it postulates the existence of an entirely self explanatory, universal system based on causal necessity. The leading representation in this value system is rather simplistic: the reality is a clock, a machine (or a computer). The hidden parts, provisionally beyond our reach, are constituents that can be analytically identified and eventually deciphered. Of course, this representation remains inherently creationist: machines are not self-informed, they rely on external formative principles for their coming into being. Chance, selection and evolution must take over the ancient role of the divine architect or engineer to explain how the machines function (Kupiec and Sonigo, 2003). Most importantly, machines are not self-sustained. They are intrinsically dependent for their maintenance, prone to failure and decay and eventually perishable. Finally, machines are not ends-in-themselves but contingent devices to satisfy others' needs and expectations.

Genetic determinism and resources: the qualification of biodiversity under liberal scientific values

The entire living world spread over the earth can be addressed with the notions of 'biosphere' coined by Teilhard de Chardin (1955), and biodiversity. The two notions embrace distinctive features of the living world, the wholeness of life as a unifying essence endowed with intrinsic teleological value, and its expression through a wealth of different particular manifestations in individuals and phyla, wild and domestic relatives. These notions are usually evaluated with an analytic, quantitative rationale based on collection and classification strategies: knowing species by their names, traits, locations, frequencies and degrees of kinship or phylogenetic relationships. The qualification of biodiversity in terms of 'genetic resources' is very revealing of the cognitive and ethical values that are implicitly applied in evaluation approaches.

The DNA machine in a gene order

According to scientific values, living organisms are sophisticated machines determined by their genetic blueprint. The current interpretation is that genetic programs control every aspect of the functioning of the machines: ontogenetic development, interaction with the environment, creation of variation available for further selection etc. DNA, a quasi-mineral substrate, is supposed to contain all the instructions of the program. This 'book of life' is thus granted special value and is meant to represent the very essence of living organisms. Genes, in spite of the difficulty in circumscribing their specific identity as DNA entities, remain the main actors in this instructionist worldview.

Resources: a commodity in a consumption model

According to liberal, utilitarian values, living organisms represent a commodity, a 'resource' to respond to our needs for consumption whether in the short term or in the distant future. Conservation and preservation of biodiversity are aimed at maintaining living organisms as a resource potential for us and future generations but not for their own good. As a commodity, they fall under the commercial regulation applicable to any other good, hence their eligibility for patenting. In this context, the inherent worth or intrinsic value of living organisms is not acknowledged.

Life auto-organisation and emergence: valuing autonomy and integrity

We all know that the essence of a book is not in the paper nor in the ink but in the spirit or intelligence which conceived its content. In fact, any cognitive endeavour must first postulate that the world is intelligible. Eventually, the essence of the living machine has to be found elsewhere than in DNA, in non material principles that constitute a form of 'intelligence'. So, who wrote the 'book of life'? The synthetic theory of evolution tells us that the intelligence is the sum total of failures (mutations) and contingency (chance and selection), i.e. external principles. But can we think of internal principles that are not eventually external? What are the processes involved in the material manifestation of this internal intelligence, its emergence into a global coherence: an organisation (organism)? To answer theses questions, a shift from mechanistic thinking ('the parts explain the whole') to emergence (systemic) thinking ('the whole explains the parts') is necessary. It is supported by formal methodological approaches and experimental data from the science of complexity. It leads to a conception in which processes can only be autonomous, with no need for genetic programs or instructions to govern them, i.e. auto-organised processes (Pouteau, 2007).

Non linear dynamic systems

The notion of auto-organisation encompasses the properties of non linear dynamic systems. All living processes are not only material (i.e. spatial) but also dynamic (i.e. temporal). The principles of these systems are based on the notion of attractors, or basins of attraction and the possible occurrence of alternative states depending on the initial conditions and history of the system. This theoretical framework can explain the emergence of spatio-temporal structures (organisations) that are 'dissipative', i.e. that allow a creation of order out of disorder (also called dissipation of entropy).

Most - if not all - biological processes are non-linear: the fact that A implies B is only a superficial observation, a mechanistic illusion. The determinist interpretation is correct at the macroscopic level but erroneous at the microscopic level, supposed to be causal for the macroscopic level. For instance, the cell cycle is a determinist process that leads to partition of one cell into two daughter cells; but every independent step in this cycle is reversible (Novak and Tyson, 2003). Bistability is probably the most common example of non-linearity. But it usually passes unnoticed because it can only be detected by applying very specific experimental methodologies. The emblematic example is the lactose operon (Laurent *et al.*, 2005). Ironically, this was initially described as the model for genetic linear determinism by Jacob and Monod (Nobel Prize in Physiology or Medicine, 1965). Prion is another example (Kellershohn and Laurent, 2001) illustrating how epistemological and ethical issues can eventually merge together. Indeed, sporadic cases of BSE in cow herds were probably not due to infection but to occasional threshold overriding in a slow non-linear transconformation of healthy into pathogenic prion protein. In the light of this epidemiological analysis it may be inferred that herd slaughter was most probably unnecessary.

Variation: the essence of life

If information is thought to be instructional, then variation must be interpreted as a failure, a mistake in proper transmission and effective realisation of a program. In this case, inter-individual variability is merely a 'noise' that blurs true biological reality and must be buffered with appropriate statistical tools. In fact, variation cannot be eliminated because it is constitutive of living phenomena. Without permanent disequilibrium and fluctuations between stationary states and unstable states, growth - hence morphogenesis, development, and evolution - could not take place. Qualitative modifications associated to changes in variability and correlations lead to emergent events and global re-organisations. Some developmental phases prove more sensitive than others to external perturbations. This may lead to multiple alternative states as in the case of salt adaptation in sorghum (Amzallag, 1999). Variability thus needs to be addressed as a specific biological property that provides specific information on the state of the system. This is necessary to qualify biodiversity in terms of process - relying on plasticity and polymorphism - and not only of the products (species) resulting from this process.

Material distributiveness and integrity

Information cannot be equated to a material substrate (DNA or RNA). Yet, its manifestation is material and involves actual physico-chemical interactions. In-formation is physico-chemically and materially distributed over all possible factors, endogenous and exogenous: an integrity. From a biophysicist's point of view, the biological matter is comparable to a growing viscous 'paste' and can be studied by hydrodynamics. This 'paste' is ruled by physico-chemical properties that are self-explanatory and do not require additional putative instructions. In other words, the 'paste' is self-instructed by its inherent constitution - its structural, material constitution and its dynamics. Because of this constitution, it folds, unfolds, is creased etc. So, finally, the development of a tetrapod embryo appears to be entirely spontaneous (Fleury, 2006). One does not need genes to make legs and arms: these are auto-catalytically organised due to the inherent material properties of the living 'paste'. Of course, genes come into action as parameters in the equations of growth since proteins, especially enzymes, are essential material factors that set these parameters (viscosity, elasticity etc.).

Evolution and metamorphosis: valuing process and plasticity

Growth, transformation and evolution are key words to qualify processes both in the natural world and in cultural society. Civilisations have emerged, culminated in some form of culture, then declined and disappeared. Biological phyla have their hour of glory to later become extinct and replaced by new successful forms of organisation. We, occidentals of the last four centuries, have taken these processes very seriously by placing faith in progress right in the centre of all our affairs. Evolution itself has become an icon of modernity. But nostalgia for the ancient created cosmos and the need for permanence is still perceptible in some concepts, even in the theory of evolution itself.

The synthetic theory of evolution

With the metaphor of the living machine determined by a central causal system (genetic program), it is necessary to invoke some form of external intelligence (selection) to explain how something 'intelligent'/ organised/ complex/ etc. can arise in the course of evolution out of a succession of failures and mistakes (mutations, noise). The instructionist or synthetic theory of evolution (neo Darwinism) still relies on the notion of fixed categories, once created and then perpetuated through reproduction. It defines their identity as an external property based on the criterion of sameness established by intellectual construction. In this context, plasticity, i.e. the capacity to express different phenotypes depending on the environmental conditions - and polymorphism, i.e. the capacity to express different features in a given environment - are seen as hindrances for proper identification of specimen.

Metamorphosis: a concept for growth, development and evolution

A major effort in evolving our concept of evolution is needed to keep up with most recent scientific advances. According to Jablonka and Lamb's analysis (2005), modern epigenetics is calling for an evolution of neo-Darwinism and the authors make a plea for a darwinism that would accommodate some of Lamarck's principles. New biophysical research is also prompting us to revise our variation-selection over-simplistic interpretation. The work by Fleury (2006) reveals that the inherent dynamic and physico-chemical properties of biological matter are sufficient to explain the spontaneous auto-organisation and 'at once-formation' of embryos.

With a distributed in-formation system developing through spontaneous organisation, variation in any constituent can affect the whole make-up through feed-back effects and lead to new unpredictable, yet deterministic organisations. Epigenetic effects (independent from alterations in the DNA sequence) exemplify how the environment can impact on genetic expression and lead to heritable phenotypic changes. Categories are plastic natures, they are continuously remodelled through interactions, communications and exchanges with the wider context. Their identity is dynamic, established in a permanent 'becoming-an-other'. Metamorphosis, the term used by Goethe in 1799 to describe the gradual transformation of plant appendages into one another, seems an appropriate concept to qualify this evolution 'in process' as a continuous flow of transformation relying on growth and developmental emergence of novel organisations (Pouteau, 2006).

Sustainability: toward an agro-ecological scientific value system

Sustaining... what... and how?

Sustainability is not a static but a dynamic concept. It needs to be thought of in terms of process, autonomy and self-propagation. The current scientific paradigm relying on the metaphor of the living machine is intrinsically unable to address these properties, hence the issue of sustainability whether this concerns biodiversity as illustrated in this paper or human activities and society. The hypothetico-deductive approach mainly used in current biosciences, especially in biotechnology, is locked in a mechanistic mode of investigation, prediction, and action. To step beyond this contradiction, the following issue must be addressed: how can we think of internal formative principles able to create a global coherence in living organisms in the same way as the intelligence of a machine inventor?

Seeing integrity and process transformation: a cognitive sense for metamorphosis

Integrity is not an integrality, the sum total of constituents in interaction. It is an emergence, a novel property emerging from underlying levels of organisation as a specific global coherence with its own intrinsic worth (Pouteau, 2006). This coherence, or intelligence, is an entelechy in Goethe's words (2003): a force that draws itself into existence out of itself. The notion of 'Urorgan' or 'Urpflanze' - translated as primordial organ/plant, archetype or type - is not a simple mental concept: the idea of organism is the entity of the entelechy itself. To comprehend living organisms, percept and concept must merge together in what may be called a phenomenology or epistemology of intuition. This requires a recognition and enhancement of inductive approaches in scientific investigation, i.e. the integration of the notions of process, dynamics and evolution within the cognitive endeavour and practice.

The issue of sustainability is not just one more research subject on the agenda. It is a call for an epistemological revision. Agro-ecology is not just one more disciplinary field. It is a new scientific value system. Part of the challenge is to gain new access to the process of moral and cognitive mediation through intuitive perception and inner evaluation. De-automation of world representations currently locked in the scientific liberal value system must be achieved by raising new faculties, both individual and social. At the individual level, process learning may be educated by specifically training a vision of phenomena that is temporal, non spatial and non linear, for instance by developing a scientific methodology inspired from Goethe (Bortoft, 2001). At the social level, process learning is embedded in participatory research confronting different sources of knowledge through partnership between scientists and communities to reach agro-ecological goals (Warner, 1997). Finding new ways of knowing should become an integral part of policy research on sustainability.

References

Amzallag, G.N. (1999). Individuation in *Sorghum bicolor*: a self-organized process involved in physiological adaptation to salinity. Plant Cell Environment 22: 1389-1399.

Bortoft, H. (2001). La démarche scientifique de Goethe. Triades, Paris, France.

Fleury, V. (2006). De l'oeuf à l'éternité: Le sens de l'évolution. Flammarion, Paris, France.

Goethe, J.W. (2003). La métamorphose des plantes. Triades, Paris, France.

Jablonka, E. and Lamb, M.J. (2005). Evolution in four dimensions: Genetic, epigenetic, behavioral, and symbolic variation in the history of life. MIT Press, USA.

Kellershohn, N. and Laurent, M. (2001). Prion diseases: dynamics of the infection and properties of the bistable transition. Biophysical Journal 81: 2517-2529.

Kupiec, J.-J. and Sonigo, P. (2003). Ni Dieu ni gène ; pour une autre théorie de l'évolution. Points, Paris, France.

Laurent, M., Charvin, G. and Guespin-Michel, J. (2005). Bistability and hysteresis in epigenetic regulation of the lactose operon. Cell Moleular Biology 51: 583-594.

Novak, B. and Tyson, J.J. (2003). Modelling the controls of the eukaryotic cell cycle. Biochemical Society Transactions 31: 1526-1529.

Pouteau, S. (2004). From agri-culture to techno-culture: the intertwined ethical and epistemological grounds of biotechnology. In: J. De Tavernier and S. Aerts (eds.) Science, ethics and society, EurSafe 2004, Belgique, pp. 105-108.

Pouteau, S. (2006). (2006) L'intégrité comme identité. Revue Cadmos 9: 57-73.

Pouteau, S. (ed.) (2007). Génétiquement indéterminé - Le vivant auto-organisé. Quae, Versailles, France.

Teilhard de Chardin, P. (2000). Le phénomène humain. Points, Paris, France.

Warner, K.D. (2007). Agroecology in Action: Extending Alternative Agriculture through Social Networks (Food, Health, and the Environment). MIT Press, USA.

Values behind biodiversity: ends in themselves or knowledge-based attitudes

Arne Sveinson Haugen
Centre for technology, innovation and culture, University of Oslo, Box 1108 Blindern, 0317 Oslo, Norway,
a.s.haugen@tik.uio.no

Abstract

The article discusses how ecological literacy influences or is part of philosophical theories of environmental ethics. Ecological literacy refers to knowledge of and care for nature. One observation is that ecological literacy seems to be of increased importance in practical ethics when inner worth as ends in themselves are expanded to an increasing spectrum of nonhuman ecological entities. This means that practical applications of the ethical theories more or less seem to rely on some kinds of moral attitudes based on ecological literacy.

Keywords: ecological literacy, environmental ethics, values, biodiversity

Introduction

The objective of the present paper is to discuss what role ecological literacy may play in practical applications of philosophical theories of environmental ethics. The basis for the discussion are anthropocentric and nonanthropocentric philosophical theories.

Armstrong and Botzler (1993) say that anthropocentrism is the philosophical perspective asserting that ethical principles apply to humans only, and that human needs and interests are of the highest, and even exclusive, value and importance. The present paper refers to anthropocentric ethics and Norton's pragmatic view.

The nonanthropocentric philosophical theories can be split into individualistic biocentric ethics and holistic ecocentric ethics (Pojman, 2001). The biocentric philosophical theories focus on criteria for inner worth as ends in themselves of individual ecological entities. Compared with the ecocentric ethics, Callicott (2002) contends that biocentric approaches face the challenge that they too narrowly distribute inner worth to individual organisms only. Unlike ecocentric ethics, they therefore do not provide moral considerability to supra-individual ecological entities like species populations, communities or ecosystems.

Ecological literacy

My use of the term 'ecological literacy' is inspired by Callicott (1998) when he, linked to Leopold's land ethics and ecological knowledge, says that a 'universal ecological literacy would trigger sympathy and fellow-feeling for *fellow members* of the biotic community *and* feelings of loyalty and patriotic regard for the *community as a whole*'. Callicott says further that Leopold supports approaches to environmental ethics that are rooted in altruistic feelings like benevolence, sympathy and loyalty. I interpret this to mean that Leopold's land ethics is based on two parts. One is an ecological knowledge part which triggers the other part consisting of sympathy, fellow-feeling and feelings of loyalty.

Building on this, I see that ethics might be based both on a knowledge part and a feeling part. I have therefore decided to use the term 'ecological literacy' as a synthesis of both these parts. Though Callicott may think of ecological literacy as constituted by the knowledge part only, I prefer to see the feeling part that follows from or is triggered by knowledge as an integrated part of the ecological literacy. In the knowledge part, I will not only include scientific knowledge but also such as contact with and observations of nature. In the feeling part, which I refer to as the care part of the ecological literacy, I include such as the way humans love, respect and care for nature. The appearance of ecological literacy, or its attributes or characteristics, will then depend on the relative importance of the knowledge part and the care part, but also of their respective total importance or extent.

I also say that ecological literacy, as a synthesis of the knowledge part and the care part, might be a basis for attitudes, world views and ethics in the field of the environment. Ecological literacy might therefore be part of or influence the attitudes, world views and ethics of philosophical theories in the field of the environment. Examples of this will be discussed in the following.

Anthropocentric ethics

Both Aquinas and Kant warn against being cruel to animals because such cruelty might damage humanity (Clarke and Linzey, 1990). Kant says further that we should be kind to animals since that will develop good character in us and help us treat our fellow human beings with greater consideration (Pojman, 2001). Though the main motivation is to create a good attitude for the relations between humans, in practice it is also said to create an attitude of not being cruel to animals. This is such as when Kant holds that the more we come in contact with animals and observe their behaviour, the more we love them (Clarke and Linzey, 1990).

In the context of ecological literacy, I perceive the references about contact with animals and observing their behaviour as representing the knowledge part, while the way we love them represents the care part. The synthesis of these parts is then a kind of ecological literacy that in practice should create a basis for the attitude of not being cruel to animals.

Biocentric ethics

Regan (1983) limits himself to assume that self-consciousness is a sufficient, but not a necessary condition for the attribution of inherent value or inner worth as ends in themselves of nonhuman organisms. He supposes that this criterion will include at least all normal mammals of one year or more, but leaves it to others to work out environmental ethics on the basis of some alternatives he has proposed for what might be necessary conditions for the attribution of inherent value. I believe therefore that there should be something in Regan's ethical theory to act as guidance for how this should be done, and that this something is a kind of attitude or world view. This might, as I see it, be an attitude based on ecological literacy, more or less the same way as the attitude of love towards animals that is created when being in contact with them and observing their behaviour.

When Taylor (1981) says that biological knowledge is an essential means to fulfil the aims set for adopting an attitude of respect for nature, it sounds like he is talking about the knowledge part of ecological literacy. Such an interpretation is strengthened when he also emphasises that adopting the moral attitude of respect for nature is both rational and intelligible. Linked to this is his trust in the moral judgements of people who he refers to as rational and factually enlightened. Additionally, he seem to refer to the care part of ecological literacy when he holds that anyone who does adopt the moral attitude of respect for nature has feelings about what is favourable and unfavourable to ecological entities. All this together points towards a conclusion, I think, that ecological literacy represents a necessary basis for the moral attitude that Taylor builds his environmental ethics on.

Taylor contends further that the biocentric outlook on nature should be consistent with all known scientific truths relevant to our knowledge of the objective of the moral attitude. It should be so that scientifically informed and rational thinkers with a developed capacity of reality awareness can find the biocentric outlook acceptable as a way of conceiving of the natural world and our place in it. He acknowledges that such a belief system cannot be proven to be true, but that it as a whole would constitute a coherent, unified and rationally acceptable picture of a total world.

Ecocentric ethics

The care part of ecological literacy is specifically addressed by Callicott's (1998) references to feelings. When Callicott in connection with Leopold's land ethic says that universal ecological literacy would trigger sympathy and fellow feeling for other members of the biotic community, as well as feelings of loyalty and patriotic regard for the community as a whole, this has similarities with Kant's belief that contact with animals will make us love them. Rolston, on the other hand, seems to be closer to Taylor's scientifically related ecological literacy, and seems thus to focus more on the knowledge part of ecological literacy. Central to Rolston's (1998) ecocentric approach is the importance of being aware of and taking into consideration the interconnection between individual organisms and ecosystems. Part of this is the need for awareness about the importance of ecosystems as support systems to secure the services and goods necessary for the survival, growth and reproduction of individual ecological entities, and for that sake also their full flourishing. This is what he refers to as the systemic value of ecosystems.

Summary so far

Within all three of the ethics discussed, i.e. anthropocentric ethics, biocentric ethics and ecocentric ethics, there are statements which, as far as I can see, relate to ecological literacy. These are both about the knowledge and the care parts of ecological literacy. An interesting observation is that ecological literacy seems to get increased importance when inner worth as ends in themselves is expanded to an increasing spectrum of ecological entities. This means that some kind of attitude based on ecological literacy seems more or less to be necessary for the practical applications of the ethical theories.

Ecological integrity

Westra (1994), who is also considered to represent ecocentric ethics, proposes an environmental ethics based on the principle of integrity. She says that much deep thinking is necessary for practical resolutions of her environmental ethics. This is reminiscent of Regan's statement about leaving it to others to work out environmental ethics. I think then especially of how I interpret Regan's idea of inherent value as representing a kind of attitude or world view that influences and decides how the environmental ethics should be worked out. I consider Westra's reference to deep thinking to be in the same vein. Westra (1998) actually seems to address this when she mentions the need for an understanding of all natural processes and laws as one of the moral implications of the principle of integrity. This refers to the knowledge part of ecological literacy, much the same way as when Taylor contends that the biocentric outlook on nature should be consistent with all known scientific truths.

Westra's environmental ethics might, as I see it, be split into a weak version, with focus on an anthropocentric functional ecosystem health aspect of sustainability, and a strong version, with focus on a nonanthropocentric structural ecosystem aspect of sustainability. It might be said that the weak version represents a kind of environmentally strong anthropocentric ethics. This is in the sense of an anthropocentric ethics added with a great extent of ecological literacy. The strong version of Westra's environmental ethics might then in practice be compared with an anthropocentric ethics where ecological literacy is even stronger and plays a necessary and crucial role.

Norton

Norton envisages a pragmatic view, which is based on careful deliberation and is compatible with a rationally adopted world view. This incorporates sound metaphysics, scientific theories, aesthetic values and moral ideals. He says that this represents a weak anthropocentric ethics which values nonhuman ecological entities for more than their use in meeting unreflective human needs. This means that they are valued for enriching the human experience (Armstrong and Botzler, 1993).

Norton (2003) is very clear in his denial of the relevance of intrinsic value or inner worth as ends in themselves in nature. He basically disagrees with his colleagues in environmental ethics who argue that we should create a distinctive language and subject matter, the intrinsic value of nature, as the subject of environmental ethics. Norton says that environmental protection is better served by a careful consideration of preferences and that we may often find that conflicting values support the same policy. As an example, Norton mentions that those who value waterfowl for hunting and those who value waterfowl for watching have common interests in supporting waterfowl habitat preservation and restoration policies. He believes that philosophers should help lay people figure that out, rather than spend time and probably create unnecessary divergence by exposing potentially different attitudes regarding intrinsic value.

Norton (*ibid.*) says that this pragmatic view, if reasonably interpreted and translated into appropriate policies, will represent a suitably broad and long-sighted anthropocentrism which will advocate the same policies as nonanthropocentric ethics. He believes that by joining all conservation interests and trusting the efforts of pragmatic environmentalists, nature will be secured optimal protection.

Conclusion

I find it interesting to note Norton's saying that weak anthropocentric ethical approaches are compatible with a rationally adopted world view. In addition to aesthetic values, he says that such a world view incorporates sound metaphysics, scientific theories and moral ideals. This has some similarities with my interpretation of the weak version of Westra's environmental ethics, which I say represents a kind of environmentally strong anthropocentric ethics in the sense of being added with ecological literacy as a core part. I even say that the strong version of Westra's environmental ethics might be compared with an environmentally exceptional strong version of anthropocentric ethics where ecological literacy plays a necessary and crucial role. What is interesting here, I think, is that these two quite opposite philosophical positions - Westra's holistic ecocentric ethics and Norton's pragmatic anthropocentric ethics - can both be interpreted in a practical context as representing a kind of environmentally strong anthropocentric ethics where ecological literacy plays a significant role.

This observation, added with the observation that some kind of attitude based on ecological literacy more or less seems to be necessary for the practical applications of the ethical theories, provides the basis for my main conclusion. Synthesising these two observations my main conclusion is that whatever the theoretical philosophical starting point of an environmental ethics might be, whether anthropocentric, biocentric or ecocentric, it seems that the differences in practical ethics will be less the more ecological literacy plays a role. This means in other words that the more people come in contact with, observe and learn about nature, and thereby come to love, respect and care for a wide spectrum of ecological entities, the less importance the different philosophical theories will have for the application of practical ethics.

This does not mean that I do not believe in the usefulness or fruitfulness of philosophical theories, such as those within the biocentric and the ecocentric ethics, which address the questions of universal inner worth of ecological entities. On the contrary, I do believe such theories have an important function in raising the questions about values behind biodiversity, and emphasising the need of being aware of such questions and the need to take a stand with regard to the valuations of ecological entities.

References

Armstrong, S.J. and Botzler, R.G. (1993). Environmental Ethics: Divergence and Convergence. McGraw-Hill, New York, USA, 570 p.

Callicott, J.B. (2002). The Pragmatic Power and Promise of Theoretical Environmental Ethics: Forging a new Discourse. Environmental Values 11: 3-25.

Callicott, J.B. (1998). The Conceptual Foundation of the Land Ethic. In: M.E. Zimmerman (ed.) Environmental Philosophy: From Animal Rights to Radical Ecology. Prentice-Hall, New Jersey, USA, pp. 101-123.

Clarke, P.A.B. and Linzey, A. (1990). Political Theory and Animal Rights. Pluto Press, London, UK, 186 p.

Norton, B.G. (2003). Searching for Sustainability: Interdisciplinary Essays in the Philosophy of Conservation Biology. Cambridge University Press, Cambridge, UK, 554 p.

Pojman, L.P. (2001). Environmental Ethics. Readings in Theory and Application. Wadsworth Thomson Learning, California, USA, 570 p.

Regan, T. (1983). The Case for Animal Rights. University of California Press, California, USA, 239 p.

Rolston, H.III (1988). Environmental Ethics: Duties to and Values in the Natural World. Temple University Press, Philadelphia, USA, 373 p.

Taylor, P.W. (1981). Human Centred and Life-Centred Systems of Environmental Ethics. Environmental Ethics 3: 197-215.

Westra, L. (1994). An Environmental Proposal for Ethics: The Principle of Integrity. Rowman and Littlefield Publishers, Maryland, USA, 235 p.

Westra, L. (1998). Living in Integrity: A Global Ethics to Restore a Fragmented Earth. Rowman and Littlefield Publishers, Maryland, USA, 269 p.

Part 2 - Theoretical, conceptual and foundational issues: assessment and models

A structuring pathway to tackling ethical problems

Michael Zichy
Institute TTN (Institute Technology, Theology and Natural Sciences), Marsstraße 19, 80335 Munich, Germany, michael.zichy@elkb.de

Abstract

Generally, ethical disputes are complex and difficult to settle. Therefore, a clear account of how to handle ethical problems successfully and an understanding of what the crucial points are, where conflicts evolve and where misunderstandings arise, would be helpful. In this paper a structuring pathway to tackling ethical problems will be presented. It is designed to support the process of ethical opinion formation and to prepare well-informed ethical judgements. In particular, it aims at alleviating the problems that severely hamper successful ethical debates, as there are: unstructured discussion, intransparent conflict situation, fuzzy facts, disorganised interdisciplinary co-operation, and unclear use of arguments. For this purpose, the pathway provides an analytical matrix and a framework to structure the ethical debate in five consecutive steps:
1. Identification and analysis of the problem.
2. Clarification of the empirical data.
3. Re-analysis of the problem in light of the findings of step 2.
4. Analysis of the ethical valuations.
5. Final ethical judgement.

Keywords: debate, methodology, analysis, facts and values

Introduction

Rapid scientific and technical developments regularly give rise to new ethical questions that are complex and difficult to answer. On the one hand, these difficulties are based on lacking knowledge about and little to no experience with these innovations (Bayertz, 1996), on the other hand, they are caused by the plurality of ethical values in modern society, leading to conflicting ethical opinions about these innovations. As a consequence, scientific and technical innovations easily generate social disputes that are frequently aggravated by emotional involvement, mutual distrust, fundamental misunderstandings and ignorance. In addition, such disputes are often severely hampered by unstructured and undisciplined discussion, unclear conflict situation, unclear facts, and incorrect use of arguments.

The structuring pathway presented in this paper aims to alleviate some of these problems by providing (1) an analytical matrix and (2) a framework to structure the ethical debate. It is designed to be used by ethical practitioners in ethical committees, conferences, debates, etc. The pathway will help:
- to structure the debate step by step, so that the right tasks are done, the right questions are asked and the right information is integrated at the right moment;
- to make the problem, the discussion and the arguments transparent by distinguishing their different aspects, in particular their normative and factual ones;
- to identify the exact cause of the conflict;
- to frame interdisciplinary co-operation.

Unlike other practical tools for ethical assessment, e.g. the ethical matrix (Mepham *et al.*, 2006, for a collection of ethical tools see Beekman *et al.*, 2006), this pathway it not primarily designed to judge problems ethically by given criteria but to facilitate the process of opinion formation and to prepare for a well-informed judgement. In this process, other ethical tools may well be incorporated.

The structuring pathway which is based on an analysis of different models for ethical assessment (Hepp *et al.*, 2002, Busch *et al.*, 2002, Busch and Kunzmann, 2006, Grimm, 2006) consists of five consecutive steps:
1. Identification and analysis of the problem: The first task is to identify and describe the problem as accurate as possible.
2. Clarification of the empirical data: Before the normative aspects of the problem can be addressed, the facts have to be clear. Therefore, the second task consists in the clarification of the facts. Possible misunderstandings, conflicting interpretations and simple errors with regard to the facts have to be corrected, open questions delegated to the concerned sciences and further information gathered and incorporated.
3. Re-analysis of the problem: If the clarifications lead to a partly different understanding of the facts, this will have an impact on the understanding of the whole problem. Therefore, the original problem needs to be re-analysed in light of the findings of step 2.
4. Analysis of the ethical valuations: Once the facts of a problem are clear and the problem itself re-analysed, only then the normative side of the problem can be addressed. For that, the arguments need first to be analysed and examined in detail and then their validity and strength assessed.
5. Based on a thorough assessment of both the empirical and normative aspects of the problem, a well-informed ethical judgement can be given or different options to solve the problem designed.

Step 1: analysis of the problem

The first step consists in the identification and the analysis of the problem causing the ethical controversy. For this purpose, the problem has to be described as accurate as possible, taking all aspects that might be relevant into account. It is particularly important to consider the specific context the problem is embedded in, because this (a) helps to keep the debate to the point, (b) helps to sharpen the arguments since they have to refer to specific and detailed information, and (c) contributes to generate a solution that in the end suits the context it has to be transferred to. What counts as relevant context certainly depends on the problem in question. However, relevant aspects of the context might be: affected country or region, main stakeholders and their interests, relevant legislation and other regulations, economy, environment, people's main attitudes, etc.

For practical reasons it is recommended to transform the problem into a specific question and to record the most important aspects of its context (see Table 1).

Table 1. Matrix with analytical dimensions.

Question:

Context:

No	Name of argument	Validity	Strength	Conclusion Moral judgement allowed/not allowed	Factual premise Facts, consequences, empirical evidence	Normative premise 1 Moral prima-facie rule	Normative premise 2 Moral principle	Stakeholder and interests
1								
2								
3								
...								

After the problem has been transformed into a clear question, and the specific context of the problem sufficiently described, the different arguments pro and con have to be collected and analysed. For this, all relevant aspects have to be considered and all the different views and opinions on the problem have to be equally taken into account. Including all peoples' opinions will contribute to the feasibility of the solution, since taking people's opinions serious in the beginning makes it easier for them to accept a possible solution in the end. In addition, considering all opinions will make it more likely that all relevant aspects have really been noticed. All together, this will help to avoid a reduction of the problem to certain of its aspects, which is likely to happen when a theory-induced approach is chosen. This would thus lead to the construction of an artificial problem (an 'ethical artefact'); its solution would not solve the original problem (Grimm and Zichy, 2006).

For the analysis, every single of the collected arguments needs to be broken down into its logical elements, i.e. its conclusion and its factual and normative premises, which are:
- Moral judgement (conclusion): Is the practice in question morally allowed or not allowed according to the argument?
- Referred to facts or consequences (factual premise): Because of which facts rsp. empirical evidence or consequences is the practice in question morally allowed or not allowed.
- Moral prima-facie rule (normative premise 1): According to which moral prima-facie rule or intuition are the referred to facts morally good or bad?
- Moral principle (normative premise 2): Because of which moral principle or rule is the moral prima-facie rule or intuition justified? Here, among others, the principles identified by Beauchamp and Childress (2004) are likely to come into play: Autonomy, nonmaleficence, beneficence, justice.

For practical reasons it is recommended to number and name all the different arguments (see Table 1).

If the debate promises to get deeper into the details of ethical reasoning, some more aspects to analyse could be added: The rule of specification of the moral principle, the highest moral principle or ethical theory, the rule of specification of the highest moral principle or ethical theory. However, as only professional ethicists will continue the debate on these abstract levels, and as they should know how their arguments work on these levels, the analysis of these aspects might also prove to be redundant.

Aspects, e.g. consequences or moral principles that seem relevant but are not yet part of any of the collected arguments should be noticed in the matrix (Table 1) for later consideration.

Although it does not influence the validity of an argument it might be enlightening to know who and what motives and interests lie behind certain arguments. Therefore, stakeholders and their interests should be indicated beside the particular arguments they hold.

For the analysis, the matrix (Table 1) that covers all mentioned analytical dimensions might be used:

One reason why ethical debates are so hard to deal with are the several forms and layers of conflicts (Moreno, 1991) which so often are intermingled and confounded. Splitting up the arguments the way described makes the arguments more transparent, the different layers accessible, and thereby serves to localise the conflict.

Step 2: clarification of the empirical data

The distinction between normative and empirical aspects in the analysis above is particularly important. Firstly, because ethical judgements are highly dependent on the facts they refer to. Empirical knowledge is therefore crucial to the ethical assessment. Secondly, because it can reveal whether the conflict is caused by differing moral values or rather by diverging understandings of the empirical facts. So step 2 serves (1) to find out if there is a controversy regarding the facts, and if this is the case (2) to analyse and possibly solve the underlying problem.

By comparing the empirical aspects and their relation to the moral judgements of all gathered arguments it becomes clear whether there is a shared view of the facts in the dispute or not. If there is a shared factual basis, i.e. the empirical data are not controversial, the conflict is necessarily a pure normative one and needs to be analysed and dealt with only in this respect. In this case, step 2 can be skipped.

However, very often ethical debates only seem to be driven by differing moral beliefs, but are in fact caused by differing understandings of the facts of a problem (Nida-Rümelin, 2005). It is of course possible that both normative as well as empirical aspects of a problem are controversial. But then the clarification of the empirical aspects needs to be done before the normative controversy can be addressed.

If the analysis reveals that the empirical basis is indeed controversial, then it has to be found out, why and in which respect this is the case. Possible reasons and their solution are:
- Ignorance or misinformation: This problem is the most easily to address; providing or correcting information should suffice, thought there might exist obstinate exceptions.
- Controversial methodology of empirical findings.
- Controversial explanatory power of data.
- Controversial scientific interpretation of data.
- Controversial relevant scientific nescience.

The last four reasons have in common that the controversy can be settled by further empirical research. What is necessary at this point of the pathway is to decide whether it makes sense to continue the pathway or not. If the controversies concern crucial questions the pathway and the discussion might have to be interrupted till further research results providing the required clarification are available. However, in any case the appropriate questions regarding the empirical data have to be formulated and assigned to the responsible scientific disciplines. Later, the respective research results will have to be integrated into the process.

There are still other reasons why the empirical basis of a moral problem could be controversial. These reasons have in common that they are normative and cannot be clarified by further scientific research. These reasons include:
- Differing estimations or bold disregard of empirical knowledge, possibly due to specific worldviews.
- Differing estimations of the extent and the relevance of nescience.
- Differing interpretations of the empirical data.

In these cases, the controversy about the empirical base is caused by controversies about normative aspects. Therefore, these controversies have to be reserved for and dealt with in step 4 where a genuine philosophical reflection has to take place.

Step 3: re-analysis of the ethical problem

Once the questions concerning the empirical facts are clarified, a re-analysis of the ethical problem with the above matrix and a reformulation of the arguments become necessary. It is very likely that in the light of the clarified empirical basis some arguments just get invalid or unconvincing and others gain or lose strength. It is also possible that the conflict turns out not to be an ethical one or to be unsolvable at the moment due to not yet available empirical data. If in step 2 a problem really turns out to be a pure empirical one, the ethical pathway can be terminated because a further normative reflection would be misplaced. A normative reflection only makes sense when the problem is a normative one, and this is the case only if the facts are not controversial, or if the controversy regarding the empirical facts can principally not be solved by further empirical investigations.

In these cases, the thorough re-analysis of the problem in light of the clarified empirical basis will reveal the real normative controversies which then have to be dealt with in step 4.

Step 4: analysis of the ethical valuations

Given that the debate is induced by differing normative beliefs, the re-analysis of the conflict will provide the starting point for the further examination. At first, the conflict's normative causes have to be identified. A clear idea of the normative conflict is essential for an adequate understanding of both the ethical dispute as a whole and of the individual arguments. Identifying the causes of conflict can easily be done by comparing the logical elements of the arguments. Conflicts can be induced by:
- Differing moral prima-facie rules.
- Differing interpretation of prima-facie rules.
- Differing moral principles.
- Differing interpretations of the same moral principle leading to different moral rules.

If the cause of controversy cannot be found at these levels, it might be necessary to extend the analysis to the other levels mentioned above, i.e. differing ethical theories, differing highest moral principles and differing interpretations of both.

After the analysis of the conflict a thorough examination of all the normative aspects has to take place. First, the wrong arguments with inconsistencies, righteous implausibility, or obvious fallacies have to be sorted out. Then, the remaining arguments have to be examined in the following respects:
- Is the moral prima-facie rule plausible and acceptable for everybody?
- How strong is the connection between facts, moral prima-facie rule and moral judgement: is it necessary, plausible or just possible?
- Is the moral principle plausible and acceptable for everybody?
- How strong is the connection between moral reason and moral principle: is it necessary, plausible or just possible?

According to the results and the overall impression of the degree of their validity and their strength, the arguments should be marked (table 1, column 'validity' and 'strength'); in cases of doubt, a comparison between the arguments might help. Possible marks could be:
- Validity: valid (3), valid under certain circumstances (depending on the circumstances 2 or 1), invalid (0).
- Strength: overriding (does not allow trade offs, excluded from calculation, therefore: +), strong (3), less strong (2), weak (1), absolutely weak (0).

The validity of an argument depends on how convincing and well justified its logical elements, the connection between the logical elements and the argument as a whole is. Only those arguments and moral beliefs which are rationally justified and therefore acceptable for everybody can legitimately request full validity.

Through this analysis the validity and the strengths and weaknesses of the individual arguments and its logical elements will become more obvious. This will prepare for the ethical judgement in step 5.

Step 5: final ethical judgement

After having identified and analysed the problem (step 1), clarified its empirical basis (step 2), re-analysed it in light of the clarified empirical basis (step 3), and after having examined the arguments in detail (step 4), either an ethical judgement on the problem could be made or different options to solve the problem could be presented. How to do this will not be discussed here, because the way judgements are made and options are designed is heavily dependent of the kind of problem, of the circumstances and on what is required.

However, it is important that the judgment should not be the result of a simple calculation of the marks previously assigned to the arguments. The marks just can give a rough idea of which arguments have to be less, and which have to be more taken into account. And they can give an idea how different answers to the original questions could look like.

In any case, steps 1-4 serve as a preparation for the judgement or the design of the different options to solve the problem by providing an excellent background for every further ethical deliberation of the problem. At this stage, other ethical tools like the ethical matrix that support ethical judging might be used if convenient.

Acknowledgements

The structuring pathway presented in this paper is based on the preliminary work done in Grimm and Zichy (2006). I'm especially indebted to Herwig Grimm, Gernot Prütz and Roger J. Busch for fruitful discussions and important remarks.

References

Bayertz, K. (1996). Einleitung. Moralischer Konsens als soziales und philosophisches Problem. In: K. Bayertz (ed.). Moralischer Konsens. Technische Eingriffe in die menschliche Fortpflanzung als Modellfall. Suhrkamp, Frankfurt am Main, pp. 11-29.

Beauchamp, T. and Childress, J. (2001). Principles of Biomedical Ethics. 4th edition, Oxford University Press, Oxford.

Beekman, V., de Bakker, E., Baranzke, H., Baume, O., Deblonde, M., Forsberg, E.-M., de Graaff, R., Ingersiep, H.-W., Lassen, J., Mepham, B., Porsborg Nielsen, A., Tomkins, S., Thorstensen, E., Millar, K., Skorupinski, B., Brom, F., Kaiser, M. and Sandoe, P. (2006). Ethical Bio-Technology Assessment Tools for Agriculture and Food Production. Final Report Ethical Bio-TA Tools. LEI, The Hague.

Busch, R.J. and Kunzmann, P. (2006): Leben mit und von Tieren - Ethisches Bewertungsmodell zur Tierhaltung in der Landwirtschaft. 2nd edition, Utz, München.

Busch, R.J., Haniel, A., Knoepffler, N. and Wenzel, G. (2002). Grüne Gentechnik - Ein Bewertungsmodell. Utz, München.

Grimm, H. (2006). Animal welfare in animal husbandry - how to put moral responsibility for livestock into practice. In: M. Kaiser and M. Lien (eds.). Ethics and the politics of food. Wageningen Academic Publishers, Wageningen, pp. 518-522.

Grimm, H. and Zichy, M. (2006). Praxisorientierung in der angewandten Ethik - eine Problemanzeige. Forum TTN 16: 56-63.

Hepp, H., Hofschneider, P.H., Korff, W., Rendtorff, T. and Winnacker, E.L. (2002). Gentechnik: Eingriffe am Menschen - Ein Eskalationsmodell zur ethischen Bewertung. 4th edition, Utz, München.

Mepham, B., Kaiser, M., Thorstensen, E., Tomkins, S. and Millar, K. (2006). Ethical matrix - manual. LEI, The Hague.

Moreno, J.D. (1991). Consensus, contracts and committees. The journal of medicine and philosophy 16: 393-408.

Nida-Rümelin, J. (2005). Theoretische und angewandte Ethik: Paradigmen, Begründungen, Berichte. In: J. Nida-Rümelin (ed.). Angewandte Ethik. Die Bereichsethiken und ihre theoretische Fundierung. 2nd edition, Kröner, Stuttgart, pp. 2-89.

Standing on the shoulders of a giant: the promise of multi-criteria mapping as a decision-support framework in food ethics

Volkert Beekman, Erik de Bakker and Ronald de Graaff
Agricultural Economics Research Institute, P.O. Box 29703, 2502 LS The Hague, The Netherlands,
volkert.beekman@wur.nl

Abstract

Whereas the ethical matrix has been pivotal for the development of food ethics from a juvenile to an adolescent sub-discipline, the dominance of this approach in food ethics forswears its development into a mature sub-discipline that could give decision-makers more practical guidance. Decision-support frameworks in food ethics face the double challenge of providing guidance in mapping and balancing values in the issue at hand. The equally brilliant and simple idea of the ethical matrix is to use a tabular format to meet the first mapping challenge. The crucial issue is the selection of appropriate headings for columns and rows in the table. The ethical matrix opts for principles and stakeholders. This is not the luckiest choice, since principles are not specified enough to result in meaningful conclusions, whereas including stakeholders is redundant in participatory applications of the methodology. It seems more fruitful to replace principles with more specified values like food security, economics, product quality, food safety, health, environment, animal welfare, fair trade and craftsmanship and to place these values in the rows of the table. Since decision-making in food ethics is a matter of choosing between options, it seems a good idea to include scenarios in the columns of the table. With these alterations of the ethical matrix, the methodology has developed all characteristics of the first stage in a multi-criteria mapping approach. Such a methodology could also provide guidance for the second balancing challenge of decision-support frameworks in food ethics.

Keywords: decision-support framework, ethical matrix, food ethics, multi-criteria mapping, participation

A word of caution

It is with the utmost respect for Mepham and his seminal work on the ethical matrix (e.g. Mepham, 1996 and 2005) that we write this paper. When we in due course explore what a decision-support framework in food ethics should be able to deliver and how the ethical matrix and other such approaches perform on that account, this should not be misunderstood as either a critique of the ethical matrix or as a suggestion that Mepham claims that the ethical matrix is constructed to do those things (Mepham, 2004). We thus humbly stand on the shoulders of a giant to do a modest proposal.

Introduction

This paper will address the paradox that, whereas Mepham's ethical matrix has been pivotal for the development of food ethics from a juvenile to an adolescent sub-discipline, the dominance of this same methodological approach in European food ethics forswears its further development into a mature sub-discipline that could guide decision-makers in policy issues that need practical solutions. Decision-support frameworks in food ethics, like the ethical matrix, face the double challenge of providing methodological guidance in both mapping and balancing relevant values in the issue at hand.

The fundamental idea of the ethical matrix is to use a tabular format to meet the first mapping challenge. The crucial issue then becomes the selection of the most appropriate headings for the columns and rows in the table. The ethical matrix opts for principles and stakeholders or affected parties respectively. This is not the luckiest choice, since these principles are not specified enough to result in meaningful conclusions and they also seem to suggest that involvement of stakeholders is not necessary to map the values at stake.

It therefore seems much more fruitful to replace the principles with more specified values or criteria like food security, economics, product quality, food safety, health, environment, animal welfare, fair trade and craftsmanship and to place these values / criteria in the rows of the table. Since public and private decision-making in food ethics is normally a matter of choosing between options or scenarios, it seems to be a good idea to include these options / scenarios in the columns of the table. With these alterations of the ethical matrix, the methodology has developed all relevant characteristics of the first stage in a multi-criteria mapping approach.

Such a methodology could also provide guidance for the second balancing challenge of decision-support frameworks in food ethics, about which the ethical matrix remains disturbingly silent. Although the ethical matrix does not preclude balancing exercises, and Mepham (2005) even mentions some possibilities in this respect, it does not elaborate upon these exercises in more practical detail. The paper will illustrate the promise of multi-criteria mapping for this second balancing stage with experiences from a pilot study about animal disease intervention strategies.

The challenge

Decision-support frameworks in food ethics face the double challenge of providing methodological guidance in both mapping and balancing relevant values in the issue at hand. More specifically, from a regulatory point of view a decision-support framework in the domain of food ethics would need to be able to do the following:
- Frame the ethical challenge;
- Map the relevant values;
- Balance the relevant values in an intelligible way; and
- Formulate action recommendations.

Such a decision-support framework would be able to facilitate structured and transparent processes of opinion-formation and decision-making about food ethical issues. Moreover, any decision-support framework in food ethics needs to include a participatory dimension. We will not further explore this participatory dimension of decision-support frameworks in food ethics but simply assume best practice in that respect.

Mapping values

The equally brilliant and simple idea of the ethical matrix is to use a tabular format to meet the first mapping challenge. The crucial issue then becomes the selection of the most appropriate headings for the columns and rows in the table. The ethical matrix opts for principles, as borrowed from medical ethics (Beauchamp and Childress, 1994), and stakeholders or affected parties respectively (Table 1).

This is not the luckiest choice from a regulatory point of view, since these principles are not specified enough to result in meaningful conclusions, whereas including stakeholders is redundant in participatory applications of the methodology. Such a participatory approach in turn is highly recommendable in ensuring societal support for the outcomes of the exercise (Kaiser and Forsberg, 2001).

Table 1. Ethical matrix (Mepham et al., 2006; p. 10).

Respect for	Wellbeing	Autonomy	Fairness
Producers	Satisfactory income and working conditions	Managerial freedom	Fair trade laws
Consumers	Safety and acceptability	Choice	Affordability
Treated organisms	Welfare	Behavioural freedom	Intrinsic value
Biota	Conservation	Biodiversity	Sustainability

It therefore seems much more fruitful to replace the principles with more specified values or criteria like food security, economics, product quality, food safety, health, environment, animal welfare, fair trade and craftsmanship (Ministry of Agriculture, Nature and Food Quality, 2006) and to place these values / criteria in the rows of the table. Such values also have a closer link to food policy issues and seem to be more suitable to get the attention of decision-makers and stakeholders. Since public and private decision-making in food ethics is normally a matter of choosing between options or scenarios, it seems to be a good idea to include these options / scenarios in the columns of the table. Finally, the impact of the different scenarios on each value / criterion needs to be mapped with the help of some further specified indicators (Table 2).

With these alterations of the ethical matrix, the methodology has developed all relevant characteristics of the first stage in a multi-criteria mapping approach (Stirling and Mayer, 2001). This mapping stage scores indicator evaluations on, for instance, a 5-point Likert scale (1 = very negative impact; 2 = negative impact; 3 = no impact; 4 = positive impact; 5 = very positive impact).

Balancing values

Such a methodology could also provide guidance for the second balancing challenge of decision-support frameworks in food ethics, about which the ethical matrix remains disturbingly silent, by assigning weights on, e.g., a 5-point Likert scale to the various indicators in Table 2. The good thing about multi-criteria mapping is that it, unlike most other multi-criteria approaches, 'focuses as much on "opening up" as on "closing down" a decision or policy process, generating a rich body of information concerning the reasons for differing viewpoints, as well as their practical implications for putting the options into practice' (Lobstein and Millstone, 2006: 10).

A pilot study

The promise of multi-criteria mapping for the second balancing stage can be illustrated with experiences from a pilot study about animal disease intervention strategies (Beekman *et al.*, 2007). In recent years, animal disease intervention strategies have given rise to societal debates in The Netherlands. The desired result of the pilot study was to map and balance ethical aspects of various scenarios for combating the next outbreak of Highly Pathogenic Avian Influenza (HPAI).The pilot study used a multi-criteria mapping approach with 23 steps (Table 3).

On a substantive level, the result of the pilot study was that in ethical terms the best option for combating the next outbreak of HPAI would be a scenario that combines sanitary slaughter of animals in infected and suspect locations with a voluntary vaccination programme that allows for non-destruction of vaccinated animals. The main bottleneck for this scenario is the number of farms opting for voluntary vaccination. The expectation is that many Dutch poultry farms will not do this, with an eye to the

Table 2. Multi-criteria map.

	Weights	Scenario$_0$	Scenario$_1$	...	Scenario$_n$
Food security					
Indicator$_1$					
Indicator$_2$					
Indicator$_3$					
Economics					
Indicator$_1$					
Indicator$_2$					
Indicator$_3$					
Product quality					
Indicator$_1$					
Indicator$_2$					
Indicator$_3$					
Food safety					
Indicator$_1$					
Indicator$_2$					
Indicator$_3$					
Health					
Indicator$_1$					
Indicator$_2$					
Indicator$_3$					
Environment					
Indicator$_1$					
Indicator$_2$					
Indicator$_3$					
Animal welfare					
Indicator$_1$					
Indicator$_2$					
Indicator$_3$					
Fair Trade					
Indicator$_1$					
Indicator$_2$					
Indicator$_3$					
Craftsmanship					
Indicator$_1$					
Indicator$_2$					
Indicator$_3$					
Balance					

economic consequences. After all, Dutch supermarkets do not appear to be willing to buy meat from vaccinated animals at a reasonable price, as they do not believe that Dutch consumers would buy such meat. It remains unclear how this spiral of negative expectations could be broken.

On a methodological level, one striking aspect of the pilot study was that participants were able to reach agreement fairly quickly with regard to changing, modifying and adding scenarios, values / criteria and

Table 3. Steps multi-criteria mapping.

Preparing
1. Formulate challenge (singular how question)
2. Formulate scenarios (unchanged policy, proposed policy, alternative policies)
3. Formulate values / criteria
4. Formulate indicators (independent and same number for each value / criterion)
5. Select type of interaction (remote or on-site)
6. Select participants (organised stakeholders or individual citizens / consumers)

Discussing
7. Explain methodology
8. Agree upon challenge, scenarios, values / criteria and indicators
9. Give individual weights to indicators (on a 5-point Likert scale)
10. Aggregate individual weights (present average and range)
11. Discuss weights (identify consent and dissent)
12. Score individual evaluations for each indicator (on a 5-point Likert scale)
13. Aggregate individual evaluations (present average and range)
14. Discuss evaluations (identify consent and dissent)
15. Combine evaluations and weights
16. Ask for conclusion (identify optimal scenario)
17. Ask for possibilities to improve optimal scenario (turning weaknesses into strengths)
18. Formulate conclusion
19. Evaluate (ask participants about their experiences with the process)

Reporting
20. Analyse process and results
21. Write report
22. Give feedback to participants (send report)
23. Communicate process and results to non-participants

indicators. In general, the participants considered the multi-criteria mapping approach to be useful and worth further development. The experience was that the structure of a multi-criteria mapping approach forces broader and more rational ways of thinking, and that the associated dialogue contributes to better mutual understanding among the various interested parties. It gives the participants a clear view about the convergence and divergence in their considerations about and priorities with respect to the mapped values in the issue at stake.

Conclusion

Multi-criteria mapping is a promising decision-support framework in food ethics that is able to meet the double challenge of providing methodological guidance in both mapping and balancing relevant values in the issue at hand. It offers a clear structure, provides a foothold in decision-support with respect to mapping and balancing various food ethical values, and thus contributes to the development of food ethics into a mature sub-discipline.

References

Beauchamp, T.L. and Childress, J.F. (1994). Principles of biomedical ethics. Oxford University Press, Oxford, United Kingdom.

Beekman, V., De Bakker, E. and De Graaff, R. (2007). Ethische aspecten dierziektebestrijdingsbeleid; Een oefening in participatieve multi-criteria analyse [Ethical aspects of animal disease intervention strategies; A pilot study in participatory multi-criteria analysis]. LEI, The Hague, The Netherlands.

Kaiser, M. and Forsberg, E.M. (2001). Assessing fisheries - Using an ethical matrix in a participatory process. Journal of Agricultural and Environmental Ethics 14: 192-200.

Lobstein, T. and Millstone, M. (2006). Policy options for responding to obesity. University of Sussex, Brighton, United Kingdom.

Mepham, B. (1996). Ethical analysis of food biotechnologies: An evaluative framework. In: B. Mepham (ed.) Food ethics. Routledge, London, United Kingdom, pp. 101-119.

Mepham, B. (2004). A decade of the ethical matrix: A response to criticisms. In: J. De Tavernier and S. Aerts (eds.) Science, ethics and society. Katholieke Universiteit Leuven, Leuven, Belgium, pp. 271-274.

Mepham, B. (2005). Bioethics: An introduction for the biosciences. Oxford University Press, Oxford, United Kingdom.

Mepham, B., Kaiser, M., Thorstensen, E., Tomkins, S. and Millar, K. (2006). Ethical Matrix - Manual. LEI, The Hague, The Netherlands.

Ministry of Agriculture, Nature and Food Quality (2006). Een goed gesprek over voedselkwaliteit [A good conversation about food quality]. LNV, The Hague, The Netherlands.

Stirling, A. and Mayer, S. (2001). A novel approach to the appraisal of technological risk: A multi-criteria mapping study of a genetically modified crop. Environment and Planning C: Government and Policy 19: 529-555.

Sustainability concept in agricultural scientific papers

Matias Pasquali
University of Minnesota, Department of Plant Pathology, Borlaug Hall, 55108 St paul MN, USA,
matias.pasquali@gmail.com, pasqu016@umn.edu

Abstract

From the original definition of sustainability of the World Commission on Environment and Development (1987) which states that 'Sustainable development meets the need of present generation without compromising the needs of future generations', the concept of sustainability has changed and has been used with different meanings, justifying opposite actions and policies.

The term has become popular also in scientific publications. The analysis of the last fifteen years of articles listed in the CAB abstract database showed that the use of the term sustainable/sustainability has increased 4 times. In order to elucidate the different meanings of sustainability, the analysis of the terms sustainable/sustainability used in agricultural scientific literature published during the year 2005 is here presented. The analysis confirms that the sustainability concept can be adopted to justify very different scientific conclusions. The lack of definition of the concept of sustainability within the majority of papers is the cause for misinterpretation of science-based conclusions. Increased attention to the communication of various facets of value laden concepts in scientific journals is advocated in order to facilitate a more transparent and efficient deliberation.

Keywords: contestable concepts, framing, scientific communication

Introduction

Framing science is a fundamental process for understanding dynamics and trends in the development of scientific discourse and its political consequences (Nisbet, 2006).

The importance of values (Kitcher, 2001) in scientific discourse (and consequently in scientific literature) is dismissed and underestimated by the majority of scientists. The value-free concept of science is still a powerful mindset. At the same time the recent transformation of science in a fund competitive enterprise requires changes in the scientific language that drives researchers to an exercise of justification of their aims and scope in a larger set of needs. This requires the adoption of value laden terminology to appeal to general interests and aims in order to obtain funds. The analysis of value laden terms is particularly interesting in the agricultural sector where the paradigm of productivity is now partially complemented by environmental and social considerations (Zimdahl, 2006).

Sustainability has become the catchword in international discussions (Dahl, 1995). The complexity and the changing definition of the concept makes the analysis of the scientific publication a valuable observation point. It is possible to verify how the scientific community has adopted different perspectives of sustainability. Secondly, outlining the different values that are considered in a publication may facilitate to put in a political context the scientific work.

In fact at least three values are interlaced in the sustainability definitions that are mainly accepted nowadays: the value of the environment is the conceptualization of nature; the economic perspective is the temporal dimension of intergenerational equity; and the social perspective is the spatial or social aspect of intragenerational equity.

In order to evaluate the use of preidentified value laden words and terminologies (sustainability/ sustainable, environmentally friendly, ethical/ ethics) in the biological/agricultural scientific literature the CAB database was interrogated for the presence of these words from year 1990 to 2005. In this paper the use of the word sustainable/sustainability is analyzed in all full text articles published during the year 2005 in order to extrapolate the ambiguity/trends and common aspects in the concept of sustainability. This analysis is then used for outlining the importance of value communication in scientific literature.

Methods

The identification of the articles that were using the word sustainability in the different years (1990 to 2005) was performed using the CAB database search tool.

The total number of articles available for each year in the database (necessary to calculate the percentage of articles that each year include the word sustainability) was provided by CABI.

For the complete analysis of the text only impact factor articles were selected. This decision was made in order to select only articles that unanimously have a scientific status.

The information obtained from the articles was classified as following: the type of article; the precision in the definition of the term sustainability and the qualitative attributes to the concept of sustainability that were declared or that can be inferred by the use of the word in the article; the values (according to Becker, 1997) that constitute the core concept of sustainability (economic, social ad environmental). In particular the economic value is defined by the explicit requirement of a temporal framework that analyses the intergenerational equity. The social value appeals to a intragenerational equity now and then. The environmental value is the transformation into values of the natural conditions.

The use of indicators was also identified. The research was performed the 02/04/2006. Other articles were deposited during the year 2006 so the analysis cannot be considered a complete analysis of the articles published in the year 2005 but it represent a good sampling and is considered sufficient for the purpose of the article.

Results

The analysis of the use of the terminology related to sustainable concepts was performed from 1990 to 2005 (Figure 1). Data show the increase in the number of articles containing the word, with a significant increase in 1994, two year after the Rio declaration, showing the impact of that event on the scientific literature.

Sixty-nine articles that were available as full text through the Ovid server were analyzed for the use of the term sustainability. The majority of them were research/experimental articles (40/69). The others were reviews. Eleven articles were using or referring to indicators of sustainability.

Approximately half of the articles had a clear definition of the term or were referring to a known definition of sustainability (34/69). Eleven over twenty nine review articles did not include a definition of sustainability. Research articles that did not have any definition were twenty seven over forty.

The concept of sustainability that was adopted in every article was evaluated from the cited definitions and from the context of the article after reading it. The majority of the articles refer to sustainability as an economic concept (34/69). Eighteen refer to the complex definition of sustainability as a multifaceted composition of economical environmental and social values, 11 specifically address the social aspect of

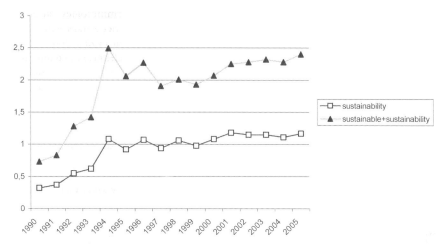

Figure 1. Percentage of articles (Y axis) listed in CAB abstract database that include the word sustainability (squared points) or sustainable/sustainability (triangular points) from year 1990 to 2005.

sustainability while only six the environmental aspects of it. Interestingly the economic perspective is widely used in research articles (23/40).

A significant example of the importance of understanding the array of concepts of sustainability used to analyze scientific data is outlined from the aquaculture research papers. The adoption of aquaculture for tuna fish is strongly supported by advocating a general environmental concept of sustainability (preserving the local sea population of pacific tunafish) in the article by Sawada *et al.* (2005). In the same journal in a different issue the feasibility of prawn aquaculture is evaluated under the perspective of a more holistic concept of sustainability that includes both environmental and social consideration by fishermans and farmers (Kutty, 2005). So while in the first case an environmental consideration is sufficient to promote aquaculture as the sustainable option in the second one a process that includes also the economic viability and the social impact on fisheries lead to a more critic approach to aquaculture implementation.

It is clear how the debate on aquaculture efficacy and acceptability depend also upon the view of sustainability that we are willing to embrace.

Another interesting example on the variability of the sustainability concept in scientific publication is the use of fertilizers in the agricultural practices. In the name of sustainable agriculture the ban of fertilizers (Rayment, 2005) and the request for increased mineral nutrition (for Zinc) are advocated (Srivastava and Singh, 2005). In the first case the need for a ban of fertilizer is due to unsustainable long term environmental condition, in the second case the need for fertilizers is advocated for granting a long term economic and social sustainability to citrus growers.

Discussion

The analysis of a value laden term such as sustainability has shown how agricultural scientific community is prone to the utilization of this terminology.

The attention towards sustainability issues is confirmed by the appearance of new publications devoted to sustainable development which include also agricultural articles such as *Environment* Science and Policy for Sustainable Development; *Annual Review of Environment and Resources;* PNAS: Proceedings of the National Academy of Sciences with the session on Sustainability: Science, Practice and Policy (SSPP); Sustainability Science; Journal of sustainability science and management.

There are therefore plenty of opportunities for publishing papers under the paradigm of sustainability, that can be intended as the sum of economic, environmental and social considerations in research outcomes. At the same time, as shown by the analysis presented here, there is a lack of scientific precision in the majority of the analyzed papers that do not state clearly which aspect of sustainability are they using for framing their research.

Because many issues are connected to the concept of sustainability it is fundamental to clarify how science is addressing the various meanings. In fact, What is to be sustained, or how long and what should be developed to implement sustainability are debatable open questions. Different views opt for sustaining Earth or biodiversity or ecosystems; other views are community centered and would sustain Cultures or groups or places.

Sustainability requires also a time scale; its definition is hardly debated: how far can we hypothesize in the future? When quality of life or economical wellbeing are advocated, what are the criteria that can be evaluated (child survival, life expectancy, equity, social capital, education, productivity)?

Thus sustainability and the related popular concept of sustainable development are defined by various essential concepts (Jacobs, 1999):
1. environmental/economy integration;
2. future perspective;
3. environmental protection;
4. equity;
5. quality of life;
6. participation of political instances.

It is a widespread opinion that sustainable development science should support common, unambiguous terminology. So the preconditions and requirements for operationalization and quantification of sustainability must be defined; and the philosophy and value system behind this concept and its translation into policies must be made explicit (Becker, 1997). This is important not only for theoretic reasons but also from very practical stand point. In fact, as shown in the analysis of publications, different scientific outcomes, that can suggest very different political decisions, can be inferred under the paradigm of sustainability. Referring to the example on the use of fertilizers in agricultural practices it is clear that different conclusions, appealing both to sustainability, can be drawn from an environmental or a social perspective. The same can be said about the example of aquaculture development. Contemporary scientific papers seem to neglect the importance of definition of the set of ideas guiding research. To help and facilitate the political discourse on the issue it would be beneficial to declare values behind research (Pasquali and Korthals, 2004) in order to facilitate the political arguments and decisions on scientific results.

Sustainability, used often as synonym of sustainable development, has been classified (Jacobs 1999) as other political terms (liberty, democracy) as a 'contestable concept' (Gallie, 1956).

Contestable concepts have two levels of meaning: the first one is vague but unitary, and often it is expressed by various definitions that complement each other; the second one is on the political argument, on how the concept should be interpreted in practice. No consensus could be ever reached by those who use it having different interests and political values.

Disagreements over the meaning of sustainable development are not semantic disputations but are substantive political arguments with which the term is concerned (Jacobs, 1999).

If public participation has to play a crucial role in the sustainable development model and if the direction of sustainability has to be defined on the political arena, scientific information should be transparent. Transparency should start from scientific publications so that scientific research can be used as tool for political debate and not as a tool to impose political solutions avoiding debate.

Acknowledgements

Work supported by the Branco Weiss Fellowship 'Society-in-Science'.

References

Becker, B. (1997). Sustainability Assessment: A Review of Values, Concepts, and Methodological Approaches. Issues in agriculture 10: 1-70.

Dahl, A.L. (1995) Towards indicators of sustainability. Scientific workshop on indicators of sustainable development (Wuppertal, 15-17 November 1995).

Gallie W. (1956). Essentially contested concepts. Proceedings of the Aristotelian Society 156: 167-198.

Jacobs, M. (1999). Sustainable development as a contested concept. In: Dobson, M. (ed.) Fairness and futurity. Oxford University Press, 21-45.

Kitcher, P. (2001). Science, truth and democracy. Oxford University Press.

Kutty, M.N. (2005). Towards sustainable freshwater prawn aquaculture - lessons from shrimp farming, with special reference to India. Research in aquaculture 36: 255-263.

Nisbet, M. (2006). What is framing. www.framing-science.blogspot.com.

Pasquali, M. and Korthals, M. (2004). New communication scheme for biotechnology innovations. 5th EURSAFE meeting 1-3 Sept., Leuven (BEL). pp. 109-111.

Rayment, G.E. (2005). Cadmium in Sugar Cane and Vegetable Systems of Northeast Australia. Communications in Soil Science and Plant Analysis 36: 597-608.

Sawada, Y., Okada, T., Miyashita, S., Murata, O. and Kumai, H. (2005) Completion of the Pacific bluefin tuna Thunnus orientalis (Temminck et Schlegel) life cycle. Research in aquaculture 36: 413-421.

Srivastava, A.K and Singh, S. (2005). Zinc Nutrition, a Global Concern for Sustainable Citrus Production. Journal of Sustainable Agriculture 25: 5-42

Zimdahl, R.L. (2006). Agriculture's Ethical Horizon. Academic Press (USA), p.272.

Part 3 - Theoretical, conceptual and foundational issues: bringing ethics into practice

Practice-oriented ethics

S. Aerts and D. Lips
Centre for Science, Technology and Ethics, K.U.Leuven, Kasteelpark Arenberg 1, bus 2456, 3001 Leuven, Belgium, Stef.Aerts@biw.kuleuven.be

Abstract

Classically two 'levels' of ethics are distinguished within normative ethics; general normative theories, which deal with fundamental issues and questions, and applied (or practical) ethics, which deals with issues in different areas of human activity.

Day-to-day questions in real life are often difficult to resolve through one of these ethical theories (or frameworks). A final decision is often difficult to be reached due to competing claims (from within different general theories or different applied theories). A third, hands-on level of ethics is necessary, and could be called 'practice-oriented ethics' (POE).

POE uses general and applied ethical theories to build POE theories (or tools) that aid in identifying the ethically best option in real-life situations. Even if any outcome of this approach is imperfect, this is better than remaining with an 'unsolvable' issue. Additionally, POE will make the underlying normative ethical assumptions explicit to the person taking the decision and to all others. It clarifies the arguments and prepositions, which are essential elements for an ethically defendable decision.

Keywords: ethics, theory, practice

Introduction

In recent years, there has been a growing attention to ethics in different parts of society. In most cases, this has been a very implicit evolution, in which 'ethics' is often translated in very concrete concepts, such as fair trade, social justice, environmental concern etc. Changes within society are sometimes advocated by referring to 'unethical behaviour' or an 'ethically unacceptable situation'. Not seldom, these claims are hailed by one part of society and demonised by another. If we do not accept a relativist approach to ethics (i.e. the view that anyone is entitled to their own ethical truth), what is the reason behind seemingly inexplicable situation?

It seems reasonable to believe that different ethical frameworks within society (and possibly even within individuals) are the fundamental underlying problem. Strictly following one line of ethical reasoning to guide daily life is impossible in the complex society we live in today. Compromises, however difficult and unsatisfactory, are unavoidable. In this paper, we propose a conceptual framework by which such decision-making (or compromise-making) can be awarded a place within the field of ethics, as a third layer of ethical theories.

Current ethical landscape

Ethics, as probably all other scholarly disciplines, is not a unified, monolithical field of work. Many subdisciplines possibly draw on the same basic principles, but are concerned with entirely different subjects. Most obvious is the difference between well-known (sub)disciplines such as medical ethics, business ethics, environmental ethics, food and agricultural ethics, but there is also the difference between normative ethics (which is what usually is meant with the term 'ethics') and the different fields of metaethics, such as moral semantics and moral epistemology. All this is visualised in Figure 1.

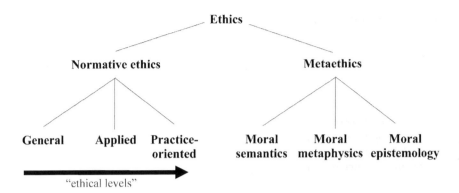

Figure 1. Overview of the main divisions of ethics, including the notion of 'practice-oriented ethics'. The arrow indicates the increasing attention to practical issues in the three ethical (sub)disciplines or ethical levels (Aerts, 2006 (adapted from Timmons, 2002).

Within this 'horizontal' differentiation between ethical subdisciplines, one could distinguish a similar diversification along 'vertical' lines. There is indeed a difference between the normative ethical work of authors such as Aristotle (350BC), Kant (1785) and Mill (1863), and that of Singer (1975) and Regan (1983). This suggests that there are multiple levels of ethical theories.

General normative ethics

The first and broadest of these levels is 'general normative ethics', which deals with fundamental issues. It is the discipline that deals with questions such as 'What is Good and Bad?' and it is centred on the creation of general theories that are relevant in a broad range of fields. General ethical theories will identify general 'truths' or 'principles' which allows one to distinguish good and bad.

The most influential general normative ethical theories are the well-known virtue ethics (Aristotelian ethics), deontology (Kantian ethics) and utilitarianism (Benthamian ethics), but others, such as Rawls' 'Theory of Justice' (1972), exist and should not be forgotten. It would be superfluous to go into further detail about these theories. More information can be found in texts such as those of Timmons (2002) and Mepham (2005).

Applied ethics

Within the framework of these general normative theories, 'applied ethical theories' exist, which constitute the second level of ethics. Applied ethics, or practical ethics as it is sometimes called, is the level dealing with the questions that arise in the different areas of human activity. This is a fairly new discipline and a plethora of subfields all have their place within this level. This has been mentioned briefly in the introduction and it is illustrated by the diversity of subjects within the *Oxford Handbook of Practical Ethics* (LaFollette, 2003). Even within such subfields, very different topics may be addressed, something that can be easily illustrated by referring to the papers in this volume, or the proceedings of previous EurSafe conferences (De Tavernier and Aerts, 2004; Kaiser and Lien, 2006).

Applied ethics is not casuistry, i.e. individual reactions to individual cases, nor the mere application of a general theory to a practical case - although Singer (1979) seems to define it in almost these exact words. Fundamental ethics will only be able to answer fundamental questions and is - usually - not apt to deal with applied and practical questions. Applied ethics builds upon these fundamental insights, but it is not a simple application of a general theory. In short, applied ethics requires the use of general theories to theorise about practical issues, not practical cases (LaFollette, 2003). Unfortunately, both names ('applied ethics' and 'practical ethics') for this level of ethics do little but add to the confusion.

Redrawing the map

General normative theories as well as applied ethical theories have their limitations. The former might be perceived as being 'too general, theoretical and therefore useless in real life', and this is indeed not without ground. An illustration could be the unsolvable conflicts between categorical imperatives within (strict) deontological ethics, or the speculative nature of utilitarian calculations.

As applied ethics essentially uses the lines of thinking propounded by general theories, much of their weaknesses are transferred to this level. Additionally, the work that has lead to the development of different subfields and the corresponding trend to specialisation can be misleading. It may well construct imaginary walls between the different fields, giving a false impression of independency between these fields. For example, 'solutions' for problems within environmental ethics that create animal welfare problems are no *real* solutions.

A second difficulty is the position of empirical evidence. One should avoid the naturalistic fallacy of directly connecting what is and what should be (see e.g. Fuchs, 1993), but empirical data can also be necessary to understand when, where and how a particular theory is relevant (LaFollette, 2003). In this way, it should guide, not determine ethical deliberation. This has been acknowledged by different authors, among them Singer (1975), Sidgwick in 1898 (see Sidgwick and Bok, 1998) and Samuel (1935), who stated 'firm philosophy [and ethics] builds on facts, not on reasoning alone'.

Practice-oriented ethics: a definition

If applied ethics is already trying to give moral guidelines to our daily life, and it is already using empirical data to build (or guide) its theories, it is reasonable to ask why yet another level of ethics be necessary. The answer is that neither general theories, nor applied theories are apt to deal with real-life dilemmas 'here and now'. The limitations we sketched in the previous paragraphs are quite fundamental when the subjects of a decision are not hypothetical, but real people, real animals, and real environments. Such real subjects suffer real consequences.

Competing claims (whether they are deontological, utilitarian, virtue-ethical or other) can block a solution to such dilemmas, and while this might be acceptable in theoretical situations, thought-experiments and other hypothetical discussions, it is not in real-life. In real-life situations, it is impossible not to take a decision, as this would in a certain sense be a decision too. Therefore, this fundamental weakness should be resolved, or remedied, by a new concept of ethical deliberation. Every option comes with different pros and cons and all will have different and often uncertain consequences, but stating 'it is impossible to conclude which option is ethically best' does not suffice. It is necessary to find a way to resolve these problems and, thereby, create a hands-on approach to ethical decision-making. This is what we propose should be named practice-oriented ethics (POE).

Goals of POE

POE is more than an institutionalised way of 'solving' specific problems or reacting to specific cases, rather like solving a mathematical equation. If this would be so, this would be the end of the general (and applied) theories we described earlier; but it is not. The practice-oriented ethicist will not jettison ethical theory, and return to casuistry, he will take the general theories, and their applied counterparts, and use these to identify the ethically preferable option in a real-life situation.

This does leave a possibility to develop POE frameworks that go beyond a mere case-by-case evaluation. Only, POE takes the interaction between ethical theory and practice (empirical data) at least one step further than applied ethics. The emphasis is not anymore on theory, but on action.

It is important to distinguish POE, its goals and methods from other forms of 'applying an ethical theory to a practical case'. POE has nothing to do with anti-vivisectionists, animal liberation movements or similar groups that - peacefully or violently - target individuals, companies or institutions. Even such pragmatic actions as those of Henry Spira - an Antwerp-born animal rights activist from New York (see Singer, 1998) - are something completely different from POE. This is 'pragmatic idealistic animal activism' (PIDA, pragmatisch idealistisch dierenactivisme; Vandenbosch, 2005), rather than POE. Such actions, however open to other considerations, are not POE because they are actions and not evaluation strategies.

Problems and solutions

Although POE is supposed to resolve some of general and applied ethics' problems, yet another difficulty arises. The differences observed between ethical theories and their proponents can also be expected in society as a whole. Discussions about decisions made can therefore be expected, even when they are based on and supported by POE. While professional ethicists can be expected to readily recognise the general and applied ethical framework(s) used in a certain decision making process, this should not be expected to be the case when working 'in the field'. Useless discussions based on assumptions and semi-misunderstandings will easily develop when people challenge each others positions without recognising their backgrounds.

This can be circumvented by making the underlying principles, assumptions and ethical theories explicit, as much and as far as possible. In some sense, this creates a common language, enabling people from within different traditions to 'meet in the middle', irrespective of their familiarity with ethical deliberation. This seems absolutely necessary if we want to invest our energy in what matters in POE: identifying the best option in the present situation. In doing so, POE does not limit itself to one theme or one applied ethical theory. As it specifically aims to deliver a decision in a real-life situation, it will take all relevant considerations into account. A simplified example is the current European legislative push towards more stringent animal welfare rules (a lengthy discussion can be found in Lips, 2004). Asking for very high animal protection level is easy, but this ignores the current economic reality. A POE evaluation aiming to answer the more realistic question of where to draw the line will have to take animal welfare considerations into account, but must also pay attention to economic, ecologic, ergonomic and communication considerations. And then propose a decision.

This is where POE's major advantage over applied ethics becomes aparent. Applied ethical theories will carry a 'prejudice' to which ethical issues are more or less important. Applied animal ethics, for example, is not apt to solve the above mentioned practical problem, because it only 'maps' the animal welfare (or animal rights) issues.

It could be argued that some applied ethical theories (e.g. Singer's animal ethics) are so well-developed and so intimately related to empirical evidence that they provide sufficient guidance for daily life and case-by-case decision making. If so, such a theory could be considered a practice-oriented ethical theory as well as an applied ethical theory. The resemblance of Singer's definition of applied ethics (cf. supra) with our notion of POE even strengthens this claim. For Singer, who can be expected to be thoroughly convinced of his theory and who accepts the premises and consequences wherever they lead, - although this has been doubted (Specter, 1999) - it may well be a sufficient basis for his day-to-day decisions. However, for others, who do not fully share (or even oppose) his believes, it will be too far-stretching or too limited to be useful in daily life. For decision-making in such a pluralist environment, an extra practice-oriented ethical layer is necessary.

Conclusion

Two remarks should be made on the discussion in this paper. The first being that it is highly improbable that one can be certain to make the *right* ethical decision in any 'scientific' sense of the word. POE is not a discipline that can be expected to deliver 'ethical certainty'. Nevertheless, in many of the difficult real-life cases in which it is to be helpful, it would be worse not to act than to act (carefully considering all aspects) but erroneously. The second remark concerns the methodology of POE. Applying an engineering approach to ethics, as in POE, seems to be a promising endeavour, one that is at the heart of the practice-oriented ethical work described in earlier work by the same authors (see e.g. Aerts *et al.*, 2006; Aerts and Lips, 2006).

In this paper, reference is mainly made to animal and agriculture related ethical issues to illustrate some of the topics discussed. Nevertheless, the theoretical background and many of the methods used in POE should be readily applicable in other fields of ethics.

General normative ethical frameworks all have their specific strong and weak points, and so do the different applied ethical theories based upon these frameworks. In part, this is the reason for the ethical pluralism that can be observed in society. This should not prevent (and might even be the very basis of) a rewarding dialogue sprouting from the development and the application of the practice-oriented ethical theories or tools.

References

Aerts, S. (2006). Practice-oriented ethical models to bridge animal production, ethics and society. Diss. Doct. Faculty of Bioscience Engineering, Katholieke Universiteit Leuven, Leuven, Belgium, 180p.

Aerts, S. and Lips, D. (2006). Stakeholder's attitudes to and engagement for animal welfare. In: M. Kaiser and M. Lien (Eds.) Ethics and politics of Food. Wageningen Academic Publishers, Wageningen, The Netherlands, pp. 500-505.

Aerts, S., Evers, J. and Lips, D. (2006). Development of an animal disease intervention matrix (ADIM). In: M. Kaiser and M. Lien (Eds.) Ethics and politics of Food. Wageningen Academic Publishers, Wageningen, The Netherlands, pp. 233-238.

Aristotle (350BC). Nicomachean ethics. (Translated by Ross, W.D. and published in English in 1923, republished in 1999 by Batoche Books, Kitchener, Canada, 182p. Available online at http://socserv2.mcmaster.ca/~econ/ugcm/3ll3/aristotle/Ethics.pdf [Accessed April 28, 2006].

De Tavernier, J. and Aerts, S. (2004). Science, ethics and society. 5th congress of the European Society for Agricultural and Food Ethics. CABME, Leuven, Belgium, 355p.

Fuchs, J. (1993). Moral demands and personal obligations. Georgetown University Press, Washington DC, USA, 232p.

Kaiser, M. and Lien, M. (2006). Ethics and politics of food. Wageningen Academic Publishers, Wageningen, The Netherlands, 592p.

Kant, I. (1785). Fundamental principles of the metaphysic of morals. (10th ed.). (Translated by Abbott, T.K.). Longmans, London, UK, 102 p. (1962 edition). Available as Project Gutenberg eBook at http://www.gutenberg.org/dirs/etext04/ikfpm10.txt [Accessed April 28, 2006].

LaFollette, H. (2003). The Oxford handbook of practical ethics. Oxford University Press, Oxford, UK, 772p.

Lips, D. (2004). Op zoek naar een meer diervriendelijke veehouderij in de 21ste eeuw. Aanzet tot het ontwikkelen van win-winsituaties voor dier en veehouder. PhD. Katholieke Universiteit Leuven, Leuven, Belgium, 184p.

Mepham, B. (2005). Bioethics, an introduction for the biosciences. Oxford University Press, Oxford, UK, 386p.

Mill, J.S. (1863). Utilitarianism. Batoche Books, Kitchener, Canada, 63p. Available online at www.ecn.bris.ac.uk/het/mill/utilitarianism.pdf (republished in 2001) [Accessed April 28, 2006].

Rawls, J. (1972). A theory of justice. Clarendon Press, Oxford, UK, 607p.

Regan, T. (1983). The case for animal rights. University of California Press, Berkeley, USA, 425p.

Samuel, H. (1935). Practical ethics. Thornton Butterworth Ltd, London, UK, 256p.

Sidgwick, H. and Bok, S. (1998). Practical ethics. A collection of addresses and essays. Oxford University Press, Oxford, UK, 142p.

Singer, P. (1998). Ethics Into Action: Henry Spira and the Animal Rights Movement. 237p.

Singer, P.A. (1975). Animal liberation. A new ethics for our treatment of animals. New York Review/Random House, New York, USA, 285p.

Singer, P.A. (1979). Practical ethics. Cambridge University Press, Cambridge, UK, 237p.

Specter, M. (1999). The dangerous philosopher. The New Yorker, September 6. Available online at http://www.michaelspecter.com/pdf/philosopher.pdf [Accessed May 8, 2007].

Timmons, M. (2002). Moral theory: an introduction. Rowman and Littlefield Publishers, Lanham, USA, 291p.

Vandenbosch, M. (2005). De dierencrisis. Houtekiet, Antwerpen, Belgium, 372p.

How do stakes and interests shape the discursive strategies for framing (multi-)causality?

Laura Maxim[1] and Jeroen P. van der Sluijs[2]
[1]*UMR C3ED n° 063 (IRD-UVSQ), Université de Versailles Saint-Quentin-en-Yvelines, 47 Boulevard Vauban, Guyancourt 78047 cedex, France, laura.maxim@c3ed.uvsq.fr*
[2]*Copernicus Institute for Sustainable Development and Innovation, Utrecht University, Heidelberglaan 2, 3584 CS Utrecht, The Netherlands, j.p.vandersluijs@uu.nl*

Abstract

Multi-causal patterns of interaction are characteristic of the behaviour of complex environmental systems under anthropogenic influence. However, in its socio-political context, multi-causality can also be an argument in discursive strategies to downplay the responsibility of a particular anthropogenic cause of an environmental problem. This paper provides theoretical insights based on a case study of the risk for honeybees of the insecticide Gaucho®. This case engendered a strong societal debate in France between 1994 and 2004 and led to the application of the precautionary principle for an environmental issue for the first time in this country. The paper explores the relationships between the way of framing (multi-)causality by experts of environmental risks and the stakes, interests, and conflicting relationships between actors in a social debate on these risks. Based on field work consisting of 32 interviews, our results show correlation between three 'variables': 1. discursive patterns on (multi)causality elaborated for explaining field observations; 2. institutional affiliation of the expert; 3. stakeholder-specific social, economic and political stakes and strategies. The links between the socio-economic stakes and the strategies of action of each stakeholder are analysed.

Keywords: multi-causality, Gaucho®, honeybee, uncertainty, post-normal science

Introduction

For complex systems such as ecosystems, changes are often caused by different combinations of factors, which may act synergistically or antagonistically. Such multi-causality may provide obstacles to understanding all the causal relationships (Gee, 2003), but it should not be used as an excuse not to directly address those causal factors that have a significant and decisive impact on the final effect(s). When research takes place in a context where economic and social stakes are high for the various actors concerned, the competence of the experts and the onus of responsibility on them become a decisive factor in framing the 'weight of evidence' (Funtowicz and Ravetz, 1990, 1993). Debates involving high stakes for the actors directly concerned by the results of the expertise may reveal a *social construction of uncertainty*, a process in which scientific fact are re-created through discourses (Michaels, 2005, Maxim and Van der Sluijs, 2007). Drawing on social constructivism and discourse analysis, Hajer (1995) describes the processes through which discourse coalitions form around story-lines that are meant to represent a particular definition of the environmental problem on which the environmental decision-making process critically depends. To exemplify these processes for the definition of (multi-)causal relationships, this paper investigates the debate on the risk for honeybees of the insecticide Gaucho® (active substance: imidacloprid) that has taken place in France for more than twelve years.

The debate started in 1994, when beekeepers observed negative symptoms in honeybees that they had never observed before such as the sudden and massive decline of honeybee populations corresponding with the sunflower honeydew. Honeybees were dying by thousands in front of their hive or seemingly 'disappeared' in that they did not return to their hive at all. In addition, a number of behavioural (sub-lethal) symptoms were observed (GVA, 1998, 1999, 2000). All these symptoms engendered massive hive depopulation and a loss of sunflower honey production between 40% and 70%. As the new symptoms coincided with the first use of Gaucho® in sunflower treatment in 1994, beekeepers demanded that Bayer, the producer of the insecticide, informs them on the potential toxicity of its active substance for honeybees. This was the start of a long series of scientific studies that involved experts from Bayer, the Ministry of Agriculture, beekeepers and public researchers. These studies resulted in arguments both for and against a causal link between seed-dressing with Gaucho® and the symptoms observed on honeybees, or simply in ambiguous conclusions.

The first risk assessment issued from public research (in 1998), as well as social pressure (e.g., press articles, manifestations), led to the first ever application in France of the precautionary principle for an environmental issue. On the 22nd of January 1999, the Minister of Agriculture suspended the use of Gaucho® in sunflower seed-dressing. However, Bayer continued to affirm its initial claim that Gaucho® has no effect on honeybees. During the public debate, the company was accused of dishonesty, corrupt behaviour including threats to other actors, and of favouring profit over environmental and public health consequences. Further to the Minister's decision, other authorities involved were: the State Council (for assessing decisions of the Ministry of Agriculture), the Commission for the Access to Administrative Documents (for allowing the access of beekeepers to documents related to Gaucho®), and courts throughout France (where Bayer prosecuted some beekeepers representatives for '*denigration of Gaucho®*').

In 2003, the Scientific and Technical Committee (henceforth SCT), which was set up by the Ministry of Agriculture for assessing the ranges of possible causal factors contributing to the decline in honeybee apiaries, published its Final Report (SCT, 2003). The report concluded that the risk of imidacloprid on honeybees is 'worrisome', both in sunflower and in maize seed-dressing. In 2004, the Minister of Agriculture also suspended the use of Gaucho® in maize seed-dressing. The apiaries started to recover after 2005, and most recent reports (in May 2007) indicate that they have vastly improved (TF1, 2007, UNAF, 2007).

In France, after the investigation began into the risk of Gaucho insecticide on honeybees, the description of the issue was later re-framed by some of the stakeholders as 'multi-causal'. Eight potential causal factors were proposed, including: 1. genetic origin of imported queens and low adaptation to local conditions; 2. unfavourable climatic conditions (during sunflower/maize flowering); 3. intoxication through pollen and/or nectar foraged in areas with Gaucho® in sunflower and/or maize seed-dressing; 4. intoxication through pollen and/or nectar foraged in areas with sunflower and/or maize which are not seed-dressed with Gaucho®, but which are grown on soils which received a Gaucho® seed-dressed crop at the year n-1; 5. honeybee diseases and viruses; 6. inadequate or illegal use of pesticides and mixes of pesticides, by farmers; 7. an insufficient quantity of pollen; and 8. changes in sunflower strains. We have synthesised the effects observed in the field to 5 symptoms: 1. unusual yield loss of sunflower honey, of 30% to 80%; 2. sudden loss of foragers uniquely during the period of sunflower/maize flowering; 3. sub-lethal symptoms following the sunflower/maize foraging: paralysis, apathy, shivering, excessive self-cleaning-up, abnormal foraging behaviour, etc.; 4. abnormal annual depopulations; and 5. winter over-mortality.

We assert that the construction of a causal link has, beyond its scientific basis, a social dimension that is strongly influenced if not directly determined by the stakeholder's self-interests. In the following, we analyse the construction of (multi-)causal relationships by different stakeholders, and their framing in relationship with their social and economic stakes.

Methods

The field work consisted of interviews with 32 people who were directly involved in the debate on the Gaucho® insecticide. They included: 2 representatives of AFSSA (French Food Safety Agency), 2 representatives of Bayer CropScience, 3 representatives of the French Ministry of Agriculture, 5 researchers and 20 beekeepers. The average length of each interview was three hours. A questionnaire was developed, based on Sir Bradford Hill's nine criteria for causation (Hill, 1965): the strength of the association, consistency, specificity, temporality, biological gradient, biological plausibility, coherence, experimental evidence and analogy). A scale from 0 to 10 was used to ascertain each interviewee's degree of conviction about an existing causal relationship between each potential cause and each symptom within the period from 1994 to 2004. This assessment was done for each of the combinations between a cause, symptom and criterion. This subjective scale was proposed by Charles Weiss (2003) and is intended as a tool to help increase the precision and rationality of discourse in conflicts in which generalists untrained in natural sciences must judge on the merits of opposing arguments in disputes among scientific experts. Each of the standards of proof of the scale, going from 'impossible' (0) to 'beyond any doubt' (10), has been drawn by Weiss from diverse branches of US law and assembled into a hierarchy of levels of increasing certainty (Weiss, 2003). For the identification of discourse coalitions, an empirical rule was applied to the mean of the results within each group. This consisted in identifying a difference ≤2 between the scores given by two stakeholders as reflecting the same discourse. The vertical axis of Figure 1 represents the percentage of answers reflecting the same discourse for a pair of stakeholders, from the total number of questions.

Results

All stakeholders acknowledged the influence on honeybees of several factors at the same time and the possible synergic effects between these factors. However, the balance between the role of Gaucho® and the role of other causes has been framed differently according to the different stakeholders.

Discourse coalitions

The first story-line is represented by beekeepers and researchers working in the public sector (see Figure 1). Based on field observations and experimental evidence, their arguments claimed that Gaucho® is the main contributor (even if not the only one) for the damage caused to honeybees colonies. Their arguments were supported by numerous studies made in France between 1997 and 2007 on the effects of imidacloprid on honeybees, as well as by the conclusions reached in the Final Report by the SCT (2003).

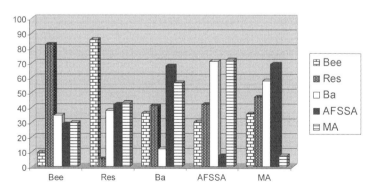

Figure 1. Discourse coalitions.

The second story-line argues for a non-causal relationship between Gaucho® and honeybees, determining that other causes are to blame. This was represented by Bayer and AFSSA and is based on their own research carried out in controlled conditions of access of honeybees to food containing imidacloprid, which does not reproduce the symptoms observed in the field,.

Finally, for the third story-line, Gaucho® is recognised as only one possible cause among others, with an unclear contribution on the final effects. This perspective is represented by the Ministry of Agriculture, whose scores on the questionnaire are ambiguous. For example, for some criteria, causes or symptoms, they indicate some influence of Gaucho® on honeybees' troubles, but for others they deny such an influence.

For beekeepers and researchers, the high levels of certainty expressed indicate a relationship between Gaucho® and all symptoms which is equally robust, consistent, coherent, specific (more or less, depending on the symptom), biologically plausible and supported by experimental evidence.

For five of the eight potential causes (i.e. genetics, climate, illegal use of pesticides, lack of pollen and sunflower strains), all stakeholders expressed low certainty levels for the consistency of the association with specific symptoms observed during sunflower flowering, mainly sudden loss of foragers only during the sunflower flowering and particular sub-lethal symptoms. However, no experimental evidence was provided supporting these causal relationships. The levels of certainty regarding the biological plausibility of the causal relationship between some of the potential causes and some of the symptoms were also low. Mixed scoring (i.e. spread on the scale) indicated that they might unequally attribute the causal factors to the honey yield loss, annual depopulations and winter mortality.

For another potential cause, namely honeybee diseases, the results were either mixed or engendered disagreement. For the beekeepers and researchers, they argued that while honeybee diseases might theoretically influence the colony in the long run (e.g., annual depopulations, winter mortality), it was not a biologically plausible explanation for the lethal and sub-lethal symptoms observed in this instance. Furthermore, the levels of certainty expressed for the other criteria of causality, related to 'what really happened from all that is theoretically possible', are low. Instead, in explanations for the symptoms and their correlation with sunflower flowering, Bayer and AFSSA proposed a combination of two causes: the lack of pollen and honeybee diseases. They based this on the following scenario: in extensive cropping areas, honeybees have no food alternative between the rape and sunflower flowerings and, therefore, are weak before sunflower honeydew and are exhausted during it because of the high work intensity. These factors would contribute to the higher susceptibility to diseases than the rest of the year. However, the basis of this scenario is not agreed on by beekeepers. For many of them, the hives are not sedentary, but movable, meaning that between the flowering of rape and that of sunflower, honeybees visit many other flowers, such as wattle, blackberry bush or chestnut trees. Despite the availability of pollen, apiaries still presented the symptoms during sunflower honeydew. No scientific evidence was given for the possible correlation between the lack of pollen, honeybee diseases and particular symptoms like those observed. In conclusion, the debate is between two possible explanations: Gaucho®, which is experimentally confirmed by researchers and by beekeepers' observations; and the combination of the lack of pollen and diseases, which is neither experimentally confirmed nor agreed by beekeepers' observations.

Socio-economic stakes

For Bayer, the new generation of systemic insecticides used in seed-dressing (which includes imidacloprid), represents an important opportunity for changing production patterns. Moreover, insecticides containing imidacloprid had a large international market, at the time.

For the beekeeping sector, the decrease of honey yield on sunflower honey between 1992 and 1999 reached 46% in some regions in France (GVA, 2000), and between 3,000 to 4,000 tonnes of the honey production at the national level between 1997 and 2004. In addition to the financial loss in revenue represented by these figures, research expenses (the sector invested European assistance funds to beekeeping, in normal conditions meant to develop the branch, for assessing the risks of Gaucho®) and attorney fees had a significant impact on French beekeepers. Between 1994 and 2004, 15 200 left this occupation.

The Ministry of Agriculture, confronted by contradictory demands from the two sectors, had a hesitant attitude. The main stake was to defend the Ministry's legitimacy, given that the debate on Gaucho® revealed important dysfunctions related to the process of authorisation of pesticides. In a letter published in the journal *Le Point* on 21 November 2003, the Head of the Bureau of Regulation of Anti-Pest Products described its lack of capacity in the following excerpt: '*three public servants for dealing with 20 000 demands of authorisation per year, a joint management of the risk assessment with industrials, lack of transparency in the procedures...it is impossible for the bureau to comply with its missions*'.

The three story-lines on multi-causality correspond to different actions demanded from decision-makers. The beekeepers' strategy focused on obtaining the support of the civil society and was characterised by cohesion among the different syndicates, perseverance, and disclosure of the social, economic and political stakes involved. They demanded a total ban on any use of imidacloprid. For the other two discourse coalitions, they suggested that more research was needed on each of the factors that possibly contributed to the negative impacts on apiaries. The discourse focuses on the uncertain aspects related to the knowledge available, rather than on the aspects which are better known, and the strategies that include action in justice and creation of 'alternative knowledge' (i.e. increasing quantity of published material on the influence of other factors than Gaucho® on honeybees).

Discussion and conclusions

The three story-lines discussed in this paper are based on different combinations of criteria of causality which are taken into account in the discourse. During the years of debate, the monitoring of symptoms by AFSSA was irregular and gave confusing results. From 1994 to 2004, the symptoms described by beekeepers were confirmed by different local or regional state services, but not systematically. The Ministry of Agriculture has repeatedly announced the creation of a national monitoring network, but trials failed due to procedural deficiencies in implementation and to the lack of trust between the Ministry's services and beekeepers. Both confusion and trust-related social deadlocks led to different descriptions of symptoms. For example, in our study, the descriptions given by stakeholders about the symptoms ranged widely with a threefold difference in some instances. In addition, no distinction is made by Bayer or AFSSA between symptoms found in areas where sunflowers were not treated with Gaucho® and areas that were treated with Gaucho®. However, beekeepers consistently asserted, based on their communication with local farmers, that honeybees showed the deleterious symptoms only on crops seed-dressed with Gaucho®, whereas the same set of symptoms did not appear where Gaucho® was absent. Whereas Bayer's experts claimed that sub-lethal symptoms are usually found in honeybees, beekeepers countered those claims stating that such symptoms are very specific and they had never seen them before 1994.

One major conclusion of our study is the need to clearly describe and differentiate the symptoms observed. Whereas some symptoms can be influenced by several factors (e.g., annual depopulations or winter mortality), others are specific (e.g., loss of foragers ONLY during sunflower flowering, repeatedly over the years). As trivial as it may seem, when a causal relationship is invoked, one should clearly ascertain precisely WHEN, WHERE and HOW the symptoms appear. Even if some causal relationship might be biologically plausible, other criteria must also be investigated (e.g., consistency, specificity, etc.). In our study, very few stakeholders (with the exception of the beekeepers) had actually been in contact with the real-life situation. This highlights the critical importance of considering local knowledge in the research process. Most importantly, all stakeholders should be involved from the very beginning to decide on the object of study and the research methods (Kloprogge and van der Sluijs, 2006), both for a good understanding of the problem and to better manage social conflicts resulting from misunderstanding and mistrust. Stakeholders can contribute knowledge about local conditions which may help determine which data is relevant or which symptoms require further investigation; provide personal observations which may lead to new foci for empirical research addressing dimensions of the problem which were previously overlooked; identify new indicators that better match the problem as experienced by the stakeholders; provide creative thinking about mechanisms regarding causal links; and scrutinize and improve assumptions made in risk assessments so that they better match real life conditions. Our findings reinforce the arguments proposed by post-normal science (Funtowicz and Ravetz, 1990, 1993) in support of 'democratising knowledge', by including all involved parties throughout the research practice on complex environmental issues.

Acknowledgements

The authors are grateful to Charlotte Šunde, for linguistic and stylistic improvements.

References

Funtowicz, S.O. and Ravetz, J.R. (1990). Uncertainty and Quality in Science for Policy. Kluwer, Dordrecht, The Netherlands, 229 p.

Funtowicz, S.O. and Ravetz, J.R. (1993). Science for the Post-Normal age. Futures 25: 735-55.

Gee, D. (2003). An approach to multi-causality in environment and health. Background paper from the European Environment Agency for the EU Environment and Health Strategy. 11 p.

GVA (Galerie Virtuelle Apicole). Apiservices - Le Portail Apiculture. Dossier Intoxications. www.beekeeping.com. Last checked on 20th February 2006.

Hajer, M.A. (1995). The politics of environmental discourse. Clarendon Press, Oxford University Press, New York, United States, 332 p.

Hill, B.A. (1965). The environment and disease: association or causation? Proceedings of the Royal Society of Medicine 58: 295-300.

Kloprogge, P. and van der Sluijs, J.P. (2006). The inclusion of stakeholder knowledge and perspectives in Integrated Assessment of climate change. Climatic Change 75:359-389.

Maxim L. and van der Sluijs, J.P. (2007). Uncertainty: Cause or effect of stakeholders' debates? Analysis of a case study: The risk for honeybees of the insecticide Gaucho®. Sci Total Environ 376: 1-17

Michaels, D. (2005). Industry groups are fighting government regulation by fomenting scientific uncertainty. Doubts is their product. Scientific American, pp. 96-101.

SCT (2003). Imidaclopride utilisé en enrobage de semences (Gaucho®) et troubles des abeilles, rapport final. Ministère de l'Agriculture, de la Pêche et des Affaires Rurales, Paris, France, 106 p.

TF1 (2007). Les abeilles de retour dans les ruches. Le journal de 20 heures, 13 Mai 2007.

Union Nationale de l'Apiculture Française (2007). Conférence de Presse, 13 Février, Paris.

Weiss, C. (2003). Expressing scientific uncertainty. Law, Probability and Risk 2: 25-46.

Depoliticizing technological decisions?

Bernice Bovenkerk
Ethics Institute, Utrecht University. Heidelberglaan 2, 3584 CS Utrecht, The Netherlands, bernicebovenkerk@yahoo.com.au

Abstract

Novel technologies, such as biotechnology, often have important social and environmental implications, are highly complex, and are characterized by great levels of uncertainty even among experts. Moreover, many disagreements about the merits of novel technologies are based on value disagreements. When such conditions are involved, decision-making on the basis of depoliticized debate might be helpful, because its outcomes would be better reflected upon and the promotion of personal interests would be avoided. In response to one particular proposal for depoliticization, that of Philip Pettit, I explore the question if and how depoliticization of decision making about animal and plant biotechnology can be achieved. I have analysed two ways in which the governments of the Netherlands and Australia have tried to do this: politically independent expert committees and lay people consensus conferences. I argue that both exercises serve to exclude certain viewpoints at the risk of leaving groups in society disgruntled and leading to more politicization rather than less. Moreover, the success of consensus conferences is dependent on the political culture - and associated view on political legitimacy - of their host country. Politicization can, therefore, not be avoided. Pettit's proposal is based on a view of politics as a struggle between interest groups and does not sufficiently take into account other views on the function of politics.

Keywords: depoliticization, biotechnology, consensus conferences, ethics committees

Introduction

Debates about novel technologies, such as the conflict about biotechnology, have certain features in common. They deal with issues that have important social and environmental implications and that are so complex that average citizens do not possess the skills to make adequate judgments about them; furthermore, they are characterized by great levels of uncertainty that exist even among experts. Moreover, as I have argued at the last EurSafe conference, many disagreements about the merits of biotechnology can be regarded as value disagreements (Bovenkerk, 2006). For example, disputes about the risks of GM food can often be traced to more fundamental disagreements about what scientific methodology is acceptable, what benefits it is worth taking certain risks for, what the status is of non-human nature, and ultimately what we consider to be the good life. Furthermore, biotechnology is a field in which many vested interests are present. When such conditions are involved, decision-making on the basis of depoliticized debate might be helpful, because it promises a more considered reflection on generalisable interests and the promotion of personal interests could be avoided. One proposal for such depoliticization is made by Philip Pettit (2004). Even though Pettit does not deal with disputes about novel technologies directly, his proposal appears an attractive one for addressing intractable disagreements in technological controversies which involve great vested interests. However, as I will argue with the help of an analysis of two ways in which governments and other organisations have tried to depoliticize decision making regarding animal and plant biotechnology, despite its promises this proposal harbours some problems and cannot be fully achieved. In fact, depoliticization is not only not feasible, but it also has undesirable effects, as it can be used to contain and truncate rather than promote public debate.

Depoliticization

Pettit (2004) argues that even if elected politicians have the good of the whole community at heart, they still have to follow their own (or their party's) interest in being re-elected. Issues that directly pertain to this interest - such as the drawing of electoral boundaries - should therefore be depoliticized. By depoliticization he means taking decisions away from the influence of politicians, by handing them over to an organisation that represents 'relevant bodies of expertise and opinion as well as the people as a whole' (Pettit, 2004: 55). Such a body should be accepted by all political parties to ensure that it does not advance sectional interests, but creates policy for the common good of society. It would ultimately remain under parliamentary control, but its decisions would not be tied to one particular politician or party. According to Pettit, depoliticization would be conducive to deliberative democracy, as according to most proponents of this model, deliberation should aim at advancing the common good and therefore should avoid decisions that unduly favour sectional interests or that are based on power imbalances. Pettit argues that the depoliticization of debate is called for particularly in three areas, because politicians could use decisions they make in these areas to gain electoral power. These areas are firstly, the popular passions, or those issues that the public is highly emotional about and which can easily be misused by politicians - such as heavy sentencing for certain crimes - secondly, aspirational morality, or those issues that involve moralistic arguments - for example about prostitution, which a politician might use to win votes, even if this might go against the common good - and thirdly, those issues that involve sectional interests, and in which lobbying and the self-interest of politicians play a large role.

Even though Pettit does not explicitly refer to novel technologies, particularly the latter two areas can be discerned in the biotechnology controversy as well. In fact, the main reasons why there is a controversy about biotechnology are exactly that moral issues are involved that people feel strongly about - and about which intractable disagreements exist - and that powerful sectional interests are at work lobbying politicians. Depoliticization of decision making in this area might be an attractive option for several reasons. Firstly, views on biotechnology do not necessarily follow party lines, which means that even if politicians could influence decision making, these decisions might go against the views of some of their own party members. Secondly, lobbying might conflate the interests of politicians and the biotechnology industry to a level where opposition is hardly possible. And thirdly, decisions about biotechnology regulation require at the same time a long term view - which is something that cannot be expected of politicians who often do not look beyond electoral cycles - and a flexibility to respond to new developments in research and development. A depoliticized body such as Pettit envisages could monitor such new developments and change policy accordingly. Even though at first sight depoliticization might appear to run counter to democracy, because decisions are not made simply on the basis of an equal vote for everyone - which is translated into the election of representatives - it is democratic in the deliberative democracy sense of the word, which seeks to move beyond the simple aggregation of preferences (which may be self-interested) to a reasoned agreement based on the force of the better argument and the fulfilment of the common good.

Committees

As the theory of deliberative democracy has been gaining more support, deliberation in depoliticized bodies has been on the rise. One way in which governments have delegated decisions about novel technologies to a depoliticized body is through the appointment of expert committees. Another instrument used to facilitate decision making on the basis of arguments rather than the negotiation of sectional interests is the consensus conference. What can we learn from an analysis of these two methods for the promise of depoliticization, particularly in the field of animal and plant biotechnology? I have examined the functioning of two biotechnology ethics committees, the Dutch Committee for Animal Biotechnology (CAB) and the Australian Gene Technology Ethics Committee (GTEC), and

Sustainable food production and ethics

two consensus conferences, those on animal cloning in the Netherlands and on gene technology in the food chain in Australia.

In short, the CAB reviews license applications for biotechnological procedures with animals and advises on these to the Minister responsible for licensing, while the GTEC has a broader mandate, namely, to provide the regulator of gene technology advice on ethical issues relating to both animal and plant biotechnology. The members have been appointed on the basis of independent scientific expertise and not as representatives of interest groups or of the opinion of particular segments of society. In my research, I looked particularly at the questions of how moral disagreements were dealt with and to what extent the diversity of views that exists in society was reflected in committee deliberations and recommendations. Considering the CAB, I concluded that certain moral viewpoints were already precluded upon the instalment of this committee. Firstly, the decision to forbid animal biotechnology save in exceptional circumstances closes off the discussion about the desirability of animal biotechnology in itself and narrows down the discussion to the acceptability of animal biotechnology in specific cases. Secondly, those with more extreme viewpoints could not identify with the terms of reference of the CAB and could not become member of the committee. These terms of reference also excluded deliberation on certain points, such as social or economic consequences of biotechnology.

The framework of the GTEC excludes less viewpoints and issues for deliberation in advance, but the members of the committee appeared to play a censuring role themselves, as they were not open to alternative viewpoints, such as religious views or views based on ecofeminism or environmental ethics. Moreover, while the GTEC has a broader mandate, it has no influence on licensing decisions and appears to have little influence on broader policy decisions. The licensing recommendations are made by another committee, the Gene Technology Technical Advisory Committee (GTTAC), which bases its decisions solely on a weighting of the risks of GMO's and thereby excludes other considerations. By separating the technical from the ethical aspects of biotechnology, and only considering the former in the licensing procedure, the latter are rendered less important. This is reflected in communiqués of the Gene Technology Regulator who seldom mentions the GTEC by name, but often refers to the GTTAC.

The composition of the CAB functions to exclude certain viewpoints as well. All of the members of the CAB are scientists, and most represent biomedical sciences or veterinary science. It is likely that this creates a bias towards a 'scientific worldview', and this might explain why technical considerations often prevail, why a majority attaches great importance to the intrinsic value of research, and why arguments that do not appear to be supported by scientific findings, such as the objection to unnaturalness, are rejected out of hand. This is exacerbated by the fact that there is no open expert controversy within the Committee. While the GTEC is also composed solely of experts, the areas of expertise are more balanced. But again, the influence of GTEC is limited and neither does its existence appear to be well-known by the public. Moreover, some members of the GTEC feel that there is a conflict of interest of some other members.

The above mentioned problems suggest firstly, that despite the independent status of its members, sectional interests have not been kept out of committee deliberations entirely and secondly, that committees have been used to contain public debate and exclude certain views. Even though concerns of the general public might have been eased by pointing to an independent committee that has 'thoroughly examined' the moral aspects of biotechnology, the more vocal critics in society - such as animal protection associations or organisations critical of genetic engineering - were left disgruntled by the limited terms of reference or influence of the committees, by the lack of transparency and the lack of public involvement. This has made them even more vocal in their criticism, increased their protests and lobbying efforts, and in the end will lead to more politicization rather than less.

Consensus conferences

In consensus conferences a panel of around fifteen lay citizens is chosen to deliberate about a certain well-defined and morally controversial topic, usually regarding a novel technology, during an intensive period, such as two weekends. Beforehand, the panel goes through a learning process, including self-study of materials selected for them by the conference facilitators and lectures given by a wide variety of experts. In the public part of the conference, the lay panel interrogates the experts and afterwards they withdraw to write a lay panel report, which aims at presenting a consensus. Doubts have been raised about the view that such a person as the 'lay person' exists at all; after people without former knowledge have engrossed themselves in literature and lectures on the topic, they cease to be real lay persons. Moreover, as Brian Wynne (1996) has successfully argued, the lay/expert divide is somewhat artificial. Lay people can often contribute experiential knowledge and experts are often familiar only with their narrow field of expertise. In these consensus conferences, the qualification of 'lay person' can therefore best be understood as a person with no personal interests in biotechnology.

However, the fact that the lay report is written by people that do not represent a sectional interest, does not mean that independence is guaranteed or power imbalances are avoided. Neither does it guarantee that all views that exist in society are represented. The lay panels in both cases I examined were expressly selected for their lack of a position on biotechnology, which ensured that the more extreme voices in society were not represented. Especially in the Australian case the independence of the organisers of the conference could be called into question, because many of the sponsors were pro-biotechnology and demanded to be on the steering committee. The steering committee influenced the lay panel at several stages, such as writing the initial briefing paper, compiling the information materials, deciding that some experts could not be heard, and limiting the topics to be discussed. On the first day of the conference, the scene was set by the introductory remarks made by the Federal Minister for Agriculture Fisheries and Forestry, which emphasised the importance of gene technology to the Australian economy and to global food security, downplayed its potential risks, and contained rhetorical remarks, such as 'we can not turn back the clocks' (Mohr, 2002). Furthermore, some of the expert speakers, most notably the Monsanto representative, deliberately omitted certain information and evaded answering the panel's questions.

Because experts could only respond to one question or a small defined set of questions, were not allowed to respond to each other, and were presented as either strongly in favour of or against gene technology, the conference had an adversarial character in which positions were simply stated, rather than a dialogue held. More neutral or middle of the road positions were not heard. Alison Mohr, who carried out an in depth study of the Australian consensus conference, likens it to a jury deliberation - in which jury members might come to a decision not so much on the basis of arguments, but of how persuasive the witnesses are - rather than a public debate: 'rather than facilitating broad public debate from a plurality of perspectives, the jury metaphor suggests that the consensus conference model facilitates debate from a duality of views' (Mohr, 2002). The lay members were under extreme time pressure and worked through the night to write the report. Some members conceded points they disagreed with because of the time pressure and because others were more dominant and had more stamina; some of those involved described it as 'survival of the fittest'. Similarly, discussions of the Dutch lay panel were dominated by those who were more highly educated. Areas of disagreement or minority positions were not described in the consensus report. Neither the Dutch nor the Australian lay panel reports had much, if any, influence on policy decisions. However, the reasons for this lack of influence differed; in the Netherlands, the conference was held at an early stage when no policy decisions yet had to be made, whereas the Australian conference was held when the main policy decisions had already been made. The Dutch conference was much less prestructured and contained than the Australian one, suggesting that the Australian organisers sought a higher level of control over the conference outcomes. This could be attributed to a difference in political culture; the Australian political climate could be characterised as an antagonistic one in which a variety of interest groups struggle for political influence, and the Dutch one as a consensual or corporatist one, in which decisions are traditionally made on the basis of stakeholder dialogue and compromise.

From this short discussion it can be concluded that even though consensus conferences are supposed to create a power free arena devoid of sectional interests, power imbalances and influences from stakeholders can not be completely avoided. In practice sectional interests do tend to come in; to what extent this happens is dependent on the political context in which the conference takes place. In defence of consensus conferences it should be noted that the lay panels turned out to be quite capable of recognizing rhetoric and self-interested motives dressed up as arguments for the common good. Moreover, the lay panels quickly picked up on the idea that science was value laden. Consensus conferences, therefore, could serve the function of countering sectional interests, but it is doubtful whether this can be interpreted as depoliticization. What also emanates from my discussion is that a consensus report gives the semblance of unity and glosses over any disagreements that exist between the panel members. Because they were faced with a polarization between experts, the panels tended to choose a position in between. These mechanisms, and the selection procedure for the panels, serve to exclude more vocal critics of biotechnology. Like in the committee system this led to a radicalisation of positions and politicization rather depoliticization.

Depoliticization or repoliticization?

Furthermore, whether or not depoliticization works appears to be dependent on the political context in which a depoliticized body functions. Especially important are the amount of control over decisions that policy makers are willing to relinquish, and associated with this, the level of influence that the decisions of such a body is allowed to have. As Peter Sandøe *et al.* (2007) argue, the success of the consensus conference model cannot be assessed independently of the political culture in which it takes place. In different countries the role of a lay panel advice will be interpreted differently. When policy decisions are regarded as legitimate because they are made by elected representatives, such as is the case in France - and which also seems to be the case in Australia - the influence of a lay panel report on policy decisions can only be minimal. Considering a consensus conference as a form of depoliticization in such a context is illusory, because the consensus conference will either not have any influence and be toothless or it will become a strategic tool of politicians who interpret its findings to suit their own agenda. In countries where lay input in policy decisions is regarded as legitimate because lay people do not have self-interested motives and are therefore more likely to deliberate about the common good - as is the case in Denmark and to a certain extent the Netherlands - the consensus conference has more chance of influencing policy decisions. However, this already assumes a certain political culture, and one can wonder whether we can really make a sharp distinction between politicization and depoliticization in such cultures. Pettit's proposal appears to be based on a view of (current, but not ideal) politics as simply a struggle between different interest groups. This does not take account of other functions of the political arena, such as facilitating debate about what is in the common interest of society. Through (re)politicization citizens in fact become more critically aware of and more likely to be involved in solving, collective problems and this could help to counter decisions made on politicians' self-interest as well.

Even though Pettit mentions deliberative polls - in which public deliberation precedes a vote - as an example of depoliticized debate, the function of public participation in policy regarding novel technologies in fact does not appear to be an instance of depoliticization, but of (re)politicization. Traditionally, important decisions in this field were left either to the industry or solely to scientific experts (such as in the committee system) and this could be regarded as depoliticization. On the other hand, 'the development and spread of the participatory consensus conference is, in effect, an attempt to 'expand the political possibilities of action' through public involvement'(Sandøe *et al.*, 2007), and hence politicize hitherto depoliticized citizens. In other words, deliberative democrats in fact want to politicize more issues rather than depoliticize them.

To conclude, Pettit's view of depoliticization assumes a political culture in which a struggle for sectional interests is central. However, in such a political culture the legitimacy of a depoliticized body itself will be called into question. When this body has to come up with concrete policy guidelines, this inevitably means that at least some people's viewpoints will be compromised. Moreover, groups representing more extreme views will either be excluded, or be involved, but in such a way that they could feel co-opted. After all, politicians can point to the depoliticized body and argue that a group of experts, stakeholders, and/or lay persons have thoroughly deliberated the issue and come up with this or that policy decision. The possibility that a minority of the body might not support the final outcome of deliberations within this body can then easily be glossed over. Finally, in an antagonistic political culture, such as Australia, different parties will try to influence the decisions of the depoliticized body and this calls into question the very possibility of depoliticization in such a culture. In political cultures that are less antagonistic, but more consensual, on the other hand, perspectives of lay people can influence policy decisions, but in such cultures the need for depoliticization is less pressing, because politics already involves more debate about the common good.

References

Bovenkerk, B. (2006). Biotechnology, disagreement and the limitations of public debate. In: M. Kaiser and M. Lien (eds.) Ethics and the politics of food. Wageningen Academic Publishers, Wageningen, The Netherlands, pp. 97-102.

Mohr, A. (2002). A New Policy Making Instrument? The first Australian consensus conference. PhD thesis, Griffith University, Australia, 216p.

Pettit, P. (2004). Depoliticizing democracy. Ratio Juris 1: 52-65.

Sandøe, P., Porsborg Nielsen, A. and Lassen, J. (2007). Democracy at its best? The consensus conference in a cross-national perspective. Journal of Agricultural and Environmental Ethics 20 (1): 13-35.

Wynne, B. (1996). May the Sheep Safely Graze? A reflexive view of the expert- lay knowledge divide. In: S. Lash, B. Szerszynski, and B. Wynne (eds.) Risk, Environment and Modernity: towards a new ecology. Sage, London, UK, pp. 44-83.

Implicit normativity in scientific advice - a case study of nutrition advice to the general public

Anna Paldam Folker[1], Hanne Andersen[2] and Peter Sandøe[1]
[1]Danish Centre for Bioethics and Risk Assessment, Faculty of Life Sciences, University of Copenhagen, Rolighedsvej 25, 1958 Frederiksberg C, Denmark
[2]Steno Department for Studies of Science and Science Education, University of Aarhus, Ny Munkegade, 8000 Århus C, Denmark

Abstract

This paper focuses on implicit normative considerations underlying scientific advice. On the basis of a case study of experts in human nutrition the paper tries to unfold two such implicit normative considerations in relation to the decision of scientific advisors to give public advice. The first concerns the aim of scientific advice - whether it is about avoiding harm or promoting good. The second consideration raises the issue of the main cause of concern in connection with the particular framing of scientific advice - the worst off members of society or the population at large? The paper also presents arguments against an apparent tendency of nutrition scientists to reduce the normative dimensions of scientific advice to individual, subjective differences and thus to bypass reflection and debate. Some advisors seem to conceive ethical considerations as merely personal preferences which are therefore not the subject of a common discussion or deliberation. It is argued that such personalization is unfortunate because it helps to conceal the normative issues that come with the decision to give public advice and accordingly serves to blur that there are genuine normative issues at play in connection with scientific advice that could be the subject of a systematic discussion. It will be concluded that the implicit normativity of scientific advice should rather be made explicit and hence debatable for scientific advisers and a wider public.

Keywords: implicit normativity, values, nutrition advice, scientific advice, scientific evidence

Introduction

Scientific advice is often called for in order to shed light on issues that seem underexposed. Regarding nutrition it is a central task of scientific advisers to evaluate the available evidence in order to decide the way in which it should form the basis for scientific advice. One element, in this kind of evaluation, is the assessment of the strength, or the probable truth, of the evidence: e.g. to ask whether the results of particular studies have been derived by proper scientific methods and by means of acceptable research designs, and to assess the extent to which any relevant results correspond with those of similar previous studies. However, these epistemic issues are not alone in guiding scientific advice: in the reflective process leading to a decision to give scientific advice, a number of *normative* considerations regarding the way the advice will effect the population are taken into account. As we shall argue, such considerations may enter the thought processes of scientific advisers, albeit implicitly (Rasmussen and Jensen, 2005).

Implicit normativity, as we understand it here, is constituted by normative questions, decisions or issues that scientific advisers and the general public are not fully aware of, but which nevertheless do and should have implications for the character of the advice given. The fact that the normativity is implicit means rather different things for the public, who receive the advice, and the scientists, who give the advice. For the public, the normativity is implicit in the sense that it is hidden: if members of the public are presented with a recommendation to do, or to refrain from doing, this or that, they stand no chance of judging the normative considerations behind. For scientists, the normativity is implicit in the sense that it is not sufficiently recognized and discussed (Molewijk *et al.*, 2003).

Nutrition advice to the general public is typically presented in the media, where nutrition scientists attempt to communicate scientific results to the general public or to give advice on issues of nutrition. Scientific advice to the general public is here understood as a broad category that may be mediated in different ways. Most often such advice involves journalists, who present the views of scientists after interviewing the scientists or after reading their papers or reports. As human nutrition involves central elements of everyday life such as eating habits, food choices, life style, and bodily appearance it makes for compelling news and stories. This means that there is often a rather rapid, automatic link between research and scientific advice: the public statements, or claims, of scientific experts are often turned into headlines in the media and transformed into recommendations for action (Hilgartner, 2000).

Normative assumptions

Based on an interview study (Folker *et al.*, 2007) we found that when they are considering whether to give advice to the public, nutrition scientists do not attend *exclusively* to an assessment of the epistemic strength of the scientific evidence. They also seem to reflect on a number of normative issues although these issues are not addressed explicitly. In what follows we try to spell out two of these issues based on a normative analysis of the findings of the interview study (Folker *et al.*, 2007). These issues are, first, whether the aim of scientific advice is to avoid harm or to promote good; second, whether the main cause of concern should be the population as a whole or just those who are worst off. We believe that the normative analysis has relevance for fields beyond nutrition - e.g. fields such as the environment, climate, energy, technology and health more generally which also involve public advice and are likewise characterized by public attention, uncertainty of knowledge and conflicting interests.

Avoiding harm or promoting good: A central issue in any decision to give scientific advice concerns the aim of the advice. This is an issue about which underlying assumptions can often be unearthed. Scientific advisers might hold different views about the aims or goals of nutrition advice. While most parties might agree that the general aim of nutrition advice is to improve people's health, this aim can be interpreted in a variety of different, more specific ways (Buchanan, 2000, 2006). For instance, it is possible to assume a narrow or a broad perspective on the improvement of public health. Within a narrow perspective, the improvement of health is seen as a matter of protection against danger or risk that is measured in terms of traditional end-points such as death and morbidity. Within a broad perspective, the same goal is interpreted as a matter of what is necessary or even optimal for good health. However, since the concept of health is itself the subject of interpretation and debate, there are in fact several broad perspectives to choose from. On one conception, health is a purely naturalistic concept describing a specific physiological state (Boorse, 1977). On an alternative conception, health is a normative concept subsumed under the value of human well-being or quality of life (Daniels, 1985).

These perspectives will have different implications regarding both the kind of evidence that should be accorded weight in a decision to give scientific advice and the character of the advice given. Take sugar, for instance. In the narrow perspective, only evidence regarding the harmfulness of sugar will be considered relevant. Within a broad perspective, on the other hand, it may also be relevant to include evidence about the interaction of nutrients in the body, including the way sugar affects the appetite, and the extent to which sugar takes the place for other important macronutrients and vitamins. Within a normative conception of health, it might also be relevant to include evidence about the role of sugar in the culture of food, and the relation of sugar intake to pleasure and social events.

The choice between perspectives on the aim of scientific advice is value-based. It involves basic views about the role of the scientific adviser vis-à-vis the public - whether the advisor is someone who protects against possible danger or someone who prescribes what is right and wrong in relation to a certain concept of health. It consequently involves a view on the extent to which scientific advisers should base their advice on more or less substantial assessments. This can be illustrated by a continuum. At one end, scientific advice is based on a minimal assessment that seeks to establish harm in relation to traditional, narrow end points. At the other, advice is based on a substantial evaluation that seeks to establish what is right in relation to wider 'end points' such as optimal nutrition, good health and quality of life. Of course, many intermediate positions lie between these extremes. This does not change the fact, however, that the position adopted is the result of a choice that can be debated.

The worst off or the population as a whole: another normative issue involved in decisions about scientific advice concerns which group in society should be thought of as the intended beneficiary of the advice. More specifically, should the advice be framed out of concern for the general population, with average health and well-being in mind, or out of concern for typical health and well-being among the most exposed or vulnerable groups, or the worst off, in society? This issue is reinforced by the general assumption among scientific experts that public advice has to be simple in order for people to understand it (Irwin and Wynne 1996). This assumption apparently excludes, or at least discourages, the giving of differentiated public advice, which in turn makes it important to consider which group public advice should be optimized for.

It might seem fair to give priority to the weakest or most vulnerable people. But this also has a price. If concern for the most vulnerable is prioritized, other groups may be alarmed without good cause. If, on the other hand, the scientific advice targets the average member of the population, the weakest in society may be disadvantaged, because they may be especially exposed to risks (Wandall, 2004) or insufficiently informed about the peculiarities of their circumstances.

To see this, consider a hypothetical case of a group of advisers who want to give public advice over the amount of added sugar in a healthy diet. There are various kinds of evidence to consider, but to simplify matters let us focus on the evidence for a corresponding, unhealthy reduction in the daily intake of vitamins. According to this evidence, the more sugar someone eats, the greater the risk of vitamin dilution and, ultimately, of vitamin deficit. This is due to two things: the fact that sugar does not itself contain vitamins; and the fact that the more you eat sugar, the less you eat other, vitamin containing foods. Assume now, further, that this risk of dilution is very low in average members of the population if the daily intake of added sugar does not exceed 25% of the total intake of energy. Now one option would thus be to give such advice to the public. On the other hand, the risk of dilution is dependent on the total amount of food consumed. Hence children and adults who only eat small amounts of food will thus be at greater risk for dilution even if they eat the same percentage of sugar as adults who eat average quantities of food. The less you eat, the greater the risk (hence, if someone eats a lot during the day, the chance that he will absorb sufficient vitamins is relatively greater, even if one third of his diet comes from sugar). Should the advisers rather base their advice to the public on the average intake of children and adults who eat less, and thus, for instance, recommend a reduction in daily intake of added sugar to the level of 10% of the total intake of energy or below? Or should they rather divide this latter group further, to account for the great variety between individuals, and therefore frame their advice on the basis of the group of most extreme individuals, e.g. children who eat very little because they are extremely fussy about their food? This option may result in general guidance to reduce added sugar to a daily level of 3% of the total intake of energy.

In the example above, there is no scientific reason for choosing one group over the other. Scientific arguments alone cannot determine which group in society should be the principal target of scientific advice. It is therefore a fundamental issue, here, whether scientific advice should be framed in order to benefit most people (using overall population averages) or the people who need it most (concentrating on the worst off). And this issue continues to arise, albeit on a smaller scale, even when the assumption that simple advice must be given is rejected, making way for more differentiated public recommendations.

At bottom, the issue, in its many alternative forms, involves values over distributive justice and equality; it therefore obliges us to engage with some central themes of modern political philosophy (Rawls, 1972).

Explicit normativity

In this article we have tried to bring out some of the normative considerations in decisions to offer public advice that are based on nutrition science: should it be the aim of scientific advice to avoid harm or to promote good? Should the main cause of concern be the population as a whole or the worst off? Considerations along these lines seem to be made by scientific advisers, although perhaps not as explicitly and systematically as they could be.

On the contrary nutrition scientists have been found to conceive ethical considerations as merely personal preferences, and as such as matters that are not (and need not be) the subject of a common discussion or deliberation among scientists (Folker *et al.*, 2007). In this way, complex normative issues are easily reduced to a matter of individual choice and different, personal thresholds regarding when it is right to communicate to the public.

In our view such personalization is unfortunate because it helps to conceal the normative issues that come with the decision to give the public advice. It serves to obscure the fact that there are genuine normative issues at play in connection with scientific advice - issues that could be the subject of systematic shared reflection and discussion. It thereby also confirms a common misunderstanding which is to go from a conception of values as subjective (in contrast with matters of fact) to the belief that values cannot be the subject of rational, systematic discussion. However, while the ultimate status of values are debatable a case can be made for saying that even if values are indeed subjective, they can still be defended or criticized with reference to reasons that can be better or worse (Blackburn, 1993, 1999).

In our view the implicit normativity of scientific advice on nutrition should, then, be recognized as an integrated part of public guidance on nutrition that has to be grappled with openly. It should be recognized that advice is a certain kind of public action, and has public implications that require normative reflection and justification. Such reflection does not lend itself to simple answers and quick solutions; nor is it a matter for specialists only. When it comes to normativity, scientists are no better equipped than anyone else (except insofar as they are in a position to supply factual information relating to the normative issues). Thus the need to consider the normative issues raised by scientific advice applies not only to the collective of scientists who communicate in the media, and to the expert policy bodies that issue recommendations, but to anyone who is prepared, and able, to think carefully about the issues.

References

Blackburn, S. (1993). Essays in Quasi-Realism. Oxford University Press, Oxford, UK, 262 p.
Blackburn, S. (1999). Is Objective Moral Justification Possible on a Quasi-Realist Foundation? Inquiry 42: 213-28.
Boorse, C. (1977). Health as a theoretical concept. Philosophy of Science 44: 542-73.
Buchanan, D. (2000). An Ethic for Health Promotion. Rethinking the Sources of Human Well-Being. Oxford University Press, Oxford, UK, 232 p.
Buchanan, D. (2006). Moral reasoning as a model for health promotion. Social Science and Medicine 63: 2715-26.
Daniels, N. (1985). Just Health Care. Cambridge University Press, Cambridge, UK, 276 p.
Folker, A.P., Andersen H. and Sandøe, P. (2007). Implicit normativity in scientific advice - An analysis of values in nutrition advice to the general public (submitted).

Hilgartner, S. (2000). Science on Stage. Expert Advice as Public Drama. Stanford University Press, Stanford CA, USA, 236 p.

Irwin, A., and Wynne, B. (eds.) (1996). Misunderstanding Science? The public reconstruction of science and technology. Cambridge University Press, Cambridge, UK, 240 p.

Molewijk, A.C., Stiggelbout, A.M, Otten, W. Dupuis, H.M. and Kievit, J. (2003). Implicit Normativity in Evidence-Based Medicine: A Plea for Integrated Empirical Ethics Research. Health Care Analysis 11: 69-92.

Rasmussen, B. and Jensen, K.K. (2005). The Hidden Values. Transparency in decision-making processes dealing with hazardous activities. Project Report 5 from the Danish Centre for Bioethics and Risk Assessment, Copenhagen, Denmark.

Rawls, J. (1972). A Theory of Justice. Clarendon Press, Oxford, UK, 607 p.

Wandall, B. (2004). Values in science and risk assessment. Toxicology Letters 152: 265-72.

Trustworthiness: the concrete task to take vague moral ideals seriously

Franck L.B. Meijboom
Ethiek instituut, Utrecht University, Heidelberglaan 2, 3584 CS UTRECHT, The Netherlands,
f.l.b.meijboom@uu.nl

Abstract

Trust in political and market institutions of the agro-food sector is on the public agenda. Consequently, methods to increase *trust* are considered the most effective to address this situation. Problems of trust, however, can better be addressed as problems of trustworthiness. From this perspective an ethical dimension surfaces. I defend that being a trustworthy agent implies that the trustee fulfils two moral conditions. First, trustworthiness ought to be motivated by a specific view on the moral status of the truster. Second, trustworthiness requires the recognition of the truster as a moral agent. In the discussion on sustainability this last condition is relevant. Consumers express their moral beliefs and ideals with respect to sustainable food production. When this moral dimension is ignored problems of trust surface. However, problems also remain when the moral dimension is taken seriously, but not accurately addressed. This is the case when the moral element of what is entrusted is addressed as if it were about the morally 'right' thing to do, while the concern starts from an ideal of the 'good life'. Concerns on sustainable food production are a good example. These concerns often start in a particular vision on the good. When this element is not addressed by the trustee problems of trust remain. However, dealing with such ideals of the good raises a practical and a fundamental problem for political liberalism. I argue that these problems are not insurmountable and that a liberal government can take the moral dimension of what is entrusted seriously.

Keywords: trustworthiness, consumer concerns, moral ideals

Problems of trust as problems of trustworthiness

Trust in political and market institutions of the agro-food sector has been on the public agenda for some years. As a result of the global character of the food market, the increased use of technology in food production and the recent food related scandals, public trust in this sector has come under pressure. Consequently, methods to increase *trust* are considered the most effective to address this situation. For instance, many elements in the current European food policy, such as a revised food safety policy, the establishment of food safety authorities, and the increase of information services all have the aim to improve the level of trust (cf. Dreyer and Renn, 2007).

Recently, we have argued that these problems of trust can better be addressed as problems of trustworthiness (Meijboom *et al.*, 2006). The observed problems of trust can be interpreted as an important signal for problems with respect to the trustworthiness of agents. This position does not deny that a decline of trust can be conceived as problematic as such, yet the definition of a decline of trust as a failure of individual agents starts from the wrong angle. To address problems of trust in a fruitful way, the question should not be 'How to increase trust?', but 'Why would an individual agent trust the other agent?' and 'Is this agent being worth to be trusted?' He who wants to be trusted should be trustworthy. Institutional agents cannot change individuals in a way that they adopt a trustful attitude. However, they can show themselves to be trustworthy. Hence, enhancing trustworthiness seems a more promising starting point in the process of regaining public trust than measures that aim to tempt consumers to adopt a stance of trust.

Two normative conditions of trustworthiness

The definition of problems of trust as problems of trustworthiness surfaces a clear ethical dimension. When we address the issue from the perspective of trustworthiness a lack of trust in the public is no longer morally neutral. When the trustee does not respond to what is entrusted this can have serious moral implications although the lack of trust itself is still morally unproblematic. The individual may suffer losses, e.g., one's freedom to choose food products is limited or one's moral concerns relating to production methods are disrespected. This has implications for the incentives of trust-responsiveness since the person who trusts is often not able to prevent this situation her/himself.

From the observation that harm can be done to individuals when they are not in the position to trust others, it would be too easy to conclude that there is an obligation for the trustee to respond to trust, however, it indicates two moral conditions with respect to trustworthiness.

First, trustworthiness ought to be motivated by a specific view on the *moral status of the truster*: the dependent and vulnerable position of the truster is not an opportunity to take advantage of, but an imperative for the trustee to act in a trust-responsive way since one is faced with an individual with inherent worth. Thus Kant's second formulation of the Categorical Imperative: 'Always treat the humanity in a person as an end, and never as a means merely' also underlies trust. As for morality in general, trusting has to start from the assumption that human beings matter. The basic assumption that other agents should be treated as an end in themselves should underlie trusting as well. Trust and trustworthiness have to start from the fundamental moral requirement to 'acknowledge each human being aright.' (Cordner, 2007: 67; Rawls, 1972; Strawson, 1974) This is not saying that the trustee may not profit from trusting, yet it implies that trust may not be used in a way that disrespect the inherent worth of the individual who will give the trust.

From this starting point we have reasons to trust and to respond to trust. With respect to the latter, the vulnerable status of the truster is crucial. This vulnerability provides reason for reacting. The reason is not based upon egalitarian arguments striving towards symmetric, power free relationships. With respect to trust, an asymmetry will always remain, otherwise there would be no need to trust. In my account one has a moral reason to respond to trust. Not responding to the vulnerable status of the truster implies a violation of the truster's inherent worth. This requirement to take the inherent worth of the other agent as a primary condition provides a direct reason to act in a trust-responsive manner. Consequently, the challenge of being trustworthy starts from the acknowledgment of the moral status of the truster as being a human being with inherent worth. This condition requires that the vulnerability of the individual agent who is confronted with an insecure position is taken seriously.

Additionally, trustworthiness includes a second moral precondition with respect to motivation: the recognition of the truster as a *moral agent*. Consumers regularly express their moral beliefs and ideals with respect to food production and consumption in the market and in the public debate. Such 'consumer concerns' reflect public uneasiness regarding a whole range of issues, such as animal welfare, sustainability, and the introduction of technological innovations (cf. Brom, 2000; Korthals, 2004). The manifestation of consumer concerns in the agro-food sector is a relevant signal with respect to trust, as it can be an indication of the vulnerable position of consumers. Neglecting a concern leads to problems as ascribed above: the disregard of the individual's vulnerability and the disrespect of the individual as a moral subject with inherent worth.

However, concerns can also be positive reactions related to consumer power (Meijboom and Brom, 2003). The positive concerns do not start out of vulnerability and thus will not automatically raise questions of trust. However, on this level they can result in problems when the moral dimension of the concern is not taken seriously. For example, this would be the case when the moral dimension in the public discussion on functional food policy is disregarded by restricting the debate to risk related issues only. As a result of this restriction, the vulnerability of the truster might still be taken seriously, thus the first condition of trustworthiness is fulfilled, yet the second one is not met: the truster is not taken seriously as a *moral* agent. Consequently, there remains a problem of trust since the moral dimension of what is entrusted is not recognised, or is not taken as important. Thus the truster is not respected as a moral agent.

We can go one step further. Problems of trust do not only occur when the moral dimension of consumer concerns is fully ignored. They also arise when the moral element is addressed as if it were about the morally right thing to do, while the concern starts from an ideal of the good life. This illustrates that the precondition that the truster has to be taken seriously as a moral agent entails that the trustee (e.g., the government) has to respond to moral ideals as well. I will focus on this specific issue in the remainder of this paper.

Moral concerns on sustainable food production: the moral good and the right

Consumer concerns on sustainable food production surface the importance of the distinction between 'the right thing to do' and responding to 'ideals on the moral good.' Following a rather old, but still concise description, 'The good, or 'the best', is the ideal pattern of the life' while the 'the right is (...), the concretion and particularisation of the good into a *hic et nun*' (Taylor, 1939: 289). The issues that trusters are concerned about comprise both the level of the 'good' and the 'right'. However, the government often responds to the subject of trust exclusively from the level of the 'right'. Consequently, the moral element of what is entrusted is only partially addressed or even remains unaddressed.

The concerns that were raised after outbreaks of BSE and the foot-and-mouth disease form a good explanation of this point. On the one hand, there were concerns that focused on the right, e.g., on how animals should be treated. This level includes concerns on the moral acceptability of the culling of healthy animals and the restricted housing conditions resulting in welfare problems for the animals. On the other hand, since the outbreaks, concerns were formulated that questioned the sustainability of the current way of husbandry and animal production. The focus here is not restricted to the actual outbreaks, but the individual takes these outbreaks as a signal of broader problems in animal husbandry. These concerns do not only focus on what we should do, but start from a particular conception of the good. From that conception the framework of animal husbandry as such has become a problem. Measures that focus on the morally right only then do not suffice. The moral dimension in this case in the trusting relationship regards the moral good, but is addressed as if it were only about the moral right. For instance, in a trusting relationship, concerns on sustainable livestock farming cannot be addressed only by drafting new standards of housing, neither can concerns on sustainable aquaculture be answered by the establishment of an ethical code of conduct. Notwithstanding the importance of such measures, they disregard an important element that is entrusted by consumers to the government, namely their fundamental view on the good life/ good society that underlies the concern for sustainable animal production.

This presents the agents that want to be trusted with an extra challenge. They have to start taking the moral value of human values as a primary condition, they should take the other as a moral agent seriously, but also have to broaden the scope of what counts as morally relevant from only the moral right to one that includes the moral good as well. This, however, raises two problems.

The problems of taking the moral good seriously

Including the element of the moral good in policy is often considered as difficult. On the one hand there is a practical problem. The ideals trusters have with respect to sustainable food production or husbandry are usually broad and vague. They form part of a particular conception of the good, but often turn out to be rather implicit. Consequently, it is difficult to grasp its practical consequences and to respond to them as a government in an appropriate way. Nevertheless, they are important since ideals are a motivating source and guide our view of what is relevant. This makes ideals alongside with other elements essential in morality and politics (cf. Van den Burg, 1997). The implicit and vagueness of ideals can be a reason to ask individuals to explicate and reflect on their ideal, yet not to disregard them from the start.

On the other hand, there is a more profound problem: the priority of the right over the good in political liberalism (Rawls, 1972; 1988). Liberalism starts from the view that sustainability is not grounded in one purpose or end. There is no such *telos* or end that fully colour the interpretation in advance. It is not that values and ends do not play any role in society, but in society there is not one particular vision of the good life that provides an authoritative interpretation of sustainability. Thus, one may argue that with respect to sustainability political liberalism asserts the priority of right and seeks principles of sustainability 'that do not presuppose any particular conception of the good.' (Sandel, 1984: 83) This, however, does not imply that a government cannot deal with concerns on sustainability that are based on particular conceptions of the good. Rawls, has convincingly argued that the priority of the right does not imply that 'a liberal political conception (....) cannot use any ideas of the good except those that are purely instrumental; or that if it uses noninstrumental ideas of the good, they must be viewed as a matter of individual choice, in which case the political conception as a whole is arbitrarily biased in favor of individualism.' (1988: 251) With respect to justice, he phrases this point as 'justice draws the limit, the good shows the point. Thus, the right and the good are complementary, and the priority of right does not deny this.' (Rawls 1988: 252)

This illustrates that a liberal government has the ability to deal with the ideas on the moral good of those who trust them. It does not mean that the government always has to act in line with the particular conception of the good of one truster in order to be trustworthy. It is, however, possible for a government to take this element into concern so that all aspects of the moral agency of the truster is taken seriously, which is a precondition for being a trustworthy agent.

References

Brom, F.W.A. (2000). Food, Consumer Concerns and Trust: Food Ethics for a Globalizing Market. Journal of Agricultural and Environmental Ethics 12: 127-139.

Burg, W. van den (1997). The Importance of Ideals. Journal of Value Inquiry 31: 23-37

Cordner, C. (2007). Three contemporary perspectives on moral philosophy. Philosophical investigations 31: 65-84.

Dreyer, M. and Renn, O. (2007). Developing a coherent European food safety policy: the challenge of value-based conflicts to EU food safety governance. In: L. Frewer and H. van Trijp (Eds.), Understanding consumers of food products. Woodhead Publishing, Cambridge, UK, pp. 534-557.

Korthals, M. (2004). Before Dinner. Philosophy and Ethics of Food. Springer, Dordrecht, The Netherlands.

Meijboom F.L.B., Visak, T. and Brom, F.W.A. (2006). From trust to trustworthiness: why information is not enough in the food sector. Journal of Agricultural and Environmental Ethics 19: 427-442.

Meijboom, F. and Brom, F.W.A. (2003). Intransigent or Reconcilable: The complex relation between public morals, the WTO and consumers. In: A. Vedder (red.) The WTO and Concerns Regarding Animals and Nature. Wolf Legal Productions, Nijmegen, The Netherlands, pp. 89-99.

Rawls, J. (1988). The Priority of Right and Ideas of the Good. Philosophy and Public Affairs 17: 251-276.

Rawls, J. (1972). A theory of justice. Clarendon Press, Oxford, UK.

Sandel, M.J. (1984). The Procedural Republic and the Unencumbered Self. Political Theory, 12: 81-96.

Strawson, P.F. (1974). Freedom and Resentment. In: P.F. Strawson (ed.) Freedom and Resentment and other essays. Methuen and Co, London, UK, pp. 1-25.

Taylor, A. E. (1939). The Right and the Good. Mind 48: 273-301.

Part 4 - Diversity, resilience, global trade

Using past climate variability to understand how food systems are resilient to future climate change

Evan D.G. Fraser[1], Mette Termansen[1], Ning Sun[2], Dabo Guan[3], Kuishang Feng[1] and Yang Yu[1]
[1]*Sustainability Research Institute, School of Earth and Environment University of Leeds, Leeds, United Kingdom, Evan@env.leeds.ac.uk*
[2]*Institute of Geographical Sciences and Natural Resources Research and Chinese Academy of Science, United Kingdom*
[3]*Judge Business School, University of Cambridge, United Kingdom*

Abstract

The literature on climate change impacts struggles to quantify how farmers will adapt to changing environmental conditions. One approach is to examine the relationship between environmental variability and crop productivity in order to understand why in some situations relatively small environmental perterbations seem to have large impacts on productivity. These cases may then be contrasted with situations where even large environmental problems did not result in production losses. Using a comparative approach, this paper will draw lessons from 'resilient' and 'vulnerable' case studies - such as the Ethiopian famines of the 1970s and 1980s where the meteorological perturbations were minor suggesting that social and economic factors created a vulnerability to drought - and apply them to theoretical work on adaptability, resilience, and vulnerability. Finally, this analysis includes a brief quantitative examination of data from Eastern China, one of the world's most vital grain producing regions, where agricultural and rainfall data from 1960-2000 suggest that factors such as urbanization are increasing vulnerability to drought.

Keywords: vulnerability, adaptation, food systems, climate change impacts

Introduction

Even in some of the world's most remote regions, we depend on complicated systems of food producers, processors and retailers to bring sustenance from farm to plate. Generally, these systems have proven quite good at reducing hunger over the past 50 years (Food and Agriculture Association, 2005) and a combination of synthetic fertilizers and pesticides, irrigation, and high yielding crop varieties have kept food production ahead of population growth in most regions except parts of Africa.

But are these systems sustainable? Some polemic and popular authors are concerned they are not, arguing that the way the globe currently produces food cannot last (For example, see: Kimbrell, 2002). These authors suggest that because the global food system depends on a stable economy, inexpensive fossil energy, and a productive environment, it may be vulnerable. Especially as we look towards a future full of climatic surprises (Intergovernmental Panel on Climate Change, 2007b), we need to know if our food systems will be able to adapt to new weather conditions, or will hunger start to grow after being in retreat for most of the last half century?

To answer this question, we must avoid seductively simplistic answers (Fraser *et al.*, 2003). For example, although we know that food production needs abundant but not excessive rainfall and sunlight, it would be easy to take the climate predictions made by global circulation models (Intergovernmental Panel on Climate Change, 2007a) and assume that productivity will be hurt if the climate changes. However, climate models suggest some areas are set to benefit from longer growing seasons and moister conditions and that this should keep global levels of productivity more or less constant (Adams *et al.*, 1995). We also know that farmers are not passive in the face of variability in the weather and will plant crops suited to new climate conditions (Smit et al., 2000). Markets should also help compensate, and production gains in one area may offset losses in other regions (Parry et al., 2004). It is important, however, not to over-emphasise the extent to which farmers may adapt. If changes are too sudden or if farmers are constrained by socio-economic factors like international trade agreements (Fraser, 2006b) or land tenure (Fraser, 2004) they may be unable to respond effectively.

As a result of this discussion, the goal of this brief paper is to review literature and propose a framework through which the adaptability of food systems can be empirically assessed and then use rainfall and agricultural data from China to test this framework.

Background

Broadly speaking, the literature on climate change, vulnerability, and agriculture falls into one of two traditions. The first is rooted in development studies and food security and draws on extensive field work to assess how people adapt to problems in their food supply. This tradition took form with Sen's 'entitlement theory' that argues food security is based on the ways that households demand food, rather than on the amount of food produced in a region (Sen, 1981). Sen arrived at this conclusion by examining a series of famines during the twentieth century (mostly from the Indian sub-continent) where market failure, poverty and political factors were far more significant at creating famines than drought or other problems in production. Sen's work highlights trends in development studies that show how individual households adapt to problems by using a range of social, natural, political, economic, and built assets (Bebbington, 1999, Watts and Bohle, 1993). This has been formalized as the 'livelihoods approach', a methodology designed to assess poverty in the developing world based on assessing these different forms of 'capital' (Scoones, 1998). An enormous number of field studies have taken place using these tools to assess vulnerability to extreme weather events and how communities cope with climate related problems (for example, see: Adger et al., 2003).

The second tradition is based on climate modeling. As noted earlier, the outputs of global circulation models have been linked to possible changes in grain yields. This approach is criticised for ignoring the extent to which farmers adapt to changing circumstances (Kandlikar and Risbey, 2000). To address this, some have built models that assume farmers will change practices and only plant those crops most suited to new climate conditions (Mendelsohn et al., 1994). Still others have linked different agro-ecological systems with both climate models and socio-economic scenarios (Fischer et al., 2002). These approaches, however, tend to miss factors that may inhibit adaptation.

These two traditions seem to be converging, with climate modelers working to include more locally relevant considerations in their impact assessments (Intergovernmental Panel on Climate Change, 2007b) while some of those working in the development community are attempting to generalize local case studies to distill broader lessons (Turner et al., 2003). To help build on this integration, some scholars have used past cases where relatively small environmental problems caused famine and compared these to cases where even large scale environmental anomalies did not cause any major problem in terms of food security (Fraser, 2003, Fraser, in press). The logic of these studies is to use historic weather variability as an analogy through which to anticipate the impact of future problems (Glantz, 1991).

For example, the rainfall record suggests the drought that triggered the Ethiopian famines of the 1970s and 1980s were minor in meteorological terms (Hulme, 2006, Comenetz and Caviedes, 2002). Social, political and agricultural changes in the decades before the famines, including war and a policy of 'village-ization' that uprooted agricultural communities and resettled them on Soviet style communes, undermined local economies and disabled drought coping strategies. These changes created a situation where relatively minor meteorological problems cascaded into a major calamity. Similarly, the Great Irish Potato Famine (1845-1849) was caused when a rainy year allowed a fungal blight to destroy the potato crop. Interestingly, fungal blights like the one that destroyed the harvest in the 1840s were common in the decades before the famine. This leads to the question, 'what was different in 1845 such that a relatively common problem caused 25% of the population to lose lives or homes?' Just as in the Ethiopian case, the decades prior to the Irish famine witnessed socio-economic changes that hurt the local economy and helped create a highly specialized agro-ecosystem that had little resilience to environmental shocks (Daly, 1986, O'Grada, 1989).

On the other side of the coin, a major drought in southern Africa in 1992 seriously affected harvests and exposed millions to starvation (Green, 1993). However, this natural disaster did not turn into a famine due to relatively diverse local economies, careful emergency planning, and weather forecasting. Similarly, work done amongst subsistence farmers in Zimbabwe shows that when weather forecasts are properly presented, farmers are able to maintain relatively high levels of food production despite poor environmental conditions (Patt and Gwata, 2004).

An interpretive framework

To synthesize these cases, Fraser (2006a) presents a heuristic framework to help illustrate vulnerability within food systems to climate change. Figure 1 plots the severity of an environmental threat, such as a drought, on the x-axis and the impact of that threat on the y-axis. (Note: as it is problematic to reduce all the multi-faceted impacts of climate change on a single axis, examples of these impacts are displayed to the left of the axis roughly ranked by their severity.) Broadly speaking, the worse the problem is, the worse we would expect the impact. However, to capture the fact that different agro-ecosystems may tolerate different environmental conditions, that different communities may have better or worse abilities to adapt, and different institutional structures may provide more effective emergency relief, three generic 'lines of defence' have been added to the middle of the figure. Seen in this way, in 'resilient' food systems the three lines of defence would be compressed along the y-axis, such that even large problems

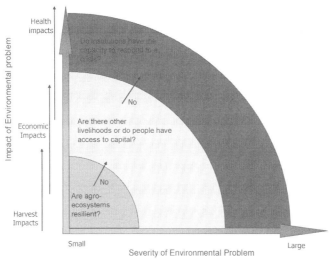

Figure 1. Heuristic framework showing the relationship between the severity of an environmental problem and its impact as medicated by the agro-ecosystem, local livelihoods and institutional capacity.

would have small impacts. Vulnerable food systems would be compressed along the x-axis suggesting that smaller problems might have big impacts.

Preliminary quantitative application of this framework

A preliminary analysis was done to see if this framework could work more quantitatively using Chinese rainfall (China Meteorological Data Sharing Service, 2007) and agricultural data (Institute of Geographical Sciences and Natural Resources Research, 2007). Data on rice paddy production (a major staple in China) and monthly rainfall were collected for all 17 eastern provinces in China between 1961 and 2000. The goal was to identify cases where anomalous rainfall was matched with normal levels of rice paddy production (a possible indication that the food system had adapted to the environmental problem) and to contrast these with the reverse situation: provinces where only slight perturbations in rainfall occurred during years where there were significant production losses (a possible indication of a failure to adapt).

To determine the size of the rainfall anomaly, for each province the long-term average rainfall between April and September (the main growing season) was divided by the amount of rainfall that had fallen during a given year. This created a simple version of the 'rainfall anomaly index' (van Rooy, 1965) that returned values of greater than one in dry years and less than one in years with above average growing season rainfall. To determine rice paddy production losses, provincial production data was de-trended over time using an 'auto-correlation' regression function (using the 'R' software package) and the yearly de-trended values were divided by the observed values for each year. Similar to the rainfall anomaly index used in step one, this produced a production anomaly index that returned a value of greater than one if the year's rice paddy harvest was below average. To determine overall vulnerability for each province in each year, the rice anomaly index was divided by the rainfall anomaly index, returning a high number for years when only small deviations from normal rainfall levels were matched with large deviations in rice paddy production. This 'paddy-rainfall vulnerability index' was calculated for every province in each year with below average rainfall. Finally, a range of demographic, population, agricultural management and economic factors were then used to divide out high and low vulnerability cases using a general linear model (using the software package SPSS version 14).

In terms of results, harvests were significantly lower in years and provinces where the rainfall between April and September was three standard deviations below the long-term average ($p<0.01$, $r^2= 0.33$), however there was also great deal of variance, especially when rainfall was between one and two standard deviations below normal. In terms of the paddy-rainfall vulnerability index, provinces with higher levels of urbanization had significantly higher vulnerability than provinces with lower levels of urbanization ($p<0.01$), while provinces with higher agricultural population were less vulnerable to droughts ($p<0.01$). One possible explanation for this is that those provinces with larger amounts of rural labour were better able to adapt to low levels of rainfall than populations with high urbanization rates. Provinces where there was a higher proportion of the agricultural land being actively cultivated were more vulnerable than those provinces with lower proportions of agricultural land in use ($P<0.05$). This perhaps indicates that agricultural potential was being maximized and there was little room for farmers to manoeuvre in the case of low rainfall. Provinces with higher fertilizer use, more fixed capital invested in agriculture, and those that cultivated two rice crops each year were significantly less vulnerable ($p<0.05$). One possible explanation for this is that that these sorts of agricultural management practices reduce the role that rainfall plays in determining crop yield, again leading to more successful adaptation. Interestingly, however, the amount of land irrigated was not found to be significant. Significant variables, and whether they are positively or negatively related to the paddy-rainfall vulnerability index, are summarized in Table 1. Overall, linear models that include different combination of these variables return r^2 values of between 0.25 and 0.4.

Table 1. Variables significantly related to the 'paddy-climate vulnerability' index (in all cases p<0.05 or 0.01). The sample was for the 17 eastern provinces in China, for the period 1961-2000, during those years when the April-September rainfall was below the 40 year average for the province (n=680). Rainfall data was from the China Meteorological Data Sharing Service (2007) and agricultural data was from the Institute of Geographical Sciences and Natural Resources Research (2007). Significance testing was done using the general linear model function of SPSS v 14.

Variables	Direction of relation
Total population in province	Negative relation
Agricultural Population in province	Negative relation
Rate of urbanization	Positive relation
Amount of rural labour	Negative relation
Amount of synthetic fertilizer used	Negative relation
Proportion of agricultural land being cultivated	Positive relation
Amount of land used to produce two rice crops each year	Negative relation

Conclusion

The specifics of these results should not be taken too literally. The data used in this analysis is too crude, and intervening variables may well be playing havoc with the statistical tests. This analysis only used a general linear modelling approach, and other more sophisticated tools are available. The rainfall anomaly index, which was based on simply averaging April to September rainfall, is too simplistic and fails to capture the repeated affect of multi-year drought, soil moisture balances or the water requirements of the specific crops planted. In addition, this analysis only looked at possible climate induced production losses for a single crop (rice paddy) and has not even attempted to quantify knock effects on food distribution or consumption more generally. Nevertheless, even with these significant limitations, intuitively plausible results have emerged. This suggests that it should be possible to use this approach, but to further refine the data and statistical tools, in order to better understand and quantify adaptation. Ultimately, it may be possible to expose underlying characteristics of regions likely to be vulnerable to climate change and to contrast these with cases where even significant climate related problems may be adapted to.

Acknowledgements

We extend sincere thanks to the Climatic Data Center, National Meteorological Information Center of China Meteorological Administration for kindly providing us with the rainfall data.

References

Adams, R., Fleming, R., Chang, C., Mccarl, B. and Rosenzweig, C. (1995). A reassessment of the economic effects of global climate change in U.S. Agriculture. Climatic Change 30: 146-167.

Adger, W.N., Huq, S., Brown, K., Conway, D. and Hulme, M. (2003). Adaptation to climate change in the developing world. Progress in Development Studies 3: 179-195.

Bebbington, A. (1999). Capitals and capabilities: A framework for analyzing peasant viability, rural livelihoods and poverty. World Development 27: 2021-2044.

China Meteorological Data Sharing Service (2007). Chinese meteorological data. Climatic Data Center, National Meteorological Information Center, China Meteorological Administration., Available from the World Wide Web at: http://cdc.cma.gov.cn/index.jsp. Accessed March 15th, 2007.

Comenetz, J. and Caviedes, C. (2002). Climate variability, political crises, and historical population displacements in Ethiopia. Global Environmental Change Part B: Environmental Hazards 4: 113-127.

Daly, M. (1986). The famine in Ireland. Dublin Historical Association, Dublin, Ireland.

Fischer, G., Shah, M. and Velthuizen, H. (2002). Climate change and agricultural vulnerability, Vienna, International Institute for Applied Systems Analysis.

Food and Agriculture Association (2005). The state of food insecurity in the world, FAO, Rome.

Fraser, E. (2003). Social vulnerability and ecological fragility: Building bridges between social and natural sciences using the Irish potato famine as a case study. Conservation Ecology 7: on line.

Fraser, E. (2004). Land tenure and agricultural management: Soil conservation on rented and owned fields in southwest British Columbia. Agriculture and Human Values 21: 73-79.

Fraser, E. (2006a). Agro-ecosystem vulnerability. Using past famines to help understand adaptation to future problems in today's global agri-food system. Journal of Ecological Complexity 3: 328-335.

Fraser, E. (2006b). International trade, agriculture, and the environment: The effect of freer trade on farm management and sustainable agriculture in South-western British Columbia, Canada. Agriculture and Human Values 23: 271-281.

Fraser, E. (in press). Travelling in antique lands: Studying past famines to understand present vulnerabilities to climate change. Climate Change.

Fraser, E., Mabee, W. and Slaymaker, O. (2003). Mutual dependence, mutual vulnerability: The reflexive relation between society and the environment. Global Environmental Change 13: 137-144.

Glantz, H. (1991). The use of analogies in forecasting ecological and societal responses to global warming. Environment 33: 10-33.

Green, R. (1993). The political economy of drought in southern Africa 1991-1993. Health Policy and Planning 8: 256-266.

Hulme, M. (2006). 'gu23wld0098.Dat' (version 1.0). Department of the Environment, Transport and the Regions (Contract EPG 1/1/85).

Institute of Geographical Sciences and Natural Resources Research (2007). Chinese natural resources database. Chinese Academy of Sciences, Available on-line at <http://www.naturalresources.csdb.cn/index.asp> Accessed Feb 15th, 2007.

Intergovernmental Panel on Climate Change (2007a). Climate change 2007: The physical science basis, IPCC, Brussels.

Intergovernmental Panel on Climate Change (2007b). Climate change 2007: Impacts, adaptation and vulnerability., IPCC, Brussels.

Kandlikar, M. and Risbey, J. (2000). Agricultural impact of climate change: If adaptation is the answer, what is the question? Climatic Change 45: 529-539.

Kimbrell, A. (ed.). (2002). The fatal harvest reader: The tragedy of industrial agriculture, Island Press Washington, D.C.

Mendelsohn, R., Nordhaus, W. and Shaw, D. (1994). The impact of global warming on agriculture: A ricardian analysis. The American Economic Review 84: 753-771.

O'Grada, C. (1989). The great Irish famine, Macmillan, London.

Parry, M.L., Rosenzweig, C., Iglesias, A., Livermore, M. and Fischer, G. (2004). Effects of climate change on global food production under SRES emissions and socio-economic scenarios. Global Environmental Change 14: 53-67.

Patt, A. and Gwata, C. (2004). Effects of seasonal climate forecasts and participatory workshops among subsistence farmers in Zimbabwe. Proceeding of the National Academy of Science 102: 12623-12628.

Scoones, I. (1998). Sustainable rural livelihoods: A framework for analysis, Institute of Development Studies, Brighton, UK.

Sen, A. (1981). Poverty and famines, Oxford, Claredon Press.

Smit, B., Burton, B., Klein, R. and Wandel, J. (2000). An anatomy of adaptation to climate change variability. Climatic Change 45: 223-251.

Turner, B.L., Kasperson, R.E., Matson, P.A., Mccarthy, J.J., Corell, R.W., Christensen, L., Eckley, N., Kasperson, J.X., Luers, A., Martello, M.L., Polsky, C., Pulsipher, A. and Schiller, A. (2003). A framework for vulnerability analysis in sustainability science. Proceedings of the National Academy of Sciences of the United States of America 100: 8074-8079.

Van Rooy, M.P. (1965). A rainfall anomaly index independent of time and space. NOTOS. Weather Bureau of South Africa 14: 43-48.

Watts, M. and Bohle, H. (1993). The space of vulnerability: The causal structure of hunger and famine. Progress in Human Geography 17: 43-67.

Integration of resilience into sustainability model for analysis of adaptive capacities of regions to climate change: the EASEY model

R. Paulesich[1], K. Bohländer[2] and A.G. Haslberger[2]
[1]University of Economics and Business Administration, Institute for Regional- and Environmental Economics and Management, Nordbergstrasse 15, A - 1090, Vienna, Austria
[2]University of Vienna, Department for Nutritional Sciences, Center for Ecology, Althanstrasse 14, A-1090 Vienna, Austria

Abstract

Global and regional management and governance will need to establish integrative models considering agro-ecological, social and economic factors to increase resilience of food production. The paper discusses an integrated concept for the analysis of adaptive capacities of regions to climate change. Data generation and analyses is carried out along three dimensions:

1. The agricultural production process has to be defined whether the cultivation only or further food processing in a sense of supply chain should be scrutinised.
2. Regional stakeholders of the production process have to be identified. Six categories are usually alleged: process holders, ecology, society, customers, market and finance.

These two dimensions are representing the state of the art in evaluating projects or programs. Here quantitative and qualitative data are used. By involving stakeholders (face to face interviews, workshops) some regional specific indicators will come up.

3. The third dimension makes the difference. It expresses three sustainability goals pre-structured by principles of multilateral organisations and European Union strategic policy intentions which can be seen as a democratically legitimated basis for a trans-disciplinary dialogue and a discourse between science and society.

This methodology can be applied as a self assessment and a benchmarking tool as well. The combination with a participative generated regional development concept could establish a feedback loop to enhance dialog between science and society. Resilience means then the availability of societal options for a flexible response to climate change effects.

Keywords: evaluation concept of regional ability to respond external shocks

Framing

Sustainability as it is most cited and once written in Brundtland-Report (1987) is defined as development, meeting the needs of the present generation without compromising the ability of future generations to meet their needs. Implementing this so to say definition, the United Nations Conference (1992) on Environment and Development in Rio de Janeiro drew up a rough structured political concept - the 3 columns of sustainable Development, challenging policy makers to weight interests of economy, society and environment equally. On that score, in Lissabon (2000) and Göteborg (2002) as well the Council of the European Union paved the way by declaring two strategies: Enhancement of Competitiveness and sustainable Development. Now the question arises how to operationalise such a political claim?

Operationalising

EASEY is the four letter word for ecological and social efficiency. The term efficiency shall provoke a debate on effectiveness and sufficiency when the model is going to be introduced. The intention is to enhance self reflection the understanding of the normative character of sustainability and development.

Concepts recently try to operationalise developing or / and assessing sustainability are to categorise as:
1. Frameworks for communities and regions (www.iisd/pdf/bellagio.pdf or manuals for LA21-projects).
2. Frameworks for business and companies (Global Reporting Initiative www.globaloreporting.org or ISO 14000 series).

Concepts in both categories lack of either comparability of results (1) or usability for all stakeholders (2). The aim is therefore to link micro-, meso- and macro level first in concept than in pilot practice to derive experience and than arguments for better understanding and decision making on each level.

This model (Figure 1) differs to conventionally used ones. The first two dimensions - (1) processes and (2) regional stakeholders - are representing the state of the art in evaluating projects or programs as well as corporate sustainability performance. The question arises whether they cover all requirements emerging from sustainability. And further more when sustainability is a normative societal issue these requirements underlie continuous changes. Which requirements a model have to comply?
- First: legal compliance is not enough as well as positive market response to a brand name or product or in political terms a qualitative democratic majority.
- Second: intergenerational justice - we have to go beyond the time horizon of a single generation at least.

We try to respond by introducing a third dimension, which expresses sustainability goals. These should be the result of an interscience and / or science and society discourse. Thus is pre-structuring the evaluation track from natural sciences [hard facts] to regional aspects and includes the normativity of preferences about social choice [soft facts].

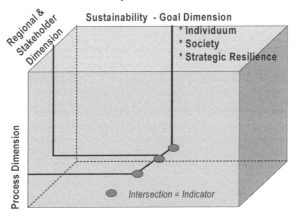

Figure 1. Model of sustainable development and evaluation.

Process dimension

We have to face up to a first decisive question here. Shall we define the food production as a part of the ecological or the socio technological system? In the first case food production is a part of ecological functions and services with less societal sovereignty than in the second case.

Food production belongs to more than one subsystem (economy, society) but is mainly part of superior system nature. Therewith food production differs clearly from subsystem economy, which is more strongly embedded in 'society' than in 'nature'- in terms of its messages. To come to the point, in their decisions and actions, food production stakeholders are not able to apply their specific system logics for other subsystems.

The process dimension of the model addresses efficiency and effectiveness in production of goods and services. Ecological efficiency of food production is measured by relation between applied material, energy, labour and extant amount in products. Effectiveness depends on need of repairs respectively applicability of products. In case of food production effectiveness is simply nutritional value or in other words contribution to health by diet. Effectiveness also reflects dynamic development of an organisation. In case of food production farmers are driven by market opportunities. It provides information about strategic investments to current achieved fruits ratio, not only in terms of fiscal and income levels.

Stakeholder dimension

The EASYE model is based on the observance that companies are always affected by stakeholders even by those assigned to the non market environment. In case of food production, especially agriculture, wide political and societal interests exist often conflictive for example relative price (bearable share of household income) and quality in ecological and physiological terms. NGOs on several levels denounce offences against ecological and health standards. All these is determining informal and formal market conditions.

In terms of agriculture as a link in the supply chain of food production family is the most important stakeholder and members of staff are narrowed to seasonal workers. Clients are on the one hand final consumers due to direct selling and on the other hand pose a strategic alliance with food industry. In order to evaluate this strategic alliance special methodology used in the management of supply chain is necessary.

Social responsibility is another critical topic, due to the kind of industrial production influencing family and local society, as quality of products with consequences for consumer´s health. In both cases choice is limited due to laws and board of control but wide range makes visible difference in impact on environment and society. Illustrative example is comparing tendency to reduce the value of diet with life-supporting expenditures in general or time exposure in relation to spare time or career (work life balance).

Goal dimension

The claim is that, in comparison with other concepts applied at the moment, this concept reflects term definitions of sustainable development more comprehensively. The present concepts mirror and assess according to benchmarks or threshold values, respectively. This leads to a merely peripheral consideration of indicators of social responsibility, at the minimal level often prescribed by law. Neither the sustainability definition proposed in the Brundtland Report nor the common 3-tier model offer starting points for an operationalisation that could do justice to the normative character and the socio-political target, respectively.

The solution for both questions lies in adding a goal dimension to the repertoire of instruments (indicator set e.g.). For this purpose, an existing three dimensioned system of goal for the conception of regional economic development is used (Table 1).

This goal system does not substitute individual goals of process participants (involved people). It points out opportunities provided by sustainable development and limits which it sets.

A system of individual process participant goals is ancillary to a system of sustainability goals, which represents multilateral agreements, international principles as well as political intentions of the European Union regarding sustainable development. In either case, agriculture, food production respectively - defined as a processing chain - is part of more than one category in the sustainability goal dimension.

The question of how this addition can be fruitfully employed for the assessment of the sustainability performance of a food production system will be answered by applying this model in a research project. The methodology includes instruments as surveys to get to know peoples preferences, priorities and assignments to who is responsible for achieving which goal (monitoring, keeping the direction etc.).

Table 1. Overview of the three goal dimensions of sustainable development.

Ensuring human existence	Maintaining the social productivity potential	Maintaining the opportunities for development and action
Health	Use of renewable resources	Equality of chances
Basic provision	Use of non-renewable resources	participation
Ensuring existence	The environment as a rubbish-	Cultural heritage and cultural
Security of revenue and assets	dump	diversity
Possibilities of making use of the	Risks	Safeguarding the cultural
environment	Material, human and knowledge	function of nature
	capital	Maintaining social resources

Resilience

The concept of resilience is widely used in ecology but its meaning and measurement are contested. While some refer to it as a new paradigm, others see it as more of an expression, complimenting use of other terms, such as vulnerability or risk. Resilience is often used in physical or ecological context, but most of the literature clarifies that the study of resilience evolved from the disciplines of psychology and psychiatry in the 1940s.

Resilience has been defined in many ways. According to C.S. Holling it is the buffer capacity or the ability of a system to absorb perturbations, or the magnitude of disturbance that can be absorbed before a system changes its structure by changing the variables and processes that control behaviour. By contrast other definitions of resilience emphasize the speed of recovery from a disturbance, highlighting the difference between resilience and resistance. Therefore, resilience can be seen as a desired outcome but also as a process leading to a desired outcome.

Social resilience is defined as the ability of communities to withstand external shocks to their infrastructure. Deficient information, communications and knowledge among social actors, the lack of institutional and community organisation, weakness in emergency preparedness, political instability and the absence of economic health in a geographic area are factors in generating greater risk. Therefore, the concept of resilience, especially social resilience is growing on importance. The direct dependence of communities on ecosystem is an influence on their social resilience and ability to cope with shocks, particularly in the context of food security and coping with hazards.

Conceptualizing and measuring economic resilience shall be illustrated by the view to economic vulnerability on a macro level. The vulnerability and the resilience characteristics can be described by four scenarios (best worst case, subsistence and prodigal). The economic vulnerability index is based on criteria as economic openness, export share and dependence on strategic imports. All vulnerability indices lead to the conclusion that small states tends to be more economically vulnerable than larger ones.

The main meaning of economic resilience is the ability to recover from the negative impacts of external shocks. The three characteristics are therefore to recover more or less quickly from a shock by appropriate counteraction, to withstand the effects by absorption and last not least the ability of avoidance.

Components of the resilience index are shock-absorbing and shock-counteracting information about macroeconomic stability, which relates to aggregated demand and aggregated supply. The macroeconomic stability aspect is based on three variables namely the fiscal deficit to GDP ratio, the unemployment and inflation rate and the external debt to GDP ratio.

Shock-absorbing and shock-counteracting from a national point of view differs to an individual company's one. Nations operate within a frame of bi- and multilateral agreements (setting a stabile democratic political system as internally given), companies has to refer to national and international business regulations. Microeconomic efficiency is a further aspect of gaining resilience.

Next steps

The Millennium Ecosystem Assessment provides a scheme which is to be classified between the two categories of sustainability concepts mentioned at the beginning. This intermediate position is characterised by:
- referring to processes - direct and indirect drivers of change; strongly based by natural sciences; macro level view dominates;
- goals to achieve - presenting human well being as strongly dependent from ecosystem services; not covering all societal needs regarding sustainable development.

It answers the question, why we should take action, but not sufficiently, who should take it and what is to do and which priorities we should care for. Comparability of results is achievable but usability for all stakeholders can not be taken for sure. Further research is needed to merge the EASEY model and the MA concept by carrying out a sub global assessment.

From a case study in Marchfeld, looking at the food production of the last 30 years (sugar beets, vegetables, maize, wheat), we conclude that there may be a number of opportunities to implement resilience strategies for a system like the food production area of Marchfeld.

In general, a portfolio approach based on managing a variety of risks and opportunities should be the most appropriate. The portfolio has to represent diversity on both levels nature (historical factors and recent recovery) and society (organisation of production, cultural heritage). The level of nature ranges from man made to intensive agriculture. The level of society extends from semi industrial farming to biological one and further more to management of protected reserves.

A portfolio approach is to be seen as a precautionary measurement because not only processes at these two levels should become robust and stabile, effects of losses are calculable, also the knowledge of society will rise by preparing for counteraction, absorption and avoidance. This knowledge provides additional economic, social, or environmental benefit.

Literature

Allenby, B. and Fink, J. (2005); Toward Inherently Secure and Resilient Societies, Science, 309;1034 -1036.

Briguglio, L., Cordina, G., Bugeja, S. and Farrugia, N.. (2004); Conceptualizing and Measuring Economic Resilience. University of Malta, Department of Economics; http://home.um.edu.mt/islands/resilience_index.pdf ,October 24th 2006 09:43.

Bruckner, B. and Paulesich, R. (Hg.) (2006); Nachhaltigkeit und Unternehmensfinanzierung. Beiträge zur aktuellen Diskussion und empirische Befunde; Hamburg Kovac.

Darnhofer, R. (2006); Resilienz und die Attraktivität des Biolandbaus für Landwirte; in: Bio Landbau in Österreich im internationalen Kontext. Band 2: Zwischen Professionalisierung und Konventionalisierung; Groier, M.; Schermer, M. (Hg.) Wien.

Fischer, G., Shah, M. and Van Velthuizen, H. (2002); Climate Change and Agricultural Vulnerability. Institute for Applied Systems Analysis, Contribution to World Summit on Sustainable Development, Johannesburg.

Hardi, P. and Zdan, T. (1997); Assessing Sustainable Development. Principles in Practice; IISD Winnipeg Canada.

Paulesich, R. (2006); EaseyX. Der HGF Ansatz in einem Modell zur Bewertung börsennotierter Unternehmen; in: Ein Konzept auf dem Prüfstand. Das integrative Nachhaltigkeitskonzept in der Forschungspraxis; Kopfmüller, J.; Berlin Sigma S 189 - 211.

Sustainability, corporate social responsibility and food markets: the role of cooperatives

Valeria Sodano
University of Naples Federico II, Department of Agricultural Economics, via Università 96 80055 Portici, Napoli, Italy, vsodano@unina.it

Abstract

The paper studies how the peculiarities of corporate governance systems of cooperatives and investor owned enterprises lead to a different attitude towards social responsibility. Stemming from the neoinstitutional theory of the firm, and particularly from the work of Hansmann, it is shown how firms are essentially political systems that make power substitute the market in the coordinating function of the economy.

An example of cooperative social responsibility is given referring to the case of the Italian tomato processing sector.

The main finding of the paper is that in contrast with investor-owned firms, cooperatives exhibit a higher 'natural' inclination to corporate social responsibility (CSR). The evidence that CSR ultimately is possible only in a socio-cultural environment rich in social capital and moral codes should suggest that when this is not the case, only command-and-control regulations (i.e the state intervention) can promote really sustainable food systems.

Keywords: corporate social responsibility, cooperative, power, trust, food system

Introduction

When market failures prevent food systems from attaining sustainability goals, a certain degree of market regulation is required. Globalization and trade liberalization have made it increasingly difficult for nation-states to regulate food markets, giving rise to a shift from public to private governance. In order for private regulation to be effective a high degree of social corporate responsibility is required.

The paper studies how the peculiarities of corporate governance systems of cooperatives and investor owned enterprises lead to a different attitude towards social responsibility. Stemming from the neoinstitutional theory of the firm, and particularly from the work of Hansmann, it is shown how firms are essentially political systems that make power substitute the market in the coordinating function of the economy.

Power corrects contract incompleteness within and outside the firm. Trust is an alternative means to correct contract incompleteness. Compared with power, trust is much more welfare enhancing in consideration of efficiency and equity, as well as economic and social welfare. While investor owned firms strongly rely on power, cooperatives are more trust-oriented and thereby more prone to 'ethical' and socially responsible behaviours.

An example of cooperative social responsibility is given referring to the case of the Italian tomato processing sector.

Sustainability, ethics and corporate social responsibility

During the last decade the food sector has undergone three important technological and institutional changes, which may strongly affect its social and environmental sustainability (Sodano *et al.*, 2007): the introduction of new bio and nano technologies and the associated environmental and human health risks; the process of deregulation and trade liberalization and the 'hollowing out' of nation states in the areas of food safety and food security policies; the consolidation processes at the downstream stages of the supply chain, and the rising concentration and power of supermarkets. These transformations led to growing moral concerns expressed by various sectors of civil society such as NGOs, consumer associations, trade unions and farmers' representatives. Delborne *et al.* (2007) have recently summarized the following values and political issues as constituting the core of food system moral concerns: food security, food safety, food quality, food sovereignty, human welfare, animal welfare, ecological sustainability, transparency and traceability.

Given the worldwide neoliberal faith, a call for corporate social responsibility (CSR) more than for market regulation and state intervention has come from civil society and public institutions; in order to face food moral concerns.

The belief that CSR may make private sector contributions to sustainable development and poverty alleviation, relies on two main assumptions: 1) that corporations are anyway able to develop a business ethics, given that their managers have the same 'moral sentiments' of other people in the society; 2) that a sound CSR can ultimately be a source of competitive advantage and thereby does not conflict with the corporation 'selfishness'. Unfortunately both these assumptions have proved unreliable. Repeated corporate scandals have shown a managerial class prone to criminal behaviours, according to the different cases that have been investigated by literature on white-collar crime. CSR can result in competitive advantages but through very non-ethical practices, such as claims made in a strategic communication policy (persuasive and deceiving advertising), and the use of fair trade and sustainable standards for heavy price discrimination practices (Bush *et al.*, 2005). The gap between corporations' claim on their CSR and their real pursuit of social goals has been proved by the results of the 'race to the top project' carried out few years ago by the IIED and other civil society organizations in order to assess the real progresses towards sustainability achieved through the good business practices claimed by UK leading supermarkets. Only three minor supermarket chains collaborated in the project, with the leading chains refusing to provide data and useful information. The finding was that for most companies concerns about their social and environmental impacts only become significant when they affect consumer trust and add to consumer value (allowing for price discrimination and market power). Instead supermarkets appear unable or unwilling to engage with a broader notion of stakeholder accountability. The general conclusion of the research was that 'there remains a conflict between business practices and sustainable development where shareholders pressure remains' (Fox and Vorley, 2004: 30).

Obstacles to true social responsible behaviours by corporations reside in their organization and ownership features. Corporations are investor-owned firms, whose board of directors, designed by shareholders, is committed to the objective of only profit maximization. Institutionally investor-owned firms have no responsibilities towards their stakeholders besides the respect of formal contracts signed with them, as settled by law. Managers are engaged by directors and are given pecuniary incentives with the precise task of profit maximization. The current phase of capitalistic development is such as to exacerbate the subordination of firm's strategies to the maximization of shares value. For examples many large corporations are currently almost all owned by institutional investors which have much more interest in financial risk than in what a particular company is doing, and therefore in its oversight. As international rating agencies reward social costly policies, such as delocalization towards countries with lower labour and environmental standards or the transfer of the burden of competition to weak customers and suppliers (using their bargaining power), managers of large multinational corporations are encouraged to behave in a very selfish and opportunistic way.

Hansmann and the ownership of firms

In contrast with investor-owned firms cooperatives exhibit a higher 'natural' inclination to CSR. This statement can be verified stemming from the theory of Hansmann on the ownership of enterprise (Hansmann, 1996).

Hansmann states that a firm can be conceived of as a nexus for contracts, meaning that the ultimate function of a firm is to be a legal entity entitled to sign contracts (under specific liability rules) with various other subjects who participate, -as suppliers, customers or workers- to firm's productive activity. Persons with whom the firm has a contractual relationship are defined as patrons. When contractual relationship are affected by high transaction costs -due to market power, asymmetric information and transaction specific investments-, the class of patrons who bear the highest costs have incentives to become owners of the firm. The owners of a firm are those persons who possess two rights: the right to control the firm (i.e. to manage it, defining its real policies and actions), and the right to appropriate the firm's net earnings. The efficient assignment of ownership therefore is such as to minimize the total costs of contracting and of ownership for all the firm's patrons combined. If ownership can be easily reassigned from one group of patrons to another, the allocation of ownership, and therefore institutional forms of firms, should be efficient.

While Hansmann claims that ownership, and therefore the boundaries of firms, can be explained on the base of an efficiency argument, his theory calls instead for a power-based theory of firm ownership. Power enters Hansmann's theory at least into three ways.

First, the ultimate reason for a class of patrons to become firm's owners is to avoid 'power abuses' by the firm. These power abuses take three forms: the classical exercise of market (buyer) power towards customers (suppliers); opportunistic behaviours associated with asymmetric information; the exploitation of exchange surplus due to a higher bargaining power, in those bilateral monopoly situations associated with hold-up problems and transaction specific investments.

Second, incentives to become owners also stem from benefits associated with the possibilities to exploit weaker contractual partners. When contractual costs due to opportunistic behaviour of a class of patrons are less than the benefits achieved by firm through its opportunistic behaviour, Hansmann theory tells that it is still efficient for this class of patrons not to be owner. What Hansmann fails to underline is that in this case power is still well present but it involves income redistribution more than efficiency losses (as in a bargaining game where the maximum exchange outcome is attained but with unequal players' pay-offs).

Third, highlighting how problems of contract incompleteness are at the core of the problem of the boundaries of the firm, Hansmann implicitly agrees with the incomplete contract theory of the firm (Hart, 1988; Grossman and Hart, 1986). This latter states that vertical integration (and hence the firm) occurs when contracts fail because of their incompleteness, then the party more able to exploit the residual claims will acquire the residual rights of control. In the model of Grossman and Hart, any process of vertical integration entails a redistribution process that will favour the more powerful party, with contradictory equity and efficiency effects. This result sharply contrasts with the Williamson explanation of vertical integration (Williamson, 1985), on which Hansmann explicitly relies on, that accounts for the shift from one organizational form (the long-term contract) to another (the firm) exclusively in terms of transaction costs minimizing; implying an efficiency improvement. The result of this third firm power-related issue is that in the case of investor owned firms capitals can be view as no longer allocated where efficiency is higher, but where there are more opportunities of exploitation through power.

Cooperatives and corporate social responsibility

The theory of Hansmann gives the same explanation to the emergence of invested-owned and cooperative enterprises. In both cases ownership is best assigned to that class of patrons which is particularly exploited by the firm, due to contract incompleteness. In the case of cooperatives the only peculiarity is the limited ownership. Hansmann says that when the cost of contracting between firms and one class of patrons is high but these patrons cannot be organized to serve as effective owners at any feasible cost, the efficient solution may be to create a non profit firm (in the form of cooperative); that is a firm without owners. In a non-profit firm there is a ban on distribution of the firm's net earnings or assets to persons who control the firm, and all the firm's revenues must be devoted to providing services. The argument of Hansmann is consistent with the theory of cooperatives given by the mainstream economics. Cooperatives are transitory and defective forms of enterprise (the traditional defects associated with the vaguely defined property rights are the portfolio, the horizon and the free-rider problems generally reported by the literature on cooperatives) and are deemed to be substituted by the more efficient proprietary enterprises as institutional innovations solve market failures and market forces select the best proprietary form, that is the investor owned firm (Ward, 1958; Cook,1995; Cook and Iliopoulos, 2000; Hendrikse and Veerman, 2001).

While Hansmann rightly points to the defensive (against power) nature of cooperatives, at the same time underlying the very nature of firms as systems of power, he underestimates the other peculiarities of cooperatives that make this kind of enterprise inclined to social responsible behaviours. Limited return on equity is only one of the Rochdale Principles of Co-operation enunciated in 1844 and updated by the International Co-operative Alliance in 1995: limited return on equity; voluntary and open membership; democratic member control; member economic participation; autonomy and independence; education, training, and information; cooperation among cooperatives; concern for community. From these principles it is clear that people engaged in a cooperative enterprise not only want to defend themselves from the exploitation of capitalistic firms, but they want to build an alternative way of organizing production and transactions, trying to offer good exchange conditions to the patrons of the firm and renouncing any form of abuse of power and exploitation. By removing incentives for the firm's managers to earn profits, the incentive to exploit the firm's patrons is reduced. By promoting cooperative rather than competitive behaviours together with a genuine interest for social objectives, cooperatives actually correct market failures and contract incompleteness, using trust instead of power. The very nature of cooperative enterprises resides in reciprocal attitudes of economic actors, grounded in social norms and ethical values. 'Cooperatives as distinct forms of business that rely upon members to work together towards collective goals; trust lies at the heart of cooperation and provides the basis for communication that is essential for members to seek mutual benefit; ethics provides the foundation for trust that must be present for cooperation to occur' (Lasley *et al.*, 2004).

Concluding, (as summarised in Table 1) requests for social responsibility in the food system are more likely to be complied with by cooperatives than by capitalistic firms.

Cooperatives and corporate social responsibility: some evidence from the Italian tomato processing industry

The Italian tomato processing industry offers a good example of positive relationships between trust-moral codes, cooperatives and SCR.

Table 1. Attitudes towards CSR: differences between cooperatives and investor-owned firms.

Requests for CSR	Responses by cooperatives	Responses by public firms
Contract incompleteness	Corrected by trust, (symmetric Nash bargaining solutions for the bargaining game associated with hold-up problems)	Residual claims exploited through power (inefficient and unequal solutions for the bargaining game associated with hold-up problems)
Avoiding negative externalities	Ethical concerns for community and fairness preferences boost the internalization of social costs	Concerns for stock values and selfish behaviours prevent from the internalization of social costs
Providing public goods	Equity and fairness preferences help to solve free-riding problems	Opportunism exacerbates free riding problems

Italy is the second producer (next to California) of processed tomato in the world. Production is localized in two agro-industrial districts, the southern and the northern district, with approximately the same size but with a very different structure and organization. The Northern district exhibits larger firms (and a higher degree of concentration at each stage of the chain), fewer brands and more collaborative inter-firm relationships, with agricultural marketing cooperatives controlling the most part of processing plants. In the South many dispersed farmers supply raw tomatoes to many independent manufacturers, in a context of highly competitive relationships at vertical as well at horizontal level in the supply chain. Results of a wide research project carried out by the University of Naples during the last three years (aimed to assess the capability of domestic firms to cope with the changing competitive international scenarios) have clearly shown that in the northern district the presence of cooperatives, backed by high levels of social capital, is associated with social responsible behaviours, and that therefore in this area public support for the sector can benefit the entire community. Opposite conclusions were reached in the case of southern district.

In the research social capital was measured on the basis of trust indicators accounting both for impersonal trust (in the form featured by Fukuyama, 1995) and interpersonal trust, this latter related to the attitude of actors towards others' trustworthiness (McAllister, 1995). Data collected through a wide set of interviews showed that collaborative agreements (such as cooperatives, inter-professional agreements, trade unions, collective quality standards) and network relationships are heavily present in the North, while are rare in the South. Farmers' (manufacturers') believing in manufacturers' (farmers') fairness and trustworthiness, and their willingness to invest in risky activities with the exchange counterpart, are as high in the North as they are low in the South.

CSR was measured looking at illegal behaviours and frauds and at policies carried out to foster social and environmental sustainability. Cases of illegal behaviours were found only in the South, namely the recruitment and exploitation of illegal African and eastern European immigrants, and the falsification of tomato purchasing records in order to access extra EU subsidies. Vice versa in the North, a lot of sustainable good practices were found: fair trade standards; organic production; traceability systems; transparent communication policies; support to civil society organizations; engagement in projects with local administrations.

Conclusions

The kind of ownership and governance structure of firms affects their social responsibility. Investor-owned enterprises are 'naturally' much less responsible than cooperatives. The higher social responsibility of cooperatives depends on ethical values and social norms on which this form of enterprise is grounded.

Traditional economic literature on cooperatives has widely denounced their economic weakness, associated with portfolio and horizon problems, while it has not paid enough attention to their role in providing positive externalities and social benefits. Taking into account these latter issues should lead legislators and economists (Cook and Chaddad, 2004) to no longer foster the revision of law on cooperatives in a way as to turn them in proprietary firms. Instead policies should be carried out in order to support the cooperative movement and to strengthen ethical attitudes within it.

The evidence that CSR ultimately is possible only in a socio-cultural environment rich in social capital and moral codes should suggest that when this is not the case, only command-and-control regulations (i.e the state intervention) can promote really sustainable food systems.

References

Bush, L., Thiagarajan, D., Hatanaka, M., Bain, C., Flores, L. and Frahm, M. (2005). The Relationship of Third-Party Certification (TPC) to Sanitary/Phytosanitary (SPS) Measures and the International Agri-Food Trade: Case Study: Eurepgap, Report 7, United States Agency for International Development, Washington D.C.

Cook, M. and Chaddad, F.R. (2004). Redesigning cooperative boundaries: the emergence of new models. American Journal Agricultural Economics 85:1249-1253.

Cook, M. (1995). The future of US agricultural cooperatives: a neo-institutional approach. American Journal Agricultural Economics 77: 1153-1159.

Cook, M. and Iliopoulos, C. (2000). Ill-defined property rights in collective action: the case of US agricultural cooperatives. In: Menard C. (ed.) Institutions, contracts and organizations. Edward Edgar.

Delbone, M., De Graaff, R. and Brom, F. (2007). An ethical toolkit for food companies: reflections on its use. Journal of Agricultural and Environmental Ethics 20: 99-118.

Fox, T. and Vorley B. (2004) Race to the top. Stakeholder accountability in the UK supermarket sector, IIED.

Fukuyama, F. (1995). Trust. The Free Press.

Grossman, S. and Hart, O. (1986). The Costs and Benefits of Ownership: A Theory of Vertical and Lateral Integration. Journal of Political Economy 94: 691-719.

Hansmann, H. (1996). The ownership of enterprise. The Belknap Press of Harvard University Press.

Hart, O. (1988). Incomplete contracts and the theory of the firm. J. Of Law, Economics and Organization. Spring

Hendrikse, G.W.J. and Veerman, C.P. (2001). Marketing cooperatives and financial structure: a transaction costs economic analysis. Agricultural Economics 26: 205-216.

Lasley, P., Baumel, C. and Hipple, P. (2004). Strengthening ethics within agricultural cooperatives. USDA. RBS Research Report 151.

McAllister, D. (1995). Affect- and cognition-based trust as foundations for interpersonal cooperation in organizations. Academy of Management Journal 38: 24-59.

Sodano, V., Hingley, M. and Lindgreen, A. (2007). The usefulness of social capital in assessing the welfare effects of private and third-party certification food safety policy standards: trust and networks. British Food Journal. Special issue: Relationships, Networks, and Interactions in Food and Agriculture Business-to-Business Marketing and Purchasing.

Ward, B. (1958). The firm in Illyria: market syndicalism. American Economic Review 48: 566-589.

Williamson, O. (1985). The Economic Institution of Capitalism. Free Press, New York.

A method for construction and evaluation of scenarios for sustainable animal production, with an application on future Swedish dairy farming

Stefan Gunnarsson[1]* and Ulf Sonesson[2]
[1]*Department of Animal Environment and Health, Swedish University of Agricultural Sciences (SLU), P.O. Box 234, SE-532 23 Skara, Sweden, stefan.gunnarsson@hmh.slu.se*
[2]*Swedish Institute of Food and Biotechnology (SIK), SE-429 02 Gothenburg, Sweden*

Abstract

Sustainability within agriculture is a complex issue. Besides the sustainability aspects generally discussed; i.e. ecological, social and economical aspects; agriculture also involves ethical issues related to animal husbandry. The aim of this paper was to develop a forecasting scenario method that would incorporate scientific knowledge and practical experiences into the scenarios. Furthermore, the method was used for analysing the sustainability of future scenarios for dairy farm production in Sweden. A methodology for working with scenarios for future agricultural production systems was developed. The scenarios can then be evaluated both quantitatively, e.g. economy and life cycle analysis (LCA), and qualitatively, e.g. animal health and welfare. Two scenarios for future dairy farming in Sweden were developed; Specialised Dairy Farming (SDF) with high production intensity and Mixed Dairy Farming (MDF) with increased crop rotations and large share of pasture. Cost of production was in SDF 3.02 SEK/kg milk, in MDF 4.34 SEK/kg milk. The contribution to eutrophication per litre milk was lower for SDF, but MDF contributed less to eutrophication per land area used. In general MDF had more positive impact on the environment than SDF. Cows in SDF had an increased risk of diseases (e.g. lameness) compared to cows in the MDF. No scenario was superior in all aspects and the goal for developing sustainable dairy farm production must be guided by analysis of values. The method offers a structured way of synthesising large amounts of research into something comprehensible and practically understandable that can be used in discussions and decisions about sustainable agriculture.

Keywords: life cycle assessment, animal welfare, economics

Introduction

Sustainability is doubtless a complex subject with many aspects. In agriculture, sustainability contains a large portion of ecological issues, if the environment is damaged you can not sustain your production since it relays on natural systems (Tilman *et al.*, 2002). Sustainability is usually described as comprising ecological, economical and social aspects. For sustainable agriculture there is also an agriculture-specific aspect of animal husbandry, which brings in ethical considerations about farm animals. In agriculture there may be several conflicting goals between theses four sustainability aspects.

Agriculture is a complex business; it consists of biological production in an environment that is neither easy measurable nor controllable (as compared to industrial production). One feasible way of elaborating sustainability issues generally, but also within agriculture, is to use scenario methodology. Scenarios originally were used for military purposes and when the method entered the civil society it was in economy and management, an example from management literature is Schoemaker (1995). The objective of this study was to develop a method to design scenarios for future agricultural systems that can be used for different production branches.

The dairy sector is one of the most important sectors in Swedish agriculture today, representing around 25% of the income to Swedish agriculture (Statistics Sweden, 2004). Within the Swedish milk sector approximately 50,000 people are working, on farms, at dairies and in transports. Dairy production is important for many positive aspects as maintained agricultural landscape, biodiversity and rural economies. The negative effects are also prominent; emissions of ammonia, energy use and nitrate leaching. The future development of dairy production influences both the environment as well as the social and economic development of the countryside. It is important to aim for a sustainable development including these issues as well as the ethical aspect of animal husbandry; which means e.g. that animal health and welfare should be considered.

The aim of this paper was to develop a scenario method that would incorporate scientific knowledge and practical experiences into the scenarios. Furthermore, the method was used for analysing the sustainability of future scenarios for dairy farm production in Sweden.

Material and methods

Scenario method

We chose a back casting scenario approach since the purpose was to develop a method for constructing scenarios that are more sustainable than today's system, not to present different outcomes of varying policies or technologies. The core principles in the method were transparency and structure. Thus, all assumptions were explicit, all choices made were clear and conflicting goals were identified.

An important background for the development of the method is the assumption that the most efficient way of incorporating experts from different fields of agricultural research as well as authorities and business is to present concrete descriptions of scenarios at meetings. This will initiate discussions about the relevance of the choices made in designing the scenarios and possibilities to develop the scenarios based on these 'expert meetings'. Hence the process described below is of an iterative nature.

1. Define and describe the value base that will guide the work.
2. Define systems boundaries and describe the system.
3. Define all relevant sustainability parameters (called 'focus parameters'). This is a list of parameters relevant for the studied system. In our work we mainly used the list of 'sustainability goals' defined by FOOD 21 (FOOD 21, 2004), but it is obvious that other definitions of sustainability can be used.
4. Describe all sub systems that make up the entire system.
5. Formulate 'focus scenarios'. A focus scenario is a scenario where the system is optimised for just one focus parameter. The focus scenario is formulated in order to describe how all functions are solved in principle and technically, e.g. by describing how the animal feed is composed and delivered to the animal or what tillage methods are used. These focus scenarios are rather extreme, taking only one aspect into consideration for every sub system. The principal solution for a sub system is described, e.g. 'the manure must be removed quickly from the house'. This 'principal concept' is then transferred to an 'implementation concept' which is a technical description how the principal concept can be achieved.
6. Identifying conflicting goals. Conflicting goals may arise from solutions for a sub system that are chosen to optimise one focus parameter that will obstruct the optimisation of another goal.
7. Describe goal visions. A goal vision is a description of what sustainability aspects are most important, as decreasing emissions, saving scarce resources or working environment and animal welfare.
8. Describe goal vision scenarios. One new scenario per goal vision is described. A goal vision scenario is a description of how the system should look if the focus parameters belonging to that goal vision are optimised.

In early steps (point 5) two levels of solutions or concepts are described, which comprise principal and implementation concepts. Conclusively two goal vision scenarios for each goal vision are designed, one principal and one implementation scenarios. On the latter it is possible to make rather detailed quantification regarding both economical and environmental impact, but for the former the accuracy of quantifications is lower.

When the goal vision scenarios for the first system are finished, the same procedure is repeated for the next system within the scope of the study. When all goal vision scenarios for the total production system are finished, the process of combining the goal vision scenarios starts.

9. Design goal vision scenarios for the total system.
10. Meeting with experts.
11. Modification of the scenarios.
12. Evaluation of the goal vision scenarios. Life Cycle Assessment combined with farm modelling was used to quantify the environmental effects. For an economic analysis it is necessary to make assumptions about agricultural policies and prices of input resources. The qualitative evaluation was based on literature review and panel discussions with groups of experts/stakeholders.

Scenarios for future Swedish dairy farming

Based on the method described above, two goal visions for milk farming were developed:
a.   Efficient production and small environmental impact per product unit ('High intensity')
This goal vision was focused on both economic and environmental efficiency. The environmental performance and resource efficiency optimised was the product oriented impact. This means that in this scenario we strived for high production per unit resource put in and per unit emission let out. The feed production was mainly based on local supply of forage feed and some grain complemented with import of high quality protein feed. The production was also concentrated on milk; it was a highly specialised enterprise, which makes it possible for the staff to become specialists on dairy cows.

b.   Focus on animal welfare, working environment and local environmental impact ('Low intensity')
This goal vision was focused on environmental efficiency, mainly on area level, but the production level was also taken into account. This means that the environmental impact per unit of land was minimised, but the impact per unit produced was also considered. The systems build on integration of milk and meat production based on local feed production, both forage and protein feed. The milk production was managed in a way that fits well into sustainable meat production. A second aspect of the integration was that in this goal vision the farm can grow more cash crops in order to optimise the crop rotation; the machinery and knowledge about crop production was a natural part of the enterprise.

Two goal vision scenarios were created from the goal visions. The goal vision 'Efficient production and small environmental impact per product' resulted in a scenario we call Specialised Dairy Farming (SDF). The goal vision 'Focus on animal welfare, working environment and local environmental impact' resulted in a scenario we called Mixed Dairy Farming (MDF).

Based on the qualitative descriptions, the quantification was done through expertise judgement based on available statistics combined with general knowledge synthesising research and extension services (Statistics Sweden, 2004; Agriwise, 2005; Swedish Dairy Association, 2005). When quantification was performed the scenarios were evaluated concerning economics, environmental effects and animal welfare, including health. The Life Cycle Assessment included investigation of eutrophication, global warming potential, acidification and toxicity. The use of resources for the system was quantified as energy use, land use and usage of phosphorus. For complete results for the dairy scenario see Gunnarsson *et al*. (2005) and Sonesson (2005).

Results and discussion

The main aim of applying the method described above was to develop scenarios in a more transparent and structured way. Since the scenarios are both rather concrete and logically constructed, they can be very valuable when the issue of sustainable agriculture is discussed. Such scenarios, and quantified results from them, can work as platforms for discussions between different stakeholders since they provide a mutual and concrete picture of different perspectives of sustainability. The explicit description of goal conflicts that are a result of the method is useful to realise where the conflicts between different interest lies. Our method deals with two time frames, principal long term scenarios and implementation short term scenarios, but it builds on the same goal visions, i.e. sustainability goals. At the same time as concrete and detailed descriptions are needed, the method also must entail discussions on a rather high systems level; otherwise the scenarios will not fulfil the aim of presenting examples of more sustainable systems. By using the method, a wide range of system levels, from definitions of sustainability to descriptions of housing for animals, are considered in a logical way.

There are certain limitations of the method. One limitation is that the method is used on farm level; e.g. how is milk production best performed. The matter of the sustainability of the food system as a whole is not addressed. This aspect includes which products should be produced and in which amounts. The question of where different products are best produced is also omitted. Other aspects not covered are 'margin effects', i.e. what will happen if the need for arable land increases or decreases. A second disadvantage is that even if the aim is to synthesise scientific knowledge into more comprehensible pictures of more sustainable systems, there is a risk that important information can be neglected. There is no absolute methodological mechanism that guarantees the completeness of the scenarios.

Cost of production was in scenario SDF 3.02 SEK/kg milk, in scenario MDF 4.34 SEK/kg milk, and in the present production 2.87 SEK/kg milk. The economic analysis shows that neither of the two scenarios was economically viable in the present economic context. The high labour cost in scenario MDF was a result of emphasis on animal welfare in dairy production. The feed in MDF was more expensive than the purchased feed used in SDF. This was not logical since the components in the feed were mostly identical, so perhaps the price of purchased feed actually does not reflect the production costs, i.e. the feed producers are not paid enough to cover their actual costs. A second explanation for the high feed costs per kg milk in scenario MDF was the relatively low milk production per cow.

The contribution to eutrophication per litre milk was lowest for scenario SDF. At the same time, scenario MDF contributes less to eutrophication per land area used. The emission per area land is important when livestock production is discussed on a regional level, in areas where the intensity is high or the receiving watershed is sensitive. The environmental assessment showed that the co-production of meat and live calves has important effects on the overall environmental impact; hence the choice of analysis method was crucial. However, the results showed the importance of including the co-products in such system analyses.

In the scenario construction, factors that are considered to improve animal welfare were integrated in both scenarios, as the legal requirements on the animal housing in Sweden have to be met. We used areas of concern found in previous research as a guideline to investigate the potential welfare differences between the scenarios we constructed. We found that a theoretical evaluation partly would be possible, considering the scientific knowledge about how housing and management is affecting health and welfare (e.g. Enevoldsen and Gröhn, 1996, Murray *et al.*, 1996; Singh *et al.*, 1994). Cows in SDF were found to have a higher risk of lameness as they have an increased risk of getting both heel-horn erosion and laminitis, compared to cows in the MDF scenario. In the SDF scenario, the cows had a higher milk yield, which has been found to be associated with an increased risk of mastitis, ketosis and abomasal displacement. Furthermore, the extended access to grazing on pasture in the MDF scenario decreases the risk of mastitis and dystocia. Comparing the two scenarios, on the long term MDF probably has more positive environmental impacts than SDF. The reason was that MDF involves a more varied crop rotation, which is beneficial for many biological aspects. Scenario MDF also uses more pasture, which can improve the biodiversity. However, the pasture was intense and hence less valuable from a biodiversity point of view.

The assessment of the scenarios is complex; there are many aspects to consider simultaneously. The fact that the evaluation of scenarios was done both quantitatively and qualitatively involves difficulties with balancing the conclusions; quantitative results often are given more weight than qualitative ones. The results show that no scenario was superior in all aspects. The implication of this is that the goal for developing sustainable dairy farm production must be guided by values, i.e. choices of sustainability goals that are more important. The choice of scenarios in this study was to some extent extreme, in reality a combination of the solutions in the two scenarios were likely to be most efficient in the quest for a sustainable development. The mainly positive environmental results for scenario MDF must be considered as rather strong, since the assumed milk yield is rather low. However, the low milk yield results in high production costs per kg of milk.

Previously, the same methodology has been applied on other branches of animal production, e.g. pig production (Stern *et al.*, 2005).

Conclusions

The method presented herein offers a structured way of synthesising large amounts of research knowledge into something comprehensible and practically understandable that can be used as a platform for further discussions about sustainable agriculture. The resulting scenarios can be subject to external assessment of all steps in the process. The method has been used on pig production, beef production, dairy production and the production of food potatoes.

No dairy scenario was superior in all aspects and the goal for developing sustainable dairy farm production must be guided by analysis of values. The economic analysis shows that neither of the two scenarios was economically viable in the present economic context. The contribution to eutrophication per litre milk was lowest for scenario SDF, but the MDF contributes less to eutrophication per land area used. Cows in SDF were found to have a higher risk of lameness as they have and increased risk of getting both heel-horn erosions and laminitis, compared to cows in the MDF scenario. No system is the sole solution and choices have to be made. Studies of this kind make it possible to assess which choices can be made and what the consequences are.

Acknowledgements

All persons within companies and organisations that gave their perspectives and comments on the scenarios are acknowledged. Furthermore, we thank Karl-Ivar Kumm, Maria Stenberg, Thomas Nybrant, Susanne Stern, Michael Ventorp and Ingrid Öborn. Finally, we thank Mistra, the Swedish foundation for strategic environmental research, which financially supported the research programme FOOD 21.

References

Agriwise (2005). Swedish University of Agricultural Sciences, www.agriwise.org.
Enevoldsen, C. and Gröhn, Y.T. (1996). A methodology for assessment of the health-production complex in dairy herds to promote welfare. Acta Agric. Scand. Section A Animal Science, Supplement 27: 86-90.

FOOD 21 (2004). Annual Report, FOOD 21, Swedish University of Agricultural Sciences (SLU), Uppsala Sweden.

Gunnarsson, S., Sonesson, U., Stenberg, M., Kumm, K.-I. and Ventorp, M. (2005). Scenarios for future Swedish dairy farming - a report from the Synthesis group of FOOD 21. Report FOOD 21 no 9/2005, 80 p.

Murray, R.D., Downham, D.Y., Clarkson, M.J., Faull, W.B., Hughes, J.W., Manson, F.J., Merritt, J.B., Russell, W.B., Sutherst, J.E. and Ward, W.R. (1996). Epidemiology of lameness in dairy cattle: description and analysis of foot lesions. Veterinary Record 138: 586-591.

Schoemaker, P.J.H. (1995). Scenario Planning: A Tool for strategic thinking, Sloan Management Rev. 36: 25-41.

Singh, S.S., Ward, W.R., Hughes, J.W., Lautenbach, K. and Murray, R.D. (1994). Behaviour of dairy cows in a straw yard in relation to lameness. Veterinary. Record 135: 251-253.

Sonesson, U., Gunnarsson, S., Nybrant, T., Stern, S., Öborn, I. and Berg, C. (2003). Att skapa framtidsbilder - En metod att utforma framtidsscenarier för uthållig livsmedelsproduktion (Creating images of the future - A Method for Scenario Construction for Sustainable Agricultural Systems, in Swedish), FOOD 21, Swedish University of Agricultural Sciences, Uppsala Sweden FOOD 21-Report no 3/2003.

Sonesson, U. (2005). Environmental Assessment of Future Dairy Farming Systems - Quantifications of Two Scenarios from the FOOD 21 Synthesis Work, SIK-Report, SIK - The Swedish Institute for Food and Biotechnology, Göteborg, Sweden.

Statistics Sweden (2004). EAA - Ekonomisk kalkyl för jordbrukssektorn 1993-2003 (Economic Accounts for Agriculture), Statistiska meddelanden JO 45 SM 0402.

Stern, S., Sonesson, U. Gunnarsson, S., Öborn, I., Kumm, K-I. and Nybrant, T. (2005). Sustainable development of food production - a case study on scenarios for pig production. Ambio 43: 402-407.

Swedish Dairy Association (2005). Web site of the Swedish Dairy Association, www.svenskmjolk.se

Tilman, D., Cassman, K.G., Matson, P.A., Naylor, R. and Polasky, S. (2002). Agricultural sustainability and intensive production practices. Nature 418: 671-677.

Protecting local diversity in scenarios of modern food biotechnology, globalised trade and intellectual property rights

A.G. Haslberger[1], A.H. Gesche[2], M. Proyer[1], R. Paulesich[3] and S. Gressler[4]
[1]Dept. for Nutritional Sciences, Center for Ecology, Univ. of Vienna, Althanstrasse 14, A-1090 Vienna
[2]Applied Ethics and Human Rights Program, SHHS, Queensland University of Technology, GP Campus, X209, 2 George Street, Brisbane, Qld. 4001, Australia
[3]Dept. for Environmental Economics, Vienna University for Economics, Augasse 2, A- 1090 Vienna
[4]Forum Austrian Scientists for Nature Protection, Mariahilfer Straße 77-79, A-1060 Vienna, Austria

Abstract

Decreased global and local diversity and a homogenization of biota is seen as a major threat to ecological and socio economic resilience. Consequences of modern food production, such as global propagation of few high yielding elite lines, declining diversity of landraces or consequences from gene flow, interact with socioeconomic drivers such as trade and intellectual property regulations in accelerating the mostly irreversible and broadening impacts of loss of biodiversity. Especially the SPS agreement under WTO prohibits any approaches to restrict trade of foods because of other reasons than sanitary and phyto-sanitary measures. Already now the reports of the UN- Millennium Ecosystem Assessment reports alarmingly increasing homogenization of biota and distribution of exotic species by trade and trans-boundary movements. This development is considered to reduce local ecological and social resilience in food production significantly. In the light of these developments trade regulations need to be reconsidered. The use of new, ethically guided structured Matrixes or Codes for an integrated assessment of safety and societal consequences and a participatory priority setting including aspects of public goods, such as conservation, seems to be mandatory.

Keywords: local biodiversity, homogenization, resilience trade, intellectual property rights

Biotic homogenization and consequences for resilience

Diversity

Damages of many ecosystems and losses of biodiversity are discussed as a major concern worldwide. The Millennium Ecosystem Assessment (MA, 2005) launched by the UN summarizes that virtually all of Earth's ecosystems have now been dramatically transformed through human actions. Between 20% and 50% of 9 out of 14 global biomes have been transformed to croplands. Biodiversity change is caused by a range of drivers. Recent and topical trends in the development of biodiversity state that current rates of change and loss exceed those of the historical past by several orders of magnitude and show no indication of slowing.

One specific aspect of the destruction of diversity detected by the analysis under the MA is a steadily increasing homogenization of the surrounding environment described as homogenisation of biota. Large parts of earth's agricultural regions are already to be characterized as monucultural. Main drivers for these developments include the removal or introduction of organisms in ecosystems which disrupt biotic interactions or ecosystem processes. The spread of exotic species respectively Invasive Alien Species (IAS, see IUCN, 2004) is promoted through worldwide trade and movement and increased global use of technologically improved high yielding races replacing local landraces. These dangerous developments are supported by socio economic drivers such as regulations for trade and intellectual property rights and are considered to be a major threat for ecological and social resilience. Resilience is now seen as an important concept which enables adaptation mechanisms and policies to dangerous global changes such as climate change.

Resilience

Although the stability of an ecosystem depends to a large extent on the characteristics of the dominant species, less abundant species also contribute to the long-term preservation of ecosystem functioning (MA, 2005). Often associated with aspects of disaster management, sustainability, vulnerability and risk (compare Manyena, 2006) the concept of (ecological) resilience could be summarized as buffer capacity or absorbing ability of an organism when facing hazards or complications. Similarly social resilience is to be defined as the capability of a social community to face, withstand, cope with and recover from outward negative impacts of varied sources, intensity and outreach. This concept of economic resilience combines aspects of shock recovery, shock avoidance and shock absorption of states or economic entities confronted with economic shocks (compare Sneddon, 2000) and integrates data derived from specific vulnerability indices, information about risks of different countries facing problems through economically 'rough' times.

Drivers of biotic homogenisation and food production

Mobility

Invasive Alien Species (IAS) are one of the most significant drivers of environmental change worldwide (IUCN, 2004) such as habitat destruction. Of all almost 2000 imperiled species in the United States, 49% are endangered because of introduced species alone or because of their impact combined with other forces. The greatest impact is caused by introduced species that change an entire habitat, because many native species thrive only in a particular habitat. For example the zebra mussel, accidentally brought to the United States from southern Russia, transforms aquatic habitats by filtering prodigious amounts of water (thereby lowering densities of planktonic organisms) and settling in dense masses over vast areas. At least thirty freshwater mussel species are threatened with extinction by the zebra mussel. Some impacts of invaders are subtle but nonetheless destructive to native species. For example the rainbow trout introduced widely in the United States as game fish are hybridizing with five species listed under the Endangered Species Act, such as the Gila trout and Apache trout (Simberloff, 2000). Intentional and unintentional distribution ('hitchhiking' of species or pathogens while transporting goods) contribute to the distribution of IAS where fast developments of globalised trading and globalised travelling significantly contributed to increased hazards.

Biotechnology enhanced breeding

In many cases products derived from modern breeding technologies resemble aspects of IAS. Modern methods of biotechnology enable the introduction of traits into recipient organisms which have not been a characteristic of the species before and which alter fitness parameters in the recipient environment. In addition few high yielding crop lines or fast growing animals derived from modern food production technologies are increasingly used in large areas worldwide competing with traditional local organisms, landraces or wild species. Gene flow between biotechnologically improved lines, conventional lines and wild organisms additionally endangers stability of ecosystems and diversity.

Trade

A biotechnologically improved organism is likely to be developed for use in large areas internationally and global trading as food and feed to endure returns of often considerable investments. Attempts become more likely that organisms are grown and traded in areas where characteristics of the product and agro-ecological characteristics do not indicate benefits, which may be existing in other areas. In consequence areas of development, production and consummation are getting more dispersed. Trade liberalization accelerates this dispersal. The SPS agreement under WTO prohibits any approaches to restrict trade of foods because of other reasons than sanitary and phyto-sanitary measures. Current market dynamics press ahead with globalised trading of foods and crops where safety instruments for protection of local diversity such as the Cartagena Protocol on Biosafety are lagging behind. Developments also result in a dispersal of areas of benefits and possible disadvantages. Claimed advantages of market liberalization and reduced subsidies cannot be realized equally by all actors on the supply chain because producers and traders are neglecting regional economic differences as market structures or information channels.

IPR

Modern methods of biotechnology in breeding can enable only an improvement of a limited number of elite lines or organism, which then should be used globally to return considerable investments, protected by intellectual property rights. This endangers the propagation of traditional local races and knowledge. There are continuing concerns about market dominance in the agricultural sector by a few powerful companies. Also in the area of the implementation of intellectual property rights considerable resistance raised internationally. Critics urge restriction of patenting possibilities to genetic material in tight combination with specified methods and uses and even a recent report of WHO identifies inconsistencies in proposed regulations and recommends reconsideration and international discussion.

Regulations

The World Trade Organization (WTO) is the only international body that sets and oversees global rules associated with trade between nations. At the core of the WTO are agreements negotiated by the majority of the world's trading nations and ratified by their governments. One of them is the Sanitary (human and animal safety) and Phytosanitary (plant safety) Agreement (SPS Agreement), overseen by the WTO. The purpose of the WTO-SPS agreement is two-fold, i) to promote free trade and ii) to protect nations against bioinvasion of unwanted pests, weeds and diseases carried by plants, animals and similar. Developing countries, which lack the capacity to implement internationally-agreed standards for food safety and animal and plant health, are supported by the Standards and Trade Development Facility (STDF) through grants, information sharing and technical cooperation. The STDF also helps developing countries 'gain and maintain market access' (STDF Secretariat, 2006). Another international body important with regards to modern food biotechnology and trade, is the Codex Alimentarius Committee. It was established jointly in 1962 by the United Nations' Food and Agricultural Organization (FAO) and the World Health Organization (WHO). Its major function is to set international food standards. The Committee is currently working on revised 'Working Principles for Risk Analysis for Food Safety for Governments'. The Committee already established 'a framework for undertaking [scientific] risk analysis on the safety and nutritional aspects of foods derived from modern biotechnology'(FAO/WHO, 2003) earlier, intended to protect human and environmental health. Although the principles have no binding effect on national legislation, they can be used in case of trade disputes. International trade in food is further guided by a code of conduct, namely the 'Code of Ethics for International Trade in Food'. Its two major aims are to protect the health of consumers and to protect consumers from 'unfair trade practices' (Article 4). There are no comments made in the Code which specify more closely unfair trade practices. The 'Code of Ethics' has been in existence since December 1979, was amended in 1985, and is currently under review once more by the Codex Committee on General Principles (CCGP; Joint FAO/WHO Food Standards Program, 2007). The aforementioned documents and guidelines are but a small proportion of international guidelines that impact on the global food market. All of these documents are directed mainly to the market. Even the 'Code of Ethics for International Trade in Food' does not address any environmental, ethical or social aspects as they relate to the production and marketing of GM foods. Even attempts to discuss a broadening of SPS criteria for an involvement of environmental or socio-economic criteria showed massive resistance. The use of international standards for traded food, focusing on food safety such as the STDF (Standards in Trade and Development Facility), a joint effort recently established between the WHO, the FAO, the World Trade Organization, the World Animal Health Organization and the World Bank will, hopefully, also focus on environmental issues in the future.

In light of the current concerns of many nations about global warming and the protection and conservation of biodiversity and the environment, we make the following recommendations with regards to a revised version of the Proposed Draft *Code of Ethics for International Trade in Foods*:
1. Reference should be made to sustainable development and the conservation of biodiversity;
2. If biodiversity is threatened to an unacceptable level, local communities and national bodies should be given the choice and be provided with the authority to protect their resources, even if this contravenes current world trade provisions. Any decisions taken should be based on an integrated risk assessment which includes scientific and normative impact assessments and is conducted by independent authorities with the mandate to invite stakeholder input and to initiate processes that ensure proper representation;
3. At the local, national, and international level, the current provision of benefit sharing should be expanded by also providing for sharing the burden of retaining biodiversity in a fair and equitable manner;
4. Reference to ethical principles should be included in order to provide an equitable and just environment within which to make decisions about trade and biodiversity.

Sustainable development and conservation of biodiversity

The Convention on Biological Biodiversity (CBD), recognised by more than 180 nations, regards the conservation of biological diversity as 'a common concern of humankind' (Article 15(1)). It also states that the collection of resources requires prior informed consent. Biodiversity is especially prevalent in the megadiverse tropical and subtropical areas of the South, where 'biodiscovery' (bioprospecting) activities have become points of conflict between local interests and trade interests, including intellectual property interests, which are heavily skewed towards the North. The term 'biodiscovery', first used in the 'Code of Ethics for Modern Biotechnology in Queensland' (2001) replaces the term 'bioprospecting' and refers to accessing and taking of biological resources. Providers can be the government or a government agency, but also individual indigenous land owners, local communities and similar. Access to these resources is required by mainly international or transnational companies, seeking valuable compounds by screening and analysing the genetic material from plants, fungi, and other biological resources in search for patentable products, such as novel pharmaceuticals, for commercial gain. These scenarios are played out on a regular basis and highlight the tension between the generally capacity poorer provider and the capacity strong producer. An ethical approach to reduce the tension and the potential disadvantage of the weaker party would be to commit stakeholders to share benefits and burdens equitably and fairly by following a process of negotiation that is situated within a framework of ethical practice and decision-making.

Sharing the benefits and the burden

The term 'benefit-sharing' was coined at the Convention on Biological Diversity (CBD) at the Earth Summit in Rio de Janeiro, Brasil in June 1992 - the first global agreement to not only conserve and sustain the biological diversity of our planet, but also to ask for the sharing of benefits that arise from the commercial or other utilization of existing genetic resources in a fair and equitable manner. The CBD obliges nations to enter into benefit-sharing agreements with the access providers of the genetic resources. This includes the valuing of knowledge within indigenous communities. These measures are a sign of how much in recent years the notion of biodiversity has shifted from a 'common heritage of mankind' to a 'resource under the sovereignty of nation states' and considerations of intellectual property rights. From a justice point of view, developing countries are increasingly asking for compensation for their assistance in providing companies with the original cultivars and their associated knowledge, which the companies then use to commercialize their research and products (Gepts, 2004). These demands have led to a number of disputes in the past. Recently, the so-called 'Bonn Guidelines on Access to Genetic Resources and Fair and Equitable Sharing of the Benefits Arising Out of Their Utilization' provide guidance about how to deal with access and benefit sharing issues (CBD Conference, The Hague, 2002). In addition, several megadiverse countries, Australia included, have devised their own guidelines to minimise disputes. Where access is sought to biological resources on indigenous peoples' land, in Australia, prior informed consent is to be obtained from access providers. While such guidelines address the commodification of resources, they do not address the other value of biodiversity, that is, the role local people play in providing healthy ecosystems, which, in turn, are vital for food security. As biodiversity declines, mainly through deliberate human interactions, and food security stressors become more prevalent, new ways must be found to encourage local communities in biodiverse areas to become guardians of that diversity or to restore past diversity. It might mean that the international community has to compensate these countries for their role in forfeiting monetary benefits in return for biodiversity protection/restauration. One such example comes from Australia, where in April 2007, a 'Global Initiative on Forests and Climate' was announced aimed at protecting the world's forests. The first beneficiary is Australia's closest neighbour, Indonesia. It will receive US $160 million to counteract illegal logging, plant new trees and find alternative income sources for people involved in the timber industry. Australia's initiative could set an example for related purposes (BBC, 29 March 2007).

Extending the current 'Code of Ethics for International Trade in Food' by including a normative framework of practice

The current 'Code of Ethics for International Trade in Food' does not pay any attention to social and ethical aspects with regards to the commercialization and trading aspects of biological resources. The question is whether it should and could be further amended to i) allow for changes in societal expectations with regards to trade plus ii) respond to recent findings regarding the relationship between biodiversity and food security. For example, if the title of the current code could be amended to 'Code of Ethics for International Trade in Food and General Principles', normative issues as mentioned above could be considered alongside trade standards. Which principles should be selected awaits further discussion. We propose a framework of ethical principles adapted from those suggested by Beauchamp and Childress (1971, 2001), which are widely used in the biomedical field. We offer the following principles for further discussions:

- Respecting persons and their communities and considering the living environment and biosphere on which life depends.
- Avoiding harm, being cautious and maximising benefits to persons, communities and the environment. This includes the sharing of benefits and burdens to maintain or restore ecosystems on which food security and trade depend.
- In trade, acting justly and equitably towards others, including other nations and future generations.
- Reducing activities that harm the biosphere and ecosystems. Taking actions that are sustainable.
- Acting with integrity in trade and development, declaring conflict of interest, and following relevant national and international guidelines and legislation designed to support both, trade and the well-being of nations and their ecosystems.

- Supporting participatory engagement and decision making, including allowing for choice and effective self-determination.

An ethical matrix (Mepham, 2000; Kaiser and Forsberg, 2001), based on the same principles, could further facilitate stakeholder deliberation when strong trade interests intersect with local, regional, or national interests, often in capacity poor regions of the globe. The suggested changes to the 'Code of Ethics for International Trade in Foods' might require a review of current trade rules, intellectual property rules and other practices to make the sharing of benefits and burdens a reality. They could set the scene for a more socially embedded and sustainable governance of international trade in modern foods that acknowledges the close relationship between food production methods, food security and the need for biodiversity conservation.

References

BBC, 29 March 2007. Australia fund to protect forests. Available from http://news.bbc.co.uk/1/hi/world/asia-pacific/6505693.stm. Accessed 7 May 2007.

Beauchamp, T.L. and Childress, J.F. (1971, 2001). Principles of Biomedical Ethics. 5th ed., Oxford University, Oxford

CBD Conference of the Parties No 6, The Hague, 2002. Decision VI/24; http://www.biodiv.org/programmes/socio-eco/benefit/bonn.asp#).

FAO/WHO (2003). Principles for the risk analysis of foods derived from modern biotechnology. CAC/GL 44-2003. Available from http://www.who.int/foodsafety/biotech/en/codex_biotech_principles.pdf. Accessed 6 May 2007

Gepts, P. (2004). Who owns biodiversity, and how should the owners be compensated? Plant Physiology 134: 1295-1307.

IUCN (2004). Policy Briefs - Trade and Biodiversity: Controlling the Movement of the Invasive Alien Species through Trade. www.iucn.org

Joint FAO/WHO Food Standards Program Codex Committee on General Principles. 24th Session. 2-6 April 2007. Proposed Draft Revised Code of Ethics for International Trade in Foods. Available from ftp://ftp.fao.org/codex/ccgp24/gp24_04e.pdf. Accessed 7 May 2007.

Kaiser, M. and Forsberg, E.M. (2001). Assessing Fisheries - Using an ethical matrix in a participatory process. Journal of Agricultural and Environmental Ethics 14: 191-200.

Manyena, S.B. (2006). The concept of resilience revisited. Disasters 30: 434-450.

Mepham, B. (2000). A Framework for the Ethical Analysis of Novel Foods: The Ethical Matrix. Journal of Agricultural and Environmental Ethics 12: 165-176.

Millenium Ecosystem Assessment MA/ Ecosystems and Human Well - Being/ Biodiversity Synthesis: http://www.maweb.org/documents/document.354.aspx.pdf

Simberloff, D. (2000). Introduced Species: The Threat to Biodiversity and What Can Be Done Action Bioscience, http://www.actionbioscience.org/biodiversity/simberloff.html

Sneddon, C. (2000). 'Sustainability' in ecological economics, ecology and livelyhoods: a review. Progress in Human Geography 24: 521-549.

STDF Secretariat (2006). Agencies agree plan for food safety, animal/plant health assistance. Press release, 18 December 2006. Available from http://www.who.int/foodsafety/biotech/en/codex_biotech_principles.pdf. accessed 6 May 2007.

Does free trade in agriculture promote 'one planet farming'?

Tom MacMillan and Neva Frecheville
Food Ethics Council, 39-41 Surrey St, Brighton BN1 3PB, United Kingdom, tom@foodethicscouncil.org

Abstract

Agricultural policy discourse in the United Kingdom (UK) is currently dominated by two key themes: on the one hand, the government is pushing heavily for thorough liberalisation of the EU Common Agricultural Policy (CAP); on the other hand, it has promoted the concept of 'one planet farming', which is a shorthand for a sustainable farming and food system that respects ecological limits. This paper begins by exploring both themes and examining how they have developed through recent policy statements. It discusses the roles played by competing government departments and interest groups, including the farming lobby and environmental groups, and the international influence of these UK policy discourses. The central part of the paper examines tensions between these two themes. For liberalisation of the CAP to promote a sustainable farming and food system, what other conditions would need to be met? What assumptions does this discursive pairing make about the responsibilities of UK/EU consumers to people in poorer countries, about countries' capacities to enforce environmental standards in production and on imports, and about the range of plausible scenarios for the CAP and international trade rules? The paper closes by discussing alternative policy themes that seek to address these tensions and to expand the opportunity for citizens and stakeholders to contribute to policy formation.

Keywords: liberalisation, CAP, free trade, sustainable development, policy

Two themes in UK agricultural policy

When David Miliband became Secretary of State at the UK's Department for Environment, Food and Rural Affairs (Defra) in the summer of 2006, he called upon farmers and the food sector to embrace the concept of 'one planet farming': 'we are living as if we had three planets' worth of resources to live with, rather than just one. So if we are to build a sustainable future... we need to cut by about two thirds our ecological footprint' (Miliband, 2006).

In the period since, the concept of 'one planet farming' - adapted from WWF's notion of 'one planet living' - has become a central theme of UK agricultural policy. It is a timely reminder that to make UK farming and food truly sustainable we need to see beyond national borders: if we tackle environmental, social and economic problems facing UK agriculture without seeing the global picture, we risk simply outsourcing them to other parts of the world.

Yet this is not the only big theme in UK agricultural policy thinking. Defra also espouses out-and-out liberalisation of the EU Common Agricultural Policy (CAP), arguing that a lean, mean CAP is better for EU taxpayers and for developing countries (HMT and Defra, 2005).

Recent official statements feature both these themes (Miliband, 2007). However, significant tensions exist between them (Porritt, 2007). This paper explores those tensions.

Policy background

Although the 'one planet' phraseology is new, the principles behind this idea have been part of agricultural policy in the UK since 2002, when the government published a '*Strategy for* sustainable farming and food' (Defra, 2002). The strategy set out eight key principles for sustainable farming and food, such as to 'respect and operate within the biological limits of natural resources' and to 'support the viability and diversity of rural and urban economies and communities'. Crucially, it aimed to support these principles not only within the UK but 'wherever our food is produced and processed'.

Similarly, UK government support for greater liberalisation of EU agricultural policy is not new. However, the joint publication by the UK Treasury and Defra in December 2005 of a 'Vision for the Common Agricultural Policy' outlined this attitude in some of the boldest terms to date (HMT and Defra, 2005). The vision included:
- Total elimination of market mechanisms from EU agricultural policy.
- Some continued public payments to land managers in return for social and environmental public goods from land management that would otherwise be uncompensated.
- Much less public money spent through the CAP on agriculture and rural development.

The vision was greeted by what the Secretary of State has described as 'a sharp intake of breath' from stakeholders, many of whom saw it as an attempt by the Treasury simply to capture resources spent on agriculture for other spending priorities (House of Commons, 2006). In particular, the Treasury's expectation that public payments to agriculture for otherwise uncompensated public goods would only cost a fraction of the current total CAP spend have been condemned by environmental and farming groups.

International relevance

These themes within UK agricultural policy discourse are of wider international relevance. Politically, the UK proved a significant player in the 2003 reforms of the CAP and the election of Nicolas Sarkozy in France is expected to strengthen the position of liberalisers such as the UK in further negotiations on the CAP in 2008-9.

Furthermore, they are instances of two larger concepts in international policy and governance: sustainable development and free trade. Exploring the tensions between them in the context of UK agricultural policy, where they are immediately juxtaposed and the relationship between them is hotly debated, offers a window onto tensions between those bigger concepts.

Tensions

The basic question is whether liberalisation can promote 'one planet farming'. If it can do so in principle, what conditions would be necessary in practice to make that happen? What assumptions does this discursive pairing make about the responsibilities of UK/EU consumers to people in poorer countries, about countries' capacities to enforce environmental standards in production and on imports, and about the range of plausible scenarios for the CAP and international trade rules?

The following three sections consider tensions between these two themes.

Trade and the environment

The first tension concerns the direct environmental impacts of agricultural trade. International agricultural trade can have major economic and social benefits, both for exporting and importing countries. When products are in season in exporting countries, but could only be grown under energy-intensive conditions where they are being imported, such trade can reduce the environmental impact of consuming those products. However, can we engage in large volume international trade in agricultural products - and eat out of season - yet also live within the planet's environmental resources?

Sustainable food production and ethics

While 'food miles' are not a simple measure of sustainability, 'in like for like systems, where food supply chains are identical except for transport distance, reducing food transport will improve sustainability' (Smith *et al.,* 2005). The climate change impacts of international food transport as a result of trade are significant with, for example, the airfreight of fresh produce to the UK registering in whole percentage points on the UK's overall emissions graph (Dalmeny, 2007). So, while reducing 'food miles' does not always reduce environmental impact, very short distance supply-chains for low-input, seasonal produce can have the lowest possible impact.

The claim that policies which promote greater international trade could also support 'one planet farming' might assume that:
- Agricultural trade currently has a negligible environmental impact.
- Transport from trade in agricultural goods is a relatively wise use of environmental resources, for example because it might support international development, so other environmentally damaging activities should be subject to greater restrictions instead.
- Improvements in transport efficiency make environmental impact of 'food miles' negligible.

However, in the absence of making such assumptions or value judgements explicit, UK policy makers appear to stakeholders not to have considered this tension at all.

Dumping problems

A second tension concerns the UK's international responsibilities to promote sustainable development. When the UK government implements regulatory measures to promote sustainable farming and food within the UK - raising environmental or animal welfare standards, for instance - there is a risk that instead of raising production standards of the food consumed in the UK, food is simply imported from places that do not have such stringent standards. This goes against the government's commitment to promote the principles of sustainable farming and food 'wherever our food is produced and processed' (Defra, 2002). Liberalising agricultural trade exacerbates this problem rather than resolving it. By contrast, producing more of the food consumed within the UK in areas that were subject to UK regulations would in theory make it easier for government to make good on its commitment to promote sustainable farming and food.

Again, however, this does not mean that liberalisation and 'one planet farming' are necessarily incompatible. Rather, the claim that they are compatible rests on assumptions that have not been made explicit in agricultural policy discourse. These might include that:
- It may be considered legitimate for a country to deplete environmental resources or compromise animal welfare in the pursuit of economic development. This argument is most plausible when it concerns localised environmental resources, such the freshwater of a single riverbasin, and least plausible when it concerns environmental resources that are generally accepted to be the common heritage of humankind, for instance biodiversity in tropical forests. However, this assumption is contentious, particularly since the concept of 'one planet farming' highlights the commonality of environmental challenges.
- It could be argued that environmental policy, not agricultural policy is the correct domain in which to address international environmental problems. Thus, a liberal approach to agricultural trade should be complemented by more robust multilateral environmental agreements. However, agreements such as the Convention on Biological Diversity are weak, in practice, since they contain few incentives or sanctions to encourage compliance.
- Another possible assumption is that the cost of meeting higher standards in the UK is supported by public payments for the provision of public goods, for example under 'Pillar II' of the CAP. One difficulty with this assumption is that, although such payments are permitted under international

trading rules, they can in practice amount to a production subsidy that leads to UK producers 'dumping' cheap produce on international markets. This raises the prospect that liberalisation could increase dumping, instead of reducing it, if it went hand-in-hand with a serious commitment to raise environmental and other production standards of food consumed in the UK.

Development conditions

The third tension concerns whether liberalisation makes good on the social aspects of 'one planet farming', to promote international development and social justice. The UK government view that out-and-out liberalisation of the CAP would support economic development in poor countries echoes the view of international financial institutions such as the World Bank. However, it raises at least two issues for debate:

- Benefits are conditional - As the Vision for the CAP notes, major benefits to poor countries from EU liberalisation depend fundamentally on additional conditions being met, including investment in the infrastructure needed to trade, help with adjusting to 'preference erosion' and protection for poor countries from forced liberalisation (HMT and Defra, 2005). These conditions are not being met and it is questionable whether they will be, since they are not integral and enforceable components of the agricultural trade framework.
- Impact within poor countries - Just as there would be pronounced winners and losers among poor countries, so there would be within them. Where poor countries benefit economically from liberalisation of the CAP, a substantial portion of those rewards may fall to large landowners and international companies.

Alternative scenarios

The World Bank view that liberalisation of US agricultural policy would help to raise world prices has been challenged by researchers who argue that strengthen some market interventions offers the best deal for people in poor countries (Ray *et al.,* 2003) The US analysis does not translate to the EU, of course, but in the absence of similar research on the CAP it raises the question of whether proposals to date have considered a sufficiently broad range of plausible policy scenarios.

World-wide, many development groups, community organisations and some national governments have called for a new framework for international agricultural trade, focused more directly on the aims of eliminating hunger and malnutrition, and promoting sustainable farming and food. These calls have centred on the notion of 'food sovereignty', championed by the international peasant movement Via Campesina (Windfuhr and Jonsen, 2005).

'Food sovereignty' raises challenging questions for the UK position in EU and international negotiations and suggests a different approach from liberalisation towards 'one planet farming'. It also challenges the way policies and international agreements are made - who is directly involved, which interest groups have access to the negotiations and what criteria underpin decisions. The provenance alone of these questions - direct from one of the key constituencies that agricultural trade reform is intended to serve - means that they deserve to be taken seriously.

References

Dalmeny, K. (2007). Low-carb diet. Food Ethics 2: 9.

DEFRA (2002). The strategy for sustainable farming and food: facing the future. Department for Environment, Food and Rural Affairs, London: 12.

HM Treasury and DEFRA (2005) A vision for the Common Agricultural Policy. HMSO, London, December: 16.

House of Commons (2006). Minutes of evidence taken before the EFRA Committee. July 12.

Miliband, D. (2006). One planet farming: speech at the Royal Agricultural Show. Defra, London, July 3.

Miliband, D. (2007). Farming 2020: speech at the Oxford Farming Conference. Defra, London, January 3.

Porritt, J. (2007). Hard to swallow. Guardian, January 4.

Ray, D., De La Torre Ugarte, D. and Tiller, K. (2003) Rethinking US agricultural policy: changing course to secure farmer livelihoods worldwide. Agricultural Policy Analysis Centre, Univeristy of Tennessee, Knoxville, TN.

Smith, A., Watkiss, P., Tweddle, G., McKinnon, A., Browne, M., Hunt, A., Treleven, C., Nash, C. and Cross, S. (2005). The validity of food miles as an indicator of sustainable development: final report. Department for Environment, Food and Rural Affairs, London, July: v.

Windfuhr, M. and Jonsen, J. (2005). Food sovereignty: towards democracy in localised food systems. ITDG Publishing, Bourton-on-Dunsmore, March: 15.

Constructing sustainable regional food networks: a grounded perspective

Dirk Roep and Johannes S.C. Wiskerke
Wageningen UR, Rural Sociology Group, Hollandseweg 1, 6707 KN Wageningen, The Netherlands

Abstract

This paper offers an empirically grounded perspective into the creation of sustainable food supply chains and networks. Based on the reconstruction of the development of fourteen food supply chain initiatives in seven European countries, we demonstrate that the process of increasing the sustainability of food supply chains is rooted in strategic choices regarding *governance*, *embedding* and *marketing*, and in the coordination of these three dimensions that are inextricably interrelated. Moreover, when seeking to scale up an initiative these interrelations need continuous coordination and rebalancing. Depending on the different starting points, the key actors, their strategies and abilities and the alliances that they are able to create, each initiative carves out its own distinct trajectory through time. Amidst all this apparent diversity three different underlying trajectories can be distinguished: *chain innovation*, *chain differentiation* and *territorial embedding*. Each reflects a specific drive and scope and a specific path towards sustainability, which balances new societal demands and marketing opportunities against new forms of dependency. All initiatives tend to start by following one of these trajectories. Some will follow the same one through their development; others may move from one trajectory to another, as part of their evolution. Finally, this paper will reflect on the relevance of the analytical framework and sustainability trajectories for practitioners, policy-makers and researchers.

Keywords: food supply chain, governance, embedding, marketing, sustainability trajectory

Introduction

During the last two decades the agro-food sector in Europe has undergone profound changes (Sonnino and Marsden, 2006). On the one hand we are witnessing processes of globalisation of the agro-food chain, the industrialisation of food production and economic concentration in the processing industry and retail sectors (Kirwan *et al.*, 2004). On the other hand one can observe the emergence of a wide variety of new food networks (in some cases these are more a re-emergence of traditional, authentic artisanal networks) that are characterised by notions of re-localisation, embeddedness and a turn to quality (Renting *et al.*, 2003; Watts *et al.*, 2005). The increase in the number and kinds of new food networks is generally understood as a countermovement to the prevailing trends of globalisation (Marsden *et al.*, 1999). Inherent in this counter-movement is the deliberate intention to create distinctiveness, for instance by producing food with distinct organoleptic qualities and/or by changing the mode of connectivity between food production and consumption, generally through reconnecting food to the social, cultural and environmental contexts in which it is produced (Kirwan, 2004).

Constructing distinctiveness: the GEM-framework

This paper is based upon an EU-funded research project entitled 'Marketing sustainable agriculture: an analysis of the potential role of new food supply chains in sustainable rural development' (SUS-CHAIN; see www.sus-chain.org for details of the project). This project was undertaken as response to the growing emergence of the issues of food quality and sustainable rural development as central concerns in discourses over the future of food and farming in Europe. As part of this project the start and evolution of fourteen food supply chain initiatives in seven European countries were reconstructed. These fourteen initiatives represent an impressive diversity with regards to the initiators, their intentions, capacities and strategies, the configuration of the food supply chain, the problems addressed, the goals pursued, the public support received and needed, their level of success and their impact on rural development. The initiatives analysed, however, also show that distinctiveness is created and realised through three dimensions:

1. Governance. This involves both structural as well as process-related aspects (Berger, 2003) of creating and maintaining a food network:
   a. The governance structure, i.e. the way in which the alliance is organised (e.g. open group, club, 'channel captain' or firm), its wider network (e.g. the kind of societal organisations and interest groups, if any, that are involved in strategic development and decision making) and its legal or formal status (e.g. association, cooperative, public-private partnership, limited company, et cetera).
   b. The governance process, i.e. the way in which the food alliance and network is governed. This includes issues such as the division of roles, decision making procedures, power relationships within the network, contractual arrangements, codes of practice, style of governance (e.g. command and control or consultation, negotiation and consensus building) et cetera.

2. Embedding. The concept of 'embeddedness' was originally based upon the idea that economic systems, such as a food supply chain, operate within a network of relationships, institutional arrangements and cultural meanings that limit the extent to which economic actors can be regarded as purely instrumentally and rational in their market orientation (Granovetter, 1985). Over recent decades this concept has gradually taken on a more specific connotation within the domain of agro-food studies:
   a. Local embedding, i.e. the extent to which (a) a food network uses local resources (e.g. soil, breeds, skills and knowledge, processing units, retail outlets) and (b) local actors and stakeholder organisations are involved in it.
   b. Societal embedding, i.e. the extent to which the values, codes and rules that represent the food product and the chain through which it is produced are shared by the wider network of stakeholders, consumers and society in general. This involves, for instance, values such as environmental friendliness, food miles, animal welfare, fair trade and health.

3. Marketing. Marketing refers to the market oriented business management of an enterprise or alliance. This dimension involves more than just 'putting a product on the market', 'enhancing sales' or 'advertising'. It is an integral part of the management of an enterprise or alliance. The marketing success of an enterprise or alliance depends on its capacity to continually understand, anticipate and adapt to market developments. Four different levels of marketing activities can be distinguished:
   a. Analysis of the market and its environment.
   b. Definition of objectives, i.e. fixing the goals that the alliance attempts to reach, e.g. market share, turnover, as well as ethical or ecological goals.
   c. Development of strategies through which these objectives are to be realised, such as degree of market competitiveness and product differentiation.
   d. Application ('marketing mix'), i.e. policies about products, prices, distribution and communication.

There is a strong and ongoing interaction between these different levels: i.e. the analysis determines the objectives which provide the framework for strategies which in turn define the applications.

Constructing a new food network always involves making conscious and strategic choices over governance, embedding and marketing and co-ordinating these three dimensions. Reconstructing an existing network will involve rethinking, reassessing and reconfiguring these dimensions. These three dimensions are interrelated and interconnected, as shown in the triangle in Figure 1. When scaling up a food supply chain these have to be continuously coordinated and balanced. In Figure 1 this process is represented by the circle with arrows. The analytical framework shown in this figure also demonstrates that a food network (being a specific combination of governance, embedding and marketing) has a specific sustainability profile and will also require specific kinds of public and/or private support to strengthen its sustainability profile.

In addressing its main objectives the SUS-CHAIN project has reconstructed the creation and development of fourteen food networks that are pursuing sustainable trajectories. These fourteen stories show how each initiative has created and pursued its own path. Although each path is unique there are clearly observable similarities and differences between them. Detailed comparison of these similarities and differences has led us to distinguish three different trajectories (see Figure 2).

1. *Chain innovation*: the construction of a new food supply chain, generally with the aim of improving the position of farmers in the food supply chain or network. This trajectory initially focuses on the design, development and implementation of new forms of food supply chain governance, such as new rules, codes of practice, division of roles and institutional arrangements. New forms of governance are realised by mobilising new strategic alliances and by building a strong support network of societal organisations, interest groups and, occasionally, also governmental authorities. This support network is of vital importance for the creation of a protected space for experimenting and learning.
2. *Chain differentiation*: the production and marketing of new, more distinctive products within an existing chain. The aim of this trajectory is to improve the commercial performance of an existing food supply chain or network by developing one, or a range of, distinctive product(s) that differ significantly from those presently available. Chain differentiation is most often initiated by chain actors such as processors or retailers. Sustainability concerns and the interrelations between governance, embedding and marketing are, therefore, primarily approached from a commercial perspective. These initiatives are frequently characterised by the presence of highly influential chain captains and their success is often dependant on the combining strategies of marketing differentiation with processes of (re)embedding distinctive food qualities.
3. *Territorial embedding*: the (re)construction of a food supply chain as vehicle for regional development. This trajectory is primarily driven by public or societal concerns over sustainable regional development and is usually initiated by public-private partnerships as a broader strategy of strengthening synergies

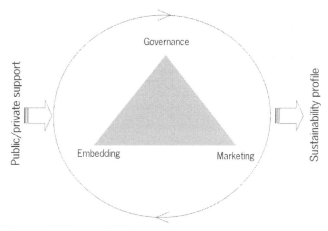

Figure 1. Analytical framework (Roep en Wiskerke, 2006).

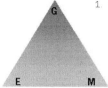

1. Chain innovation
   - Key objective: to strengthen the bargaining power and commercial position of farmers in the food supply chain.
   - Initial focus: constructing a new food supply chain through designing, developing and implementing new forms of chain governance
   - Initiators: often farmers aiming to improve their livelihood.

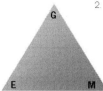

2. Chain differentiation
   - Key objective: improving the commercial performance of an existing food supply chain.
   - Initial focus: developing and marketing more distinctive products alongside existing, well established products.
   - Initiators: often highly influential chain captains or directors (usually processors or retailers) aiming to improve the competitive position of their firm.

3. Territorial embedding
   - Key objective: to (re-)construct a food supply chain as vehicle for sustainable regional development.
   - Initial focus: strengthening interlinkages and creating coherence and synergies between food supply chains and other economic activities in the region.
   - Initiators: often public-private partnerships aiming to address public/societal concerns regarding sustainable regional development.

Figure 2. Three trajectories for constructing sustainable food supply chains (Roep and Wiskerke, 2006).

between food production, consumption and regional economic development. Initiatives have met with varying levels of success in mobilising and actively involving regional food chain and institutional actors in the reconstruction of governance, embedding and marketing interrelations in order to enhance regional sustainable development.

It is important to emphasise that a sustainability trajectory always combines three components: governance, embedding and marketing (G+E+M). Different types of trajectories involve different configurations of these three components. However, each type also has its specific focal or starting point (see Figure 2). The first type, chain innovation, departs from the dimension of governance, while chain differentiation and territorial embedding depart from marketing and embedding respectively. Yet, regardless of the original starting point, the more developed and balanced these three dimensions are, the better the performance of an initiative.

Relevance of framework and trajectories

To conclude this paper we will briefly reflect on the relevance of the analytical framework and the different sustainability trajectories for practitioners, policy-makers and researchers. For a more elaborate reflection we refer to Roep and Wiskerke (2006).

Differences in sustainability profiles

An indicative, integral assessment of the contribution of such initiatives to sustainable rural development demonstrates that their effects differ significantly. This is illustrated in the sustainability profiles within each case study (see Roep and Wiskerke, 2006). The differences may partly mirror the success of the development to date, but they are also related to differences in drive and in scope, which are underpinned by differences in values and trade offs between objectives. Just as there is no single measure for sustainability, so there is no single road to sustainability. Different trajectories result in different profiles and different contributions to sustainable rural development.

One of the findings in this respect is that direct and regional marketing initiatives do generate additional income and employment for rural areas, although the degree to which they do so differs. In addition they enable synergies with other rural development activities, such as rural tourism. In more marginal areas, these benefits can help counter the abandonment of agriculture, out-migration and 'greying' populations. Furthermore, they often contribute to an increase in job satisfaction and organisational capacity within rural communities, greater consumer trust in food systems, and reductions in food miles or waste.

Policy tool

Policy is about making choices: who and what to support, and how to provide this support in the most effective way. A great range of instruments is available for creating a more favourable environment for the development of FSCs. While public support is often important, it is also crucial to note that not all initiatives *depend* on public support. It is also important to remember that support is not only financial, but can also come in other forms. We can identify a number of different types of support: financial, (e.g. through investment or start-up finance); marketing, information and public relations; advocacy and public legitimisation of the initiative, brokering; training and consulting; and technical and legal support for innovative and experimental approaches. Finally, it is not only public bodies that can act as a source of support; social organisations, communities, individuals and (actual or potential) trading partners are also potential sources of support.

The question of how to provide effective support in the most efficient way comes back to issues of identifying the type of support needed, and providing it in the right amount and at the right time. The GEM-framework (see Figure 1) allows for a better understanding of development opportunities, constraints and risks faced by different types of alternative FSCs at different stages in their development. This framework provides a tool that could prove of use in helping improve the targeting of support.

Analytical tool

The conceptual framework (the GEM framework) allows a better understanding of how sustainable chains are constructed. By using this conceptual framework it is possible to better conceptualise different types of alternative FSCs at different stages of their development. It posits that a sustainability trajectory always involves a combination of Governing, Embedding and Marketing (GEM). Different types of trajectories can be formulated that reflect different configurations of these three aspects. The analytical framework also intends to demonstrate how each type of sustainability trajectory has a specific performance in terms of sustainability, in terms of its impact on rural development as well as commercial performance, marketing and communication, etc. Particular types of trajectory require specific kinds of public or private support to enhance their sustainability performance and enable them to meet their full potential.

In initiatives that have their starting point in *chain innovation* the key objective is to strengthen the bargaining power and commercial position of farmers in the food supply chain. The focus of related research and advisory work should be on the development of the most suitable forms of chain governance. Key questions are how to mobilise strategic alliances and to build strong support networks that create a protected space, or niche, for experimenting and learning. In initiatives where *chain differentiation* is the most characteristic feature, emphasis is typically on improving the commercial performance of a particular organisational configuration. The key questions then are those of how to develop and market more distinctive products (or a range of products) alongside existing, well established ones. Initiatives that are mostly characterised by a high level of *territorial embedding* often aim to (re-)construct a food supply chain as a vehicle for sustainable regional development. An important question in related research and advisory work is how to strengthen the inter-linkages and to create coherence and synergies between food supply chains and other regional economic activities. The role of public-private partnerships that contribute to a sustainable development of 'their' region is often a key issue that needs to be addressed.

Reflexive tool

The framework can also be used as reflexive tool for practitioners and their supporters, one that can help them to position themselves, develop a clear strategy, find the right allies, develop their skills and build the capacities that they need. The framework can not only help practitioners to find the right road, but also to travel along it well equipped. It also has great relevance as a policy tool for politicians and policy makers, to improve the strategic choices that they have over what needs enhancing and how that can best be done through developing better and more targeted policy instruments.

References

Berger, G. (2003). Reflections on governance: power relations and policy making in regional sustainable development. Journal of Environmental Policy and Planning 5: 219-234.

Granovetter, M. (1985). Economic Action and Social Structure: The Problem of Embeddedness. American Journal of Sociology 91: 481-510.

Kirwan, J. (2004) Alternative strategies in the UK agro-food system: interrogating the alterity of farmers' markets. Sociologia Ruralis 44: 395-415.

Kirwan, J., Slee, R.W., Foster, C. and Vorley, B. (2004). Macro-level analysis of food supply chain dynamics and diversity in Europe - synthesis report. www.sus-chain.org

Marsden, T., Murdoch, J. and Morgan, K. (1999). Sustainable agriculture, food supply chains and regional development: editorial introduction. International Planning Studies 4: 295-301.

Renting, H., Marsden, T.K. and Banks, J. (2003). Understanding alternative food networks: exploring the role of short food supply chains in rural development. Environment and Planning 35: 393-411.

Roep, D. and Wiskerke, J.S.C. (2006). Nourishing networks: fourteen lessons about creating sustainable food supply chains. Reed Business Information, Doetinchem, 176p.

Sonnino, R. and Marsden, T.K. (2006). Beyond the divide: rethinking relationships between alternative and conventional food networks in Europe. Journal of Economic Geography 6: 181-199.

Watts, D.C.H., B. Ilbery and D. Maye (2005) Making reconnections in agro-food geography: alternative systems of food provision. Progress in Human Geography 29: 22-40.

Ethical use of Andean tomato germplasm

Bart Gremmen
Wageningen UR, Laboratory of Plant Breeding, Droevendaalsesteeg 1, 6708 PB, Wageningen, The Netherlands, Bart.Gremmen@wur.nl

Abstract

During many centuries, the development of new local varieties was a national affair in the public sector. Farmers used the abundant agricultural biodiversity they found in their own region. Starting in the sixteenth century, many crop varieties were taken by explorers to their own countries, and there has been also a world-wide loss of agro-biodiversity. Crops like maize, potato and tomato were taken, and used by national private sectors. In the last century this was considered to be bio-piracy. International treaties, like the Seed Treaty were developed to protect biodiversity, and international Intellectual Property right, like TRIPs, was developed to protect private property of plant genetic resources for food and agriculture. In 2005 these protective frameworks collided in the case of a proposed genomics research project involving a tomato collection from Ecuador and the Dutch Centre for BioSystems Genomics. The aim of this paper is to explore the ethical background of the protective legal frameworks. Special attention will be paid to Corporate Social Responsibility as a possible way to align the protective frameworks of biodiversity and intellectual property rights.

Keywords: biodiversity, intellectual property rights, corporate social responsibility

Introduction

Since the dawn of agriculture 12,000 years ago, humans have nurtured plants to provide food. Careful selection by farmers resulted in a myriad diversity of varieties of the relatively few plants we use for food - our agricultural biodiversity. This activity is now under threat by two kinds of loss of the necessary access to plants: on the one hand by loss of legal access to plants, on the other hand by the huge loss of food crop varieties. The European tradition of honouring intellectual property rights (IPR's) was installed internationally to prevent bio-piracy (Shiva, 1997). This protective framework intends to prevent people from commercially exploiting ideas or inventions without fair compensation to the originators. However, it not only caused a loss of legal access to plants for the poor farmers, it also clashes with the protective framework of the Biodiversity Treaty that tries to limit the loss of biodiversity.

The aim of this paper is to explore the ethical background of these protective frameworks. Questions are:
- What are the international legal frameworks on agricultural biodiversity and IP about?
- How are the legal frameworks on agricultural biodiversity and property rights connected?
- What are the main bottlenecks in the commercial use of germplasm in genomics research?

Two kinds of losses

In only a few years time farmers from all over the world have experienced a loss of legal access to plants. In the past centuries plant breeders and researchers, mainly from the Western world, took many plants from the centres of origin, and placed them in private gene-banks, what is nowadays condemned as bio-piracy. Recently, as part of an ongoing process of globalisation, IPRs slowly bring about a shift from public governance of seeds to private governance of seeds. Patents on seeds would restrict farmers' ability to grow the food that underpins food security and environmental integrity. Patents also provide the mechanism by which agribusinesses are able to control these resources.

IPRs are secured both under national legislation and under international law, primarily, the Agreement on Trade-related Aspects of Intellectual Property Rights Agreement (the TRIPs Agreement) concluded as part of the Uruguay Round of the General Agreement on Trade and Tariffs (GATT). The TRIPs Agreement requires countries to enact national legislation which, among other things:
- extends patenting to micro organisms and 'modified' forms of life;
- provides patents or other forms of property rights on plant varieties;
- extends patent rights over pharmaceutical products;
- increases the duration of protection for patents to 20 years from the date of the application.

The TRIPs Agreement is seen by many in the Third World, and especially by indigenous peoples, as a 'strategy to foster a form of technological protectionism', or a method of 'freezing' the comparative advantages of the developed countries vis-à-vis the developing world (Gerster, 1998). Another major objection to TRIPs is that the Agreement fails to recognize and protect the contribution of indigenous peoples, farmers and local communities to the innovation or product patented.

The second kind of loss is the fact that 95% of food crop varieties of the locally developed agricultural biodiversity have been lost from farmers' fields in a relatively very rapid pace. The loss of the wide diversity of varieties of these crops will have severe consequences for food security for all mankind given environmental degradation, pests, epidemics and climate change. Moreover, future crop species that could be used directly or modified by biotechnology are lost when entire ecosystems are wiped out. Adressing these threats, international and local actions were taken in the second half of the last century.

The Convention on Biological Diversity (CBD) of 1991, commonly referred to as the Biodiversity Treaty, tries to limit the loss of biodiversity. The treaty defines biodiversity as 'the variability among living organisms from all sources including, inter alia, terrestrial, marine and other aquatic ecosystems and the ecological complexes of which they are part; this includes diversity within species, between species and of ecosystems.' The Convention on Biological Diversity is an international agreement governing the conservation and use of the world's biological resources, whether or not they are endangered or threatened. Its stated goals are threefold:
1. the conservation of biological diversity;
2. the sustainable use of its components; and
3. the equitable sharing of the benefits arising out of the utilization of genetic resources, including appropriate access to resources and transfer of relevant technologies.

The CBD has consequences for access to natural resources, use, handling, and genetic modification of resources, and dictates certain key financial components of business arrangements between countries and their citizens concerning use and ownership of benefits deriving from such resources. One of the most controversial issues addressed by the Biodiversity Treaty is intellectual property right related to biological and genetic resources. The main aim of the CBD was to negotiate environmental agreements, not trade issues. Thus it is not altogether surprising that the negotiators were able to address the issue of intellectual property rights in only a superficial way.

Linking intellectual property rights and biodiversity

The loss of legal access to plants and the loss of biodiversity were connected in 2004, when the International Seed Treaty (International Treaty on Plant Genetic Resources for Food and Agriculture - ITPGRFA) came into force. Its goals are to conserve and sustainably use the genetic resources of the world's food respecting national and international systems of intellectual property rights. It will implement a 'Multilateral System' (as opposed to the existing CBD 'Bilateral System') of access to a list of some of the most important food and fodder crops essential for food security and interdependence for those countries that ratify the treaty. It will ensure that benefits from the commercial use of the public genetic resources of these crops are returned to farmers in developing countries, the original source of most of the resources. Importantly, the treaty will implement farmers' rights to access public genetic resources, to use, save and sell seeds and participate in decision making, although these rights are currently subordinate to national laws (Correa, 2003).

The Treaty on Plant Genetic Resources focuses on patents, while the national systems of 'Breeders' Rights' are left intact. This may become problematic in cases when the private sector doesn't take a patent, and, by referring to a national system of 'Breeders' Rights', refuges to pay for the use of plants of developing countries. But these countries will only give 'their' plants when they will receive a financial compensation.

Shared ethical background

Although ethical concerns and values are prominent in the debate about the world-wide loss of agricultural biodiversity and the globalization of agriculture, there seems to be no consensus about these underlying suppositions. A shared ethical background will help to develop legal and institutional arrangements that will enable the countries by participating in research to protect their germplasm, and, at the same time, make commercial use of this germplasm. To build an ethical background for the analysis of (agricultural) biodiversity and its relation to IP, it is possible to use general ethical theories about justice, responsibility, fairness and equity. Also literature on the societal values of sustainability and biodiversity has to be considered. This ethical background has to ensure that implementation of international treaties, like the Seed Treaty, is:
- just - ensure a level playing field on access rules without any threat of privatisation and bio-piracy, and full international recognition of Farmers' Rights;
- equitable - provide reasonable benefits to poor farming communities in developing countries;
- comprehensive - contribute to keeping the germplasm of all crops and their 'wild' relatives in the public domain.

Corporate Social Responsibility is another possible way to align the protective frameworks of biodiversity and intellectual property rights. In this approach it is argued that companies have a moral obligation to be good citizens and to 'do the right thing' (Porter and Kramer, 2006). It is expected that the commercial success of companies is achieved by honouring ethical values, but in some areas '... the moral calculus needed to weigh one social benefit against another, or against its financial costs, has yet to be developed.' (Porter and Kramer, 2006, 3). Companies may not only be held responsible for poor nutrition, but also for a strategic switch from patents to breeders' rights. In 2005 this was the case of a proposed genomics research project involving a tomato collection from Ecuador and the Dutch centre Centre for BioSystems Genomics (CBSG).

The Loja tomato collection case

The National University of Loja (UL) in Ecuador has a collection of tomatoes (between 600-800 lines) which has been collected in the past in a region extending from the northern part of Chile till the southern part of Colombia. This collection is protected by different international treaties on biodiversity because it is a public collection. CBSG wanted to use this collection for genomics research activities, and insisted on a legal and sound procedure which takes all international treaties on biodiversity into account before taking the tomato germplasm to the gene bank in Wageningen (CGN). CBSG also explored the possibility to make this collection available to the Dutch vegetable breeding companies, represented by Plantum, the Dutch association for breeding, tissues culture, production and trade of seeds and young plants.

All parties involved tried to reach an agreement, in which the companies will get the right to use results in commercial breeding programmes. In case of patenting by CBSG and/or participating companies, UL will share the revenues. The Ecuadorian Ministry of Environment will only give a permit for the use of Ecuadorian germplasm for taxonomic research purposes. In case of commercial genomics research, Ecuador, as the rightful owner of the germplasm, insists on receiving financial compensation under Ecuadorian law to be able to protect their agricultural biodiversity.

Although the companies were aiming at using the results in commercial breeding programmes, they switched from a focus on patents to breeders' rights. In that case the companies do not have to pay any financial compensation to Ecuador according to Dutch law on breeders' rights, which the Treaty on Plant Genetic Resources has left intact. Breeders' rights will allow Ecuador to use the new cultivars of the Dutch companies, but this is of limited value because Ecuador has no infrastructure to do this, and using seed of hybrid plants is not profitable. So far the possible transfer of technologies, an instrument to bridge the difference in capacities, has not received its due attention. An extra complication is that only part of the tomato germplasm originates from Ecuador.

Conclusion

Scientists insisting on legal and sound procedures, which take all international treaties on biodiversity into account, are trying to enable a responsible commercial use of genomics research in a context of expanding globalization. However, they are confronted with colliding protective legal frameworks and strategically acting companies. Development and exploration of Corporate Social Responsibility in relation to the International Seed Treaty will help to protect the germplasm of countries in the South, and, at the same time, make commercial use of this germplasm in genomics research. However, the following issues will require clarification:

- Will the International Seed Treaty allow new crop varieties or genes from food crops, if extracted, transformed or modified and included in new varieties, to be patented and be subject to other intellectual property rights claims?
- The spread of patented genes in the environment could undermine Farmers' Rights. Will the 'Material Transfer Agreement' (MTA) that has to be developed be equitable and protect crop genetic resources from privatisation?
- Will the International Seed Treaty be recognised as the competent authority to deal with plant genetic resources for food and agriculture by the World Trade Organisation and especially its Agreement on Trade Related aspects of Intellectual Property Rights (TRIPs) with respect to these resources?
- Will the International Seed Treaty provide benefits and funding commensurate with the contribution that farmers in different countries have made over past centuries to the development of the diversity of crops?

References

Correa, C. (2003). The access regime and the implementation of the FAO International Treaty on Plant Genetic Resources for Food and Agriculture in the Andean Group countries. The Journal of World Intellectual Property, vol. 6, No. 6, November.

Gerster, R. (1998). Patents and development. The Journal of World Intellectual Property, Vol. 1, No. 4.

Porter, M.E. and Kramer, M.R. (2006). Strategy and Society, The Link Between Competitive Advantage and Corporate Social Responsibility. Harvard Business Review, December.

Shiva, V. (1997). The Plunder of Nature and Knowledge. Research Foundation for Science, Technology and Ecology, New Delhi.

A legal examination of the requirements relating to developing countries' exports of organic products to the European Union

Morten Broberg
University of Copenhagen, Faculty of Law, Studiestraede 6, DK-1455 Copenhagen K. and Danish Institute for International Studies, Strandgade 56, DK-1401 Copenhagen K., Denmark

Abstract

When producers of organic agricultural products in the developing countries attempt to export to the EU they meet considerable legal obstacles in the form of the EU's Organic Regulation. This paper provides a legal analysis of these obstacles. First it examines the legal basis of the Organic Regulation. Next it examines whether the requirements of the Organic Regulation conform to international law. Following this, the paper examines the Regulation in the light of the proportionality principle, the Community's environmental obligations and its obligations in the field of development policy. While it is not possible to establish that the Regulation is unlawful under EU law, the paper recommends further examination of two points, namely the Regulation's legal basis and its conformity with the principle of proportionality.

Keywords: organic food, European Union, import requirements, lawfulness, developing countries

Purpose of this paper

Organic food products have evolved as a substantial new market over a short period (European Commission, 2001). A product sold as 'organic' attains a price premium compared with non-organic food products (Gibbon, 2006a). The EU is one of the world's largest import markets for organic agricultural products (OECD, 2002). In the EU, agricultural products may only be sold as 'organic' if they comply with Council Regulation (EEC) 2092/91 on organic production of agricultural products and indications referring thereto on agricultural products and foodstuffs (the 'Organic Regulation'). This is also the case for organic agricultural products originating outside the EU. Under the Organic Regulation, third country products may be imported either under a 'compliance guarantee' or under a significantly more burdensome 'equivalence guarantee'. The importation regime that has hitherto been in force lists only seven countries, none of them African (Commission Regulation 94/92). It is clear that African producers of organic agricultural products are faced with very real barriers if they want to export to the EU and attain the price premium that comes with the use of the organic label (Forss and Sterky, 2000; Courville, 2006).

This paper provides a legal analysis of whether the requirements of the Organic Regulation which developing countries' exporters must meet when selling to the EU conflict with superior legal principles of Community law. The limits of the paper do not allow for a detailed evaluation of the EU requirements relating to organic production.

The following section contains a brief presentation of the Organic Regulation. In Section 3 there is a more detailed definition of the problem. Section 4 provides a legal examination of the import requirements. Finally there is an identification of those superior principles of Community law which it may be argued run counter to the import requirements.

The Organic Regulation

The Organic Regulation was adopted in 1991 and has since been frequently amended. A proposal to revise the Regulation was published in December 2005, and a new regulation is expected to be adopted sometime during 2007. The 1991 Organic Regulation provided a special scheme for imports from third countries. This scheme was to expire by the end of 2006, but on 21 December 2006 a new scheme was introduced as an amendment to the Organic Regulation. The new scheme took effect from 1 January 2007 and is expected to be carried over in the revised regulation, or at least its main features (European Commission, 2005).

Under the new scheme, in order for agricultural products from third countries to be marketed as organic, the amended Organic Regulation provides two alternative routes. One is based on a 'compliance guarantee', and the other is based on an 'equivalence guarantee'.

The compliance guarantee provides the more direct access to the EU market for organic food products. The requirements of the guarantee may be fulfilled either by the export being made from a country that has been recognised by the Commission as providing a system of production that complies with rules equivalent to those in the Organic Regulation, or they may be fulfilled by the product complying with the provisions of the Organic Regulation and all operators submitting their activities to an inspection body recognised by the Commission and accredited to a specific European standard and/or a specific ISO standard.

If it is not possible to qualify under the compliance guarantee, a producer may still export food products under the organic label, provided the products have been produced in accordance with production standards equivalent to the production rules laid down in the Organic Regulation. 'Equivalence' means that the production may have been carried out according to standards that differ from EU standards, provided these other standards are considered equivalent (Borowsky-Kyhos, 2003). There are stringent requirements for the inspections to which the operators must submit. The formal requirements associated with exporting under the equivalence guarantee are considerable. For example, organic products exported to the EU must be accompanied by an original certificate of inspection issued by the competent authority or inspection body of the exporting country. The importer must keep this certificate available to the inspection body or inspection authority for at least two years.

Comparing the EU's new approach to equivalence with the one hitherto in force, Gunnar Rundgren has characterised it as 'a step in the right direction, but still not enough' (2006).

The problems relating to imports of organic products

In developing countries agricultural products are often organic by default, simply because the producers lack the necessary means to apply non-organic production methods such as artificial fertilisers and pesticides (De Vylder, 2007; Rundgren, 2003). On the face of it, the evolving European market for organic food products could mean that what at first has been a disadvantage could instead be turned to advantage if developing countries' producers could sell their products as 'organic' on the EU market (Twarog and Vossenaar, 2003). First, they would be able to claim the organic premium, thereby improving their standard of living. Second, the premium would act as an incentive to continue to use organic production methods to the benefit of the environment.

However, as noted above, the EU requirements for imports of organic products are a significant barrier. The requirements are far-reaching. For example, the Organic Regulation, Annex III, point 5, in its amended version provides that '[t]he inspection body or authority must make a full physical inspection, at least once a year, of all operators ...'. To a Ugandan smallholder the cost of such an annual inspection is a significant burden (Gibbon, 2006b; Rundgren, 1999). However, alternative systems have been created to overcome this problem, such as group certification (Axelsson Nycander, 2000; OECD, 2002), but the Organic Regulation does not allow for these.

Put differently, it may be argued that the EU's requirements tend to confound the sustainable economic and social development of the developing countries and frustrate their integration into the world economy.

Taking a legal approach

Non-lawyers are often surprised to learn that it is possible to challenge the legality of a legal measure such as a Community regulation. This is nevertheless the case if, for example, the legal measure conflicts with a rule of superior rank in the legal hierarchy (e.g. a regulation that conflicts with a Treaty or a fundamental principle of Community law). If a legal measure does not conform with superior law, it may be challenged before the European Court of Justice which may decide to overturn it in whole or in part. In the following there is a review of the possible conflicts between the Organic Regulation and superior law.

The legal basis

Under Community law it is of considerable importance to establish the basis for a legal measure such as a regulation. First, Article 5, first sentence, of the EC Treaty provides that '[t]he Community shall act within the limits of the powers conferred upon it by this Treaty and of the objectives assigned to it therein'. In other words, the Community may not act if there is no legal basis for doing so. Second, choosing the right legal basis is important since the different legal bases provide for different procedures for the adoption of legal acts. This means that choosing the wrong legal basis may result in using the wrong procedure for adoption.

The Organic Regulation was adopted on the basis of Article 37 of the Treaty which forms part of the Community's provisions for the common agricultural policy. This is not surprising since the Organic Regulation is concerned with organic agricultural products. However, the question is whether regulating agriculture is the Regulation's only main objective. It is clear from the Regulation's preamble and from its Article 12 that another important objective is to harmonise the rules on organic production so as to ensure the free movement of organic products within the Community (Scharpé, 2003). In this context it is worth pointing to the objectives which the Commission has identified for the proposed new organic regulation (European Commission, 2005). In the explanatory memorandum accompanying this proposal, the Commission points to the protection of consumer interests, the development of organic production, the protection of the environment and respect for animal welfare.

Thus, both the original Organic Regulation and the current proposal clearly go beyond the scope of the common agricultural policy, and both the objective of approximating laws in order to establish the internal market and the objective of protecting consumer interests must normally be pursued on the basis of Article 95 of the Treaty. Hence, there are solid reasons for arguing that the existing Organic Regulation should have had both Article 37 and Article 95 of the Treaty as its legal bases, and that the same is true with regard to the current proposal for a new regulation (European Parliament, 2006). The inclusion of Article 95 of the Treaty as legal basis would vest important powers in the European Parliament - powers which it does not have under Article 37. Applying the wrong legal basis can justify annulment by the European Court of Justice.

Conformity with international law

The European Community is bound by international law (Broberg, 2000). This means that the Community may only impose conditions on parties outside its territory to the extent that this does not conflict with the international law on jurisdiction. The question therefore arises as to whether the fact that producers in third countries may only sell into the Community if they fulfil the requirements laid down in the Organic Regulation constitutes an extraterritorial procedural requirement that contravenes international law? In the view of this author, it does not. The Community does not require third country producers to fulfil specific requirements, it merely requires that if third country producers want to sell products in the Community under the label 'organic' they must comply with the Organic Regulation. This is not an extraterritorial application of the Regulation.

A different question is whether the requirements conform to WTO law. Some high profile WTO rulings have given rise to debate as to whether, under the WTO's GATT rules, it is illegal to distinguish between products on the basis of their production methods. The matter remains unclear (Howse and Regan, 2000), but it seems unlikely that a distinction between organic and non-organic products will be considered inadmissible under WTO law.

The principle of proportionality

Article 5, third sentence, of the Treaty establishes proportionality as a general principle of Community law. The principle has three cumulative conditions. Only if all three conditions been met, will a Community measure be proportionate - and thus lawful. The three conditions are:
- The measure must be appropriate for attaining the objective.
- The measure must be necessary, meaning that no other less restrictive measure is available.
- The measure must not be disproportionate to its aims.

In order to establish whether the Organic Regulation complies with the principle of proportionality in respect of imports from developing countries, the Regulation's objective must first be established. The Regulation itself states that it shall first of all contribute to the reorientation of the common agricultural policy, to the protection of the environment and to securing a common market with fair competition between organic producers.

The question is whether the Regulation's requirements that apply to producers in developing countries meet the three conditions. There is little doubt that, in a legal sense, the measures are appropriate. But are they really the least restrictive? And are they not disproportionate if their consequence is that many developing countries' producers are barred from applying the organic label to products that undoubtedly are organic? In order to answer these questions it is necessary to consider what weight may be attributed to the interests of non-EU producers. If the interests of third country producers may be given weight, the fact that substantially less restrictive but efficient control systems do exist (Axelsson Nycander, 2000; Rundgren, 1999) will be a strong argument that the Organic Regulation contravenes the proportionality principle.

Protecting the environment

In Community law, protection of the environment is an overriding aim, as reflected in Article 6 of the Treaty. Promoting organic production is in itself a fine example of protection of the environment, but it is clear that the EU rules are not designed to support the development of organic farming globally (Axelsson Nycander, 1999). Indeed, one could argue that the Organic Regulation's very strict requirements - not least with regard to imports under the equivalence guarantee - discourage producers in developing countries from producing organically. Even if this is the case, taking a general view of the Regulation there is no doubt that it works to the benefit of the environment. In this context, the problems of organic producers in developing countries when exporting to the EU are likely to be considered unimportant.

The Community's development policy

Articles 177 and 178 of the Treaty together require that in its policies the Community must take account of the sustainable economic and social development of the developing countries, and more particularly of the least developed countries. Moreover, the Community must foster the smooth and gradual integration of the developing countries into the world economy as well as the campaign against poverty in the developing countries.

According to a strict reading of these two articles, the Community must pay particular attention to the needs of the developing countries when, for example, laying down the requirements for selling organic products in Europe. In practice, outside the field of development policy these two articles appear to have been almost completely overlooked by the Community and it seems unlikely they will be given much prominence in the foreseeable future.

As part of the Community's development policy, a scheme of generalised tariff preferences (European Council, 2005) has been established giving the developing countries preferential access to the Community market (i.e. the developing countries may export their products to the Community at a reduced customs tariff). Part of this scheme provides that developing countries that have ratified and effectively implement a number of international treaties on human rights, employment rights, good governance and on the protection of the environment may be given particularly favourable treatment. In other words, the scheme encourages the developing countries to protect their environment. These efforts could have been furthered if the Organic Regulation had encouraged organic production by making it easier to sell into the EU. The fact that, rather than encouraging organic exports, the Regulation creates obstacles to them, reveals an inconsistency between the two policies. Whilst inconsistencies should be avoided, they do not in themselves mean that either of the policies is unlawful.

Conclusion

As has been shown above, the Community's conditions for imports of organic agricultural products cause considerable problems to exporters, especially those in developing countries. Whilst this analysis has shown that the Organic Regulation is at variance with a number of superior legal principles, there is no basis for concluding that the Regulation could be successfully challenged for a breach of these principles. However, two problems have been singled out as meriting further consideration, namely: (1) the Organic Regulation's legal basis, and (2) the Regulation's conformity with the principle of proportionality.

References

Axelsson Nycander, G. (1999). Council Regulation 2092/91: Consequences for the import of organic products from developing nations, a report produced on behalf of Sida, Swedish International Development Cooperation Agency, (English version 2000).

Axelsson Nycander, G. (2000). Toward a Level Playing Field in Organic Farm Trade, Bridges - Between Trade and Sustainable Development 4: 15-16.

Borowski-Kyhos, H.-G. (2003). Imports under the regulatory system of EEC-regulation No 2092/91 - Contribution from a German länder authority (Baden-Württemberg). In C. Westermayer and B. Geier (eds.): The Organic Guarantee System - The need and strategy for harmonisation and equivalence. IFOAM, Germany, pp. 118-121.

Broberg, M.P. (2000). The Court of First Instance's Judgment in Gencor v Commission. International and Comparative Law Quarterly 49: 172-182.

Courville, S. (2006). Organic standards and certification. In P. Kristiansen, A. Taji and J. Reganold (eds.): Organic Agriculture - A Global Perspective, CSIRO Publishing, Australia, pp. 201-219.

Forss, K. and Sterky, E. (2000). Export Promotion of Organic Products from Africa - An evaluation of EPOPA, Sida Evaluation 00/23, Sweden.

Gibbon, P. (2006a). Decoding Organic Standard-Setting and Regulation in Europe (1991-2005), United Nations Industrial Development Organization, Vienna, Austria.

Gibbon, P. (2006b). An Overview of the Certified Organic Export Sector in Uganda, Danish Institute for International Studies Working Paper No 2006/13.

Howse, R. and Regan, D. (2000). The Product/Process Distinction - An Illusory Basis for Disciplining 'Unilateralism' in Trade Policy. European Journal of International Law 11: 249-289.

OECD (2002). The Development Dimension of Trade and Environment: Case Studies on Environmental Requirements and Market Access, COM/ENV/TD/2002)86/FINAL, France.

Rundgren, G. (1999). The Challenge for Developing Countries to Establish an Organic Guarantee System, paper given at the 6th IFOAM Trade Conference in Florence, October 1999.

Rundgren, G. (2003). Organic - Making markets work for developing countries, May 2003, available at www.grolink.se.

Rundgren, G. (2006). Organic certification today and in the future, paper presented at the IFOAM Certification Conference in Rome, November 2006.

Scharpé, A. (2003). The EU Regulation. In C. Westermayer and B. Geier (eds.) The Organic Guarantee System - The need and strategy for harmonisation and equivalence, IFOAM, Germany, pp. 24-29.

Twarog, S. and Vossenaar, R. (2003). Obstacles Facing Developing Country Exports of Organic Products to Developed Country Markets. In C. Westermayer and B. Geier (eds.) The Organic Guarantee System - The need and strategy for harmonisation and equivalence, IFOAM, Germany, pp. 122-128.

De Vylder, S. (2007). The Least Developed Countries and World Trade, 2nd edition, Sida Studies No 19, Sweden.

EU legislation and reports

Council Regulation (EEC) No 2092/91 on organic production of agricultural products and indications referring thereto on agricultural products and foodstuffs, OJ 1991 L198/1.

Commission Regulation (EEC) No 94/92 of 14 January 1992 laying down detailed rules for implementing the arrangements for imports from third countries provided for in Regulation (EEC) No 2092/91 on organic production of agricultural products and indications referring thereto on agricultural products and foodstuffs (with later revisions), OJ 1992 L11/14.

European Commission (2001): Organic Farming - Guide to Community rules, Office for Official Publications of the European Communities, Luxembourg.

Proposal for a Council Regulation on organic production and labelling of organic products - Proposal for a Council Regulation amending Regulation (EEC) No 2092/91 on organic production of agricultural products and indications referring thereto in agricultural products and foodstuffs (presented by the Commission), (COM(2005) 671 final).

Council Regulation (EC) No 980/2005 of 27 June 2005 applying a scheme of generalised tariff preferences, OJ 2005 L169/1.

Council Regulation (EC) No 1991/2006 of 21 December 2006 amending Regulation (EEC) No 2092/91 on organic production of agricultural products and indications referring thereto on agricultural products and foodstuffs, OJ 2006 L 411/18.

Part 5 - Sustainability and animal welfare: animal welfare and basic value

Vegan agriculture: animal-friendly and sustainable

Tatjana Visak
Ethiek Instituut, Utrecht University, P.O.Box 80103, 3508 TC Utrecht, The Netherlands

Abstract

There is a broadly shared idea that animals have some moral status and that inflicting unneccessary suffering to them is wrong but that this is compatible with keeping and killing them for food. The idea that animal husbandry is sanctioned when so-called welfare considerations are taken into account can not be supported by any plausible normative theory. Every plausible theory in the field of animal ethics comes to the conclusion that animal husbandry is not compatible with the moral status that animals are due. If a vegan agriculture is indeed the most animal-friendly option, it is important to check how it scores in terms of sustainability, because sustainability is another moral aim or prerequisit in relation to agriculture. I will argue that vegan agriculture is not only animal-friendly, but also sustainable.

Keywords: animal welfare, animal husbandry, vegan agriculture, animal ethics

Introduction

In the Dutch societal debate about the future of agriculture, it is broadly agreed that animal-friendliness and sustainability must be major aims or prerequisites in the transformation of the sector. However, when concrete policy proposals are discussed, it becomes obvious that people have only vague and/ or contradicting ideas about how those aims should be understood and what they would mean in practice.

In my PhD-thesis, I turn to normative theories in order to reveal the normative basis and practical implications of the moral aim of animal-friendliness. After having determined what an animal-friendly agriculture would be, I turn to investigate whether that kind of agriculture would also be a sustainable one. In this paper, I will discuss some questions and preliminary results of my research. I will first focus on animal-friendliness, then on sustainability.

Animal-friendly animal production: a contradiction in terms

'Animal-friendliness' is the literal translation of a term ('diervriendelijkheid') that is used in the Dutch debate. To some, including myself, 'animal-friendly animal production' might sound as a contradiction in terms. Yet, it indicates the broadly shared idea that we have certain moral duties towards animals, such as not causing unnecessary suffering, but that what we owe to animals is compatible with keeping and killing them for food. A huge majority of Dutch people, for instance, claims to oppose factory farming. Recently, 102.726 signatures have been collected in a citizen's initiative under the heading of 'stop wrong meat' in order to force Parliament to reconsider the issue of intensive animal husbandry and hopefully speak out against it. A new political party, the Party for the Animals, found its way into Dutch parliament. In its radio spots it presents meat consumption as a personal choice, but people should take welfare considerations into account when choosing where to buy their meat. In Dutch law, inflicting unnecessary suffering to animals is forbidden and animals are even acknowledged to have an intrinsic value, and at the same time animals have the status of property, and all kind of animal use, including animal husbandry, is sanctioned. Many people say they have no problems with meat eating, as long as it is produced responsibly, that means without 'unnecessary suffering' and as long as the animals have had a good life. Even this sub-theme of the conference, which is 'Sustainability and animal welfare in animal production' fits in this line of thinking: there is a concern with animal welfare which does not question the practice of animal husbandry as such. The central question of my PhD-thesis is whether this position is defensible. I want to find out whether animal husbandry can be morally sanctioned at all.

I discuss diverse approaches in animal ethics on that issue, because in that discipline, the question what we owe to animals is systematically inquired. I wonder whether any plausible approaches in animal ethics would sanction the keeping and killing of animals for food. I have not yet finished this task, but my preliminary conclusion is that the idea that animals do have moral status in combination with the idea that we can use them in animal husbandry can not be supported. For instance, the animal rights theories of Regan (2004), Nussbaum (2006), Pluhar (1995) and Wolf (2004), as different as they are, all condemn the use of animals for food. The utilitarian theory of Peter Singer, as well as the 'reflective equilibrium approach' of De Grazia (1996), also oppose animal production. There are some minor approaches that seem to reconcile the moral status of animals with their use as food products, such as Korthals' (2002) deliberative approach and some relational approaches, but those can be quite easily dismissed for weakness of argumentation. If it is true that no plausible normative theory about what we owe to animals sanctions the practice of animal husbandry, then I see three options:

1. If we cannot seriously accept animals as moral objects and still use them for food, then we might conclude that it is wrong to accept them as moral objects in the first place. This option seems to be indefensible, because it has successfully been argued that speciesism is not tenable, for instance in the argument from marginal cases (Singer, 1995 and Pluhar, 1995).
2. We come up with a plausible theory that accepts animals as moral objects and sees no fault with using them for food. I invite someone else to try this, but not me, because it does not seem possible to me.
3. We draw the conclusion that animal husbandry is unacceptable, at least as far as our duties towards animals are concerned. There might be other morally relevant considerations which should be included in an overall moral judgment. In my research, I will shortly discuss other morally relevant considerations, such as the effects of stopping animal husbandry on other practices, on public health, on the economy and on international relations. I give special attention to the effects that an animal-free agriculture would have on sustainability. That is because In the Dutch debate on the future of agriculture, sustainability, along with animal-friendliness is explicitly brought forward by the public and by politicians as an aim in the transformation of the sector (and I think rightly so).

If the argumentation that I have only been able to sketch above is successful, I have at least shifted the burden of proof to those who think that the moral status of animals is compatible with keeping and killing them for food. I have shown that the broadly shared attitude that animal husbandry as such is morally acceptable as long as so-called 'animal welfare' is taken into account, lacks a sound theoretical basis. Therefore, my preliminary conclusion is that an animal-friendly agriculture is a vegan agriculture. Note that I could also have reached that conclusion more directly by applying one of the above-mentioned theories that reaches it. In a way that would have been stronger, if successful, because it once and for all would determine what our duties to animals are and that all who think otherwise are wrong. In another respect, however, the method of argumentation that I have chosen is stronger, because it does not hinge on the acceptance of one particular theory. Yet it succeeds in showing that a broadly shared belief lacks any plausible normative foundation and that until that foundation is given, we should refrain from using animals in agriculture.

Vegan agriculture: a sustainable option

Let us assume, therefore, that we have strong reasons, provided by our consideration for animals, to adopt a vegan agriculture. As animal-friendliness is not our only concern, I will now investigate how vegan agriculture scores in terms of sustainability. Again: I have not fully studied this subject yet, but I will share some tentative considerations.

Vegan agriculture is agriculture without animal production, such as meat, dairy or eggs and without the use of manure that results from animal production. Vegan agriculture is already successfully practiced. I personally know examples of vegan organic agriculture in Austria, Germany and the Netherlands. Vegan agriculture produces plant food, such as grains, vegetables, legumes, herbs, fruit, nuts and beans.

In order to determine how vegan agriculture scores in terms of sustainability, we first need to know what sustainability is. Sustainability is a complex and controversial concept. Roughly, it is concerned with intergenerational justice (what we owe to future people), intragenerational justice (concerned with a just worldwide distribution and eventually redistribution) and ecological justice (whether we have direct duties towards nature). In the 1980's there have been fundamental discussion of those issues and scholars tried to base their ideas on normative theories. Recently, while fundamental discussions have not been solved, the focus is more on pragmatic principles, norms and measurement tools for sustainability. Those pragmatic principles might be linked to underlying theories, at least in the sense that they are compatible with them. Furthermore, despite some underlying controversies, the practical norms that are brought forward show reasonable overlap and are probably more likely to be broadly endorsed. The greatest practical problem concerning sustainability is not that there are no plausible ideas about what is sustainable, but that there seems to be a lack of motivation to act accordingly.

The idea of sustainability is about striking a balance between conservation and development. It is acknowledged that there is a growing consensus among natural and social scientists that sustainability depends on maintaining natural capital (Wackernagel *et al.*, 1999). It is about making use of natural capital in such a way that a practice can in principle be sustained endlessly. If something is called 'capital' that means that it is conceived as having (also) a certain *use-value*. This is acceptable in the case of natural capital, as long as the use-value is conceived broadly enough. The idea is that a certain 'stock' causes a 'flow' of goods that can be used. Natural capital can be defined as the stock of environmentally provided assets such as soil, atmosphere, water, and forests, which provide a flow of useful goods and services.

Useful goods and services that are provided by our natural environment have been categorized in different ways. For instance, the *Total Economic Value* concept includes the following categories of values (Ott and Döring, 2004):
- The Actual Use Value comprises the traditional view of nature as raw material, as well as ecological functions and the so-called 'experience value', referring, amongst other things to the value of nature experiences and the recreational function of nature.
- The Option Value refers to all potential future use-value, for instance the potential future medicinal use of a certain plant.
- The Existence Value refers to the value that humans attach to the mere existence of something (such as animal species or special ecosystems), independent of the use they could make of it. This value is conceived as being contingent on human preferences and is thus different from moral ideas about intrinsic value.
- The Bequest Value refers to the wish people might have to leave natural goods intact for the next generations to inherit. This, again, is not based on moral ideas about intergenerational duties, but on mere preferences of existing people.

This categorization is problematic, because it is only based on actual preferences and not on normative considerations. Despite this shortcoming, the *Total Economic Value* concept illustrates the broad range of use-values that natural capital has.

Use-values are also described as *functions*. Functions refer to the capacity of natural processes and components to provide goods and services that satisfy human needs (directly or indirectly). Functions are also specified under the labels of 'source', 'sink' and 'service'. As further functions 'scenery', 'site' and 'life support' can be added. It is also common to simply talk about the 'ecological capacity' as a summary of all possible functions. The following distinction of four functions will be sufficient for my purpose of offering an idea of what the use-value of the natural capital comprises (Ott and Döring, 2004):

- The Production Function refers to the use of the products of natural eco-systems. Examples are food, raw materials and genetic resources.
- The Regulation Function is a broad heading under which 'climate regulation', 'water regulation', 'soil formation', 'biological control', 'nutrient recycling', 'sink functions' and others can be grouped.
- The Habitat Function refers to ecosystems as a habitat of wild plants and animals and as such is also related to global biodiversity on different levels and as such they have a functional value with respect to maintaining certain species on a sustainable level.
- The Information Function is also a broad heading which refers to nature as a basis for scientific research, discoveries and development, as well as to the possibilities nature provides for recreation and cultural and historical information.

Thus, when I talk about natural capital, I refer to the use that can be made of nature in all those different ways. 'Use', as I made clear, has to be understood in a very broad sense.

Carrying capacity refers to the capacity of the earth or biosphere to provide resources and to absorb waste. In order to be sustainable, the impact of human activities must not exceed nature's carrying capacity. With other words, that means that humanity must maintain the planet's natural capital stocks in order to achieve (strong) sustainability. Maintaining the capital stock means using up not more than the interests without diminishing the stock itself. This amounts to staying within the carrying capacity of the earth. Whether the impact of human activity exceeds the carrying capacity of the earth, can be determined on a global level, but also on a smaller scale, such as national or local level. Now, it should already be clear that agricultural practices can be evaluated with regard to their impact on the carrying capacity or, with other words: with regard to their use of natural capital. In this way it can be determined how sustainable an agricultural practice is.

What about vegan agriculture then? First, I will mention some unsustainable aspects of non-vegan agriculture. Worldwide, two thirds of the arable soil is used for animal husbandry. Meanwhile, the scarcity of arable soil is estimated to become a problem within 50 years from now, when the growing world population needs to be fed. At the moment, there is in principle enough food for everyone and even more, but it would already be impossible to feed everyone with a typical Western diet. There is simply not enough arable soil for that. If the grains and other plant food would be eaten directly by humans, much more humans could be fed. Dairy production, the most efficient animal production, wastes at least 33 to 45 percent of the energy from the food.

A varied vegan diet needs 700 square meters per person. If one third of the calories would be replaced by dairy products and eggs, twice as much land would be needed. The typical Western diet needs five times as much land. Where does all this land come from? Europe imports 70 percent of the proteins for animal feed. People in poor countries are thereby pushed to grow so-called cash-crops instead of food for themselves. This happens in huge monocultures, which leads to depletion of the soil. Worldwide, about 3.4 billion hectares of land is used for pasture. Nearly all of this land could be used instead to grow plants that could deliver fruit and nuts and would have other advantages for nature and wildlife.

Animal husbandry contributes to climate change, pollution, deforestation, soil degradation and loss of biodiversity. The large amounts of manure lead to emission of ammonia and to acid rain and destruction of forests. Methane emissions lead to climate change. Deforestation and land degradation lead to a loss of habitat for wild life which results in a loss of biodiversity. Animal production also needs much more water and energy than vegan agriculture. Meat is rightly considered to be the part of the food parcel that is most harmful to the environment. The production of one kilo of meat needs a hundred times as much energy as the production of one kilo of potatoes.

Vegan agriculture makes less use of nature in terms of 'raw material' and leaves more possibilities for the maintenance and creation of natures other values which I described above as experience value, option value, existence value and bequest value. Vegan agriculture seems to be promising with regard to the sustainable maintenance of the different above mentioned functions of nature. Vegan agriculture is a promising option for feeding the world, now and in the future. It puts less pressure on the earth's carrying capacity in terms of sinks and resources. It needs less arable soil and makes it possible to leave more space for nature and non-human species.

In conclusion, vegan agriculture has important advantages with regard to sustainability as compared to the typical Western agriculture. Furthermore, vegan agriculture scores well in terms of animal-friendliness. My conclusion is a preliminary one and the issue deserves further investigation.

References

De Grazia, D. (1996). Taking Animals Seriously. Cambridge University Press, New York.

Korthals, M. (2002). Voor het Eten. Filosofie en Ethiek van Voeding. Boom, Amsterdam.

Nussbaum, M. (2006). Frontiers of Justice. Disability, Nationality, Species Membership. The Belknap Press of Harvard University Press, Cambridge, Massachusetts, London, England.

Ott, K. and Döring, R. (2004). Theorie und Praxis starker Nachhaltigkeit. Metropolis-Verlag, Marburg.

Pluhar, E. (1995). Beyond Prejudice. The Moral Significance of Human and Nonhuman Animals. Duke University Press, Durham and London.

Regan, T. (2004). The Case for Animal Rights. University of California Press, Berkeley, Los Angeles.

Singer, P. (1995). Animal Liberation. Random House, London.

Wackernagel, M., Onisto, L., Bello, P., Callejas Linares, A., Susana Lopez Falfan, L., Mendez Garca, J., Isabel Suarez Guerrero, A. and Guadalupe Suarez Guerrere, M. (1999). National natural capital accounting with the ecological footprint concept. Ecological Economies 29: 375-390.

Wolf, U. (2004). Das Tier in der Moral. Klostermann, Frankfurt am Main.

Understanding farmers' values

Carolien de Lauwere[1], Sabine de Rooij[2] and Jan Douwe van der Ploeg[2]
[1]*Agricultural Economics Institute of Wageningen UR, Location Leeuwenborch, P.O. Box 35, 6700 AA Wageningen, The Netherlands, carolien.delauwere@wur.nl*
[2]*Rural Sociology Group, Wageningen University, Hollandseweg 1, 6706 KN Wageningen, The Netherlands*

Abstract

This paper deals with the differentiated approaches by means of which pig breeders and dairy farmers try to operate in the highly complex contradictory and contested arena of animal welfare. An empirical and mainly explorative sociological study among 41 pig and dairy farmers was carried out to give insight into farmers' values in relation to animal welfare, sustainability and biotechnology. The study is part of a larger project, which aims to bridge the gap between farmers and society by making farmers' values more explicit in the public debate on issues as animal welfare. As a first step, in depth - sometimes paired - interviews were hold with a varied group of male and female farmers. Farmers' standards for animal welfare vary from 'meeting minimum legislation' and 'good health and good production levels' to 'the ability to express species specific behaviour' and 'respect for the intrinsic value of an animal'. Farmers refer to the significance of different dimensions of sustainable farming; nonetheless, the majority stresses the importance of 'economic results'. Many farmers also consider food safety as a key responsibility (a social aspect). Their judgement of ecological aspects varies: from very important to having no priority. Conventional farmers tend to emphasize more on social wellbeing at micro level, organic farmers on ecological aspects of sustainability and on social responsibility at macro level. Farmers appear to be rather unanimous about the use of animal biotechnology: although animals or their organs may be used for human purposes one should be very careful with genetic modification (and rather not use it) because of unknown and possible risky consequences for human beings and the natural world.

Keywords: farmers' values, animal welfare, sustainability, biotechnology

Introduction

Farm animal welfare is of increasing social concern, especially in the Western world. Farmers are confronted with ever higher standards and pressed to reflect on their production methods Modern agriculture also raises other concerns as its environmental implications, its impact on nature and landscape, food safety, the cost-price squeeze putting pressure on farmer's incomes, consumer prices, working conditions in agriculture, international trade, the growing dependency on external inputs. The concept that links most issues is that of sustainable agriculture (McGlone, 2001) for it aims at an ecologically sound, economically viable, and socially responsible agriculture. Core of the sustainability concept is life quality and the possibilities to maintain this quality, also in the future. Sustainability is thus dependent on social definitions of quality of life, its distribution worldwide (normative standards) as well as on scientific insights in the functioning of human beings and the natural system (RIVM, 2004). Consequently, the concept is used in many different ways ranging from simply 'a new term for responsible environmental and labour management practices' to 'a vast, diverse set of goals, such as poverty elimination and fair and transparent governance' (Marshall *et al.*, 2005). Sustainability is also a complex concept. It is multi-dimensional involving an economic, ecological and social/institutional dimension. Each dimension is furthermore composed of various elements. Ecological sustainability for instance comprises environment (soil, water, and air), biodiversity (animals, nature and landscape), and non-renewable resources. Farmers can choose to focus on one element or on more elements simultaneously. Beyond this, sustainability is a multi level (farm/ community, regional and global level) as well as a multi actor concept: it refers to the involvement of many different actors and institutions. Finally, sustainability is a dynamic concept - it is subject to changes.

Sustainability translates amongst others into animal welfare. The complexity of the concept - multidimensionality, dynamics, etcetera - is, as it were, reflected in the concept of animal welfare. Although at first sight a concept that has a far more limited domain of application, animal welfare is as richly chequered as the wider notion of sustainability. It is equally having several dimensions, different levels, many actors - amongst which the animals- and it is subject to ongoing change. And probably more explicit than is the case with sustainability in general, animal welfare refers to moral values and, especially, to a wide range of moral dilemmas. Whilst the problem of sustainability may have the aura of being resolvable (albeit in a far from easy way), animal welfare seems to coincide with the persistence of ambiguities and ongoing contestations. In the first place because the varying and often contrasting interpretations (among farmers as well as between farmers, consumers, citizens, activists, scientists, politicians, etcetera) of what constitutes farm animal welfare are highly problematic. Secondly, a multidisciplinary approach of farm animal welfare which includes levels of productivity, behaviour, physiology, health and immunity, may seem a good approach (McGlone, 2001), but it cannot avoid emerging frictions between farm animal welfare and other concerns and will remain to cause moral dilemmas. Thirdly, there is the complication that animal welfare can only be assessed in an interpretative way, indirectly. Animals, put in the centre stage in the notion of animal welfare, cannot directly communicate with human beings.

The differentiated approaches (or strategies) by means of which pig breeders and dairy farmers try to operate in this complex contradictory and contested arena of animal welfare entail both discursive elements and specific practices. The first are used to pattern, as it were, the relations with wider society (i.e. offering explanations, justifications or adapted problem definitions), while the former (the specific practices) pattern the process of production, i.e. the interaction with the animals in a specific way. Discursive elements and specific practices are not necessarily congruent. They neither enclose necessarily all relevant aspects, levels and effects. That is, in practice, animal welfare is as much characterized by spheres of visibility and discursiveness, as it is by lack of transparency and unwillingness to dialogue. Although the focus in the paper is on individual farms, it should not be forgotten that institutions play an important, and maybe even decisive (although not yet fully recognized) role in the social construction of animal welfare or the lack of it.

Gap between farmer's values and public perceptions

In today's agriculture great emphasis is laid upon sustainability and social responsibility. Farmers are blamed by society that they subordinate animal welfare, environment and nature to economic interests, and therefore hamper sustainable agriculture. Farmers tend to react defensively by trivializing the problem - 'there is nothing wrong with animal welfare as long as the animals produce' - or shirking their responsibility - 'I would improve animal welfare if consumers would want to pay for it' (Te Velde *et al.*, 2002). This does not bridge the gap between farmers and society. Legislative measures by European governments, imposing handling and husbandry procedures on farmers, do not seem to bring together the different interests either. Farmers argue that regulations affect their competitive position. The rules also seem to be partially in conflict with their view on doing good to animals (De Greef *et al.*, 2006[a]). In the project 'A new ethics for livestock farming: towards value based autonomy in livestock farming?' it is studied whether moral empowerment of farmers can reduce the gap between farmers and society. The underlying thought is that farmers do have moral views underlying their farm management, but that these values are divergent from urban values (Stafleu *et al.*, 2004, De Greef *et al.*, 2006[b]) These differences, it is assumed, could be reduced by making farmers' values more explicit in the public debate on animal welfare. If a mutual understanding would be translated into increased integration of farmer's values into rules and regulations, farmers would be more willing to comply with governmental directives. The project combines sociology and ethics with an interactive stakeholder approach (De Greef *et al.*, 2006ᵃ).

The sociological study is explorative and so far based on literature review and in-depth interviews with respectively 20 and 21 male and female farmers in the pig and dairy sector. Differences in farming styles (amongst others organic and conventional farming) and farm scale were important criteria for selection. The farmers were asked about their thoughts and ideas on animal welfare, good farming practices, how to treat animals well, how to deal with nature, environmental issues, biotechnology and areas of tension between farmers, society, market and technology and the visions behind.

Farmers' values as regards animal welfare

The interview data show that the farmers' views on animal welfare show a considerable variation that ranges from 'meeting minimum legislation' and 'good health and/or good production levels' to 'the ability to express natural behaviour' and 'respect for the intrinsic value of an animal'. This is comparable to findings of Van Huik and Bock (2007, in press). Within this section we will describe the five main value orientations that might be distinguished. The interesting point is that these orientations are to be encountered both among dairy farmers and pig breeders, although the distribution of value orientations differs within the two groups.

1. In a first value-orientation it are non-moral, that is economic and commercial values that are central in both the discussion about and the treatment of animals. The farmers concerned, consider animals first and foremost as means of production that are serving human interests. The production is based on control and modern technology strongly determines the treatment of animals (e.g. housing system). Possible negative effects of the production system are denied, trivialized or imputed to consumers. Some farmers do not comply with minimum standards if these interfere too much with economic goals. This value orientation is wide spread among the majority of conventional pig breeders; it is equally widely shared among (large) dairy farmers.

2. A second value-orientation is partly identical to the first one but simultaneously more ambiguous. It is recognized that current productive systems and technologies might be detrimental for animal welfare. Hence, it is argued that it would be preferable to treat animals different than is currently the case, but this, it is argued, does not 'fit' in the production system and/or is excluded by reigning market relations and price levels. As compared to the first orientation this position opens some room for dialogue and ethical reasoning. This orientation is to be found partly among conventional farmers (both dairy farmers and pig breeders) and partly among those who are more than others oriented - through different institutional arrangements as e.g. nature and landscape conservation, environmental programs, and classification schemes for improved food quality - to sustainability. An important borderline that at the moment separates the first and the second value orientation within dairy farming is summer grazing.

3. In a third value orientation animal welfare is a starting point for the definition of farm management, but it is a somewhat limited and neatly-delineated starting point. Animal welfare is, within this orientation, important in as far as it sustains a specific nice and a premium price for the products. This position is frequent amongst organic and free range farmers for whom economic and commercial values dominate their choice for these styles of farming: within this overall choice they accept and defend animal welfare as being instrumental for their niche market and as marketing instrument. Emphasis is on creating space for animals to express natural behaviour and on the animal's individuality ('let the pig be a pig'; dairy farmers on their turn translate this position in somewhat lower milk yields and in an extended longevity). These values are based on the demands of the production system rather than on ethics.

4. In a fourth value orientation farmers go beyond instrumentalism in as far as animal welfare is concerned. They fully recognize animal welfare as an independent field that might enter into contradiction with e.g. economic requirements or environmental aims. Hence, moral dilemmas and ambiguities do appear here. For example, keeping the bull together with the cows is more natural, but it might be in conflict with the safety of family members or visitors. Family life may also be affected by a higher workload that goes together with the choice for more natural production systems. For some farmers tail docking, castrating male piglets or dehorning of cows is in conflict with respectful treatment of animals and their integrity. Summer grazing may furthermore not go with environmental aims.

5. In and through a fifth value orientation farmers opt in an integral way for an all encompassing sustainability and for the associated animal welfare. Economic values are only considered to be relevant in as far as they are *derived* from an approach that is build on starting points and values that basically differ from the market. A substantial part of the income is therefore raised from other sources. Provision of maximal animal welfare is the starting point and the production process is fully adapted to the needs of the animals. Apart from emphasis on naturalness, respectful treatment of animals, the possibility to express natural behaviour, the animal's integrity and intrinsic value.

An intriguing finding in our research concerned farmers running two units of production (often on different locations). We encountered cases in which one unit was inspired and organized by one value orientation, while the other one corresponded to another orientation. Typical cases are farmers who run a conventional farm and, additional to that, another 'animal-friendly' farm. Thus they apply different animal welfare standards for similar animals within the enterprise units they control. In this case, economic and commercial values are decisive: the main motive for improved animal welfare is the added value of the meat (animal welfare is considered a niche market and used as marketing instrument).The involved professional challenge and social acceptance play a role as well. The production system is partly adapted to animals' needs, that is, as long as it goes with the financial and economic demands.

Farmers in all orientations mention 'good care, good and sufficient feeding, guarding their health, optimal housing, making them feel well, and handling them calmly' as elements of treating animals well. Additionally, farmers of the other orientations mention 'offering them an as good as possible life (within the 'demands' of the production system), no unnecessary suffering and personal attention'. Values as compassion and dedication are considered important. 'Treating animals respectfully, the possibility to express natural behaviour, naturalness and the animal's individuality and integrity' are in particular emphasized by farmers in the organic and biodynamic orientations. Conventional farmers refer to good production results as an important indicator for animal welfare more often than farmers in other orientations, who more frequently mention physical indicators such as the animal's appearance (colour, posture, and glance) and behavioural parameters.

Sustainability

The farmers stress similar as well as different dimensions of sustainable agriculture as leading principle in their farming strategy. They all underline the importance of an economically viable farm except for the farmers who are motivated by ideological reasons. Social sustainability at the level of the farm (work quantity, working schedule, division of tasks, balance work/ family life/ social life) is in particular mentioned by large(r) conventional farmers. At a higher level (e.g. food safety, fair trade, fair prices, poverty reduction), organic and biodynamic farmers feel more responsible to contribute to social equity.

Many farmers (but certainly not all) recognize their responsibility towards the ecosystem. Their support of ecological sustainability is however diverse and differently motivated. They may focus on sustainable cattle, engage in mineral management, nature and landscape conservation and development and/ or energy production. Dairy farmers are more engaged in these activities than pig farmers due to their land based farming practice. It is obvious that organic and biodynamic farmers with basic values as respect for nature, the aim of sustainability and a holistic view (Lund, 2000) are involved in more aspects of ecologically sound agriculture than conventional farmers do.

Biotechnology

Though a lack of knowledge may hinder a grounded judgement, most farmers say they are not in favour of biotechnology in agriculture. They especially reject it when it is driven by profit making or used for raising production levels. Farmers are most critical of animal biotechnology. Only a number of the bigger dairy farmers don't consider it as problematic. Their moral concerns may refer to the intervention in itself or to the effects of the intervention i.e. intrinsic and extrinsic concerns (Kaiser, 2004). The former refer to moral problems with interventions in the basis of nature itself (e.g. one should not touch God's Creation; it is unnatural; one should not violate an animal's integrity). The latter to the unknown consequences and possible risks - for people's health, animal's health, the biodiversity, the ecosystem, the lack of control on what is being created - also in the long run. Except for the organic farmers, farmers seem less reserved to the use of GMO's (Genetically Modified Organisms) for 'plants are a different species, having no feelings' and 'people are already used to GMO's in their foods without visible negative effects'. As possible advantages they mention environmental improvement (reduced use of pesticides) and solving the hunger problem in the world. The lack of free choice ('GMO is forced onto farmers'), the increased dependency on multinationals as well their increasing influence on agricultural production and the negative effects for small farmers (especially in developing countries), are considered as worrisome consequences by part of the farmers.

Concluding remarks

These preliminary results give insight in the diversity of farmers' values as regards animal welfare, sustainability and biotechnology. The values will be validated in an interactive process with farmers (focus groups) and used as input for the ethical analysis of the project mentioned before ('a new ethics for livestock farming: towards value based autonomy in livestock farming?' De Greef *et al.*, 2006a). That material, in turn, will serve as input to answer the question whether it is desirable and possible to acquire a kind of 'professional autonomy' for farmers. Such a mandate would assume acceptance and trust of involved farmers' values by society.

References

Greef, K.H. De, De Lauwere, C.C., Stafleu, F.R., Meijboom, F., De Rooij, S., Brom, F.W.A. and Van der Ploeg, J.D. (2006a). Towards value based autonomy in livestock farming. In: Mattias Kaiser and Marianne Lien (eds.) Ethics and the politics of food. Preprints of the 6th Congress on the European Society for Agricultural and Food Ethics. Eursafe Oslo, Norway, pp. 61-65.

De Greef, K.H., Stafleu, F.R. and De Lauwere, C.C. (2006b). A simple value-distinction approach aids transparency in farm animal welfare debate. Journal of Agricultural and Environmental Ethics 19: 57-66.

Huik, M.M. Van and Bock, B.B. (2007; in press) Attitudes of Dutch Pig Farmers towards Animal Welfare. Forthcoming in British Food Journal.

Kaiser, M. (2005) Assessing ethics and animal welfare in animal biotechnology for farm production. Revue Scientifique et Technique 24(1):75-87.

Lund, V. (2000) Is there such a thing as 'organic'animal welfare? In: Hovi, M. and Garcia Trujillo, R. (eds) Proceedings of the Second NAHWOA Workshop, pp.151-160, Cordoba 8-11.

Marshall, J. and Toffel, M. (2005). Framing the elusive concept of sustainability: a sustainability hierarchy. Environmental Science and Technology 39(3): 673-82, p. 673.

McGlone, John, J. (2001). Farm animal welfare in the context of other society issues: toward sustainable systems. Livestock Production Science 72: 75-81.

RIVM (2004). Kwaliteit en toekomst. Verkenning van duurzaamheid. Milieu en Natuurplanbureau RIVM, Bilthoven.

Stafleu, F.R., De Lauwere, C.C. and De Greef, K.H. (2004). Respect for functional determinism. A farmer's interpretation of respect for animals. In: J. de Tavernier and S. Aerts (eds). Science, Ethics and Society. 5th Congress of the European Society for Agricultural and Food Ethics. CABME, Leuven, 2004, 355p.

Velde, H. te, Aarts, N. and Van Woerkom, C. (2002). Dealing with ambivalence: Farmer's and consumers' perception of animal welfare in livestock breeding. Journal of Agricultural and Environmental Ethics 15: 203-219.

Consumers versus producers: a different view on farm animal welfare?

Filiep Vanhonacker[1], Els Van Poucke[2], Frank Tuyttens[2] and Wim Verbeke[1]
[1]*Department of Agricultural Economics, Ghent University, Coupure links 653, B-9000 Ghent, Belgium*
[2]*Institute for Agricultural and Fisheries Research, Animal Sciences, Animal Husbandry and Welfare, Scheldeweg 68, B-9090 Melle, Belgium*

Abstract

Animal welfare is more than ever a hot issue throughout the entire livestock production chain. Consumers, in their role of citizen, emphasize on the importance of animal welfare in their food decision process; retailers and producers anticipate the more and more by means of product differentiation based on animal welfare; and also governmental actions are taken (e.g. Community Action Plan on the Protection and Welfare of Animals). To facilitate the debate and prevent different stakeholders to talk at cross-purposes, it is important to assess how the concept of farm animal welfare is interpreted by the different stakeholders. A quantitative study was carried out in Flanders, April 2006, including consumers (n=459) as well as producers (n=204). Perceptions and beliefs of consumers and producers towards farm animal welfare were measured using two 5-point interval scaled questions: '*Could you indicate the importance of the following aspect/action, according to your personal opinion, in obtaining an acceptable level of farm animal welfare?*' (perceived importance); '*What is your personal opinion concerning this aspect/action; do you think that this aspect/action entails a problem towards the animal welfare in the present Flemish agriculture?*' (evaluative belief). The interpretation of farm animal welfare by consumers and producers was rather analogous although the producers' perception was more fine-grained. The largest differences concerned aspects related to the expression of natural behaviour (higher relative perceived importance at consumer level). Furthermore, producers did not report major welfare problems with current Flemish livestock production practices, while consumers' perception of animal welfare in current livestock farming was generally negative.

Keywords: farm animal welfare, consumer, producer, perception, Flanders

Introduction

The issue of farm animal welfare is more than ever a topic under concern throughout the livestock production chain: at consumer level, an increasing demand for food products, perceived by consumers as more animal friendly, is noticed, together with a growth of the number of vegetarians and the call for more stringent regulation of welfare in livestock production (Harper and Henson, 2001; Harper and Makatouni, 2002; European Commission, 2005); producers anticipate to maintain their market share and to raise their revenues; retailers react to entice consumers (Phan-Huy and Fawaz, 2003); the European Commission responds with the off-going of the *Community Action Plan on the Protection and Welfare of Animals (2006-2010)*. However, much of the efforts loose part of their efficacy due to the different stakeholders interpreting the concept of animal welfare differently. This originates from the perception being shaped against a background of different visions, interests and knowledge concerning farm animal welfare. Science has an important task through visualizing the perception of the different stakeholders in detail, such that the themes subject to the highest degree of disagreement between the different stakeholders are uncovered. In this study, focus was put on the actors at both ends of the chain, i.e. the producer and the consumer in its role of citizen.

Recently, perception of farm animal welfare has been subject to consumer research all over the world (Harper and Henson, 2001; Te Velde *et al.*, 2002; Verbeke, 2002; Kanis *et al.*, 2003; Phan-Huy and Fawaz, 2003; Quintili and Grifoni, 2004; European Commission, 2005; Frewer *et al.*, 2005; Boogaard *et al.*, 2006; Lassen *et al.*, 2006; Maria, 2006; Van Poucke *et al.*, 2006; Vanhonacker *et al.*, 2006b; European Commission, 2007), while much less research is devoted to producers perception of farm animal welfare (Te Velde *et al.*, 2002; Dockes and Kling-Eveillard, 2006; Morgan-Davies *et al.*, 2006). Similar results appear throughout this literature, with consumers indicating a negative overall perception of animal welfare, and producers expressing a more positive perception of the current state of farm animal welfare. These opposite perceptions are explained by Te Velde *et al.* (2002) through consumers and producers having different convictions, values, norms, knowledge and interests with respect to farm animal welfare.

In the present study, a cross-sectional survey was conducted among Flemish consumers and producers in order to obtain a detailed insight in the perceptual difference concerning the concept and the current state of farm animal welfare between both stakeholder groups. For the 72 farm animal welfare aspects listed by Vanhonacker *et al.* (2006a), interviewees were asked to indicate firstly how important they feel the aspect is for the welfare of farm animals in general, and secondly to what extent this aspect is believed to be problematic with respect to farm animal welfare in current livestock production practices in Flanders.

Material and methods

Study design and subjects

Cross-sectional survey data were collected through questionnaires in Flanders during April 2006. Consumers were selected using a quota sampling procedure with gender, age and living environment as quota control characteristics, via a primary distribution of web-based questionnaires, and supplemented with a more targeted distribution of paper questionnaires to realize the predetermined quota. In total, the sample comprised 459 consumers (Table 1). Producers on the other hand were selected through purposive sampling. Questionnaires were distributed electronically through the use of websites frequently visited by producers, yielding a sample of 204 producers. Respondents were defined as a producer if they or their parents had a farm.

Table 1. Socio-demographic characteristics of the sample; the consumer sample is compared to the characteristics of the Flemish population (NIS, 2002).

		Consumer		Producer (n=204)
		Sample (n=459)	Population	
Gender (%)	Men	48.5	49.3	64.5
	Women	51.5	50.7	35.5
Age (years)	mean (S.D.)	37.8 (14.1)	40.2	35.9 (13.4)
Living environment (%)	Urban	38.9	35	8.9
	Rural	61.1	65	91.1

Questionnaire and scales

The key questions were related to the perception of farm animal welfare. The perception was pictured using an exhaustive list of 72 aspects concerning farm animal welfare, derived from four citizen focus group discussions and a profound literature review (Van Poucke *et al.*, 2006). Factor analysis using principal components subdivided these 72 aspects into seven key dimensions (Housing and Climate; Transport and Slaughter; Feed and Water; Human-Animal Relationship; Animal Suffering and Stress; Animal Health; Expression of Natural Behaviour). For each aspect, both consumers and producers were asked to indicate its perceived importance for obtaining an acceptable level of farm animal welfare (*'perceived importance'*), and whether they believe the aspect poses a potential problem with respect to animal welfare in present Flemish livestock farming ('evaluative belief'). Both statements were scored on a five-point Likert scale.

Statistical analyses

Data were analysed using SPSS 12.0. Bivariate analyses through correlation and comparison of mean scores, i.e. independent samples t-tests were used to detect differences in perceived importance, evaluative belief and knowledge between farmers and citizens. Linear regression analyses was performed to search for relationships between producers' and consumers' responses.

Results

Perceived importance of farm animal welfare

For the consumer sample, average perceived importance for the 72 aspects ranged from 3.14 (distraction material) to 4.56 (availability of water). At the top of the list, mainly aspects relating to the dimensions Feed and Water, Human-Animal Relationship and Animal Health emerge, together with some specific aspects from the dimensions Housing and Climate (stocking density, available space and air quality), Animal Suffering and Stress (pain through human intervention and stress), and Transport and Slaughter (slaughter without pain or stress). Among producers, average perceived importance ranged from 2.40 (size of livestock herd on farm) to 4.35 (availability of water). Producers considered aspects relating to the dimensions Feed and Water, Animal Health and Human-Animal Relationship as most important in obtaining an acceptable level of farm animal welfare.

When the results from both samples are compared to one another, a positive linear relationship is found between perceived importance scores of producers (Y) versus consumers (X) (Y= 1.09x - 0.84; R^2=0.55). The negative intercept indicates that, in general, consumers gave higher scores than producers. This holds for most aspects, but particularly for aspects relating to the Expression of Natural Behaviour dimension. In addition, singular aspects within Housing and Climate (stocking density, outdoor access, size of livestock on farm) and Transport and Slaughter (thirst and hunger during transport, stunning before slaughter) also yield (relatively) large differences. Exceptions are genetic selection, taste of feed, preventive medication, frequency of visual inspection and attention of farmer for animals. Other aspects had rather similar mean values. These are aspects within Feed and Water (availability of water and feed, freshness of feed, balanced feed), Human-Animal Relationship, Animal Health (with the exception of lifespan animal), plus some singular aspects within Housing and Climate (barn temperature, flooring type) and Transport and Slaughter (lairage time). The correspondence regarding the first three dimension originates from the particular aspects being perceived as top priorities for farm animal welfare both at consumer and producer level, while the similarity regarding the latter aspects is due to a rather average perceived importance at producer level and a more moderate score at consumer level. The positive slope close to unity reflects a systematic and consistent difference, indicating a rather similar view on most aspects' relative importance in obtaining an acceptable level of farm animal welfare. A similar positive linear relationship is found for all the dimensions separately, except for the dimension Expression of Natural Behaviour, where a slope of nearly zero suggests no relationship between the scores assigned by the two stakeholder groups (Table 2). Hence, apart from this latter dimension, a similar interpretation of the different dimensions is shared by the citizen and the farmer sample, at least in relative terms.

Table 2. Linear relationship for perceived importance and evaluative believe of the animal welfare dimensions between producer data (y) and consumer data (x).

Dimension	Perceived importance		Evaluative belief	
	Equation	R^2	Equation	R^2
Housing and Climate	Y= 1.04x - 0.69	0.55	Y= 0.32x + 2.76	0.32
Expression of natural behaviour	Y= -0.05x + 3.20	0.00	Y= 0.41x + 2.50	0.55
Animal Health	Y= 1.02x - 0.28	0.75	Y= 0.50x + 2.19	0.10
Human-Animal relationship	Y= 1.03x - 0.19	0.88	Y= 1.15x + 0.45	0.58
Transport and Slaughter	Y= 0.78x + 0.29	0.47	Y= 0.23x + 2.86	0.15
Feed and Water	Y= 1.15x - 0.92	0.74	Y= 0.55x + 2.20	0.83
Animal Suffering and Stress	Y= 1.07x - 0.77	0.94	Y= 0.36x + 2.55	0.47

Evaluative belief of farm animal welfare

Among consumers, average evaluative beliefs for the 72 aspects ranged from 2.11 (stocking density) to 3.34 (feed on fixed moments), with only eight aspects receiving a mean score above the mid-point score of 3. The aspects evaluated most positively mainly relate to the dimensions Feed and Water (with the exception of growth hormones) and Human-Animal Relationship. The lowest evaluative beliefs were found for aspects related to the availability of space (stocking density, available space, space during transport and outdoor access), to the ability to express natural behaviour and also to a high number of the aspects within Transport and Slaughter and Animal Suffering and Stress. For producers, average evaluative beliefs are scored above the mid-point score for all aspects and ranged from 3.27 (thirst during transport) to 4.02 (availability of feed). Aspects related to Feed and Water (with the exception of growth hormones) and Human-Animal Relationship are perceived as the least problematic by producers.

When the results are compared between both stakeholders groups, a positive linear relationship is found between evaluative belief scores of producers (Y) and consumers (X) (Y= 0.56x + 2.09; R^2=0.63). The positive intercept indicates that the current level of animal welfare is generally evaluated more positively by producers as compared to consumers. Also, the evaluative beliefs are in the same direction (positive slope), with a broader range in the answers of the farmers (slope < 1). A similar equation is found for all dimensions with the exception of Human-Animal relationship, where a slope around unity and a small positive intercept is found (Table 2).

Discussion and conclusion

The perception of the concept of farm animal welfare by both consumers and producers is described and compared, based on their perceived importance and evaluative belief of 72 aspects which are considered relevant for the welfare of farm animals (Vanhonacker *et al.*, 2006a). In general, a rather analogous interpretation of the concept was found, given the roughly similar ranking of the perceived importance scores of these aspects. However, consumers generally attributed higher absolute importance scores as compared to producers. Possibly, acquiescence bias is at the base of this. This category of response bias corresponds with a tendency to report a positive connotation and especially relates to opinion or attitudinal questions (Watson, 1992). Aspects with low differences in mean value between consumers and producers pertain either to the aspects perceived most important by both groups or to aspects which require a practical knowledge base of livestock rearing conditions. The former may be due to a ceiling effect, since mean values are close to the maximum score, while the latter reflects the more fine-grained perception of the farmer, which is in accordance with the findings from Beekman *et al.* (2002) and Verbeke (2002). The exception on the overall similarity in perception is found in aspects related to the ability to express natural behaviour. These aspects received a very high perceived importance score by consumers and yield a relatively strong difference, in spite of a relatively high score at producer level. This perceptual difference is in correspondence with the two additional values that consumers take into account compared to producers: freedom to move and freedom to fulfil natural desires. Especially the latter is relevant in this perspective (Te Velde *et al.*, 2002; Lassen *et al.*, 2006; Marie, 2006; Morgan-Davies *et al.*, 2006; Milne *et al.*, 2007).

Concerning the evaluative beliefs, several opposite opinions between consumers and producers were detected. Whereas producers did not perceive problems for any of the aspects, a much more negative picture emerged from the consumer sample. Only the evaluation of issues relating to feed and water, together with the relationship between the farmer and his animals were judged positively by the citizens. Aspects related to space availability, expression of natural behaviour, transport and slaughter, and suffering and stress were evaluated most negatively at consumer level.

The findings from the present study can help to anticipate societal debate relating to farm animal welfare and can be valorized for diminishing the perceptual gap between livestock producers and consumers. Furthermore, these insights could offer business and market opportunities for stakeholders by means of for instance product differentiation based on prominent aspects in the perception of animal welfare at consumer level, or by paying attention particularly to aspects scoring high in terms of consumer concern.

Acknowledgements

The partial financing of this research by the Ministry of the Flemish Government through the project ALT/AMS/2005/1, and by IWT Flanders through the project 50679 is gratefully acknowledged. (Vanhonacker *et al.*, 2006a)

References

Beekman, V., Bracke, M., Van Gaasbeek, T. and Van der Kroon, S. (2002). Begint een beter dierenwelzijn bij onszelf? [Does a better animal welfare starts with ourself?]. LEI/ID. Rapport 7.02.02, 94 p., Den Haag/Lelystad.

Boogaard, B. K., Oosting, S. J. and Bock, B. B. (2006). Elements of societal perception of farm animal welfare: A quantitative study in The Netherlands. Livestock Science 104: 13-22.

Dockes, A. C. and Kling-Eveillard, F. (2006). Farmers' and advisers' representations of animals and animal welfare. Livestock Science 103: 243-249.

European Commission. (2005). Attitudes of consumers towards the welfare of farmed animals. Spec Eur 229, http://ec.europa.eu/food/animal/welfare/euro_barometer25_en.pdf

European Commission. (2007). EU consumers willing to pay for better animal welfare. Press release 22/03/2007. Available on http://europa.eu/rapid/pressReleasesAction.do?reference=IP/07/398&type=HTML&aged=0&language=EN&guiLanguage=en (last consulted: 18/04/2007).

Frewer, L. J., Kole, A., Van De Kroon, S. M. A. and De Lauwere, C. (2005). Consumer attitudes towards the development of animal-friendly husbandry systems. Journal of Agricultural and Environmental Ethics 18: 345-367.

Harper, G. and Henson, S. 2001. Consumer concerns about animal welfare and the impact on food choice. EU FAIR CT98-3678 Final Report, 38 pp. Available on http://europa.eu.int/comm/food/animal/welfare/eu_fair_project_en.pdf (last consulted: 18/04/2007).

Harper, G. and Makatouni, A. (2002). Consumer perception of organic food production and farm animal welfare. British Food Journal 104: 287-299.

Kanis, E., Groen, A. F. and De Greef, K. H. (2003). Societal concerns about pork and pork production and their relationships to the production system. Journal of Agricultural and Environmental Ethics 16: 137-162.

Lassen, J., Sandoe, P. and Forkman, B. (2006). Happy pigs are dirty! conflicting perspectives on animal welfare. Livestock Science 103: 221-230.

Maria, G. A. (2006). Public perception of farm animal welfare in Spain. Livestock Science 103: 250-256.

Marie, M. (2006). Ethics: The new challenge for animal agriculture. Livestock Science 103: 203-207.

Milne, C. E., Dalton, G. E. and Stott, A. W. (2007). Integrated control strategies for ectoparasites in Scottish sheep flocks. Livestock Science 106: 243-253.

Morgan-Davies, C., Waterhouse, A., Milne, C. E. and Stott, A. W. (2006). Farmers' opinions on welfare, health and production practices in extensive hill sheep flocks in Great Britain. Livestock Science 104: 268-277.

Phan-Huy, S. A. and Fawaz, R. B. (2003). Swiss market for meat from animal-friendly production - Responses of public and private actors in Switzerland. Journal of Agricultural and Environmental Ethics 16: 119-136.

Quintili, R. and Grifoni, G. 2004. Consumer concerns for animal welfare: from psychosis to awareness. Global conference on animal welfare: an OIE initiative, Paris.

Te Velde, H. T., Aarts, N. and Van Woerkum, C. (2002). Dealing with ambivalence: Farmers' and consumers' perceptions of animal welfare in livestock breeding. Journal of Agricultural and Environmental Ethics 15: 203-219.

Van Poucke, E., Vanhonacker, F., Nijs, G., Braeckman, J., Verbeke, W. and Tuyttens, F. (2006). Defining the concept of animal welfare: integrating the opinion of citizens and other stakeholders. In: H. M. Kaiser and M. Lien (eds.) Ethics and the politics of food. Wageningen Academic Publishers, Wageningen, The Netherlands, pp.555-559.

Vanhonacker, F., Van Poucke, E., Braeckman, J., Verbeke, W. and Tuyttens, F. 2006a. Definiëring van het begrip dierenwelzijn. Rapport ALT/AMS/2005/1. Pages 103.

Vanhonacker, F., Van Poucke, E., Nijs, G., Braeckman, J., Tuyttens, F. and Verbeke, W. (2006b). Defining animal welfare from a citizen and consumer perspective: exploratory findings from Belgium. In: H. M. Kaiser and M. Lien (eds.) Ethics and the politics of food. Wageningen Academic Publishers, Wageningen, The Netherlands, pp.580-582

Verbeke, W. (2002). A shift in public opinion Pig Progress 18: 25-27.

Watson, D. (1992). Correcting for Acquiescent Response Bias in the Absence of a Balanced Scale - an Application to Class-Consciousness. Sociological Methods and Research 21: 52-88.

Market segmentation based on perceived importance and evaluation of farm animal welfare

Filiep Vanhonacker[1], Els Van Poucke[2], Frank Tuyttens[2] and Wim Verbeke
[1]*Department of Agricultural Economics, Ghent University, Coupure links 653, B-9000 Ghent, Belgium*
[2]*Institute for Agricultural and Fisheries Research, Animal Sciences, Animal Husbandry and Welfare, Scheldeweg 68, B-9090 Melle, Belgium*

Abstract

There is a general tendency in the Western society towards a stronger influence of post-materialistic values. For livestock production, this entails that ecological and socio-cultural considerations occupy an increasingly important role in the food purchasing decision process. One of those emerging issues susceptible to consumer concern is animal welfare. As not all consumers take animal welfare into account, it is important from a marketing perspective to uncover potential target segments of consumers with specific interest in welfare improvement. Also, with respect to a goal-oriented communication strategy, it is important to consider how different consumer segments evaluate the current state of animal welfare in livestock production. In this study, a cluster analysis using a cross-sectional dataset (n=459 citizens) collected in Flanders, April 2006, is performed, with the perceived importance attached to animal welfare in the food purchasing decision process, relative to other product attributes (relative importance; RI) and the subjective evaluation of the current state of farm animal welfare (evaluation; EV) as grouping variables. Six clusters are obtained: Cluster 1 with moderate RI and positive EV (21.1%); Cluster 2 with very low RI and strong positive EV (12.9%); Cluster 3 with low RI and moderate EV (18.7%); Cluster 4 with moderate RI and low EV (12.6%); Cluster 5 with high RI and moderate EV (23.5%); Cluster 6 with very high RI and very negative EV (11.1%). The clusters have been characterised in terms of meat consumption behaviour, knowledge about animal welfare, interest in information and socio-demographics.

Keywords: farm animal welfare, consumer, segmentation, Flanders, survey

Introduction

The growing importance of farm animal welfare throughout the livestock production chain, together with the strong position of the consumer in this consumption society, has recently led to numerous studies about consumer behaviour and consumer attitudes towards farm animal welfare (Harper and Henson, 2001; Te Velde *et al.*, 2002; Verbeke, 2002; Kanis *et al.*, 2003; Phan-Huy and Fawaz, 2003; European Commission, 2005; Frewer *et al.*, 2005; Boogaard *et al.*, 2006; Lassen *et al.*, 2006; Maria, 2006; Vanhonacker *et al.*, 2006). These studies largely confirmed findings by Verbeke and Viaene (1999), stating that animal welfare and acceptable production methods emerged as key attention points for the future. Subsequent studies only affirmed the increasing consumer concern about the subject. Besides, there has been a lot of discussion about the duality between consumer and citizen. People tend to respond to questionnaires as citizens and claim to pay more attention for animal welfare, but when they make a choice in the store as 'consumer', they are not equally willing to pay for such products (Aarts and Te Velde, 2001). Consequently, it is concluded that consumers do not prioritise animal welfare considerations while shopping for food. In the literature, too often these conclusions are very general and based on sample averages, without acknowledging for different segments that might exist with respect to the importance of animal welfare in the food purchasing decision process.

Little research has been performed with regard to segmenting consumers based on the product attributes importance when purchasing food in general, and to our knowledge the segmentation of consumers based specific on the relative importance of animal welfare when purchasing food has not yet been studied. Hansman (1999) found four consumer segments based on the food consumption pattern: the 'cooperating consumer', with a traditional food pattern; the 'responsible consumer', with a mainly vegetarian and ecological consumption pattern; the 'competitive consumer', who likes to eat exclusive; and the 'rational consumer', a mainstreamer that cannot be differentiated from other consumers. Meuwissen and van der Lans (2004) identified six consumer segments in The Netherlands: Environmentalists, Ecologists, Animal Friends, Health Concerned, Unpronounced and Economists. Both studies clearly prove that market segments which take into account animal welfare when purchasing food exist.

Several studies have also tried to characterise consumers who are more concerned about animal welfare and who take animal welfare more into account. Hansman (1999) stated that the impact of socio-demographical variables is negligible. Verhue and Verzeijden (2003) to the contrary found only a small effect from life-consideration and value-orientation on the evaluation of animal welfare, whereas they did find a more negative attitude among young people, townspeople, high educated women and vegetarians. Also Burrell and Vrieze (2003) found a higher importance for the welfare of chicken among women, elderly, high educated people and vegetarians. The same effect of the urbanisation degree was confirmed by Frewer *et al.* (2005), indicating a more positive evaluation of pig welfare among rural people. Finally, the studies by Verbeke and Viaene (2000) and Harper and Henson (2001) related the impact of animal welfare as a product attribute to socio-demographics. The former found a higher attention among men, while the latter stated that especially young people are more likely not to eat meat for ethical reasons.

The aim of this study is to perform a market segmentation, based on both perception of farm animal welfare, in terms of the evaluation of farm animal welfare in current livestock production, and on the relative importance of farm animal welfare as a product attribute taken into account during purchasing decisions. Since a representative sample is used for the Flemish population, this segmentation exercise should enable us to gain insight in the representative size of the segments. The segments will be characterised in terms of meat consumption behaviour, knowledge about animal welfare, information need and interest, and socio-demographics.

Material and methods

Survey design and subjects

Survey data were collected through self-administered questionnaires during April 2006 in Flanders. A quota sampling procedure with province, gender, age and living environment as quota control characteristics was applied. Respondents were selected in a first phase through a wave of web-based questionnaires and supplemented with a more targeted distribution of paper questionnaires such that the predetermined quota were realized. This resulted in a total sample of 459 respondents (Table 1).

Segmentation variables

Relative importance of animal welfare as a product attribute
In the questionnaire, 13 product attributes were probed for their perceived importance (PI) in the food purchasing decision process of animal food products on a five-point interval scale ranging from 'totally unimportant' to 'very important'. Product attributes were: safety; quality; reliability; taste; origin; health; price; appearance; freshness; environmental friendliness; availability; animal welfare; and production method. This extended list allowed to compute a relative perceived importance score for each of the thirteen attributes, using the following formula:

Table 1. Socio-demographic characteristics of the sample (% of respondents, n=459).

		Consumer	
		Sample (n=459)	Population
Gender (%)	Men	48.5	49.3
	Women	51.5	50.7
Age (years)	Mean (S.D.)	37.8 (14.1)	40.2
Living environment (%)	Urban	38.9	35.0
	Rural	61.1	65.0
Province (%)	West-Flanders	25.2	19.0
	East-Flanders	27.5	23.0
	Antwerp	21.7	27.7
	Limburg	8.9	13.3
	Flemish Brabant	16.8	17.0

$$RI_i = \frac{13 * PI_i}{\sum_{i=1}^{13} PI_i}$$

Where:
RI = relative perceived importance
PI = absolute perceived importance

A RI-score below 1 indicates that the specific product attribute ranks among the less important product attributes, while a score above 1 corresponds with a relatively important product attribute. As the focus will be on the relative perceived importance of animal welfare, we will use the abbreviation RI_{AW} in further discussion as reference for the relative perceived importance score assigned to the attribute animal welfare. RI_{AW} ranges from 0.27 to 2.60 with a mean score of 0.98 (SD=0.23) (Table 2).

Evaluation of the current state of animal welfare in Flemish livestock production (EV)
The second segmentation variable was used in the analysis as probed for in the questionnaire: '*Do you consider the current state of farm animal welfare in Flanders in general...*' measured on a seven-point scale anchored at the left pole by 'very bad' and at the right pole by 'very good', with 'moderate' as central score and ranges from 1 to 7, with a mean score of 4.13 (SD=1.49) (Table 2).

Table 2. Profile of consumer segments on segmentation variables.

	S1	S2	S3	S4	S5	S6
Segment size (% of sample)	21.1	12.9	18.7	12.6	23.5	11.1
Perceived importance (PI)	4.14	2.51	3.15	3.95	4.55	4.98
Relative importance (RI)	1.03	0.66	0.81	0.95	1.12	1.30
RI_{AW} z-score (segmentation variable)	0.22	-1.40	-0.75	-0.15	0.59	1.39
Evaluation (EV)	5.44	5.95	4.53	2.36	3.76	1.67
EV z-score (segmentation variable)	0.88	1.22	0.26	-1.18	-0.24	-1.65

For the clustering itself, there is opted to work with the standardized score (z-score) of both variables rather then with the actual scores, in order to obtain a segmentation that better brings the relative relationship between the segments into vision. In further discussion, absolute perceived importance score and evaluation of farm animal welfare in the current Flemish livestock production refer to the mean scores of the non-standardized variables RI and EV.

Results and discussion

Cluster analysis

A combination of hierarchical and K-means cluster analysis was used to determine the amount of clusters (so-called segments) yielding the highest degree of differentiation, resulting in a six-cluster solution. Table 2 summarises the respective segment size and the scores of the segmentation variables. The z-scores as well as the actual scores are given. Figure 1 shows a projection of the cluster centres in a two-dimensional graph.

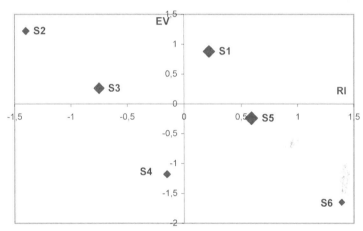

Figure 1. Projection of the cluster centres according to RI z-score and EV z-score; size of marker reflects cluster size.

Profiling the clusters

Meat consumption behaviour
The highest meat consumption was noticed for segment 1 (S1) and segment 2 (S2), closely followed by segment 3 (S3). Segment 6 (S6) on the other hand reported a very low meat consumption frequency; segment 4 (S4) and segment 5 (S5) reported a meat consumption in between. Hence, meat consumption seems to be strongly related and inversely proportionate to the evaluation of farm animal welfare.

Knowledge
Both subjective and objective knowledge with regard to farm animal welfare was probed for in the questionnaire. The six consumer segments differed significantly on both measures (resp. $F=10.89$, $p<0.001$; $F=5.21$, $p<0.001$). For both measures, S2 and S6 showed up as most knowledgeable, while S3, S4 and S5 had the lowest knowledge level.

Information need and interest
The consumer segments show clear differences with regard to information need and interest. S2 clearly expresses the least interest. This low interest is also reflected by a very low expected impact of receiving more information on their meat consumption. In addition, S2 was the sole segment which disagreed to pay more for food produced with attention for animal welfare. S1 and S3 are somewhat more positive towards information interest, although only a neutral attitude towards the expected impact of receiving more information on their meat consumption and a limited willingness to pay for food produced with attention for animal welfare was found. In contrast, the remaining segments (S4, S5 and S6) express a high information need and interest. While this need is high for S4 and S5, it is very high for S6. Despite this strong request for information, only a moderate expected influence of receiving more information on the meat consumption is indicated by these segments. Explanations pertain to a segment (S6) being largely vegetarian, hence not willing to eat meat, or to segments consuming already a large amount of meat (S4, S5), hence hardly leaving room for further increase. Regarding willingness to pay (wtp), a high wtp is found for S4 and S5 and a very high wtp is found for S6.

Also, big differences are found between the segments with respect to preferred information source concerning animal welfare. In accordance to the low information interest, S2 expressed low preferences for most of the information sources probed for. Only the farmer as a source was quoted positively, while very low preference was attributed to animal welfare organisations, television and radio. S1 and S3 expressed similar preferences, both in degree of interest as in type of source preferred. Higher preference values were found for S1 and S3 as compared to S2, but the values were clearly lower as compared to the preferences of the remaining segments. The difference between S1 and S3 concerned S3 attributing a somewhat higher preference to animal welfare organisations and a somewhat lower preference to farmers. Government, labels and educative sources were the most preferred sources of information for S4 and S5. Finally, S6 attached the highest preference to animal welfare organisations and welfare labels, while the farmer was the least preferred source.

Socio-demographics
With regard to gender, significantly more men are found in S2, whereas women rather belong to S5 and S6. With regard to age, S6 was the youngest (32.6 years on average), followed by S4 (35.2 yrs), S5 (36yrs), S3 (37.4 yrs), S2 (41.0 yrs) and S1 (42.5 yrs). No significant differences were found between the segments with respect to living environment, although there was a tendency to a more rural composition of S1 and S2. Finally, the segmentation seems to be strongly influenced by the involvement in agricultural practices. The respondents which have themselves a farm or have parents with a farm are equally distributed over S1 and S2.

Conclusions

Six distinct market segments have been identified based on the relative importance of animal welfare as a product attribute in the food purchasing decision process and the evaluation of the current state of animal welfare in Flemish livestock production. These segments differ significantly with respect to meat consumption behaviour, knowledge about animal welfare, information need and interest, and some socio-demographical variables. The fact that distinct market segments can be identified, imposes some limits on pursuing a generic marketing and communication strategy with respect to farm animal welfare, though also offers specific opportunities for targeted information provision to meat consumers.

References

Aarts, H. and Te Velde, H. (2001). Eten, maar niet willen weten. In: N. Poll, L. Sterrenberg, E. Bozkurt, and I. Miedema (eds.). Hoe oordelen we over de veehouderij? Rathenau Instituut, Den Haag, 192pp.

Boogaard, B. K., Oosting, S.J. and Bock, B.B. (2006). Elements of societal perception of farm animal welfare: A quantitative study in The Netherlands. Livestock Science 104: 13-22.

Burrell, A. and Vrieze, B. (2003). Ethical motivation of Dutch egg consumers. Tijdschrift voor Sociaal Wetenschappelijk Onderzoek voor de Landbouw 18: 30-42.

European Commission (2005). Attitudes of consumers towards the welfare of farmed animals. Spec Eur 229, http://ec.europa.eu/food/animal/welfare/euro_barometer25_en.pdf

Frewer, L.J., Kole, A., Van De Kroon, S.M.A. and De Lauwere, C. (2005). Consumer attitudes towards the development of animal-friendly husbandry systems. Journal of Agricultural and Environmental Ethics 18: 345-367.

Hansman, H. (1999). De consument gevangen in cijfers: Zoektocht naar het bestaan van consumentenbeelden. LEI, Den Haag.

Harper, G. and Henson, S. (2001). Consumer concerns about animal welfare and the impact on food choice. EU FAIR CT98-3678 Final Report, 38 pp. Available on http://europa.eu.int/comm/food/animal/welfare/eu_fair_project_en.pdf (last consulted: 18/04/2007).

Kanis, E., Groen, A.F. and De Greef, K.H. (2003). Societal concerns about pork and pork production and their relationships to the production system. Journal of Agricultural and Environmental Ethics 16: 137-162.

Lassen, J., Sandoe, P. and Forkman, B. (2006). Happy pigs are dirty! conflicting perspectives on animal welfare. Livestock Science 103: 221-230.

Maria, G. A. (2006). Public perception of farm animal welfare in Spain. Livestock Science 103: 250-256.

Meuwissen, M. and Van der Lans, I. (2004). Trade-offs between consumer concerns: an application for pork production. *International conference on chain and network management in agribusiness and the food industry* Ede, pp 27-28.

Phan-Huy, S.A. and Fawaz, R.B. (2003). Swiss market for meat from animal-friendly production - Responses of public and private actors in Switzerland. Journal of Agricultural and Environmental Ethics 16: 119-136.

Te Velde, H.T., Aarts, N. and Van Woerkum, C. (2002). Dealing with ambivalence: Farmers' and consumers' perceptions of animal welfare in livestock breeding. Journal of Agricultural and Environmental Ethics 15: 203-219.

Vanhonacker, F., Van Poucke, E., Nijs, G., Braeckman, J., Tuyttens, F. and Verbeke, W. (2006). Defining animal welfare from a citizen and consumer perspective: exploratory findings from Belgium. In: H.M. Kaiser and M. Lien (eds.). Ethics and the politics of food. Wageningen Academic Publishers, Wageningen, The Netherlands, 580-582.

Verbeke, W. (2002). A shift in public opinion Pig Progress 18: 25-27.

Verbeke, W. and Viaene, J. (1999). Beliefs, attitude and behaviour towards fresh meat consumption in Belgium: empirical evidence from a consumer survey. Food Quality and Preference 10: 437-445.

Verbeke, W.A.J. and Viaene, J. (2000). Ethical challenges for livestock production: Meeting consumer concerns about meat safety and animal welfare. Journal of Agricultural and Environmental Ethics 12: 141-151.

Verhue, D. and Verzeijden, D. (2003). Burgeroordelen over de veehouding, uitkomsten publieksonderzoek. Onderzoeksrapport. In: Veldkamp (ed.), Amsterdam.

Part 6 - Sustainability and animal welfare: different practices

Stacking pigs: Dutch pig tower debates and the changing nature of ethical livestock production

Clemens Driessen
Wageningen UR, Applied Philosophy, Department of Social Sciences, P.O. Box 8130 NL-6700 EW, Wageningen, The Netherlands, clemens.driessen@wur.nl

Abstract

In the Netherlands heated debates have been spawned by designs of industrial scale pig farms in the shape of 'pig towers'. This new concept of integrated livestock production was a government plan created within the approach of system innovation. Ethical concerns regarding livestock farming such as emissions, animal welfare and transportation were optimised in the design of the new system. The public reactions nevertheless were strongly dismissive. In order to understand the arguments brought forward as more than irrational and emotional outbursts, the pig tower in this paper is analysed as a modest proposal in the tradition of Jonathan Swift. Then it appears that the debate that has arisen with the pig tower design has moved the issue of intensive pig farming from a market development following an autonomous industrial logic, to a subject of political decision making and moral deliberation. Finally this paper discusses an alternative approach for dealing with ethical issues in the livestock sector that does not require complete control and an integral design of large scale systemic change, but instead departs from farmers as ethical actors.

Keywords: pig tower, system innovation, ethical room for manoeuvre

Introduction

Over the past decade in the Netherlands twice a debate has flared on the design of a high rise pig farm. What became known as the 'pig tower' (or 'varkensflat' in Dutch) started out as a plan for a 6 story building, a 1000 meters in length. It was projected to contain 300,000 pigs, 1.2 million chickens, a slaughterhouse and a salmon nursery, and was to be located somewhere in a large seaport such as Rotterdam, near the supply of cheap grains and usable wastes of the food industry. The facility was projected to be largely self-supportive and produce no environmentally damaging emissions. The manure would be used for generating electricity, heating greenhouses and growing mushrooms in the basement. Cattle food in at the entrance, pork out at the exit, windmills on top, and even the pigs were better off in this way: They are not being transported, live in stable groups, and have more space than in their current stables, even including a balcony.

Protest

A storm of protest gathered in the national media in autumn of the year 2000. The Dutch minister of agriculture Brinkhorst, a non-Christian-Democrat minister of agriculture and by many in the field regarded as a technocrat, had aired the scheme for the pig tower on the 4th of October, world animal day. A heated debate ensued in which all kinds of actors came to the front. Farmers, rural organisations, public figures, newspaper editorials, animal welfare organisations, all denounced the project on their own terms. Allusions both to fascism and Stalinism where made in the national media. The minister rushed to say it never had been a serious proposal, or at least a very immature one. Eventually that year the pig city became an art project by the Dutch architectural firm of MVRDV. And it was not until recent, after several years, publicly heard of again as a serious plan. In order to understand how the design came about, including the role of the government in developing it, the approach to policy issues in terms of system innovation is relevant.

System innovation

The pig tower design first of all can be understood as a reaction to the Classic Swine Fever and Foot and Mouth Disease crises. In this respect it is a design ordering the chaotic practices and ending the untraceable movements of animals, by reducing the complexity of the system and making it easier to control (Bos, 2004). In it there are no more farmers taking regulations for granted, no more contacts with wild animals spreading diseases. (A modernisation effort resembling the moving of humans from the crowded and chaotic inner city slums to ordered and spacious apartment dwellings in the previous century.) The systemic approach to pig production, even integrating it with other agricultural sectors, can be regarded as an instance of a way of dealing with the undesirable side effects of modernisation processes by further modernisation. By taking up the entire agricultural production sector as a system with inputs and outputs, and dealing with ethical concerns by closing substance cycles and optimising system relations, an attempt is made at creating an 'industrial ecology'. In order to make this assembly of natural and technological processes work, a radical shift away from the previous system is to be created. This is not just a technological challenge but also requires an institutional change and the adoption of new ways of thinking, which could even be regarded as a paradigm shift. (Grin *et al.*, 2004)

The pig tower proposal, and especially later efforts to design and implement new integrated agricultural production systems, can be understood as an example of a certain kind of system innovation project. With it the ethical concerns are taken up as parameters in processes that can be optimised. In the end there are no ethical dilemmas; decisions are made in an integral way by technologists. There appears to be no real political decision even. The role of the government is just one of creating networks and facilitating societal actors and businesses to cooperate in bringing about the new industrial ecology.

As long as the industrial agro-parks remain plans on paper, the pig tower might be considered a failed government project, which can at best be the subject of an inquiry into the reasons for its failure and then is likely to become a classic case study in the policy sciences. Or it can still be regarded as the unavoidable destiny of pig farming in the Netherlands, the final stage of a process of decades of specialisation, increasing scales and integration of businesses in the sector. The vehement protests then appear as emotional and irrational reactions of the public that denies the character of current pig production practices and refuses to see the advantages of an integrated approach. Or, a more complex moral dynamic has been spurred by the pig tower proposal, analysis of which can help understand the role of ethical arguments in system innovation processes, and that could lead to another approach to these processes.

A modest proposal

In 1729 Jonathan Swift published (at first anonymously) a satirical pamphlet entitled: 'A Modest Proposal for preventing the children of poor people in Ireland, from being a burden on their parents or country, and for making them beneficial to the publick' [sic]. In it he proposed to deal with poverty and famine in Ireland by feeding children of the poor as a delicacy to the rich. (Swift, 1984) With detailed calculations of the amount of children that were born and the amount of food available, Swift demonstrated logically this to be the best solution for the problem of poverty and famine. He suggested recipes for children's ragouts in the polite and careful style in which gentlemen in learned societies usually proposed technical improvements to society. (Downie, 1984) In this way a systematic weighing of aspects of a problem was put forward in which the best solution for all involved was determined.

Sustainable food production and ethics

The children would die anyway at an early age or lead lives of misery and crime, and now at least their parents would be able to feed them and themselves.

The public was shocked by this proposal. But with the re-description of the pamphlet it was forced to regard the issue of poverty and famine as pressing and morally important. Implicitly responsibilities were redistributed for a situation that had developed over the years and previously was considered a given. Those in power in England lost their moral innocence with their first indignation for the scheme. It was hard to reject the proposal without taking on a moral perspective, without specifying where exactly the flaw is in the line of reasoning, and without sensing some need to formulate an alternative. While at the same time the detached style in which these kinds of issues usually were approached had been disqualified. A neutral, technical description of the current situation and meticulous abstract methods for solving problems were made ridiculous. Those in power could no longer hide behind this kind of morally neutral analyses. Rather the issue was made into a moral and political one, calling for the taking up of responsibilities and radical measures to put an end to the miserable situation of the Irish poor.

Pig tower moral dynamics

Those proposing the 'agro-production-parks', as the pig towers officially were called, are not known to have meant their schemes as satire. Nevertheless it can be claimed that its effects on public opinion were as if it concerned a proposal in the genre in which Jonathan Swift had specialised himself. Especially the design of pig towers by architectural firm MVRDV can be seen to operate as a modest proposal; Offering a kind of policy analysis in the pictorial style of Otto Neurath, with the towers as a three-dimensional depiction of current pig production in the Netherlands. Exactly 76 of the sketched towers would be enough to contain the current pig population on the square meters required for organic farming. (Betsky, 2003) Rejoicing in the extremity of the ensuing designs while sticking to their moral neutrality, the designers have drawn pigs in colourful computer images of their endless dwellings; The towers being just a depiction of current amounts of livestock, and an improvement on current practices. Satire in the classic tradition always contains a moment in which the solutions the writer really advocates are brought forward, but as opposed to Swift the architects nowhere reveal the irony, sustaining it throughout their plans.

The moral outrage over the pig tower scheme brought pig farming to the centre of debate. Of course animal activists and environmentalists have criticised intensive livestock farming for years. But with the pig tower the situation that has grown gradually appears suddenly anew, and this time completely intended. What previously had been considered mostly a market issue was made political. With the design, the rearing of animals on an industrial scale was suddenly turned into an officially chosen situation, propagated by the state and represented by the minister of agriculture.

As part of the pig tower plan, specific characteristics of current livestock production were proposed to be dealt with in innovative ways, such as CO_2 and methane emissions, the elimination of transporting live animals, the use of land and scarce resources, and the energy efficiency of production. And as the new design was set off against current pig production practices, these practices were implicitly evaluated on these terms. Even though the proposal was almost universally derided and dismissed, the parameters it highlighted were considered ethically important by actors in the field. With the pig tower design these actors are forced to redefine themselves and reassess their practices. After the first publicity of the pig tower plans, pig farmers' organisations advertised in magazines stressing their hi-tech professionalism and sound ethical practices towards animals and the environment. Their independence, knowledge, and care for animals make for the ideal characteristics of keeping pigs professionally. In a recent 'town hall debate' on a possible location for a pig tower in the harbour of Zaanstad, the dynamic could be seen at work. The representative of the agro-production-park stressed the need for improvement in the sector with respect to environmental concerns and the reduction of transporting live animals. The conventional pig farmers were found struggling to gain legitimacy in the new arena of debate and with the new benchmark production standard. 'But we can do that too!', exclaimed the pig farmers' representative when all the advantages of the pig towers over current farming were put forward. 'But you don't!', retorted the proponent of the pig tower.

Integrated system design and ethical room for manoeuvre

Out of the description of the dynamics, broadly speaking two models for pig farming can be seen to evolve. The current pig farmers attempting to incrementally improve on their practices, and the new integrated system design of the pig tower. The first model is a continuation of the loose coupling of the still to some extent independent elements of the sector. The second model a complete and systemic re-design of livestock production. (Bos, 2004) The models imply two different approaches towards ethical issues. The 'we can do that too' of individual pig farmers, requiring room for manoeuvre in order to ethically improve on practices; and the technological optimisation of parameters and solving of potential dilemmas by centralised decision making on agro-production-parks. Next to that the models imply different kinds of knowledge, as well as different ways of developing technology. Farming at the moment is one of the few sectors left in a modern economy in which manual labour and strategic decision making are often done by the same person, offering opportunities for dealing with unexpected situations and for improving technologies based on practical knowledge. While individual farmers often do not change their entire facilities at once as in the large scale industrial systems, they improve on their farming technologies by piecemeal engineering in that way spreading investments; Helped in their innovations by the organisation of peer group networks. Both of these models have pros and cons. Where the current pig farmers are in danger of inherent conservatism with regard to ethically improving their practices, they can be expected to have a higher degree of flexibility when dealing with practical difficulties in reconciling different values, such as requirements for animal welfare and environmental emissions.

Conclusion

The combination of the two approaches to innovation in pig farming accompanied by public deliberation and legitimating efforts of the actors involved, could lead to a development of livestock agriculture in sustainable directions, as the models seem able to reinforce each other in critical competition. Whether or not this new need for ethical justification and morally inspired innovation will take on solid shape and lead to substantial improvement of livestock production in practice still largely remains to be seen. As it is always possible that the moral re-description offered by the pig tower loses momentum, and both conventional farmers and pig towers will end up in a race to the bottom of cheap pork production with little care for ethical concerns. As with Swift's modest proposal, it might be centuries and many disasters later until things really change for the better.

References

Betsky, A. (ed.) (2003). Reading MVRDV. NAI Publishers, Rotterdam, the Netherlands. 152p.

Bos, B. (2004). Een kwestie van beheersing, over de rol van planten, dieren en mensen in technologische systemen. De Vliegende Beer, Amsterdam, the Netherlands.192p.

Downie, J.A. (1984). Jonathan Swift political writer, Routledge and Kegan Paul, London, UK. 391p.

Grin, J., Felix, F., Bos, B.and Spoelstra, S. (2004). Practices for reflexive design: lessons from a Dutch programme on sustainable agriculture. International Journal of Foresight and Innovation Policy 1: 126-149

Swift, J. (1984). A Modest Proposal for preventing the children of poor people in Ireland, from being a burden on their parents or country, and for making them beneficial to the public. (originally published: 1729) In: A. Ross and D. Woolley (Eds.) *Jonathan Swift* Oxford University Press, Oxford, UK. pp.492-499.

Towards sustainable livestock production systems

John J. McGlone and Mhairi Sutherland
Pork Industry Institute, Texas Tech University, Lubbock, TX 79409-2141 USA

Abstract

We define sustainable as production methods that are friendly to the animals, the people, the environment, the community, and that are economically competitive. If all of these features were simultaneously met, one could claim the production and processing methods were sustainable. If pork was produced in a sustainable manner, the production methods may or may not look quite different than what we see today. Imagining a sustainable pig farm may bring up an image of a pastoral setting on a green hillside. Adapting that image to a modern day scenario, this production system is unlikely to be sustainable. In the USA, this style of farm would probably not have health insurance or an occupational health and safety program for its workers, protection of waterways, and it may not be economically competitive with larger, more efficient pig farms. When economies of scale save money and that money is put into worker health and safety, and environmental and animal protection, and minimization of zoonoses, the larger farm may be considered more sustainable. Furthermore, if a nation puts non-competitive restrictions on its livestock producers in the name of animal welfare or environmental protection, it makes their livestock production systems less sustainable than those in other countries. Standards are needed for the farm as a whole so that when changes are made towards a single sustainable goal, other elements of sustainability are not lost. Each element of the model should evolve over time as more is learned.

Keywords: sustainability, pig, animal welfare, environment

Introduction

Livestock production can only be considered sustainable if it lasts far into the future without negatively impacting people, animals, or the environment as well as other components. For the purposes of this paper, we will define sustainable livestock production as production methods that are friendly to the animals, the people, the environment, the community, and that are economically competitive. However, some of these features of a sustainable production system are at cross purposes. Understanding where we are at the moment and where we would like to be in the future will help us set a path towards improvement of production systems. We are not proposing that this is the only model of sustainable production systems; however, the approach outlined in this paper is one potential approach.

In this paper, we will consider a specific example of livestock production. Pork production is growing around the world as the population grows and as incomes increase in pork-eating populations. Some estimates have pork demand continuing to increase for decades to come; hence, the world would be a better place if all livestock production was sustainable. Consumers currently have a poor image of pig production systems and the associated production practices, which is not compatible with a sustainable demand for pork production (Petit *et al.*, 2003); hence this image needs to improve as a socially important aspect of sustainability of this production system.

For a given production system to be sustainable, it must pass several tests. The aesthetic appearance of a farm to the casual observer is not one of these tests. Still, natural intuition may hold that certain systems are more sustainable than others. Let us consider the two production systems shown in Figure 1. Figure 1a shows an outdoor system while 1b shows an indoor system. As the paper progresses, we will examine elements of these diverse production systems and how these elements contribute or detract from a sustainable goal. Science and reason should prevail when deciding the extent to which a given system passes tests of sustainability.

Due to space limitations, only 2 examples will be given of how individual components of sustainable pork production are at cross purposes. Making progress in one component of sustainable production system development while losing ground on another component is not true progress towards sustainable livestock production system development.

Food safety and animal welfare

The outdoor production system shown in Figure 1a brings to mind an idyllic, pleasant pastoral setting while the image of a building containing animals as in Figure 1b may not conjure up a positive mindset. But is the production system shown in Figure 1 sustainable?

The outdoor system works very well in mild weather. However, when the weather is very warm, very cold, snowy, or wet, the outdoor system can fail. Failure may involve having no access to the animals during severe weather. In the winter of 2006, a series of large snow falls befell southern Colorado. One large outdoor pig unit could not get to the sows for several days. Workers were so frustrated with the working conditions and the unpleasant weather that they walked off the job. This system failed and large losses of animals resulted in a loss of sustainability. Figure 2 illustrates weather-related challenges for an outdoor pig production model. The parallel problem for an indoor production unit is to have a power failure. For disaster planning, the risk to pig and people welfare is the critical factor. A sustainable production model would be required to minimize risks to animal and human welfare.

Food safety is an important component of a sustainable production model. For a livestock production model to be valuable in providing nutritious animal protein for human consumption, it must do so while minimizing the risk of introducing food borne illness.

Indoor production systems on slatted floors or on bedding create a microenvironment that allows pathogens to grow. An example pathogen is *Salmonella*. *Salmonella* is one pathogen among others that can be transmitted from pig to pig via faeces. When indoor-kept pigs have *Salmonella*, they can spread it from pig to pig and pen to pen by nose-to-nose contact (such social behaviours are positive for animal welfare). Outdoor-kept pigs have the luxury of large spaces which may disperse and minimize

Figure 1. Sow breeding/farrowing/weaning production systems. Figure 1a shows an outdoor production system and 1b shows an indoor production system.

Figure 2. Seasonal pictures of outdoor production units. In 2a, the weather and conditions are ideal for pigs and people. In 2b, a large volume of snow causes significant challenges. In 2c, rain and mud presents a risk to pig welfare and it makes for an unpleasant working condition. In 2d, some snow and cold are not a particular problem for this sow and litter.

respiratory pathogens, but enteric zoonotic agents are a different matter. Calloway *et al.* (2005) examined generic *E.coli*, coliforms and *Salmonella* in two production systems (indoor and outdoor) and found that outdoor wallows contained *Salmonella* and these bacteria were cycling between the sows and its environment (Figure 3). Wild birds seemed to contribute to the continued re-seeding of wallows and soil with *Salmonella* and other enteric pathogens.

The very nature of the outdoor production system which promotes natural pig behaviours and positive animal welfare (wallows and soil) is actually a major source of pathogen maintenance in the production system. This is a clear example of how animal welfare and food safety are at cross purposes which makes striving for a sustainable model that accommodates both animal welfare and food safety (and other issues) so critical.

A major concern in the area of animal and human health is the spread of viral respiratory pathogens. Influenza moves among birds, pigs and people while mutating and changing its pathogenicity. Outdoor units by design would have a large risk of influenza movement among animals and people. Creative solutions to the zoonotic risk of the outdoor system are needed to make it sustainable. As is, outdoor units provide an unacceptable zoonotic risk.

Figure 3. The wallow provides critical water for drinking and cooling during warm weather. The wallow accommodates the natural behaviour of the pig while providing an incubator for pathogens associated with food borne illness.

Animal welfare and the environment

Sows are much better able to express natural behaviours when outdoors. Outdoors sows spend considerable time rooting and chewing on soil and plants as do their wild counterparts (Graves, 1984; Stolba and Wood-Gush, 1989; Rachuonyo and McGlone, 2006). Soil and moisture are essential aspects of the environment in which wild and feral pigs prefer to habitat (Graves, 1984).

An environmentally sustainable livestock production unit would not pollute the soil, air, or water, and it would not contribute to greenhouse gases. Therefore, what would an environmentally sustainable outdoor system look like? A sustainable livestock production unit would have a cycle of nutrients that include the following:
Input:
- Feed (N, C, P, H_2O, etc.)
- Replacement animals (N, C, P, H_2O, etc.)
Output:
- Pigs
- Manure nutrients (C, N, P, H_2O, etc.)
- Gases (CO_2, Oxides of Nitrogen, etc.)

Rachuonyo and McGlone (2006) examined the changes in nitrate and phosphorus over time in an outdoor production unit. Nitrate cycled over time with the changes in plant growth while phosphorus levels increased in the soil over time, especially in the feeding area of the outdoor pen ($P < 0.05$).

Therefore, for an outdoor unit to be sustainable, nutrients and gases should be prevented from escaping from the farm into the air or water. The best way to do that given present-day technology is to keep the carbon, nitrogen and phosphorous in the form of animals and plants as much as possible and not in liquid or gas form. In addition, plants grown on the site must be harvested to remove nutrients (esp., phosphorus) to be recycled in other systems (for example ruminant feeding systems).

Figure 4. Aerial view of an outdoor gestation system (4a) showing the difference in ground cover between 2 stocking densities (7 or 14 sows per 0.4 ha). Note that when the stocking density is too high, ground cover is removed and nutrient flow over the land are not checked. When an irrigation system is used (4b) in a warm, dry climate, irrigation water can be added as needed to maintain ground cover while preventing issues with mud and snow.

Nations making their industries less sustainable in the name of animal welfare

When a single nation introduces laws or regulations that are intended to improve animal welfare in the absence of consideration of other components of sustainable principles, then the farmers and the system are put at risk. We can use the United Kingdom as an example.

Animal welfare laws and marketing drives of grocery store chains drove the UK pig industry towards outdoor production models (among other models). This was done without environmental and food safety safeguards in place. Putting sows outdoors in a crop rotation with wheat and potatoes (as was and is common) results in pigs being placed on pastures after a crop is harvested, or in other words, on nearly bare ground. This poses an environmental risk. Plant density is low or non-existent, preventing the capture of nutrients from manure.

There is concern that the introduction of increased animal welfare standards in the UK will increase production costs and potentially drive pig farmers out of business (Bornett *et al.*, 2003). Less has been said about how this move contributed to increased risk of food safety and environmental challenges, but these are surely the case.

Balance is more important than individual components

Making progress on one component of a sustainable model of livestock production must be considered in the light of an entire sustainable system. Making apparent progress in animal welfare, while introducing a food safety or environmental risk, makes no sense. All of the components of a sustainable production system should be identified and considered when any change in production methodology is proposed to be introduced. This is the only way we can prevent one component of sustainable production to cause loss of progress on other components.

The need for models and standards

Towards the end of developing models of sustainable production systems, new models and standards are needed. New production models should start with an eye towards accommodating as many components of a sustainable production system as is possible. To accomplish this, standards are needed for each component of a sustainable production model. By the very nature of the diverse components that go into a sustainable production model, an inter-disciplinary team is needed to address this complex model development.

Because new production systems are being established in less developed countries as they develop their economies, the need for standards and new models of sustainable production systems is urgent. Standards must come first. Then, a review of current production systems should be conducted to determine if systems that are out there now meet all the goals of a sustainable livestock production system. And certainly new sustainable production system models will evolve.

References

Bornett, H.L.I., Guy, J.H. and Cain, P.J. (2003). Impact of animal welfare on costs and viability of pig production in the UK. Journal of Agricultural and Environmental Ethics 16: 163-186.

Callaway, T.R., Morrow, J.L., Johnson, A.K., Dailey, J.W., Wallace, F.M., Wagstrom, E.A., McGlone, J.J., Lewis, A.R., Dowd, S.E., Poole, T.L., Edrington, T.S., Anderson, R.C., Genovese, K.J, Byrd, J.A., Harvey, R.B. and Nisbet, D.J. (2005). Environmental prevalence and persistence of Salmonella spp. in outdoor swine wallows. Foodborne Pathogens and Disease 2: 263-273.

Graves, H.B. (1984). Behaviour and ecology of wild and feral swine (Sus scrofa). Journal of Animal Science 58:482-492.

Petit, J., Hayo, S. and van der Werf, H.M.G. (2003). Perception of the environmental impacts of current and alternative modes of pig production by stakeholder groups. Journal of Environmental Management 68: 377-386.

Rachuonyo, H.A. and McGlone, J.J. (2006). Impact of outdoor gestating gilts on soil nutrients, vegetative cover, rooting damage, and pig performance. Journal of Sustainable Agriculture 29: 69-87.

Stolba, A. and Wood-Gush, D.G.M. (1989). The behaviour of pigs in a semi-natural environment. Animal Production 48: 419-425.

Economic, ecological and societal performance of organic versus conventional egg production in the Netherlands

Eddie A.M. Bokkers[1], Sophia van der Ploeg[1], Erwin Mollenhorst[2] and Imke J.M. de Boer[1]
[1]*Animal Production Systems Group, Wageningen University, P.O. Box 338, 6700 AH, Wageningen, the Netherlands, eddie.bokkers@wur.nl*
[2]*Farm Technology Group, Wageningen University, P.O. Box 17, 6700 AA, Wageningen, the Netherlands*

Abstract

Due to the ban on traditional battery cages in the EU egg producers have to change to another production system. The organic egg production system may be such an alternative. To assess the contribution to sustainable development of this production system economic, ecological and societal performance was quantified and analysed according to a formerly established method. Results were compared to other egg production systems studied before. Eleven organic laying hen farms were visited to collect data of the last flock in 2005. Regarding economic performance farm income per full time equivalent was computed. For ecological performance, a life cycle assessment was used to express the contribution of acidification, eutrophication, global warming, energy use, and land use. Regarding societal performance animal welfare and health as well as product quality were studied. Farm income and contribution to acidification and eutrophication was higher for organic egg producers compared to other egg production systems. Contribution to global warming and energy use, however, were lower for organic egg producers. No difference was found for land use. Animal welfare level was high as well as the use of anthelmintics in organic egg production. The compelled outdoor run affected these two indicators strongly. Variation was high between egg producers, but organic egg producers tended to have higher mortality levels and lower percentages of second grade eggs. Although organic egg production has a strong sustainability image, the results showed they do not contribute only positively to sustainable development. Improvements can be made especially regarding mortality and the contribution to acidification and eutrophication.

Keywords: laying hens, sustainability analysis, organic, battery cage, deep litter, aviary

Introduction

Animal welfare is the reason for banning the traditional battery-cage system for laying hens in the EU in 2012. Farmers are forced to choose a more hen-friendly production system. In the Netherlands, many egg producers with battery cages still have to make their choice for the future. In order to choose the best egg production system for a specific farm, it is important to investigate the economic, ecological and societal (EES) performance for each alternative. The EES performance gives an indication about the contribution to sustainable development of a production system. A support system will be developed to help a farmer to gain insight into the most sustainable solution. This support system is based partly on the method and data of Mollenhorst *et al.* (2006). They used a method to assess the contribution of egg production systems to sustainable development. This method implied four steps: (1) description of the situation; (2) identification and definition of relevant EES issues; (3) selection and quantification of suitable indicators; (4) final assessment of the contribution to sustainable development (Mollenhorst and de Boer, 2004). Mollenhorst *et al.* (2006) compared the EES performance among the battery-cage system (BC), the deep litter system with (DLO) and without (DL) outdoor run, and the aviary system with outdoor run (AO).

For a proper support system the data base needed to be extended. In this study, we investigated EES performance of the organic egg production system (OR) in the Netherlands, which is compelled to have an outdoor run, and compared it to the earlier studied systems. For that we needed to perform step 3 and 4 of the described method for the organic egg production system.

Material and methods

Eleven organic egg producers with more then 1500 hens were visited to collect data about their farm and the last finished flock in 2005. The same questionnaire as formerly used by Mollenhorst *et al.* (2006) was used containing detailed questions about, e.g., the farm, housing system, feed and feeding, production, manure management, etc. Data were used to compute outcomes for the different indicators. For the other egg production systems data of Mollenhorst were used. Economic data were updated to the prices of 2005. Farm income per full time equivalent (FTE) per year was used as indicator for economic performance. For ecological performance, a life cycle assessment was performed (Haas *et al.*, 2000). The life cycle assessment assessed the impact of all relevant production processes to produce one kg of eggs. In that way the contribution of a production system to acidification, eutrophication, global warming, energy use and land use was calculated. Societal performance included the issues animal welfare and health. Animal welfare was assessed by using the Animal Needs Index (ANI, maximum 200 points). The ANI is mainly an environmental based method to gain insight into the potential animal welfare level on farm (Sundrum *et al.*, 1994; Bokkers and Koene, 2001). The ANI scores a production system at eight categories: locomotion, feeding and drinking, social, resting, comfort, nesting, hygiene, and care. Medicines used per laying round and mortality between 20 to 68 weeks of age were used to quantify health status. Another societal issue, product quality, was quantified by using the percentage second grade eggs and *Salmonella enteritidis* status. In the Netherlands, each flock is tested for *S. enteritidis* status. Results of these tests were used for this assessment.

Data were analysed with an analysis of variance for the effect of production system. When a significant effect was found a post hoc analysis was conducted with least square means adjustment for multiple comparisons with Bonferroni correction. Discrete data were analysed with the Fisher's exact test.

Results and discussion

OR had the smallest farms with on average 5673 ± 2648 laying hens per farm. Farm income per FTE per year differed between production systems ($F_{4,58}=20.24$, $P<0.001$) but a large variation in farm income was found in all egg production systems. Although OR realised the highest farm income (45784 ± 59762 €/FTE/year) and BC the lowest (-105355 $\pm$ 79475 €/FTE/year), results have to be interpreted with care. Data earlier obtained by Mollenhorst *et al.* (2006) were updated to the prices of 2005 which includes risks because technical results and prices differ per farm per year. Nevertheless, Mollenhorst *et al.* (2006) found also the lowest farm income for BC. The other indicators are less sensitive for short term effects.

For all ecological indicators a significant production system effect was found ($F_{4,58}=5.24$ to 49.27; $P<0.01$). OR contributed more to acidification than BC, DL and AO ($P<0.01$), and more to eutrophication than all other production systems ($P<0.001$) (Table 1). The higher acidification potential for OR was caused by a higher ammonia emission from manure due to the fact that manure was not dried in contrast to the other production systems. Eutrophication potential was higher for OR due to the combination of a higher ammonia emission and leaching from manure in the outdoor run. OR contributed less to global warming and energy use than the other systems ($P<0.001$) and no effect was found for land use. Global warming potential and energy use were lower for OR because artificial fertilizers are not allowed in growing crops for concentrates, less transport (shorter distances) and grains are less processed.

Table 1. Environmental impact for different egg production systems (mean ± s.d.), i.e., organic (OR), deep litter (DL), deep litter with outdoor run (DLO), aviary with outdoor run (AO), and battery cage (BC).

	OR (n=10)	DL (n=13)*	DLO (n=14*)	AO (n=11)*	BC (n=15)*
Acidification potential (SO_2-eq. / kg egg)	0.07 ± 0.01^a	0.06 ± 0.01^b	$0.07 \pm 0.00^{a,b}$	0.04 ± 0.01^c	0.03 ± 0.01^c
Eutrophication potential (NO_3^--eq. / kg egg)	0.62 ± 0.15^a	0.31 ± 0.04^b	0.41 ± 0.03^c	$0.35 \pm 0.02^{b,c}$	0.25 ± 0.04^b
Global warming potential (CO_2-eq. / kg egg)	2.91 ± 0.74^a	$4.31 \pm 0.47^{b,c}$	$4.59 \pm 0.29^{b,c}$	$4.19 \pm 0.30^{b,c}$	3.90 ± 0.29^b
Land use (m^2 / kg egg)	$5.03 \pm 1.44^{a,b}$	4.81 ± 0.54^a	$5.66 \pm 0.56^{a,b}$	$5.11 \pm 0.38^{a,b}$	4.51 ± 0.28^a
Energy use (kJ / kg egg)	0.87 ± 0.24^a	1.34 ± 0.19^b	1.39 ± 0.15^b	1.37 ± 0.11^b	1.30 ± 0.14^b

*Data originate from Mollenhorst *et al.* (2006)
[a,b,c]Data with different superscripts differ significantly (P<0.05)

ANI scores differed between systems ($F_{4,66}$=261.24, P<0.001). ANI score of OR (109 ± 6.7) was higher compared to BC (37 ± 4.5), DL (60 ± 9.2) and DLO (97 ± 9.0) (P<0.01) and did not differ from AO (114.4 ± 8.9). As expected, minimum demands for organic production as determined in the regulation had a positive effect on animal welfare. OR, however, did not score higher than AO, which was related mainly to the favourable scoring for number of animals per square meter, and the better feather condition of the birds in AO.

Mortality differed between the systems ($F_{4,51}$=2.98, P<0.05), but post hoc analysis showed that mortality in OR (12.5 ± 7.3%) was not different from other systems (5.8 ± 2.2% for BC to 12.5 ± 9.7% for DL). Data showed that mortality varied tremendously between farms and probably also between laying rounds. In general mortality is seen as a high risk in OR because beak trimming is not allowed. Birds can damage each other more easily due to severe feather pecking which can lead to cannibalism.

Medicine use was higher in all three production systems with outdoor run compared to the production systems without an outdoor run due to the use of anthelmintics (OR 73%, DLO 59%, and AO 69% versus DL 27% and BC 19% of the farms; P<0.05). No difference was found regarding the use of other medicines. Although farmers said that the use of anthelmintics differed per laying round, in general they agreed that anthelmintics are necessarily in egg production systems with outdoor run. The percentage second grade eggs tended to differ between the production systems (P=0.06) with OR with the lowest (3.8 ± 2.4%) and BC with the highest percentage (7.6 ± 2.2%). Except for one DLO flock, none of the flocks was contaminated with *S. enteritidis* because most flocks were vaccinated against *S. enteritidis*.

Conclusion

Although OR has a strong sustainability image, the results showed that OR did not contribute only positively to sustainable development. Improvements have to be made especially regarding mortality and the contribution to acidification and eutrophication.

References

Bokkers, E.A.M. and Koene, P. (2001). Activity, oral behaviour and slaughter data as welfare indicators in veal calves: a comparison of three housing systems. Applied Animal Behaviour Science 75: 1-15.

Haas, G., Wetterich, F. and Geier, U. (2000). Life cycle assessment framework in agriculture on the farm level. International Journal of Life Cycle Assessment 5: 345-348.

Mollenhorst, H. and de Boer, I.J.M. (2004). Identifying sustainability issues using participatory SWOT analysis. Outlook on Agriculture 33: 267-276.

Mollenhorst, H., Berentsen, P.B.M. and De Boer, I.J.M. (2006). On-farm quantification of sustainability indicators: an application to egg production systems. British Poultry Science 47: 405-417.

Sundrum, A., Andersson, R. and Postler, G. (1994). Tiergerechtheitsindex -200. Köllen Druck und Verlag GmbH, Bonn, Germany, 211p.

Considering the farmer-animal relationship in the development of sustainable husbandry systems for cattle production

X. Boivin[1], S. Waiblinger[2], A. Brulé[3], N. L'hotellier[4], F. Phocas[5] and G. Coleman[6]
[1]Unité de Recherche sur les Herbivores-Adaptation et Comportements Sociaux, Institut National de recherche Agronomique, UR1213 Herbivores, Site de Theix, F-63122 Saint-Genès-Champanelle, France, xavier@ clermont.inra.fr
[2]University of Veterinary Medicine, Institute of Animal Husbandry and Welfare, Veterinärplatz 1, A-1210 Wien, Austria
[3]Institut de l'Elevage, 35652 Le Rheu, France
[4]Caisse Centrale de la Mutualité Sociale Agricole, 40 rue Jean Jaurès, 93 547 Bagnolet Cedex
[5]Station de Génétique Quantitative Animale, Institut Nationale de Recherche Agronomique, 78 352 Jouy-en-Josas Cedex, France
[6]Animal Welfare Centre, Departement of Psychology, Monash University, P.O. Box 197, Caulfield, Vic. 3145, Australia

Abstract

European husbandry systems for cattle are subjected to considerable changes: herd sizes are increasing; loose housing stables and outdoor husbandry systems replace tie stalls; farmers' working time and organisation skills are particularly concerned due to the decrease in the number of workers on farm (e.g. less overlapping between generation) and due to the multi activity of the farmers. Numerous consequences occur through these changes that should be considered in the development of sustainable husbandry systems. What is often neglected is the conflict between the aim to keep up a high productivity level while saving time and the necessity of having enough time to properly watch and care for the animals and to build up a positive human-animal relationship. Our paper aims at highlighting this human factor when considering animal behaviour and welfare. Based on recent studies, our presentation will particularly discuss the consequences of a poor human-animal relationship in terms of animal welfare, human work and safety, as well as production. It will also present different research programs running in Europe that aim to develop practical solutions (farmers training, animal selection) to help farmers to solve occurring conflicts and to reach their targets. Regarding our understanding of the human-animal relationship, we will shortly discuss the need and the possible way to maintain a positive human-animal relationship, based on knowledge about animal behaviour, moral concerns and aspects of job satisfaction. By improving animal welfare, farmers' safety and job satisfaction these programs can contribute to the sustainability of animal husbandry/agriculture in Europe.

Keywords: animal welfare, stockmanship, human-animal relationship, sustainability

Introduction: sustainability and animal welfare

Sustainable agriculture should combine both the social and economic interests of the farmers, and ethical concerns about animal welfare and nature. The question of animal welfare addresses the respect of the animals' needs which is nowadays a strong societal concern and demand. Animal welfare can be defined as a state of (dis)harmony between an individual and its environment (Veissier *et al.*, 2007). If an animal is not able to cope with husbandry conditions its welfare is reduced and this progressively leads to animal pathology and probably death of this animal (Broom, 1986). Without considering the animals' needs as living beings, animals get transformed to 'sophisticated machines of production', being managed in intensive systems. Besides a negative impact on animal welfare those systems often are critical also from an ecological or social point of view (e.g. high use of energy and chemicals such as pesticides, antibiotics; high use of food imported from developing countries). However, farm animals are not machines, but living beings with physiological as well as psychological needs, which are in focus of welfare research since decades. However, the necessity to include these aspects into decision making was or is still not always recognised. Actually, the question of animal welfare directly addresses ethical questions on animals' husbandry conditions for various human purposes. The question of animal welfare consequently addresses also the question of responsibility of the first actors, the farmer and stockperson, in taking care of their farm animals. The quality of the human-animal relationship is nowadays particularly challenged, with the empirical observation (in absence of long-term studies) of more and more animals being fearful of humans. This tendency could strongly affect animal production and welfare on one side and safety, comfort and job satisfaction of people working with the animals on the other side (for review, Hemsworth and Coleman, 1998, Waiblinger *et al.*, 2006a). We will focus on cattle husbandry in Europe, in France in particular, as huge changes are presently occurring in this production sector. Traditionally, cattle were often kept in tie stalls and the calves separated from the dam immediately or soon after birth even in beef cattle (Raussi, 2003, Boivin *et al.*, 2005). In these systems, particularly in early age, calves received many human contacts, in association with food provision being a positive reinforcement. Farms with less then ten animals were common 30 years ago. However, these characteristics have strongly changed, particularly over the 10 last years. Adult cows are more and more kept in loose housing stables (Raussi, 2003, Boivin *et al.*, 2007). Very frequently, in the beef systems, calves are reared permanently with their dam. These changes are accompanied by a huge increase in herd size and herds of more than hundred cows are found in farms (Raussi *et al.*, 2003, Pichereau *et al.*, 1995). Farmers also tend to increase the mechanisation of their husbandry practices, in particular for feeding, identification of animals, or changing littering. We will question some reasons of these changes, their consequences and the problems both cattle and farmers face as a result. We will shortly discuss the need and the way to maintain a positive human-animal relationship, based on knowledge about animal behaviour, moral concerns and job satisfaction

The human point of view

When the society requires the development of new durable production systems, farmers need to adapt their working organisation and working time (Dedieu and Serviere, 2001). The stronger societal demand for control of food safety, impacts on environment, and animal welfare has multiplied the number of quality certification and control and is costly in working time. Farmers want to be more efficient and aware about the last innovations, and often want to participate in the commercialisation of their products. More and more farmers multiply their activities (different productions, additional job) to secure their income. They also want to have more leisure time (hobbies, holidays) in addition to their working time (Pichereau *et al.*, 2005). One direct consequence of these tendencies is the increase in productivity per working hour and the often reduced time farmers spend in close contact with their individual animals. On the other hand, they share their work with other persons available in their environment (other family members such as parents, children or other relatives). The family is still the basic structure of the farm (Pichereau *et al.*, 2005) but these unpaid help is decreasing with the decrease in overlapping of generations. There is a tendency towards increasing development of farmers' associations and the use of external services or employees. Another indirect consequence is found on their perception of their daily work as farmers. Is the care for their animals and contact with the animals still the heart of a farmer's job?

Recently, we investigated farmers' attitudes (e.g. in Austria or France) towards their animals in dairy and beef suckler cows. Attitudes towards their animals and especially beliefs about the effects of practices and the effects of human behaviour on animal behaviour vary among stockpeople. This variability has been related to cattle behaviour and stress, ease of handling and production (for review Hemsworth and Coleman, 1998, Waiblinger *et al.*, 2006a). What is especially interesting is that these attitudes towards animals, as well as tactile human contact are also related to the attitudes towards work and cleanliness of the animals (observed in veal calves, Lensink *et al.,* 2001), as well as to housing and management decisions (Waiblinger *et al.,* 2006b). Nevertheless, all of the European studies reported positive attitudes of many farmers towards their animals. There are still names given to the cows by a majority of farmers. The value of each animal is particularly high and loss should be limited. The majority of farmers also declared to avoid aversive interactions with their animals and considered that regular contact with the animals is the major factor to have animals easy to handle. Thus, they agree with the classical statement: 'The herd behaviour is the reflection of the stockperson behaviour'. So, many stockpeople face a paradox: they know that a good human-animal relationship needs repeated contact and declare to give importance to the quality of their human-animal relationship. But at the same time they develop practices and work organisation that could limit the possibility of positive interactions. Today, an important need for improved safety and efficiency during handling is expressed among cattle handlers. This could be caused by more difficulties linked to animal behaviour. In a recent survey in France on beef cattle breeders, 48% of the farmers asked stated that cattle are dangerous and 27% expressed difficulties when handling them (Boivin *et al.,* 2007). In another survey, 74% of Austrian dairy farmers stated that catching their cows is at least partly difficult (unpublished data). These attitudes are confirmed by statistics from a survey of the French agricultural Insurance company (MSA, 2006). 15 to 20% of farmers' accidents occurred when handling cattle. MSA believes that these indicators are the sign of a lack in farmer's knowledge about their animals' behaviour. These work dimensions could certainly affect job satisfaction and working organisation which in turn can also affect farmers' practices, behaviour towards the animals and then animal welfare (Hemsworth and Coleman, 1998).

The animal point of view

Solving the problems of animal behaviour and animal welfare in relation to changes in animal husbandry and human behaviour needs to consider the animals' point of view (Boivin *et al.,* 2003, Waiblinger *et al.,* 2006a). This idea is based on different considerations:

1. The phylogeny of the cattle species that modulates the genetic potentiality of each individual. Despite millennia of domestication, it is particularly important to remember that cattle are prey animals and that they have developed sensory abilities and behavioural strategies along their 'natural history' to limit predation. In particular, they could be easily characterised by their gregariousness, their propensity to suddenly flee a perceived danger and, under special conditions, their active defence against a detected predator, specifically in presence of their calves. Cattle are large, heavy and powerful, and so, associated with anti-predator strategies, they sometimes are difficult to handle. These predispositions may be of particular importance when husbandry practices limit human contact.

2. The living history of each individual, in particular during sensitive periods of their life (early age, weaning, calving...). Cattle are able to communicate with their stockpersons; they can recognise them, and remember easily the issue of their interactions with humans. Estep and Hetts (1992), using the concept of inter-individual relationship considered the human-animal relationship rather as a learning process. Based on intra-specific social mechanisms, cattle build a relationship with their stockperson that allows them to predict the issue of future interactions. Fear provoking stimuli (shout, hits with a stick, quick movements) often are used when handling cattle. However, animals can associate these stressful cues with the human in general. They can become more and more fearful, difficult and dangerous to handle with high risks of injuries. Regular negative interactions can affect animal welfare and production through chronic stress (Hemsworth and Coleman 1998). By contrast, positive contact, or the higher ratio between positive and negative contact, during sensitive periods of the animal's life and during daily interactions, can even transform the human to become a social-like partner to them, allowing the establishment of a confident relationship between animals and their human partner (Boivin *et al.,* 2003, Waiblinger *et al.,* 2006a). Many other events

can affect the human-animal relationship: Since birth, technical actions (e.g. sanitary treatments, dehorning...) that are perceived as aversive by the animals can increase their fear of humans. In contrast, maternal or automatic feeding reduce the rewarding association between humans and feeding. Social factors can be important. Rearing the calves permanently with their dam in loose housing stables or outdoors made the calf more focused on its dam and more easily influenced by her possible fearful or defensive reaction towards humans. The increase of herd size will also limit the familiarity and the individual interaction between each animal and the stockperson. Fearful animals can more easily hide among the others, thus limit their chance to receive positive contacts and so to change their negative perception of the human. Therefore, even if group housing has many beneficial effects on animal welfare, it also may increase the risks of negative perception of the human by the animals (Raussi, 2003). This strongly depends on the farmer's implemented behaviour and practices regarding human-animal interactions and thus his/her attitudes and decisions - often in conflict with other necessities and priorities as mentioned above.

The possible solutions

The possible solutions for maintaining or reach a good human-animal relationship need a multi-factorial approach based on knowledge about animals' behaviour, moral concerns and job satisfaction. The first step to find solutions to these problems is to make the actors (farmers, breeding organisations, animal scientists...) aware of the fact that animal behaviour and animal welfare will directly reflect husbandry changes. The need to maintain a good human-animal relationship seems not obvious at first sight. The difficulties can probably be found in the multi-factorial determinism (genetic, ontogenetic) of the human-animal relationship and the important role played by the attitudes and behaviour of the farmer. In addition, evaluations of animal behaviour and welfare are still perceived as subjective by many farmers and animal scientists and many prefer to ignore them.

Three research programs are presently running to propose solutions to maintain a good human-animal relationship in future husbandry systems. The first program named 'Handling stress' is a part of an integrated European program entitled 'Welfare Quality'[4]. The 'Handling stress' program investigates the variability among farmers with respect to farm characteristics, farmers' practices and attitudes that can influence the human-animal relationship and by consequence stress during handling. Its applied goal is to propose a training program targeting attitudes and behaviours of the stockpeople towards the animal in order to improve the human-animal relationship and consequently animal welfare. The second research program is a European Leonardo project called 'CAFRAT'[5]. It is involving agricultural insurance companies and scientific institutes and addresses the question of handlers' safety in relation to animal behaviour during handling. Handlers do generally not consider the knowledge on the animal's behaviour towards humans when such accidents during handling occur. The project aims at defining common guidelines for the training of professionals working with cattle and horses on the basis of animal behaviour. The project will help professionals to establish a diagnosis of their own competencies and promote their knowledge. The third project, a French program called 'COSADD'[6] addresses the question of the development of genetic selection criteria, in particular for improving cattle docility and animal welfare and their acceptability by farmers, breeding organisations and representatives of the society. We believe that these programs can contribute to the development of sustainable agricultural

---

[4] Welfare Quality® Research Project co-financed by the EC 6th Framework Programme, contract No. FOOD-CT-2004-506508.

[5] 'Comportement Animal dans la Formation afin de Réduire les Accidents du Travail', Leonardo di Vinci European program, FR/05/B/P/PP-152034.

[6] 'Critères et Objectifs de Sélection Animale pour un Développement Durable'. Programme Agriculture et Développement Durable.

systems by improving animal welfare, animal production, farmers' safety and job satisfaction, without imposing new control and time loss for the farmers. This question of the minimal working time necessary for good-human animal relationship and good human care is really a critical point in developing new sustainable system that are respectful of both humans and animals. However, as the programs target different factors playing in interaction, they challenge different moral questions:

Should we genetically select animals, even for improving animal welfare, when it is possible to improve their reaction to handling by adapted taming practices? In the beginning of the seventies, in New-Zealand, an easy care system was applied or evaluated (Fisher, 2003). Sheep were living freely on pasture with minimal human care. Flocks of sheep benefited with respect to animal welfare (freedom and grass feeding). This system was also beneficial for stockpeople, having to spend a lot less time for taking care of their numerous animals. Very quickly, a natural selection process occurred, with high mortality among lambs and many ewes culled. But the population survived with the more robust animals and with one lamb per ewe, leading to an increase in the number of animals and an increase in farm productivity. This abrupt change in the philosophy of the human-animal relationship is now progressively occurring. Researchers are developing experimental protocols to select animals that are more autonomous considering human care and more robust to harsh environments (Boissy *et al.*, 2002). If this approach is not different from the traditional domestication process (selection of the best adapted animals) and could help farmers with their work, we should be aware that increased autonomy will nonetheless not decrease farmer's responsibility for taking care of their animals.

What role and what competency for the present and future stockpeople? We believe that for many European cattle farmers, the contact with their animals and the interactions they exchange is a part of their competencies and work motivation. It is necessary to decrease hard working conditions and to allow farmers to save time by helping them to achieve a better work organisation. Saved time may at least partly be invested in positive contact to the animals: all what we demonstrate in this paper indicates that the human-animal relationship should not be neglected, but by the contrary should be considered as central point. The human competencies to take care of the animals, based on adequate knowledge and attitudes towards animal and human behaviour, should be more recognised and maintained or improved if necessary. Stockpeople would then have a better job satisfaction, feel more responsible for their animals and their welfare and show a better image of their profession. Rather than neglecting these aspects, would it not be interesting for the professionals to compare stockmanship taking care of their animals to medical professionals taking care of their patients? Indeed, Vaarst *et al* (2004) compared stockpeople to nurses. Larrère and Larrère (2000) imagined a 'social contract' for mutual benefits between domestic animals and their human caretakers. With the strong responsibility of the stockperson in these questions, the professional organisation thinking about sustainable husbandry systems could elaborate professional ethic codes and structures of control. A professional order, under public control, could be allowed to deliver licences to stockpeople after training courses on such aspects. It could also have the competence/right to remove the licence for those who would not respect this code and help them to improve or change for other jobs. This could be a way to practically and socially build the recognition of these competencies essential for durable husbandry systems. However, this would need large changes in the whole ethics and legislation of society.

References

Boissy, A., Le Neindre, P., Gastinel, P.L., Bouix, J . (2002). Geneticque et adaptation comportementale chez les ruminants: perspectives pour améliorer le bien-être en élevage. INRA Productions Animales, 15: 373-382.

Boivin, X., Lensink, J., Tallet, C. and Veissier I. (2003). Stockmanship and Farm Animal Welfare. Animal Welfare 12: 479-492.

Boivin, X. and Le Neindre, P. (2005). The stockperson as a social partner to the animal? A stake for animal welfare. In: M. Marie, S. Edwards, G. Gandini, M. Reiss and E. von Borrel (eds.) Animal Bioethics: Principals and teaching methods. Wageningen Academic Publishers, pp. 113-132.

Boivin, X., Marcantognini L., Boulesteix, P., Godet J., Brulé., A. and Veissier, I. (2007) Attitudes of farmers towards their Limousine cattle and their handling. Animal Welfare: 16: 147-151.

Broom, D.M. (1986). Indicators of poor welfare. British Veterinary Journal 142: 524-526.

Dedieu, B. and Serviere, G. (2001). Organisation du travail et fonctionnement des systèmes d'élevage. Rencontres Recherches Ruminants 8: 245-250.

Estep, D.Q. and Hetts, S. (1992). Interactions, relationships, and bonds: the conceptual basis for scientist-animal relations . In: H. Davis and D. Balfour (eds.) The Inevitable Bond: Examining Scientist-Animal Interactions. Cambridge University Press, Cambridge, pp. 6-26.

Fisher, M. (2003). New Zealand farmer narratives of the benefits of reduced human intervention during lambing in extensive farming systems. Journal of Agricultural and Environmental Ethics 16: 77-90.

Hemsworth, P.H. and Coleman, G.J. (1998). Human-Livestock interactions: the stockperson and the productivity and welfare of intensively farmed animals. CAB International, Bristol, UK, 152p.

Larrère, C. and Larrère, R (2000). Animal rearing as a contract? Journal of Agricultural and Environmental Ethics 12: 51-58.

MSA, 2006. http://www.cafrat.eu

Lensink, B.J., Veissier, I. and Florand L. (2001). The farmers' influence on calves' behaviour, health and production of a veal unit. Animal Science 72: 105-116.

Pichereau, F., Becherel, F., Farrie, J.P., Legendre, J., Veron, J., Lequenn, J, Mage, C., Servière, G., Cournut, S. and Dedieu, B. (2004): Fonctionnement des grands troupeaux de vaches allaitantes: analyse des déterminants structurels et techniques de l'organisation du travail. Rencontres Recherches Ruminants 11: 129-136.

Raussi, S. (2003). Human-cattle interaction in group housing. Applied Animal Behaviour Science 80: 245-262.

Vaarst, M., Wemelsfelder, F., Seabrook, M., Boivin, X. and Idel, A. (2004). The role of humans in the management of organic herds. In: M. Vaarst, S. Roderick, V. Lund and W. Lockeretz (eds.) Animal Health and welfare in organic agriculture. CABI Publishing, CAB International Wallingford Oxon, pp. 205-225.

Veissier, I., Beaumont, C. and Levy, F. (2007). Les recherches sur le bien-être animal:buts, méthodologie et finalité. Inra Production Animal 20: 3-10.

Waiblinger, S., Boivin, X., Pedersen, V., Tosi, M.-V., Janczak, A.M., Visser, E.K., and Jones, R.B. (2006a). Assessing the human-animal relationship in farmed species: A critical review. Applied Animal Behaviour Science 101: 185-242.

Waiblinger, S., Mülleder, C., Menke, C. and Coleman, G. (2006b). How do farmers' attitudes impact on animal welfare? The relationship of attitudes to housing design and management on dairy cow farms. In: Amat, M. and Mariotti, V. (eds.) Proc. of the 15th Annual Conference of the International Society for Anthrozoology, 5-6 October 2006, Barcelona, Spain, pp. 55-56.

From the backyard, through the farm, to the laboratory - changes in human attitudes to the pig

Reinhard Huber[1], I. Anna S. Olsson[1, 2], Mickey Gjerris[2] and Peter Sandøe[2]
[1]IBMC - Instituto de Biologia Molecular e Celular . Universidade do Porto, Rua do Campo Alegre 823, 4150-180 Porto, Portugal
[2]Danish Centre for Bioethics and Risk Assessment, University of Copenhagen, Rolighedsvej 25, DK-1958 Frederiksberg C, Denmark

Abstract

Over its at least 9000 years of common history with humans, the pig has moved out of the pigsty, into the farrowing crate or fattening pen of the pork production industry, and more recently into the laboratory, where it is a standardized model animal. Likewise, human attitudes have shifted: pigs were once regarded almost as family members, then as production units, and then as pivots in the development of new medical treatments. The way we view animals doubtless influences the way we treat them. The intensification of farm animal production has lead to a devaluation of the individual animal. After an introduction to the different roles of the pig, in this paper we discuss whether the increased economic and scientific value of the research model pig may result in its acquisition of elevated status and better treatment. Possible arguments are that humans will treat pigs better the more they contribute to our welfare (out of gratitude), the closer we see them as being to us, or simply to compensate for the wrongs we do to some of them. However, given the human tendency to fit animals to our own purposes, it is questionable whether an increased perception of individuality will improve the treatment and welfare of pigs in general.

Keywords: human-animal relationship, animal ethics, animal welfare, reverence, pig

The history of the human-pig relation

All domestic pigs descend from the wild boar, *Sus scrofa*, which is still a widespread species in many parts of Europe (Clutton-Brock, 1999). Recent genetic data suggest that the domestication of pigs occurred in various localities across Eurasia at least 9000 years ago (Barrios-Rodilles *et al.,* 2005). In the past, people and pigs lived in a very close relationship, the pigs being fed leftover food by their owners. It has been argued that it is feelings of guilt and shame evoked by killing these almost household members that brought about the large repertoire of pejorative expressions connected with pigs in everyday language (Stibbe, 2003).The traditionally close relationship is also reflected in the frequent appearance of the pig in the arts (e.g. literature), myth, legend and religion, often but not always in a negative context (Meyer, 1992).

Small-scale pig farming is still widespread in developing countries (Lemke *et al.,* 2006, Wabacha *et al.,* 2004), above all in Asia. It also seems to retain importance - if not economically, then at least culturally - in some European countries (e.g. the Czech Republic, France, Italy, Poland, Portugal and Slovakia), and this has prompted the European Union to grant special legal regulations governing holdings with no more than one pig (Commission Decision 2006/80/EC).

In general, however, pig husbandry and meat production in industrialized countries (although not only there: Devendra, 2007) has been characterized by specialization and intensive stock rearing, or factory farming, during the last few decades; and this has led to a shift of focus, from the individual animal to groups of animals (Hendriks *et al.,*1998). Unlike other domestic mammals, such as cattle and small ruminants, pigs are not required by EU legislation to be individually identified and registered (Council Directive 92/102/EEC), with the exception of breeding animals. Individual health problems and their origins are given very little attention in intensive pig production (Olsson and Sandøe, 2006). With its low input of labour, energy and space per individual animal, and with its handling of animals in bigger production units (herds), and its specialization and standardized production procedures, intensive animal production reaches strikingly high levels of biological and economic productivity.

People working in the pig industry tend to speak about pigs in rather technical language. Here productivity is the main concern, and expressions such as 'herd health' dominate the discourse at the expense of attention to the individual animal (Stibbe, 1996). Efforts to improve intensive pig production focus on increased competitiveness, a potential reduction of environmental pollution and the potential for more stringent quality management during the production process.

However, across society as a whole, particularly in Europe, there is a growing awareness of animal welfare issues (Mennerich-Bunge, 2003), and this has prompted several organizations involved in meat production and trade to adopt codes of conduct on animal welfare (Lassen *et al.*, 2006). Again, both at EU and national level, legal instruments safeguarding minimum standards of welfare for pigs have now been introduced.

The ethics of keeping animals for food production has been questioned, and several of the most influential writers on animal ethics defend the abolition of most (Singer) or all (Regan) animal production. The issues at stake here are principally animal welfare and the killing of animals, but also resource use and environmental impact (see Lund and Olsson, 2006 for an overview). The pig occupies a special position where two of these issues are concerned: killing and resource use. While it is possible to obtain some products (eggs, milk or wool) from the other common European farm animals without killing them, the farmed pig is kept exclusively for its meat. As opposed to the ruminants, the monogastric and omnivorous pig feeds on grain rather than forage and therefore competes with humans for feedstuff. On the other hand, pigs produce less of the greenhouse gas methane than ruminants (Steinfeld *et al.*, 1996). Moreover, pigs are not always kept to be killed: they are valued as pets and used for searching for drugs and truffles (Meyer, 1992).

Pigs were used in biomedical research by early anatomists and physicians such as Erasistratus of Chios (310 BC- 250 BC; vascular system) and Galen (AD 129-200/216; anatomy, nervous system). Later, during the renaissance, they were employed in often cruel experimental vivisection (Guerrini, 2003).

In the last few decades domestic pigs have become increasingly valued as animal models in biomedical research. This is because they are widely available and have anatomical and physiological similarities with humans - a comparable organ size and a similar digestive system (AWIC, 2000; Bassaganya-Riera *et al.*, 2004; Domeneghini *et al.*, 2006; Simon and Maibach, 2000; Swanson *et al.*, 2004). Published studies of domestic breeds and miniature breeds now range widely, covering xenotransplantation, cardiovascular studies, agricultural applications (meat quality and disease resistance), nutritional studies, skin research, immunological studies, and much else besides (Bassaganya-Riera *et al.*, 2004; Domeneghini *et al.*, 2006; Wheeler and Walters, 2001; Whitelaw and Sang, 2005). Pigs are easier both to obtain and to keep than non-human primates and stir fewer emotions in the public mind. They are also more suitable than rodent models in studies of arteriosclerosis and myocardial infection, and in general in cardiovascular studies, as well as in studies of ataxiatelangiectasia, where their advantages are connected with gene homology with humans (Bassaganya-Riera *et al.*, 2004; Swanson *et al,* 2004).

Controversial techniques like cloning and transgenesis are already used for research purposes on pigs in order to obtain viable disease models, gather knowledge about the genome and advance various other fields like xenotransplantation research (Niemann *et al.*, 2003; Niemann *et al.*, 2005). Genetically engineered animals have been, and continue to be, produced with the aim, among other things, of providing valuable models to study disease progression and disease control (Paterson *et al.*, 2003; Whitelaw and Sang, 2005). Questions about the welfare of pigs used in biomedical research remain to be investigated, as data are currently available only for farm pigs (Bollen and Hoitinga, 2004).

Changing perceptions of pigs

The pig now seems to have come full circle in its relationship with humans. Things began with the keeping of individual animals, or small groups of animals, which were crucial to the survival of the human communities that bred them. The pig then became a smallholder's animal, as dependent on humans as humans were on the meat they could gain from keeping it. Over the last half century of modern industrial farming the significance of individual animals has diminished almost to nothing. Individual pigs have come to be seen exclusively as production units. With modern biomedical research, however, it may be that pigs are on their way back to being reinstated as individual animals, owing to their high economic value and to their importance in the understanding, and possible treatment, of serious human disease.

It has been suggested that this return to a focus on the individual animal will lead to better treatment of the pigs. The argument goes like this. As we (again) become more aware of our dependence on animals, in this case pigs, and understand how crucial their contribution is to our welfare, we humans shall re-evaluate the way we treat pigs; and out of the sense of gratitude we develop, we shall ensure that the general welfare of pigs improves. At the same time, our growing understanding of the complex social behaviour of pigs, and of the similarities between our biology and that of the rest of the animal kingdom, will enhance our feelings of reverence for nature in general - but especially for animals that we can see are closest to us, both from a biological and a practical perspective.

Furthermore, it has been argued that humans ought to compensate for the wrongs they do to some animals by increasing the welfare of others. Thus if we choose to sacrifice the welfare of certain pigs in the name of medical research, we ought to increase the welfare of other pigs to compensate for this. This approach is not based on a hazy notion that the relevant 'payment' somehow balances the calculation. The idea is rather that compensation gives us a way of expressing our sense of reverence[7] towards the animals that provide us with means to increase our own welfare (Sørensen, 2001).

This suggestion, however, that an increased focus on the individual animal will lead to improvements for the animals is questionable. First of all, it appears to presuppose that in earlier times, when the individual animal was focused upon more, animals fared better than they do in industrialized farming. This is not self-evident. It has to be shown that the welfare of animals in less intensive production systems was (and is) higher than it is in industrial systems. But this claim certainly hinges on which concept of animal welfare one uses (Gjerris *et al.*, 2006). And even if a definition can be agreed upon, it remains to be shown that this standard of welfare was (or will be) brought about by the higher individualization of the animals. An alternative explanation would be that human powers over animals were once limited to the extent that humans had, in some sense, to work together with the animals. That is, animals that were not faring well would not have provided the services needed (in terms of labour, protein, wool and so on). Thus adequate animal welfare was a question of prudence. In fact it can be argued that as human understanding of, and power over, animal biology grew after the second world war, as a result of scientific and technological developments, the idea of a sort of mutual dependence between humans

---

[7] It should be noted that the term 'reverence' is used without religious connotations here. It refers simply to the feelings that the experience of our lives with animals induce in us. A feeling of reverence can in this context be understood as a feeling of gratitude for the involuntary and unconscious 'sacrifice' of the animal to our benefit (Woodruff 2001).

and animals was replaced by a notion of animals as production units that could be put under production pressure and more or less 'fitted' into more efficient production systems. As Bernard Rollin remarks, the framework went from putting square pegs into square holes to putting round pegs through square holes with the help of medicine, breeding technologies and the presumed low value of the individual animal (Rollin, 1995).

The connection between human appreciation of animal welfare and the individualization of animals is, therefore, no guarantee that increased use of very valuable pigs in the biomedical sector will lead to a greater reverence for pigs in general and thus rising levels of pig welfare in general. It seems just as possible that, as we diversify the number of ways that pigs can be utilized, and as we use increasingly advanced biotechnologies to change the pig's basic biology, humans will become even more accustomed to an essentially instrumental view of the pig.

It is already widely believed that it becomes more acceptable to harm animals as the reason for inflicting the harm becomes more substantial. Thus using animals for research into serious diseases is seen as much more acceptable than using animals for research into more efficient meat production (Lassen *et al.*, 2006). The fact that animals have, for hundreds of years, played an important role in science and biomedicine does not seem to have brought about increased awareness of the importance of animals, at any rate of the sort reflected in a greater reverence for animals. So although the importance of avoiding animal suffering is appreciated in western culture today, this could very well be counter-balanced, where pigs are concerned, by the perceived usefulness of pigs. Obviously, that would lead to the acceptance of more suffering even in animals that are more valuable and individualized than they were previously. (It is true, however, that this would be unlikely to occur if the reasons for the increased suffering were not taken to justify it, e.g. if the goal were merely improved efficiency in meat production.)

As pigs move out of the pigsty and into the laboratories, the likelihood that this will have a positive impact on the species, or on individual animals, thus seems very small. History suggests that although the value of the individual animal might increase, this increased value will be mainly economic in nature. Again, while the individual pig in a biomedical research facility may receive more attention and veterinary care than the average farm pig, the positive consequences thereby secured may be offset by the cost, to the animal, of invasive research.

References

Animal Welfare Information Center (2000). Information Resources on swine used in biomedical research. In: Smith C P (ed) AWIC Resource Series No. 11 www.nal.usda.gov/awic/pubs/swine/swine.htm

Bassaganya-Riera, J., King, J. and Hontecillas, R. (2004). Health benefits of CLA - lessons from pig models in biomedical research. European Journal of Lipid Science and Technology 106: 856-861.

Bollen, P. and Ritskes-Hoitinga, M. (2004). The welfare of pigs and minipigs. In: Kaliste E. (Ed.) The Welfare of Laboratory Animals. Kluwer Academic Publishers Dordrecht, The Netherlands, pp. 275-279.

Clutton-Brock, J. (1999). A natural history of domestic mammals Cambridge University Press 238p.

Devendra, C. (2007). Perspectives on animal production systems in Asia. Livestock Science 106: 1-18.

Domeneghini, C., Di Giancamillo, A., Arrighi, S. and Bosi, G. (2006). Gut-trophic feed additives and their effects upon the gut structure and intestinal metabolism. State of the art in the pig, and perspectives towards humans. Histology and Histopathology 21: 273-283.

Gjerris, M., Olsson, A.,and Sandøe, P. (2006). Animal biotechnology and animal welfare. Ethical eye - Animal welfare. Council of Europe Publishing.

Guerrini, A. (2003). Experimenting with humans and animals. From Galen to animal rights. The Johns Hopkins University Press Maryland, USA 165p.

Hendriks, HJM., Pedersen, BK., Vermeer HM. and Wittmann, M. (1998): Pig housing systems in Europe: current distributions and trends. Pig News and Information, Cabi Publishing.

Lassen, J. and Sandøe, P., (2006). Happy pigs are dirty! - conflicting perspectives on animal welfare. Livestock Science 103: 221- 230.

Lassen, J., Gjerris, M. and Sandøe, P. (2006): After Dolly - Ethical limits to the use of biotechnology on animals. Theriogenology 65: 992-1004.

Lemke, U., Kaufmann, B., Thuy, L.T., Emrich, K. and Zarate, A.V. (2006). Evaluation of smallholder pig production systems in North Vietnam: Pig production management and pig performances. Livestock Science 105: 229-243.

Lund, V. and Olsson, I.A.S. (2006). Animal agriculture: Symbiosis, culture or ethical conflict? Journal of Agricultural and Environmental Ethics 19: 47-56.

Meyer, H. (1992). 10,000 years 'high on the hog': some remarks on the human-animal relationship. Anthrozoös 5: 144-159.

Niemann, H., Kues, W. and Carnwath, J.W. (2005). Transgenic farm animals: present and future Revue Scientifique et Technique-Office International des Epizooties 24: 285-298.

Niemann, H., Rath, D. and Wrenzycki, C. (2003). Advances in Biotechnology: New tools in future pig production for agriculture and biomedicine. Reproduction in domestic animals 38: 82-89.

Olsson, I.A.S. and Sandøe, P. (2006). Ethical considerations in Danish pig breeding - a report for Danske Slagterier. 13 pp.

Paterson, L., DeSousa, P., Ritchie, W., King, T. and Wilmut, I. (2003). Application of reproductive biotechnology in animals: implications and potentials Applications of reproductive cloning. Animal Reproduction Science 79 : 137-143.

Rollin, B.E. (1995). The Frankenstein Syndrome. Ethical and social issues in the genetic engineering of animals. Cambridge University Press, Cambridge.

Steinfeld, H., de Haan, C. and Blackburn, H. (1996). Livestock - Environment interactions: Issues and options. 59p. Available at: http://www.fao.org/docrep/x5305e/x5305e00.HTM#Contents

Sørensen, M. (2001). Xenotransplantation i et dyreetisk perspektiv, i Jørgensen, Eva og Nordentoft, Eva (red.): Hjerte for svin - Hjerte fra svin. Det økumeniske Center, Århus, pp. 29-36.

Stibbe, A. (2003).As charming as a pig: the discursive construction of the relationship between pigs and humans. Society and Animals 11: 375-392.

Swanson, K.S., Mazur, M.J., Vashisht, K., Rund, L.A., Beever, J.E., Counter, C.M. and Schook, L.B. (2004). Genomics and clinical medicine: rationale for creating and effectively evaluating animal models. Experimental Biology and Medicine 229: 866-875.

Wabacha, J.K. (2004). Characterisation of smallholder pig production in Kikuyu Division, central Kenya. Preventive Veterinary Medicine 63: 183-195.

Wheeler, M.B. and Walters, E.M. (2001). Transgenic technology and applications in swine. Theriogenology 56: 1345-136.

Whitelaw, C.B.N. and Sang, H.M. (2005). Disease resistant genetically modified farm animals. Revue Scientifique et Technique - Office International de Epizooties 24: 275-283.

Woodruff, P. (2001). Reverence. Renewing a Forgotten Virtue. Oxford University Press.

Logistics at transport to slaughter: food and environment-optimised animal transport

Sofia Wiberg[1], Anne Algers[2], Bo Algers[1], Ulrika Franzén[3], Magnus Lindencrona[3], Olof Moen[3], Sofia Ohnell[3] and Jonas Waidringer[3]

[1]*Swedish University of Agricultural Sciences, Department of Animal Environment and Health, P.O.Box 234, 532 23 Skara, Sweden*
[2]*Swedish University of Agricultural Sciences, Department of Food Science, P.O.Box 234, 532 23 Skara, Sweden*
[3]*WSP Analys och Strategi, Rullagergatan 6, 415 26 Göteborg, Sweden*

Abstract

Transport to slaughter affects animal welfare and has a negative impact on the environment.

The aim of this pilot study was to investigate the possibilities to improve logistics of transport to slaughter in medium scale abattoirs, with a simultaneous decrease in environmental load and improvements in transport conditions. Data on animal welfare and transport routes and routines were collected for animal welfare analysis, transport simulations and environmental impact calculations. The analysis shows that it is common to collect single animals from the farms which contribute to multiple stops and an increase in transport time. Mixing of animals is common both in the transport and at the abattoir. The transports are in general relatively short. The time from when the animals arrive at the abattoir to the time of slaughter varies greatly; the time in lairage varies between 1 minute and 27 hours. Almost 50% of the pigs and 20% of the cattle are kept in lairage overnight. The transport simulation shows good possibilities to improve logistics by collecting the same amount of animals in shorter time with reduced distance driven; 58% reduction of number of vehicles, 23% reduction of total time and 30% reduction of distance driven. Calculations on the optimized scenario show a 20% reduction of HC-emissions and a 12% reduction of CO_2. The conclusion from this pilot study is that transport optimization can result in simultaneously increased animal welfare, reduced costs and reduced environmental impact.

Keywords: animal welfare, animal transport, emissions

Introduction

There is an increasing consciousness of animal welfare in food production and societal demands on the transport system are high regarding animal welfare and environmental impact. Research shows that transports can be detrimental to animals leading to reduced welfare. The transports can cause stress and injuries on the animals which also can affect the meat quality (Atkinson, 2000). The profitability in the Swedish meat industry is low and the slaughter industry is moving in a direction toward fewer and larger abattoirs with increasing areas of service.

The work presented here was conducted as a pilot study. The aim was to investigate the possibilities to optimise transport to slaughter in medium scale abattoirs and to improve transport conditions for the animals with a simultaneous decrease in environmental load and transport costs.

Material and methods

Data on animal welfare and transport routes and routines were collected from a medium scale abattoir in Sweden in June-July 2006 through questionnaires and visits to the abattoirs. At the abattoir distance, times, number of animals, etc. was recorded. This data was used in both an animal welfare analysis and in a transport simulation. The latter also served as basis for environmental impact calculations where a comparison between the performed transports and optimised scenarios was made.

Records on 439 of 507 slaughtered cattle and 4353 of 4422 slaughtered pigs from 178 collections divided in 85 rounds were obtained. Of these, 101 were collections of cattle, 76 of pig and 1 of sheep. Since there is only one recorded collection of sheep no optimisation was made on transport of sheep. The transports were conducted with 17 vehicles. Of the vehicles, two belonged to the abattoir, four to farmers and the rest to private hauliers.

In the questionnaire, farmers, transporters and representatives from the abattoirs were asked about the attitudes towards a number of different changes that could make the animal transport system more optimal.

Results

Vehicle 1-4 each performed between 17 and 26% and together 81% of the collections. Group size at collection of cattle to the abattoir was low; one single cattle was collected in 31% of collections. In 34% of the collections, the groups were of more than three animals. Pigs were collected in groups of 1-10 animals in 36% of the cases (see Figure 1).

The mean total travelling time was 2 hours and 36 minutes for cattle and 2 hours and 11 minutes for pigs.

The proportion of animals lairaged overnight was high, especially for pigs; 114 cattle (23%), 1990 pigs (45%) and 5 sheep (10%) (see Figure 2).

Mixing of animals was recorded on the transport and upon arrival to the abattoir. Transporters answered that 19% (n=75) of cattle and 78% (n=46) of the pigs were mixed in the transport vehicle. Mixing in lairage was recorded in 82% (n=22) and 100% (n=25) for cattle and pig, respectively.

Optimized transports

The traditional way of transport planning is done manually with pen and paper. During the last 2-3 years, in-car navigation with digitalized maps and driving restrictions has been deployed nationally in

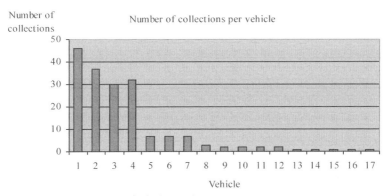

Figure 1. Number of collections per vehicle, (n=178).

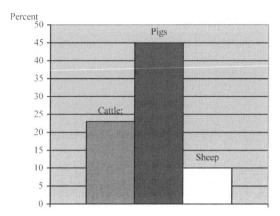

Figure 2. Animals kept in lairage overnight.

Sweden. Digitalized maps in combination with route optimizing computer programs are tools that can revolutionize transport logistics.

For animal transports, there are several conditional elements that need to be considered to accomplish a good transport; e.g. legislation, overnight lairage, transport conditions, access to animals and access to vehicles and transporters. But there is also an advantage compared to many other parts of industry - a large 'time window' for the collection of goods. A pig ready for slaughter has about a five day interval where slaughter can be done before it gets too heavy. For healthy cattle, the interval is about three weeks, which is about the time the farmer accepts to wait. This means there are five days and 3 weeks, respectively, to plan the transport to slaughter. In the slaughter industry, the planning can be done during regular working hours with good planning in advance.

At present, transports to the abattoir are performed using 17 vehicles with different starting points; some start at the abattoir, others where the transporters live.

The first scenario, 'Present' was done to recreate the performed transports as they were done in reality, deleting the five transport vehicles which only marginally contributed to the transports. Some information from the forms that were filled in at the visits to the abattoirs was used, and some basic conditions were set up, for example: Vehicles started at their present starting points and returned to the same place at the end of the day, the number of collections of animals was 178 and the time to unload and wash the vehicles was estimated to 50 min. In the second scenario, 'Optimised I', a computer program was used to optimize the transports. The program could choose what day and what vehicle to use for each round. The basic conditions were, for example: Only the largest vehicles were used, the vehicles started at their present starting points and returned to the same place at the end of the day, a maximum transport time of eight hours and no animals kept in lairage over night. Another optimization was done, 'Optimised II'. Here, the basic conditions were still maximum transport time of eight hours and no animals kept in lairage over night, but also that only five of the largest vehicles were used and that the vehicles all had the same starting and return point (the abattoir) (see Table 1).

Environmental and economic impact

Calculations on emissions have been made for the scenarios 'Present' and 'Optimized II'. These calculations are based on the transporters' data on loading capacity, fuel consumption and driving

Table 1. Different scenarios after optimisation.

	Present	Optimised I	Optimised II	% reduction present -optimised II
Vehicles, n	12	7	5	- 58%
Distance, km	14153	12167	9894	- 30%
Time, min	26720	22988	20653	- 23%
Rounds, n	85	51	58	- 32%
Time/round, min	326	390	356	+ 9%
Distance/round, km	167	239	171	+ 2%

distances. Data on emissions from vehicles of different Euro classes are collected from 'nätverket för transporter och miljö' (2007; [Network for transports and environment]).

The optimisation shows large potential to reduce emissions (see Table 2). The CO_2 -emissions are reduced in relation to driving distance. Thus what is effective from a commercial viewpoint also reduces the environmental impact. To decrease emissions further (NOx, HC and particles), replacement of old vehicles was shown to have the largest effect. Euro 2 and Euro 3 show the results from vehicles with current EU-standards. Subsidies or legal demands are ways to speed up such replacements.

The optimised scenario indicates that the abattoir can reduce the number of vehicles and the amount of time used for transport as well as the lairage area.

Table 2. Emissions as% reduction of scenario'Present'.

	Optimised route	Euro 2	Euro 3
Distance	11%	11%	11%
HC	20%	34%	42%
NO_x	16%	4%	27%
PM	20%	40%	42%
CO_2	12%	12%	12%

Attitudes

The studies on attitudes show that it is a general belief that picking up animals earlier in the day to avoid overnight lairage is positive for animal welfare and that the use of pre loading facilities is viewed as being positive for animal welfare and labour situation. Producers of slaughter animals believe that animal welfare can be improved by the use of mobile slaughter facilities. There is a positive attitude to transporters given their own geographical region irrespective of the receiving abattoir and a negative attitude to several transporters working in the same area to increase flexibility.

Discussion and conclusion

The main part of the collections was performed by only a few vehicles, and the animals were often collected in small groups. To make use of vehicles with large loading capacity, collection of small groups of animals means more stops per round. Many stops are negative for animal welfare (Gebresenbet and Eriksson, 1998). Small groups also increase transport time and the risk of mixing animals.

It was common to mix animals in the transports and in lairage. Studies have shown that animals that are mixed in the transport vehicle are exposed to stress (Bradshaw *et al.*, 1996). To transport small groups of animals under conditions with good animal welfare and at the same time use the vehicle capacity, the equipment for fencing the animals in the transport needs to be flexible.

Time from arrival at the abattoir until slaughter varied greatly because some animals were slaughtered immediately after arrival, while others were kept over night. A large part of the animals are kept in lairage over night. After a well performed transport there are no advantages in keeping animals in lairage and a long time in lairage increases the risk of spreading contagious diseases (Warriss, 2003). Animals exposed to stress can benefit from around two hours in lairage, under good circumstances (Santos *et al.*, 1997). Time in lairage is apart from that considered a factor of stress and should be avoided (Geverink *et al.*, 1998; Santos *et al.*, 1997).

In the optimised scenarios, mean travel distance per animal increase compared to 'present'. Further research is needed to analyse the effect of an increased number of stops and longer journeys compared to long and overnight lairage on animal welfare. Further research is also needed to study the effect of flexible interiors in the transport vehicles to reduce mixing and thus the stress load on the animals.

Emissions are reduced as the distance decreases. To further reduce emissions it is the exchange of vehicles that induces the largest effects. The cost for animal transport will be lower due to reduced number of vehicles, driven distance, handling times and lairage area.

The conclusion from this pilot study is that transport optimisation can result in simultaneously increased animal welfare, reduced costs and reduced environmental impact. Farmer attitudes show openness to such changes.

References

Atkinson, S. (2000). Farm Animal Transport, Welfare and Meat Quality. Thesis, Rapport 4, Department of Animal Hygiene and Health, Swedish University of Agricultural Sciences, Skara, Sweden.

Bradshaw, R.H., Parrot, R.F., Goode, J.A., Lloyd, D.M., Rodway, R.G. and Broom, D.M. (1996). Behavioural and hormonal responses of pigs during transport; effect of mixing and duration of journey. Animal Science 62: 547-554.

Gebresenbet, G. and Eriksson, B. (1998). Effects of transport and handling on animal welfare, meat quality and environment with special emphasis on tied cows. Department of Agricultural Engineering, Swedish University of Agricultural Sciences, Uppsala, Sweden.

Geverink, N., Bradshaw, R.H., Lambooy, E., Wiegant, V.M. and Broom, D.M. (1998). Effects of simulated lairage conditions on the physiology and behaviour of pigs. The Veterinary Record 143: 241-244.

Nätverket för transporter och miljö (2007): www.ntm.a.se (accessed 2007-03-25)

Santos, C., Almeida, J.M., Matias, E.C., Franqueza, M.J., Roseiro, C. and Sardinha, L. (1997). Influence of lairage environmental conditions and resting time on meat quality in pigs. Meat Science 45, 253-262.

Warriss, P.D. (2003). Optimal lairage times and conditions for slaughter pigs: a review. Veterinary Record 153: 170-176.

The ethics of automatic milking systems and grazing in dairy cattle

Leonie F.M. Heutinck[1] and Clemens Driessen[2]
[1] Division of Animal Production, Animal Sciences Group, Wageningen UR, P.O. Box 65, NL-8200AB, Lelystad, The Netherlands, leonie.heutinck@wur.nl
[2] Applied Philosophy, Department of Social Sciences, Wageningen UR, P.O. Box 8130, NL-6700EW, Wageningen, The Netherlands, clemens.driessen@wur.nl

Abstract

Over the last decade an increasing number of automatic milking systems (AMS) have been adopted, especially in The Netherlands. The appraisal of this new technology in ethical terms appears to be a complex matter. Compared to a conventional milking parlour the use of an AMS entails in several respects a different practice of dairy farming, the ethical implications of which are not self evident. In this paper these potential implications are approached in terms of the co-evolution of technology and ethics. The approach starts from the assumption that ethics and technologies both are subject of change and that there is a relation of mutual influence. From this perspective it is argued that an AMS in a number of ways entails a different practice of dairy farming. Cows health and wellbeing are monitored to a large extent by means of data generated by the AMS management system, instead of in direct interaction during milking in a conventional system. There even is evidence that ethical criteria of cow welfare co-evolve with the new system and its behavioural implications. Furthermore the implications of AMS towards grazing are discussed. The AMS and the co-evolving ethical concerns can be regarded as part of a wider process of co-evolution of dairy technologies and ethical concerns on cow welfare, towards a more monitoring role for the farmer. Directions for further research on farming practices and ethical decision making are indicated.

Keywords: dairy farming, automatic milking system, grazing, co-evolution

Automatic Milking Systems

In The Netherlands as well as in other countries worldwide the number of automatic milking systems (AMS) replacing a conventional milking parlour on dairy farms has been increasing substantially ever since their commercial introduction in the 1990s. With an increase of 800 AMS farms the total number of farms using automatic milking was about 4000 worldwide in 2006, most of them located in North-Western Europe (Scandinavia, The Netherlands, Germany, France). Currently in The Netherlands over 800 farmers operate an AMS, which covers about 3.5% of the Dutch dairy farms nowadays. Some prognoses by the Dutch dairy industry indicate that in 2020 over 50% of the Dutch farmers will use an AMS.

The AMS is a system able to milk dairy cows without (direct) supervision of the farmer. Automatic milking strongly relies on the cow's motivation to visit the AMS voluntarily. The main motive for this is the supply of concentrates dispensed in a feed manger in the milking box during milking. The system operates the milking device with special software, including the storage of data and a management program to control the settings and the conditions for cows to be milked. Sensors are used for electronic cow identification, and also for the detection of abnormalities in milk (Koning and Rodenburg, 2004).

When a cow visits the AMS and the interval between 2 milkings exceeds a minimum threshold level the cow is accepted and will be milked. When the cow does not visit the AMS and the interval between 2 milkings exceeds the maximum threshold level (i.e. the cow does not visit the AMS for a period set by the farmer) the cow's identification number will be on a so called 'attention list', which the farmer can use to collect all cows with long milking intervals. Most farmers have a small area surrounded by fences near the AMS device (also called waiting area) for those cows, which can only leave this area by passing through the AMS. With the management program the farmer keeps track of all the milking parameters (like daily average per cow, amount of milk, interval between milkings) and milk quality parameters.

Ethical technology assessment?

How could and should this technology and the changes it brings for the relevant stakeholders be considered ethically? Simply listing the advantages and disadvantages of the new system compared to the conventional could of course be attempted. The investment costs can be set off against the amount of labour saved. The welfare of the cows can be measured in the previous and the new situation. Impacts on milk quality and production can be traced. Nevertheless a more complex dynamic seems to be involved here. The AMS implies in many respects different farm management, requiring new skills of farmers and new behaviour of cows.

When studying the ways in which in practice is dealt with ethical issues, when formulating ethical arguments, and when judging the application of new technologies from an ethical point of view, it is important to understand how ethical considerations can develop along with technological change.

Co-evolution of ethics and technology

Ethical issues in various ways can be part of technological developments. On the one hand moral concerns, such as regarding animal welfare or environmental degradation, can motivate the search for specific innovations. At the same time new technologies can be seen to bring along new ethical issues, and give rise to new concerns. Ethics in this paper is not considered as something completely outside of these developments. Moral judgments of situations can in many ways develop along with the practices they criticize or endorse.

Of course with new technologies new possibilities and new choices arise. 'Ethics' is then called in to consider whether these new possibilities do not cross moral boundaries or bring along unacceptable risks. But a more complex dynamic can be seen at work as well. Technologies often can be found to contain so called 'scripts', by which implicitly certain actors, roles, and behaviour is prescribed with the use of an artefact or system (Akrich, 1992). Together with new machines new actors can enter the scene, or old ones change in character. Who is to decide on a technology and its use, and who is involved in its development can differ from the actors previously involved in a practice. Power structures can be rearranged. Different needs for justification arise, together with different arenas of argument and deliberation. Responsibilities are redistributed, expectations redirected, and what is considered an autonomous development or rather a subject of choice can be seen in a new light.

In order to make the system work, also (what previously was considered) its context needs to be adapted to it. For instance cows udders needed to be adapted to the machines for milking them as much as vice versa (Bieleman, 2000). Developments that sometimes even make it difficult to meaningfully distinguish a technology and its co-evolving context. Applying the new system can bring along a new perspective on the world it is made part of. Different parameters can come up for measurement in the new system, new aspects can become comparable, new processes can be controlled. Attention gets redirected, dominant metaphors switched. Normative concepts are redefined, while previous moral vocabularies are considered a hindrance to dealing with new problems adequately (Keulartz *et al.*, 2004). A notion such as 'naturalness' loses its seemingly solid ground on which to found arguments, for instance when breeding is used to change the welfare requirements of livestock animals. Issues can get reframed and agendas set with different priorities. What is contested and what taken for granted can shift. New experiences are generated and practical knowledge changes in character. And what is considered good practice together with the terms in which it is discussed can all become part of the innovation process.

This description of the in many ways changing (nature of) ethics is not meant to endorse a naive ethical relativism in which moral judgments can only be seen as cynical strategies brought forward for the sake of (powerful) interests. Nor is it meant to simply unmask these ethical concerns for their being enmeshed in technological (and other cultural) developments.

In order to analyse the ways in which technological innovations come about, as well as to deal ethically with these developments, it is important to acknowledge and understand the ways in which ethics and technologies co-evolve. By having an open eye for the dynamics of co-evolving ethics and technology, new insights for the ethical assessment of a new technology and its implications might be developed. In this paper the AMS is analysed as a subject of co-evolving ethics and technology.

AMS dynamics

An AMS can be considered to generate a number of advantages compared to a conventional milking parlour (Koning and Rodenburg, 2004). As there is no longer a fixed milking routine with 2 or 3 milkings a day, the AMS means more flexible and in general reduced labour for the farmer. On the one hand this advantage is taken up in monetary terms and set off against the higher investment costs. Then the AMS is attractive in countries with high labour costs. On the other hand the flexibility of labour means the farmer has more time for social and family contacts, in this way allowing for increasing scale in family farming.

For an AMS to function adequately, a number of aspects of farm management need to be aligned with the system. First of all the housing is to be changed: attention should be given to the location of the AMS in the barn which must be favourable for cow traffic. And it is to be decided whether the access to the milking unit is designed free or forced, in which case access to the feeding area is only granted after first visiting the AMS. But also the cows have to adapt: the farmer may have to cull cows with udders that are too extreme in size or teat positioning to be milked by the AMS. And the cows will have to get used to the new milking routine, in which she is to some extent free in her choice to visit the AMS. It is found that the new system requires a new look at cow and herd behaviour in order to accommodate and use these to 'fit' with the AMS (Heutinck *et al.*, 2004).

The farmer has to adapt as well. He or she has to find a management strategy in collecting cows with long milking intervals which fits him/her best: collecting those cows strictly may save some milk but may turn the cows 'lazy' because of habituation to being collected vs. giving those cows some extra time may cost some milk, but may save labour of collecting in the long run.

An AMS brings with it the individual management of each cow during the entire lactation period, with possibilities for optimising individual milking frequencies to increase milk production. The AMS can be seen to amount to a new style of dairy farming, as it requires a different way of relating to the cows. Where previously farmers would spend up to half of their time in the milking parlour, with an AMS the assessment of the cows' health and wellbeing is done in large part by means of the data generated by the management program. Apart from milking the cows the system also detects milk quality deviations, it registers the milk volume and body temperature, and tracks instances of disease and heat (Bieleman, 2000). This means the farmer will need to adapt to a different way of visual control of his cows, and to learn to deal with 'management by exception' (Ouweltjes and Koning, 2004). Instead of being disciplined in rising early every day for milking, the dairy farmer utilising an AMS has to be disciplined and plan to check up on cows of which he or she interprets the data to say she does not perform well. Farmers' motivations for dairy farming may vary from direct interaction with animals, towards working with hi-tech systems and challenging technologies.

Interestingly, the ethical discussion of the new system has brought forward new concepts and aspects of dairy farming. The newly generated ability of cows to more or less choose when to be milked is by some regarded as an (important) ethical improvement. Dairy farming by means of an AMS in this way includes a shift of focus with regard to ethical indicators that went previously unmentioned. Another aspect that was mentioned was that a robot milked cow would be able to lie more comfortably on udders that where emptied more often. Even new concepts of naturalness were introduced, as initially the AMS was compared to and proposed as an emulation of the natural situation of a calve suckling the cow at will (which of course in reality is still quite a different situation from an AMS).

A British ethical technology assessment of 2000 was critical of early promises surrounding the AMS for more freedom for the cows. Certainly as the desire to be milked in itself was found to be not enough motivation to enter the AMS. Some kind of reward or forced routing is necessary in order to prevent the need for recurrent fetching of cows. 'Cow autonomy' in the analysis nevertheless became further expanded upon as criterion of ethical assessment of dairy systems (Millar, 2000).

The new circumstances of dairy farming furthermore gave rise to new ethological, stress physiological and veterinary research into animal welfare of dairy cows (Meijering *et al.*, 2004).

Grazing

It is often assumed that grazing does not combine well with automatic milking, since it may lead to a decrease in use of the capacity of the AMS and an increase in labour required for fetching cows with long milking intervals from the pasture site. The labour freed by the AMS could of course be put to use here. But the whole rationale behind the AMS and its justification as an investment is focused on saving labour.

However, cow health and welfare as well as the 'green' public image of the dairy industry are seen to benefit from giving dairy cattle the opportunity for grazing.

Previous research in The Netherlands showed that grazing in combination with automatic milking can be successful under various management conditions, although some farmers were less successful (Ruis-Heutinck *et al.*, 2001). In a 3-year study from 2001-2003 it was found that with daytime grazing on average almost 17 cows from a 60-cow-herd had to be collected to visit the AMS at the end of grazing each day (Heutinck *et al.*, 2004). In the same study it was found that grazing resulted in a significant lower occupancy of the AMS (a 2-box-system) during daytime grazing compared to the occupancy at the end of the winter season (each unit 15 vs. 17 hr occupied per day). From an economic point of view it is important for the farmer that his investment (the AMS) is used optimally and does not stand idle waiting for cows to return from pasture.

On the other hand implementing an AMS, together with the inadequacy of the home pasture site (distance, size), are brought forward as justifications for refraining from pasturing. The importance of grazing is an issue on which the ethical concerns are still not settled, but which at least in The Netherlands is increasingly coming to the fore, if only with recent Dutch dairy company commercials.

An earlier co-evolution process in livestock farming

Earlier innovations in dairy farming have also shown ethical co-evolution processes. For example the switch from the tie-stall to loose housing in cubicle barns around the 1970s was motivated by various factors and considerations. Not only was there a development towards increasing the scale and decreasing the amount of labour, also the working conditions of farmers became more important and the hard manual labour required by the tie-stall increasingly unacceptable. The change in housing system next to that saw a shift in what was considered desirable and allowable practices with regard to the animals.

For instance the dehorning of the cows was deemed necessary in loose housing, but had been considered highly controversial a decade earlier (Bieleman, 2000; Ploeg, 2003). Reluctant farmers considered the new loose housing as messy aberrations to good farming practice. While tying up cows, which was common practice in the old tie-stalls, got to be considered undesirable from an animal welfare point of view. In organic farming it is mostly ruled out and with respect to calves tying up is now prohibited in the EU. Nevertheless, in a number of European countries the tie-stall still is common, even in countries that have seen extensive debates on animal welfare, such as Nordic and Alp countries.

In line with the development from tie-stall to loose housing, the development of the AMS can be seen as part of a process of (the automation of) dairy farming in which 'good farming' is changing from *caring for the animals*, towards *allowing the animals to take good care of themselves*. The concept of 'recursive control' is a description of this approach, a technical term originating in mechanical engineering, as a design paradigm following the change in understanding of the role of farmers and the functionality of housing and other management (Bos *et al.*, 2003). A development in the role of the farmer and a different view of animal welfare that can also be regarded to amount to a shift in ethics.

Conclusion

While the introduction and widespread adoption of automatic milking systems entails significant changes in the practice of dairy farming, how to assess these changes ethically is a complicated matter (though at least with regard to environmental concerns its character in terms of increased energy use seems to be a clear disadvantage). Not just is it uncertain whether the AMS will live up to some of its promises, but also how it will work out with regard to serious ethical issues in dairy farming, such as disease proneness, grazing, claw health and lameness. The broader development towards 'recursive control' could for instance generate a need for more robust animals that are less disease prone.

It is interesting on what parameters and aspects farmers' decisions regarding AMS are taken, as it is clear that these are not based on economic calculations alone. Understanding the considerations of farmers as well as the implications of AMS with regard to ethical aspects can be instrumental in improving AMS technology and its accompanying farm management. Further research will consist of detailed analysis of the discussions of farmers working with an AMS and trying to apply grazing.

The kind of co-evolution processes are a challenge to research, as important changes largely take place in unregistered and little reflected daily practices of farmers. The aim of the ethical research effort is nevertheless not so much at producing solid evidence of causal relations between ethical and technological development, but rather to make the dynamic visible and an input to ethical debates. Then wider implications of new technologies from the perspective of co-evolution can become apparent and contested. Also the granting of ethical priorities can then be scrutinised, and the ethical disadvantages of new technologies be a subject of debate and redesign in an early stage. Instead of only after its economic lifespan, as seems to be the case with the loose stall housing in cubicle barns.

References

Akrich, M. (1992). The De-scription of Technical objects. In: W. E. Bijker, and J. Law (eds.) Shaping technology/building society. MIT Press, Cambridge MA, USA, pp. 205-224.

Bieleman, J. (2000). Landbouw en Voeding. In: J.W. Schot (ed.) Techniek in Nederland in de Twintigste Eeuw, Volume III. Walburg Pers, Zutphen, The Netherlands, 442p.

Bos, B., P.W.G. Groot Koerkamp. P.W.G. and Groenestein, K. (2003). A novel design approach for livestock housing based on recursive control - with examples to reduce environmental pollution. Livestock Production Science 84: 157-170.

Heutinck, L.F.M., van Dooren, H.J.C. and Biewenga, G. (2004). Automatic milking and grazing in dairy cattle: effects on behaviour. In: Meijering, A., H. Hogeveen and C.J.A.M. de Koning (eds.) Automatic Milking: a better understanding. Wageningen Academic Publishers, Wageningen, The Netherlands, pp. 407-413.

Keulartz, J., Schermer, M., Korthals, M. and Swierstra, T. (2004). Ethics in a Technological Culture. A Programmatic Proposal for a Pragmatist Approach. Science, Technology and Human Values 29: 3-29.

Koning, C.J.A.M. de, and J. Rodenburg, J. (2004). Automatic milking: State of the art in Europe and North America. In: A. Meijering., H. Hogeveen and C.J.A.M. de Koning (eds.) Automatic Milking: a better understanding. Wageningen Academic Publishers, Wageningen, The Netherlands, pp. 27-37.

Meijering, A., Hogeveen, H. and de Koning, C.J.A.M. (2004). Automatic Milking: a better understanding. Wageningen Academic Publishers, Wageningen, The Netherlands, 544p.

Millar, K. (2000). Respect for animal autonomy in bioethical analysis: the case of automatic milking systems. Journal of Agricultural and Environmental Ethics 12: 41-50.

Ouweltjes, W. and de Koning, C.J.A.M. (2004). Demands and opportunities for operational management support. In: Meijering, A., H. Hogeveen and C.J.A.M. de Koning (eds.) Automatic Milking: a better understanding. Wageningen Academic Publishers, Wageningen, The Netherlands, pp.433-443.

Ploeg, J.D. van der (2003). The Virtual Farmer. Past, present and future of the Dutch peasantry. Van Gorcum, Assen, The Netherlands, 444p.

Ruis-Heutinck, L.F.M., van Dooren, H.J.C., van Lent, A.J.H., Jagtenberg, C.J. and Hogeveen, H. (2001). Automatic milking with grazing on dairy farms in The Netherlands. In: J.P. Garnet, J.A. Mench and S.P. Heekin (eds.) Proceedings of the 35th International Congress of the ISAE. The Centre for Animal Welfare at UC Davis, Davis (CA), USA, 188 p.

Part 7 - Sustainability and animal welfare: implementation and legislation

Conflicting areas in the ethical debate on animal health and welfare

Albert Sundrum
Department of Animal Nutrition and Animal Health, Faculty of Organic Agricultural Sciences, University of Kassel, Germany, Sundrum@wiz.uni-kassel.de

Abstract

There is a far-reaching agreement within western society, that a high level of animal health and welfare can be regarded as an ethical value. An increasing number of consumers are interested to purchase products that derive from animal-friendly housing conditions. Often animal-friendly and organic products are affiliated. Although the prescriptions of the EU-Regulation on organic agriculture clearly exceed the minimal standards of conventional livestock production, they do not automatically lead to a high status of animal health and welfare. Animal health status emerges from complex interactions within a farm system and thus is not primarily related to the production method or to minimal standards. Literature reviews indicate profound discrepancies between the claim and reality of animal health and welfare on organic farms. Conflicting areas on different levels within and between the groups of retailers, producers, and consumers are responsible that efforts with regard to improvments in relation to animal health and welfare have come to a standstill. In order to preserve the credibility of organic agriculture and the confidence of the consumers in organic products there is a need for more transparency and for a change in the paradigm from a standard-oriented to an output-oriented approach. Credible information about animal health and welfare has to be provided, which consumers are able to understand. Simultaneously, a high level of animal health and welfare has to be honoured by premium prices to cover the additional costs and efforts that are needed to improve the current situation.

Keywords: organic livestock, consumer expectations, inconsistencies

Introduction

The existence and nature of subjective experience in animals such as suffer and pain - although denied for a very long time in history - nowadays is without question within the scientific community. The ability for subjective experience is a central question in the debate about animal health and welfare, leading into the general demand to accept animals as fellow creatures. Animals should be kept in such a way that their body functions and their behaviour are not disturbed and that they are capable to cope with the environment. Thus, there is a far-reaching agreement within western society, that the realisation of a high level of animal health and welfare can be regarded as an ethical value.

Long-standing activities of animal protectionists and the effects of scandals in connection with food production have led to a development in which people are becoming increasingly aware of welfare problems within livestock production. Consequently, an increasing number of consumers expect their food to be produced with greater respect for the needs of farm animals. Animal welfare has become an important component of consumer motivation to purchase products that derive from specific brand label programmes or from organic farms, claiming to provide animal-friendly living conditions for farm animals (Harper and Makatouni, 2002). In this way consumers have the possibility to buy food that is closely linked to the ethical value of animal health and welfare. However, the implementation of animal health and welfare into livestock production systems and the use of this item as a process quality trait in marketing strategies lead to various questions. Different dilemmas can arise when an ethical approach must be applied to very down-to-earth choices in practice. Some of the conflicting areas are discussed below with special emphasis on organic livestock production.

How to define and assess animal health and welfare?

Animal health and welfare has different meanings to different people, and the interests of one particular group may conflict with those of others. The attributes included in the concept of animal health and welfare primarily depend on who is making the definition. Typical actors participating in the evaluation of animal health and welfare are scientists, veterinarians, farmers, food producers, government officials, marketing people and consumer groups. In the literature, there is a great variety of definitions of animal health and welfare, thoroughly discussed by Rushen and de Passillé (1992) and Stafleu *et al.* (1996). Hence, there is no generally accepted definition of animal health and welfare within the scientific community. Three types of animal welfare definitions are often distinguished, depending on what is considered important for the well-being of the animal (Fraser *et al.*, 1997):

- The natural living approach: the welfare of an animal depends on its being allowed to perform its natural behaviour and live a life as natural as possible.
- The biological functioning approach: animal welfare is related to the normal functioning of physiological and behavioural processes.
- The subjective experience approach: the feelings of the animal (suffering, pain and pleasure) determine the welfare of the animal.

Depending on the basic opinion about life, different conclusions can be drawn regarding what is important for animal health and welfare and how to assess this issue (Verhoog *et al.*, 2004).

The assessment of animal health and welfare raises several methodological problems while the results are often hard to interpret (Webster and Main, 2003). In general, legislators and brand label programmes are using technical indicators which refer to single aspects of housing conditions (e.g. space allowance, laying surface, feeding regime), to describe different levels of minimum standards in relation to the appropriateness of housing conditions in terms of animal health and welfare. The prescriptions of the EU-Regulation (EEC-No. 2092/91) on organic livestock production clearly exceed the minimum standards of conventional livestock production or other brand label programmes, thus clearly enhancing the options for the farm animals to act out their normal behaviour.

Literature reviews, however, indicate that organic standards do not automatically lead to a high status of animal health that exceed the level in conventional production (Hovi *et al.*, 2003; Sundrum *et al.*, 2004). Comparable high rates of mortality and morbidity in livestock production interfere with the well-being of farm animals and indicate that the animals are not able to cope appropriately with their environment. Hence, in many cases consumer expectations are not met to an acceptable level.

This is, among others, due to the fact that the meaningfulness of technical indicators used in basic standards with regard to animal health and welfare is very limited. Technical indicators represent only a small section of the complex interrelationship among farm animals and their living conditions, thus lacking validation when the responses of the animals are not assessed directly in the specific situation (Sundrum, 1999). Animal health and welfare emerges from very complex interactions between the individual animals and the environment and within a farm system, and thus is not primarily related to minimal standards but to the farm management in the first place (Sundrum *et al.*, 2006).

Consumers' demands and perceptions

While a number of food scandals (BSE, foot-and-mouth disease, nitrofen) have stimulated consumer concerns about the safety and quality of food, consumers are also becoming increasingly interested in foods produced according to ethical aspects of animal welfare principles (Verbeke and Viaene, 2000). Sometimes the concept of animal welfare is mixed with items of food safety and sensory quality. Different individuals show different preferences and subjective perceptions in relation to the different features (Harper and Makatouni, 2002). The interests of one particular group may even conflict with those of others. Hence, consumers are neither a uniform group, covering common interests nor are they experts who can decide on how to evaluate animal health and welfare and what is needed to provide housing conditions and management that are appropriate to farm animals.

Products with attributes of process quality have in common that their unique selling proposition is not directly visible to the consumer. Only additional information will identify the nature of the origin, or the production process of these foods. Correspondingly, perception of the consumers is to a high degree influenced by information through media and advertising. However, advertising campaigns often are neither announced with any clear labelling nor do they define their view on animal health and welfare or by which criteria the status is assessed. High risks of misuse and unfair competition are the consequences.

The change from confinement to more free-range systems has been e.g. one of the tools to evoke positive associations with the product and to sell stories (Andersen *et al.*, 2005). The promotion of products from such production systems is based on anecdotal information rather than on real facts regarding the obtained process quality. Introduction of the wholesomeness concept in livestock production, most often represented by organic production, is mainly due to a wish for re-establishing a positive image of food safety and animal welfare aspects (Verbeke and Viaene, 2000). Consumers make a whole range of positive inferences from the label 'organic'. Surveys of consumers' attitudes show that regular consumers of organic products expect organic livestock products to be safer, have a high sensorial quality, contain less additives, were produced with the use of natural and healthy fodder, less drugs and hormones, as well as a higher level of animal welfare on the farms (McEachern and Willock, 2004). Many consumers directly associate organic farming with enhanced animal welfare and conflate organic and animal-friendly products (Harper and Makatouni, 2002).

On the other hand, positive interferences do not necessarily lead to a purchase if consumers do not think that the trade-off between give and get components is sufficiently favourable. According to Von Alvensleben (2003), consumers often are in a conflict situation where it is difficult for them to decide whether it is worth or they should be willing to pay a premium price for a premium product. In order to avoid dissonances and incoherencies when confronted with inconsistent information and uncertainty, consumers often ignore or actively mask out specific aspects that do not fit in their world view or justify their doubt in relation to the credibility of the information or deny being involved.

Some consumers appear to delegate responsibility for ethical issues in meat production to the meat retailer or the government as many consumers do not seem to wish to be reminded about issues connected with the animal when choosing meat (Bernués *et al.*, 2003). Moreover, knowledge of production systems often appears of little consequence in terms of any food market potentials as consumer groups often freely remark that there is no link between the negative images of production methods and their purchase behaviour (Ngapo *et al.*, 2003). According to the authors, consumer groups are often confused and mistrust the limited information available at the point of purchase. In contrast, price is an extremely visible attribute of products related to quality by the notion of value (McEachern and Schröder, 2002).

Conflicting areas on the farm level

Between organic livestock farms, a huge variation exists with regard to the living conditions, and the implementation of animal care and hygiene management. The causes of variation are diverse and encompass among others differences in financial and labour resources, education, awareness, and traditional habits. It is reasonable that there is always a history and a background why the specific status of animal health and welfare on each farm is as it is. This does not exclude possible improvements, but makes it difficult to identify the most relevant constraints within the farm system without evaluating the farm by a system approach (Sundrum, 2007). In general, the capacity for changes is limited, and high pressure on the production costs is a relevant but not the only reason for deficits in relation to animal health and welfare.

As resulting costs of production for most organic farm types are higher than for conventional systems, price premiums are urgently needed to achieve an appropriate income (Offerman and Nieberg, 2000). In order to increase the level of animal health and welfare on livestock farms, there is a need for enhanced efforts, encompassing improvements in housing and feeding conditions as well as animal care and hygiene management (Sundrum *et al.*, 2006). Those efforts are expected to clearly increase the costs of production. On the other hand, organic farmers do or will not adequately benefit from those efforts as an increased health status is not honoured through the market by extra prices. Often, prices for organic animal products do not even cover the previous additional expenditures of organic livestock farming. On the other hand, those producers who aim for a high level of animal health and welfare by increasing their labour and management efforts compete with their products on the same markets as those who widely ignore the issue of animal health and welfare. Thus, producers that follow the organic principles strictly and aim for a high level of animal health and welfare are at a clear disadvantage, due to higher production costs compared to those producers who are just following the minimum demands providing them a competitive advantage. In this way, the approach to use minimum standards prevents clear improvements in relation to animal health and welfare and at the same time causes unfair competition between organic farmers.

Cognitive dissonances

Ethics is the study of values and habits of a person or group and covers the analysis of concepts such as right and wrong, good and evil, and responsibility. It has been indicated above, that to deal with an ethical value does not ensure that the debate, the implementation into practice or the stakeholders involved always meet ethical demands, e.g. in relation to transparency, fairness and truthfulness. There is reason for the concern that the current handling of the issue of animal health and welfare in organic livestock production by the different stakeholder groups provide huge discrepancies between the claim and reality and cause various conflicting areas on different levels:

- Retailers and/or producers claim to offer products that derive from animal-friendly living conditions, without providing adequate health and welfare standards.
- Retailers want to increase the turnover by offering organic food with comparable low prices and at the expense of the possibilities of the farmer to investigate in substantial improvements of animal health and welfare.
- Producers who strive for a high status of animal health and welfare by using appropriate management concepts and encountering higher production costs are confronted with unfair competition when competing with their products on the same markets as those who widely make use of minimum standards and produce on a low cost base.
- A high percentage of consumers announce their special interest in the issue of animal health and welfare and their willingness to pay premium prices, but hesitate to do so when corresponding food is offered and instead are miserly and are greedy for cheap food.

- Many consumers prefer to delegate responsibility for ethical issues when choosing animal products to the retailer or the government and are by their ignorance jointly responsible for the severe deficits in animal health and welfare within livestock production.
- Consumers partly wish to buy animal products related to the issue of animal health and welfare but are not willing to pay premium prices that cover the higher expenditures.

One of the main reasons of those inconsistencies can be traced back to the selective perception and single minded perspective members of the different stakeholder groups may have on this issue, causing cognitive dissonances and mistrust. Thus, it does not make sense to point with a finger to single stakeholder groups and blame them for being responsible for the discrepancies between the claim and reality, as all stakeholder groups are a part of the complex field of interests. In this context, the issue of animal health and welfare seems to be a cover without content which serves as a projection surface for partly contrasting interests within and between the different stakeholders.

While farmers are responsible in the first place for the well-being of their farm animals, they are very limited in their freedom of decision-making as they possess little financial scope that can be used for improvements. In contrast, consumers are able to make a choice between large ranges of products. Expenditures for food in relation to the total budget of a household have dramatically decreased during the last decades. Hence, consumers in general would have money to spend more on food if they would get the priorities right.

For organic livestock production, consumers' interests and expectations are very important as they are closely linked to their willingness to pay premium prices which are an essential precondition to cover the higher productions costs in comparison to conventional production. Hence, it is of essential importance for organic farming to clarify on how to cover consumers' interests, to ensure consumer confidence and to avoid misleading labelling. Thus, the organic movement is challenged to ensure that its credibility and the confidence of the consumers does not get lost due to discrepancies between different expectations.

Conclusions

Retailers and producers of organic food seem to be the victim of their own announcements, which, on the one hand, has increased consumer expectations in relation to process quality but which was not accompanied by the claim for the real price needed to cover all costs for an enhanced level of process qualities. The current framework conditions of the food market contribute to a situation in which the existing potential for a high level of animal health and welfare in organic livestock production is not fully realised and the further development of quality production is hampered by contradicting expectations and perceptions and by unfair competition within organic and between organic and conventional agriculture.

Organic agriculture is challenged to ensure that its credibility and the confidence of the consumers in organic products do not get lost due to conflicting areas on different levels. There seems to be no alternative in increasing transparency and to provide credible information about animal health and welfare, which consumers are able to understand, and which can serve as a prerequisite for launching such products in the market. Furthermore, there is a need for a change in the paradigm from a standard-oriented to a result- and output-oriented approach. Thus, reliable monitoring systems for assessing the animals' health and welfare status are urgently required to accommodate societal concerns and market demands. Simultaneously, retailers have to make sure that a high level of animal health and welfare will be honoured by adequate premium prices to cover the additional costs that are needed to provide food that derive from farm animals with a high status of animal health and welfare.

References

Andersen, H.J., Oksbjerg, N. and Therkildsen, M. (2005). Potential quality control tools in the production of fresh pork, beef and lamb demanded by the European society. Livestock Production Science 94, 105-124.

Bernués, A., Olaizola, A. and Corcoran, K. (2003). Labelling information demanded by European consumers and relationships with purchasing motives, quality and safety of meat. Meat Science 65, 1095-1106.

Fraser, D., Weary, D.M., Pajor, E.A. and Milligan, B.N. (1997). A scientific conception of animal welfare that reflects ethical concerns. Animal Welfare 6, 187-205.

Harper, G.C. and Makatouni, A., (2002). Consumer perception of organic food production and farm animal welfare. British Food Journal 104, 287-299.

Hovi, M., Sundrum, A. and Thamsborg, S.M. (2003). Animal health and welfare in organic livestock production in Europe - current state and future challenges. Livestock Production Science 80, 41-53.

McEachern, M.G. and Schröder, M.J. (2002). The role of livestock production ethics in consumer values towards meat. Journal of Agricultural and Environmental Ethics 15, 221-237.

McEachern, M.G. and Willock, J. (2004). Producers and consumers of organic meat: A focus on attitudes and motivations. British Food Journal 106, 534-552.

Ngapo, T.M., Dransfield, E., Martin, J.F., Magnusson, M., Bredahl, L. and Nute, G.R. (2003). Consumer perceptions: pork and pig production. Insights from France, England, Sweden and Denmark. Meat Science 66, 125-134.

Offermann, F. and Nieberg, H. (2000). Economic performance of organic farms in Europe. Organic farming in Europe: Economics and policy. Vol 5. University of Hohenheim.

Rushen, J. and de Passillé, A.M. (1992). The scientific assessment of the impact of housing on animal welfare: A critical review. Canadian Journal of Animal Science 72, 721-743.

Stafleu, F.R., Grommers, F.J. and Vorstenbosch, J. (1996). Animal welfare, evolution and erosion of a concept. Animal Welfare 5, 225-234.

Sundrum, A. (1999). EEC-Regulation on organic livestock production and their contribution to the animal welfare issue. In: KTBL (ed.) Regulation of Animal Production in Europe, KTBL-Schrift 270, p. 93-97.

Sundrum, A. (2007). Obstacles towards a sustainable improvement of animal health. In: Zikeli, S., Claupein, W., Dabbert, S., Kaufmann, B., Müller, T., Valle Zárate, A. (Eds.), Beiträge zur 9. Wissenschaftstagung Ökologischer Landbau, p. 577-580.

Sundrum, A., Benninger, T. and Richer, U. (2004). Statusbericht zum Stand der Tiergesundheit in der Ökologischen Tierhaltung - Schlussfolgerungen und Handlungsoptionen für die Agrarpolitik. http://orgprints.org/5232/.

Sundrum, A., Padel, S., Arsenos, G., Kuzniar, A., Henriksen, B.I.F., Walkenhorst, M. and Vaarst, M. (2006). Current and proposed EU legislation on organic livestock production, with a focus on animal health, welfare and food safety: a review. Proceedings of the 5th SAFO Workshop, 01.06.2006, Odense/Denmark, 75-90.

Verbeke, W.A. and Viaene, J. (2000). Ethical challenges for livestock production: meeting consumer concerns about meat safety and animal welfare. Journal of Agricultural and Environmental Ethics 12, 141-151.

Verhoog, H., Lund, V. and Alrøe, H. (2004). Animal welfare, ethics and organic farming. In: M. Vaarst, R.L. Roderick, V. Lund and W. Lockeretz (eds.) Animal Health and Welfare in Organic Agriculture. CABI Publishing, Wallingford, Oxon, UK, pp. 73-94.

Von Alvensleben, R. (2003). Gesellschaft und Tierproduktion. In: E.-J. Lohde and F. Ellendorf (eds.) Perspektiven in der Tierproduktion, Landbauforschung Völkenrode SH 263, 15-21.

Webster, A.J. and Main, D.C. (eds.) (2003). Proceedings of the 2nd International Workshop on the assessment of animal welfare at farm and group level. Animal Welfare 12, 429-708.

The concept of sustainable agriculture necessitates a duty of stewardship in contemporary animal welfare legislation

Ian A. Robertson

Definitions

Sustainable agriculture integrates three main goals-environmental health, economic profitability, and social and economic equity. This paper focuses on the animal welfare component of those goals. More specifically, this paper considers how legislation does, or does not, address issues of sustainability in contemporary animal welfare legislation.

Law has been described as a 'house of words'. Legal definitions are critical to clarifying legislative application of concepts such as rights, duties, and enforcement. Legal definitions may also be substantially different in scope and application to common or traditional, interpretations.

Defining agriculture

A common definition of agriculture, for example, may be simply the science or process of farming or cultivating the soil for the production of plants and animals that will be useful to humans in some way. The legal definition, in comparison, may vary according the circumstances, and document, that it applies to. A simple search of legislation referencing the term 'agriculture' reveals everything from no definition at all to very detailed definitions such as found in the 2003 Horizontal Regulation which defines 'agricultural activity' as 'the production, rearing or growing of agricultural products including harvesting, milking, breeding animals and keeping animals for farming purposes, or maintaining the land in good agricultural and environmental condition'

For the purpose of this paper, the focus of attention is on the animal and its welfare in agriculture, defined as 'a system that produces animals to be used primarily as a source of food for humans'.

Defining stewardship

Sustainability is generally agreed as a concept that denotes the ability to meet the needs of the present without compromising the needs and abilities of future generations. The term captures a concept that considers both short *and* long-term needs and effects.

Central to notions of sustainability is the concept of stewardship. Stewardship has been defined as a responsibility for taking good care of resources entrusted to one for the benefit of another. In legal terms, this may be viewed as analogous to a trustee, a person or entity that holds and manages the assets (corpus) for the current and future benefit of beneficiaries. Importantly, this is a defined legally enforceable obligation, where the steward is obligated to provide a higher duty of which is much more than simple cursory consideration. If the steward's actions do not meet this standard of care, then the acts are considered negligent, the steward may be removed and potentially held accountable for relevant damages.

It may be argued that today's decision makers and caretakers, including legislators, are fiduciaries for the interests of future generations of animals and humans alike.

The human animal relationship defined in animal law

Whilst issues regarding animals have been dealt with under various legal headings, Animal Law is a developing, separate discipline within law. At a general level, Animal Law may be interpreted as issues of law that have three elements: they deal with an *animal*; take into account the *unique nature* of animals, and additionally affect the *relationship* between humans and animals. The concept of stewardship also denotes a relationship between one party and/to another. Animal welfare legislation outlines the duties, and thereby the relationship, 'regarding' current animals (and human interests) and *future* animals and human interests on the grounds that animal and human interests are inextricably linked. Similarly, the objectives of sustainable agriculture extend contemporary animal welfare legislative concepts of the human-animal relationship, by incorporating duties of stewardship to the human interests invested in the animals, the environment, and the animals themselves.

Defining welfare

This paper does not purport to be a discussion of animal rights v animal welfare. Nevertheless, given that this paper considers how contemporary animal *welfare* legislation does, or does not, incorporate the objectives of sustainable agriculture, a brief clarification is provided to avoid the confusion that often results in respect of these terms and concepts.

A 'right' may be defined as a 'justified claim or entitlement, validated by moral principles and rules' (Orlans *et al.*, 1998: 28). The use of the term 'animal rights' has proved problematic and there are a number of different, and conflicting, ways that the term has been used. One perspective holds that animals are legally classified as property, and given that property cannot hold rights then animals cannot have rights. Philosophers like Immanuel Kant, have argued that individuals can only be rights-bearers if they understand the 'social contract' of duties and obligations that accompany rights - which animals arguably cannot. Some have said the term 'animal rights' has been 'hijacked' from the concept of rights as applied to fundamental human rights, or basic human rights. Additionally the term welfare has been interpreted by some as a right on the basis that if humans have a duty to treat animals well, then animals have a legal expectation be treated accordingly. Finally, the term 'animal rights' has been ambiguously and interchangeably used to denote concepts as animal welfare, interests, and protection.

For the purpose of this paper, Joel Feinbergs definition has been used to distinguish the terms. Animal welfare, and animal protection, are used to refer to the *legal use* of animals, which involves legal duties *regarding* them, as opposed to animal rights which involves legal duties *to* the animals.

How does contemporary animal welfare legislation currently address objectives of sustainable agriculture?

Legal issues regarding animals have been addressed in multiple pieces of legislation, but legislators have specifically addressed 'animal welfare' in legislation of the same, or equivalent, name. Animal Welfare legislation could appropriately be viewed as the 'first stop' and umbrella of animal welfare law and principles. Accordingly, this paper focuses on the leading legislative developments in animal welfare. The hallmark of contemporary animal welfare legislation is the incorporation of positive duties of care broadly known as the Five Freedoms. These may be stated as:
1. Freedom from hunger and thirst - by ready access to fresh water and a diet to maintain full health and vigor.
2. Freedom from discomfort - by providing an appropriate environment including shelter and a comfortable resting area.
3. Freedom from pain, injury or disease - by prevention or rapid diagnosis and treatment.

4. Freedom to express normal behaviour - by providing sufficient space, proper facilities and company of the animal's own kind.
5. Freedom from fear and distress - by ensuring conditions and treatment that avoid mental suffering.

Although the Five Freedoms provide a useful paradigm, they are not absolutes. Animal welfare is a developing field and there have already been concerns expressed about the potential limitations of the Five Freedoms. On the well founded legal understanding that a change of words often imputes a change of duty, this author has illustrated discrepancies between jurisdictions having 'contemporary' animal welfare legislation, that arguably result in Five 'inconsistent' Freedoms. Nevertheless, the Five Freedoms currently distinguish contemporary animal welfare law.

Successful realization of sustainability objectives requires a multidisciplinary holistic approach. In respect of the animals, particularly those used in agricultural, this requires attention not only to the factors surrounding and impacting the animals, but also necessitates the continued presence and well being (welfare) of the animals themselves, in terms of the individual animals, the breed, and issues that extend beyond the breed itself. Given the impact of agriculture on wider issues, especially international trade and food safety, the longer term vision of sustainability objectives imputes a necessity for responsible stewardship towards a range of animal welfare sustainability issues including the maintenance of a suitable genetic pool to maintain biodiversity, responsible breeding programmes, and appropriate care, monitoring and enforcement of standards in respect of the well being of current and future generations of animals. The concern is that contemporary animal welfare legislation fails to adequately address sustainability issues. This may be illustrated by examining, for example, legislation's lack of clarity in respect of a pivotal issue such as genetic selection. Consider the recently updated animal welfare legislation of Scotland and England. The legislation refers to breeding establishments, specifically those breeding animals used in a scientific procedure establishment, the Breeding of Dogs Act 1973, and Transmissible Spongiform Encephalopathies (TSE's) - but nothing in the modern welfare legislation directly imputes an obligation of responsible animal breeding. Although there has been discussion about the possibility of including such an offence, contemporary animal welfare legislation does not make it an offence to predispose animals to suffering, whether through defects in management, breeding, or both.

At a stretch, it might be said that an argument could be made for responsible breeding. According to England's Animal Welfare Act 2006, for example, a person commits an offence if:
4. a. an act of his, or a failure to act, causes an animal to suffer.
4. b. he knew, or ought to reasonably have known, that the act, or failure to act, would have that effect, or be likely to do so.

It is arguable that section 4b could apply to breeders of certain genetic lines of chickens. If it has been shown historically that breeding with certain genetic strains results in offspring that suffer from chronic pain, then arguably the breeders ought reasonably to recognize that continued breeding with these strains will continue to result in offspring that suffer from chronic pain. With regard to the special case of leg weakness in broiler chickens, authors like veterinarian and scientist John Webster argue that there is overwhelming evidence that the animals suffer both physical and emotional distress from chronic pain. There are those who believe that breeding a strain of animals in such a way that knowingly renders the offspring liable to suffering should constitute an offence under legislation. Rather than leaving such an issue open to judicial interpretation however, clear and direct legislative attention to this matter would be an obvious preferable alternative.

If we examine where contemporary animal welfare legislation addresses the issue of sustainability or stewardship, the short answer really is - it doesn't. Counter argument that suggests that these issues are addressed, to a degree, in other legislation, misses the point that the focus is on animal welfare, and that legal issues of animal welfare should logically and appropriately be addressed in the leading piece of animal welfare legislation.

Given that sustainability is a key animal welfare issue, it follows that clear directives should be given in primary animal welfare legislation.

Issues of sustainability have been addressed in certain European legislation, although it doesn't go so far as to refer to the requirements as a 'stewardship'. Nevertheless, required standards are listed which are a pre-requisite to farmers receiving relevant subsidies and payments. It is noteworthy that the EU continues to lead the way in much of the animal welfare issues. A directive due to come into force in 2010 means that chickens reared for meat production will be covered by strict regulations governing the condition in which they are kept. In addition to measures including stocking density, training, labeling and enforcement, the initiative also seeks to establish systems of data collection and 'scientific monitoring of impacts on welfare such as genetics'. The initiative is a start and, although contained in secondary legislation rather then primary animal welfare legislation, perhaps the most important feature of this initiative to this discussion is that it is an acknowledgement that genetics, and the well being of future animals, is an animal welfare issue that warrants legislative attention. Additionally, this illustrates that matters of genetic selection affecting pain, suffering, or death of future generations of animals is not simply a measure of good stockmanship, or an economic consideration to minimise wastage to improve profit. With legislative direction, it becomes a matter of law that clearly states that you do *not* breed from animals with heritable traits that compromise the welfare of future animals - *and* it is a prosecutable offence to do so.

Existing animal welfare legislation has the means for implementation. Primary animal welfare legislation sets out the broad principles, and secondary legislation, through welfare codes, provides the flexibility to enunciate the details, which may evolve with input from science and other disciplines. If legislators turn their minds to the issue of sustainability and it is included in primary legislation, then the importance of sustainability and stewardship is arguably elevated to a position equivalent to other issues such as transportation and slaughter. Furthermore, it is hardly a huge leap to then ensure that the EU initiatives and requirement such as data collection to assist decision-making in respect of related matters of welfare and genetics, could then be incorporated into secondary legislation.

The temptation is to simply formulate the wording for a section regarding sustainability. The reality is that it is not quite that simple - but the idea of capturing the concept in legal terms so that it can be included in legislation is indeed thinking along to right lines that could result in sustainability objectives being included in primary and secondary animal welfare legislation. Legislation that clearly communicates the definition, responsibilities, liabilities and enforcement procedures of sustainability as it pertains to animal welfare, is likely to be a time consuming and complex. task that requires multidisciplinary input and consideration.

And there is a difference between drafting legislation, and having it enacted in one state, let alone globally. Indeed, the concept of unified global animal protection is an objective that has so many political, social, cultural and economic hurdles that it could appear unrealistic and unattainable. Yet the evidence would suggest that in spite of these apparent obstacles, there are people, organisations and states who, although sharing a variety of motivations, are actively involved in trying to achieve it.

A global legislative model of animal welfare based on 'best practice'

While there is no single international organisation that has a standard-setting role in the field of animal welfare, there are a large number of international bodies with obvious interest in the field of animal welfare. One such example is the OIE. David Bayvel, Chair of that organisation's Permanent International Animal Welfare Working Group, has stated, 'the need for international leadership in respect of animal welfare policy and standards has been evident for some time and is likely to be an expanding core role for the OIE in the decades ahead'. The stated tasks of the OIE include the 'development of policies and guiding principles to provide a sound foundation from which to elaborate draft recommendations and standards'. While skeptics may raise questions of implementation of global standards in the face of perceived barriers such as trade restrictions and state sovereignty, this has not deterred key organisations such as the OIE from exploring how global consistency might be achieved. Illustrating this fact, in October 2008, the OIE have planned a conference in Cairo, appropriately entitled 'Global Implementation of Animal Welfare Standards'.

In a world where there is increasing globalization of trade there are likely to be significant risks and costs unless there is clarity and consistency in respect of how animals, trade and relevant technology are governed. At an animal law conference held at Harvard Law School in 2007, this author proposed that, given the international recognition of issues related to animal welfare, it was timely to develop a global animal welfare legislative model based on standards of 'best practice'. In addition to identifying elements of a best practice model by examining common features of existing animal welfare law, it was suggested that a living model would also address animal welfare issues not currently addressed in contemporary animal welfare legislation. This would include, for example, sentience, the incorporation of indigenous and cultural perspectives of animals that currently exist outside of traditional western concepts, and the issue of sustainability.

There are those that believe that as long as state sovereignty, and rules based trading systems which ignore animal welfare as a trade criteria, remain, then true harmonization will be an elusive goal. While there is no question that the existence of such systems poses certain obstacles, there is increasing recognition and attention being directed toward the legislative opportunities that come with state sovereignty. Accordingly, it does not necessarily follow that obstacles are necessarily fatal to the implementation of global initiatives. The WTO, for example, currently prohibits member states from restricting imports on the grounds that the goods to be imported have not been produced to the same animal welfare standards as those required of the home producer. To allow such action, it is claimed, would be allowing importing countries to impose their own moral view on other countries who may not share the same attitude regarding animals. Importantly however, there is increasing awareness that such an approach ignores the counter argument that the imposition of moral values can be equally argued the other way.

The multifunctional nature of global animal welfare issues means that implementation will require a multidimensional approach to accommodate the required economical, political and social adjustments. In the words of Michael Cardwell, this will logically require 'adjustment by degrees'. Mirroring historical agricultural models and developments, the inclusion of sustainability principles in animal welfare legislative model may be accompanied by an assortment of 'carrot and stick' initiatives, including, for example, labeling, levies, incentives, special systems of aid, and compensatory allowances. As with previous effective models, the involvement of a suitably internationally respected and authoritative organisation is likely to be necessary. The impact however, could be substantial. In addition to facilitating the provision of consistency and accountability which is vital in a global marketplace; a living model of best ethics, science and law; would also provide a global blueprint for the future of animal welfare, and serve as a benchmark for current and future animal welfare initiatives for regions, states and NGO's alike.

Conclusion

Contemporary animal welfare law focuses predominantly on the physical and mental well-being of existing animals, but fails to clearly address animal welfare issues of future generations, which are central to the animal welfare portion of sustainability objectives. Issues of sustainable agriculture have enormous breadth and relevance, but if the objectives are to be achieved then it is necessary to include a legal duty of stewardship in animal welfare legislation in respect of the welfare of current *and* future animals.

Delays, or failure, to incorporate the stewardship principle into animal welfare legislation will equally result in delays, or failure, in achieving sustainable agricultural objectives. Delays due to differences of opinion are often bound up with differing religious, cultural, economic and political perspectives which, when combined with inconsistency and ad hoc approaches to an issue that has enormous impact, locally and globally, risks enormous costs to both humans and animals alike. It is therefore timely to develop a global animal welfare legislative model which addresses the issue of sustainability and other gaps within animal welfare legislation. Such a living document would provide a blueprint for the future of animal welfare that implement current animal welfare initiatives, and serve as a standard by which established and developing domestic and international law can be assessed and reviewed.

References

Turner, J. and D'Silva, J. (Eds) (2006). Animals, Ethics and Trade: The Challenge of Animal Sentience, Earthscan, London.

Webster, J., (2005), Animal Welfare Limping Towards Eden, Blackwell, Oxford.

Brooman, S. and Legge, D., (1997). Law Relating to Animals, London; Sydney: Cavendish.

Cardwell, M., (2004). The European Model of Agriculture Oxford University Press, Oxford.

Sunstein, Cass R. and Nussbaum, M.C., (Eds), (2004). Animal Rights Current Debates and New Directions. Oxford University Press, New York.

Animal welfare rules for broilers (2007) (Reference: IP/07/630). Brussels: European Commission. Available on the World Wide Web: http://europa.eu/rapid/pressReleasesAction.do?reference=IP/07/630&format=HTML&aged=0&language=EN.

Orlans, R.B., Beauchamp, T.L., Dresser, R., Morton, D.B. and Gluck, J.P., (1998). The Human Use of Animals, Oxford University Press, New York.

The Australian Animal Welfare Strategy: sustaining food production drivers within broader societal agendas

H.R. Yeatman
School of Health Sciences, University of Wollongong, Northfields Avenue, Wollongong, NSW 2522, Australia,

Abstract

The Australian Animal Welfare Strategy (AAWS) was accepted in 2004 by the Primary Industry Ministerial Council, with implementation commencing in 2005 (Product Integrity Unit, 2005). The Strategy covers all aspects of animal welfare in Australia - livestock and production animals; animals in the wild; companion animals; animals in research and teaching; aquatic animals; and animals used in work, sport, recreation or on display. While considerable effort by professional and industry advocates was undertaken in the lead up to the acceptance of the AAWS, political drivers were strongly linked to primarily international public criticism of Australia's treatment of livestock and production animals and hence threats to Australia's international trade. These economic drivers are reflected in the positioning of responsibility for the implementation of the Strategy within the national primary industry department (Agriculture, Fisheries and Forestry, Australia). The AAWS offers an interesting contrast to the initiatives in Europe. The AAWS is very broad in its scope. Embedding animal welfare concerns across a wide societal framework offers potential for sustainable impacts in the area of food production, perhaps more than just focusing on animal welfare in food production in isolation of animals in society more generally. However, the strategies undertaken have a presumed vision of what constitutes animal welfare goals, perpetuated within professional and government networks. The challenge is to sustain an appropriate and encompassing animal welfare agenda, when the impetus is primarily driven from one sector, primary industry, and in the absence of a strong (domestic) public positioning of the issues.

Keywords: animal welfare, policy agenda, policy implementation

Background to the Australian Animal Welfare Strategy

Animal welfare issues have been on the Australian government's agenda since 1989 when a National Consultative Committee on Animal Welfare (NCCAW) was established to advise the Australian Minister for Primary Industries (Shiell, 2006b). The Consultative Committee comprised representatives from industry groups such as the National Farmers Federation, Aquaculture Industry and Australian Racing Board, the national research agency (National Health and Medical Research Council, NHMRC), non-government groups such as the RSPCA and the Australian Veterinarian Association, together with various government groups. Even at its early meetings, the breadth of animal welfare issues under its purview was debated, quickly leading to inclusion of non-production animals as well as production animals and broadening the areas under consideration wider than originally intended. Its primary role was to provide policy advice to the Minister for Primary Industries. When advice related to activities in other portfolios, it was provided via the Minister for Primary Industries, thus maintaining the same reporting mechanism.

In keeping with the broad approach to animal welfare issues taken by the NCCAW, subsequent work by Australian governments to develop a national animal welfare strategy included all animal circumstances. Five years of development and consultation resulted in the Australian Animal Welfare Strategy (AAWS), which was accepted by the Australian Primary Industries Ministerial Committee in 2004 (Product

Integrity Unit, 2005). The Australian Animal Welfare Strategy (AAWS) covers the humane treatment of all sentient animals in Australia including:
- Animals used in research and teaching
- Companion animals
- Animals in the wild
- Aquatic animals
- Livestock/production animals
- Animals used for work, sport, recreation and on display (Product Integrity Unit, 2005)

In terms of Kingdon's concept of 'policy window' (Kingdon, 1995), a 'policy solution' was available to place on the table when two significant events occurred. In 2003, a live sheep cargo ship the Cormo Express was refused entry into Saudi Arabia, resulting in several additional weeks of travel for over 50,000 sheep until an alternative country, Eritrea, was prepared to berth and unload the sheep. This event occurred after a series of problems in the live export trade during 2002, and it resulted in a high level of international criticism of the live sheep trade and of the way in which the Australian government handled the event. (de Fraga, 2004; Gratton, 2003). At a similar time, the international action group PETA (People for the Ethical Treatment of Animals) ran an international campaign against the practice of mulesing - removing folds of skin around a sheep's tail and anus, usually without anaesthetic, to prevent the occurrence of flystrike which causes a high level of distress and pain for the animal (Hogan, 2004; People for the Ethical Treatment of Animals). Together, these two events focused the Australian government's attention on animal welfare issues, in a federal election year. The AAWS subsequently was adopted with $6 million to support its implementation over a 4 year period.

Implementing the AAWS

Having adopted the AAWS, the Primary Industry's sector then focused on the 'mobilisation' (Cobb, 1976) of effort to implement the strategy. On the one hand the Minister of Primary Industries represented the strategy as a 'shared vision', on the other hand he emphasized that it was 'vital in the agricultural sector - healthy animals and sound welfare practices, contribute significantly to improving production and meeting consumer expectations'. He clearly identified that 'implementing this strategy will require the commitment and resources of all stakeholders - it is very much a shared responsibility' (Product Integrity Unit, 2005).

A strategy of broad representation and involvement was adopted for the implementation of the AAWS. A high level Advisory Committee was appointed, with representation from a range of government and non-government agencies and including a person with wider, public interest expertise (author). A two day workshop was run to develop and refine the Strategy's Implementation Plan. This involved almost 100 representatives from across Australia, involved in all aspects of animal welfare, and all engaged in the determination of the Implementation Plan. This Implementation Plan was subsequently endorsed by the Primary Industries Ministerial Council (Department of Agriculture Fisheries and Forestry, 2006). In addition, sectoral working groups were formed to progress each area of the AAWS. Independent chairs were appointed and again, wide representation from interested groups and government departments were involved. A working group was established for each of the six areas of animal welfare identified in the AAWS and subsequently three additional working groups were established to focus on cross-sectoral issues - research, communication and education.

Public engagement with AAWS

One of the key strategies identified by the AAWS Advisory Committee was greater public engagement in animal welfare issues. The Australian situation is quite different from that in Europe, where animal welfare, particularly as it relates to food production processes, is at the forefront of public interest. In Australia, action in introducing the AAWS could be considered to be a government response to international pressure, not the result of domestic constituency pressure. One indication of this is the lack of public demand for the declaration of food production methods to be declared on food labels, as is under consideration in Europe. In one location, the Australian Capital Territory (where the country's capital, Canberra, is located) there has been some public interest, as reflected in an attempt to ban the sale of battery hen produced eggs in 1996. This legislation was not implemented due to national constitutional problems regarding trade between states and territories (Productivity Commission, 1998). Also, although a range of government departments and agencies have been involved in the AAWS processes (e.g. primary industries, environment, national parks, heritage, local government), health departments and the food regulation agency have not been involved at any stage. This reflects a government position that animal welfare does not impact on human health. Little public discourse has challenged this position.

The AAWS Advisory Committee was mindful that it needs to engage the public. Market research has focused on communicating animal welfare concepts. It has identified that the public readily engaged with the concept of animal welfare but were somewhat complacent about the issue (e.g. 'I just know how to care for my pets') and were not aware of government action in the area (Southwell, 2006). Key messages were identified as 'collective responsibility' and 'the social benefits of animals'. Work is currently underway to develop and implement communication initiatives. However, actions which aim to engage in public debate around controversial animal welfare topics, such as live exports, or in areas not considered within the AAWS, such as considering the mandating of food labeling declaration of food production processes, have not been contemplated.

Opportunities and threats

Controlling the animal welfare agenda within the framework articulated within the AAWS has both advantages and disadvantages. It has raised the profile of animal welfare issues across a full range of animal-related areas within the one framework for action. This provides a vehicle for many different agencies and organizations to engage with each other and learn from shared experiences. For example, all sectoral working groups undertook reviews of legislation and initiatives in their area. These reports were then collated into a single review, identifying gaps and opportunities shared by all groups, while also communicating needs specific to each sector (Shiell, 2006a). Collective actions in the areas of research, legislative reform, communication and education were identified and actions commenced. However, limited public involvement and a primarily single sectoral (primary industry) responsibility may expose the AAWS to criticism or conflict in the future. Communicating about the AAWS and animal welfare 'shared responsibilities' needs to be considered as an important but preliminary step. To enable public trust and engagement with animal welfare issues in Australia to develop in a meaningful manner, issues of public interest need to be identified and openly debated. Animal welfare needs to be linked to issues that people contend with on a daily basis, such as purchasing food, or with issues that underpin their support for their government agents, such as ensuring trade or wildlife management issues are supported by appropriate animal welfare standards. Robust public discussion, accompanied by government actions perceived by the public to be supportive of their, not industries', interests, is yet to occur. Thus the development and implementation of the AAWS should be considered to be an important first step in achieving appropriate animal welfare action in Australia.

References

Cobb R. (1976). Agenda building as a comparative political process. American Political Science Review 70: 126-138.

de Fraga, C. (2004). The Cormo Express: Australia's latest live export shame. Animal Welfare Institute, http://www.awionline.org/pubs/Quarterly/04-53-1/531p7.htm.

Australian Government Department of Agriculture Fisheries and Forestry (2006). National implementation plan of the Australian Animal Welfare Strategy: Australian Government, http://www.daffa.gov.au/__data/assets/pdf_file/146745/aaws_implementation_april06.pdf.

Gratton, M. (2003, September 24). Sheep on board a national shame. The Age, http://www.theage.com.au/articles/2003/09/23/1064082991895.html.

Hogan, J. (2004, October 15). Farmers ridicule US wool ban. The Age, http://www.theage.com.au/articles/2004/10/15/1097784011310.html.

Kingdon, J. (1995). Agendas, alternatives, and public policies. New York: HarperCollins College Publishers.

People for the Ethical Treatment of Animals. An examination of two major forms of cruelty in Australian wool production: mulesing and live exports. http://www.livexports.com/mulesing.txt.

Product Integrity Unit, Australian Government Department Agriculture, Fisheries and Forestry (2005). The Australian Animal Welfare Strategy. In DAFF (pp. 34): Commonwealth Government of Australia.

Productivity Commission. (1998). Battery eggs sale and production in the ACT, Research report. AusInfo, Canberra, http://www.pc.gov.au/study/batthen/finalreport/batthen.pdf.

Shiell, K. (2006a). Australian Animal Welfare Strategy. Final summary report on priorities for action from inventories of animal welfare arrangements. Animal Welfare Unit, DAFF, Australia http://www.daffa.gov.au/__data/assets/pdf_file/152111/aaws_stocktake_summary.pdf

Shiell, K. (2006b). Australian Animal Welfare Strategy. Report on the Review of the National Consultative Committee on Animal Welfare (NCCAW). DAFF, Australia. http://www.daffa.gov.au/__data/assets/word_doc/146750/nccaw_review_feb07.doc.

Southwell, A., Bessey, A., Barker, A. (2006). Attitudes towards animal welfare. A research report. In: Agriculture Fisheries and Forestry (ed.) TNS Consultants, http://www.daffa.gov.au/__data/assets/pdf_file/146748/tns_aw_research.pdf.

Sheep welfare in the welfare state: ethical aspects of the conventionalisation of Norwegian organic production

Marianne Kulø and Lill M. Vramo
SIFO, the National Institute for Consumer Research, P.O. Box 4682 Nydalen, 0405 Oslo, Norway,
marianne.kulo@sifo.no

Abstract

In the western part of the world the organic sector is going through a process of expansion and commercialisation. Norway currently lags behind other European countries in the production and consumption of organic products. In order to improve the sustainability of the Norwegian agricultural sector, the Ministry of Agriculture has decided on a policy of increasing organic production from 4 per cent of total agricultural production in 2006 to a goal of 15 per cent within 2015. Sheep production is given particular attention in this policy. This paper explores the agricultural sector's focus on animal welfare in this process, discussing how the conventionalisation in organic sheep production may affect sheep welfare. The empirical analysis in this paper is based on in-depth interviews with key informants from central organisations in the agricultural sector. We discuss how the informants perceive the conversion to an organic production system. A common view is that such a conversion will not benefit sheep welfare, and that it instead may lead to worse conditions for the sheep. Based on cultural norms and conditions in Norway, it is explored why key informants in the agricultural sector are negative to animal welfare being improved through the market differentiation that organic certification implies.

Keywords: sheep production, organic, conventionalisation, animal welfare

Introduction

In the western part of the world the organic sector is going through a process of expansion and commercialisation. This process involves several challenges and dilemmas, connected to conventionalising of organic agriculture, which means that the special attributes associated with organic products diminish or even disappear in the conventional food system (Kratochvil and Leitner, 2005). Organic attributes include intrinsic values such as striving for the assurance of animal welfare and long-term sustainable production systems. This paper discusses why these attributes may diminish in the Norwegian conventional food system, and how the conventionalisation of organic sheep production may affect sheep welfare.

Conventional farming can be characterized as specialised and intensive production systems where the emphasis is put on maximizing productivity and profitability, and the animals are seen as more or less passive production units. Organic agriculture developed partly as a reaction to the negative effects of the industrialisation of agriculture, with the central idea that nature as a whole must be kept in balance. Animals are understood as active beings that contribute to this balance (Vaarst *et al.*, 2004). The two different agricultural systems can be seen as part of different rationalities. While conventional farming has an inherent economic and productivist rationality, organic farming has a visionary rationality where the goal is a system in balance.

The Norwegian government recently made a political statement of increasing organic production from originally being 4 per cent in 2006 to become 15 per cent within 2015. Sheep production is one of the areas of commitment in this policy. Norwegian consumers increasingly demand organic products. They already perceive conventional sheep farming as nearly organic, and also as having the best level of animal welfare in Norwegian livestock production (Berg, 2002). Traditionally, however, stakeholders in the organic sector have not focused on animal welfare as a central value. Instead organic farming has been presented as a food production system that is more sustainable for the environment. The Agricultural Minister recently pointed out that animal welfare is among the areas where organic agriculture should take the lead, through breeding, knowledge about animal behaviour and improvement of practical production methods (Ministry of Agriculture, 2006). One might expect that the conversion of Norwegian sheep farms to an organic production system would lead to improved sheep welfare in Norway. Yet several stakeholders claim the opposite. According to them, the conversion to an organic production system will not benefit sheep welfare, and may even lead to worse conditions for the sheep. We discuss how these conditions are part of the conventionalisation of organic sheep production. We argue that the policy of converting sheep production to an organic production system will affect the mind-set of the organic sector towards a more rational marketing perspective. In-depth interviews have been conducted among stakeholders in the agricultural sector as the basis for this study.

Background

Sheep welfare in Norwegian organic agriculture

In Norway organic farming was initiated in the 1930s inspired by Rudolf Steiner's anthroposophy. The biodynamic method focuses on the farm as a holistic ecosystem where animal health and vitality is central in the ensurance of the sustainability of farm production. Later, in the 1960s, ecophilosophy became a central political reference for the organic movement. As a part of the emerging environmentalism, this philosophy is based on respect for the intrinsic value and diversity of non-human life forms and has a precautious approach to the management of natural resources. The opportunity for all species to realise their natural needs is emphasized (Vittersø *et al.*, 1994).

Until recently, Norwegian organic agriculture has been dominated by private initiatives and has constituted a very small part of the agricultural sector. During the 1990s an expansion process began. The processing industry owned by conventional farmer cooperatives got involved in the organic sector, and parallel to this process the regulations for organic production became subject to governmental control and harmonised with EU regulations. Organic regulations represent a supplement to other laws and regulations being in force. In order to set a common uniform standard and simplify its implementation, several articles concerning animal welfare have been removed. Organic sheep are no longer necessarily entitled to access outdoor areas in the winter period, the transportation period is now the same as in conventional production and expanded metal is now allowed as flooring. There is a general agreement that the removal of these articles is a setback for animal welfare. What is left of additional mandates that affect sheep welfare in organic regulations is the demand for a lower stocking density in pens.

The Norwegian welfare state

Norway has several unique features compared to other European countries. The modern society is built on values of the social democratic welfare state. One of the basic pillars of the Scandinavian welfare states are strong collective values, leading to an emphasis on the elimination of diversity (Vike, 2001). In Norway, equality is understood as sameness: Difference between people is easily perceived as unwanted hierarchy and thus as injustice. Hierarchical relations are being under-communicated through focusing on common traits of individuals and groups, or quite simply by not cooperating with those who are perceived as too different (Gullestad, 1989). The agricultural sector is characterised by high subsidies, supported also by import restrictions. This strongly protected industry is believed to be beneficial for Norwegian society in a number of ways, including good animal welfare. There is a widespread view both within the agricultural sector, among retailers and in the general public that Norwegian regulations on animal welfare are stricter than in other European countries (Dulsrud and Vramo, 2006). The high level of trust in Norwegian institutions and society in general applies even to domestic food production (Kjærnes *et al.*, 2005).

Method

The discussion in this article is based on findings in 13 in-depth interviews with central stakeholders in the agricultural sector. The interviews were carried out in December 2006- April 2007, covering sheep production and sheep welfare in Norway. The study is part of the interdisciplinary project 'Sheep welfare in the food chain', which is a collaboration between The Norwegian School of Veterinary Science (NVH, the coordinator), the National Institute for Consumer Research (SIFO) and the Norwegian Meat Research Center (Animalia). In the analysis we have divided the actors into three categories:

- Conventional agricultural organisations:
  a. Public: Public food authorities, the Agricultural University, the Ministry of Agriculture.
  b. Private: The meat processing industry, the organisation for sheep farmers.
- Marketing organisations:
  a. Public: The national company for marketing and innovation.
  b. Private: The marketing organisation for the meat processing industry.
- Organic organisations:
  Private: The auditing and certification organisation for organic production, the national organisation for organic producers and consumers, the national agricultural extension service.

The interviews

Sheep welfare in Norway

When asked about the state of sheep welfare in Norway, all informants claimed it to be generally good. At the same time, all of them mentioned several critical animal welfare issues in Norwegian sheep production. Common issues were loss on pasture due to predators, breeding, losses during the lambing period and long distance transport. Generally the informants were talking about animal welfare problems in two ways. Either they conceived problematic issues as integrated and inevitable parts of sheep production, such as lambing problems and shortage of labour, or they characterised them as unacceptable. In the latter case they either categorised them as exceptions due to human error or as caused by conditions in society such as economical limitations.

Organic agriculture

Generally the informants associated organic agriculture with restrictions on artificial additives such as the use of inorganic fertilizers and pesticides. A distinction can be made between a *product* and a *process* perspective. The representatives from the sheep organisation and the meat processing industry focussed more on the final product and emphasized qualities such as 'pure' or 'local' food. This can be categorized as a product perspective. Both the representatives from the processing industry and the marketing institutions emphasized the market potential for organic mutton. Those having a process perspective, which were the representatives from the organic organisations, associated organic farming with protection of the environment, sustainable use of resources and food safety. Generally the informants did not perceive sheep welfare problems as relevant for improvements through organic farming. This is reflected in the answer from a representative from an organic organisation, when asked why the difference in welfare conditions in conventional and organic sheep production is so small:

'There isn't much related to animal welfare that can be improved in sheep production, I don't think that.[...] I assume so, because otherwise for sure those issues would have been taken into account when the regulations were formulated.'

This statement reflects the general trust in public regulations. It also indicates an attitude of animal welfare being 'good enough' at the expense of the visionary perspective in the organic philosophy. However, in addition to existing animal welfare problems in conventional sheep farming, several potential animal welfare problems in organic sheep farming were mentioned. There was a general concern regarding the possibility of malnutrition due to the prohibition on inorganic fertilizers, and concerns about insufficient formal capabilities of sheep farmers who convert to an organic production system.

Market differentiation

Although the informants perceived citizens as having important influences on the development of animal welfare conditions, the citizens' role as consumers was not emphasized. Instead the informants focused on how public opinion may improve the level of animal welfare through the general political development. The informants believed that Norwegians trust sheep production to have high quality, irrespective of whether it is from an organic or a conventional production system. The representatives from the conventional agricultural organisations were resistant to differentiate on animal welfare in the markets. The statement from the representative from the food authority can be seen as a typical presentation of this view: *'Animal welfare in sheep production should generally be on such a high level that it should be irrelevant whether it is an organic or a conventional production system'*. A representative from the sheep organisation did not see any point in having different production regimes and stated: *'We have to be careful not to call attention to organic production at the expense of conventional production.'* In this informant's opinion there is not really any point in having different production regimes, because the general high ethical quality of products from sheep farming makes choice between alternatives irrelevant.

Discussion

We will in this last section discuss the findings from the interviews in relation to cultural norms and conditions in the agricultural sector. The interviewed stakeholders had different perspectives on organic sheep production. Several of the representatives from the marketing organisations and the conventional organisations emphasized the market potential for organic mutton, which can be categorized as a rational marketing perspective on organic agriculture. The organic organisations, on the other hand, focused on aspects of the production. However all informants perceived restrictions of artificial additives to be the main idea of organic agriculture and animal welfare was not mentioned as a central factor. This view has often been seen as the main defining feature of organic farming by the conventional sector, and is known as the no-chemical approach (Vaarst *et al.*, 2004). The fact that this view is shared by representatives from both organic and conventional organisations reflects to which degree the gap between organic and conventional sector is diminishing. The inclusion of a rational marketing perspective on organic production is also an indication of a process of changing visions in organic agriculture.

The informants pointed out several critical animal welfare factors relevant for improvement in Norwegian sheep production, such as breeding and long distance transport. However, these problems were not perceived as relevant to upgrade through organic production. Instead the level of animal welfare in the agricultural sector should be improved as a part of the general evolvement of the state, as a responsibility of the public authorities. As shown in the quotes, the informants were reluctant to differentiate on animal welfare between organic and conventional production systems. The representatives from conventional agricultural organisations were also negative to differentiation because it would put conventional agriculture in an unfavourable light. This can be seen as a consequence of the historical emphasis on homogeneity in the development of Norwegian agricultural sector and as a general trust in ethical values to be upheld by the state.

The focus on state regulations at expense of organic philosophy is acknowledged as a feature of the conventionalisation of organic agriculture, as well as the involvement of stakeholders with interests beyond organic agriculture (Vittersø *et al.*, 2005; Kratochvil and Leitner, 2005). These factors seem to inhibit the possibilities of focusing on animal welfare improvements in organic sheep production. The conversion of sheep farming to an organic production system is a central part of the policy of expanding the organic sector. Yet, a general opinion among our key informants is that the conversion will not benefit sheep welfare and may possibly even affect it negatively. This is an indication of a change of rationality in the organic sector, and can be seen as a consequence of stakeholders from the conventional sector being increasingly involved in organic production.

As a result of the rationalisation of organic regulations, the differences between the production systems regarding sheep welfare mandates have become minimal. This process decreases the possibilities of organic sheep production to be a pioneer in the general development of animal welfare, and also radically contrasts the ideas from the original organic philosophy. At the same time, the government points out that animal welfare is an important value in the promotion of organic farming. This disagreement between the presentation and the actual situation of sheep welfare in organic production might be a risk for citizens' distrust in organic production. However, public discourse about ethical aspects of food production is minor due to the general trust in agricultural stakeholders.

Stakeholders in the agricultural sector have, in their presentation of organic production, more often called attention to the sustainability of the environment than to the assurance of animal welfare. This can be seen as an explanation to the low focus on improving animal welfare in organic production. As conventionalisation continues, the role of animal welfare in the basic principles of organic production may be weakened further and in that way contribute to a further diminishing gap between the two production systems. The possibility of organic agriculture to function as a pioneer in the development of farming will hence decrease. This can promote the scenario of organic production to become just another niche production.

As a consequence, the weakening of attention towards animal welfare can reduce the sustainability of organic production as a whole.

References

Berg, L. (2002). Dyr er ikke bare mat. Project Report No. 10, The National Institute for Consumer Reasearch, Oslo, Norway.

Dulsrud, A. and Vramo, L. M. (2006). Deliverable 1.2.3. National Report Norway. Welfare Quality Reports. The National Institute for Consumer Reasearch, Oslo, Norway.

Gullestad, M. (1989). Kultur og hverdagsliv. Universitetsforlaget, Oslo, Norway.

Kjærnes, U., Poppe, C. and Lavik, R. (2005). Trust, distrust and food consumption. A study in six Euopean countries. Project Report No. 15. The National Institute for Consumer Reasearch, Oslo, Norway.

Kratochvil, R. and Leitner, H. (2005). Organic farming between vision and reality. Paper presented at the XXI ESRS Congress August 22-25, 2005, Keszthely, Hungary.

Ministry of Agriculture (2006). Avspark for øko-kampanjen 'Naturlig bortskjemt mat'. Speach September 21, 2006, Oslo, Norway.

Vaarst, M., Roderick, S., Lund, V. and Lockeretz, W. (2004). Animal health and welfare in organic agriculture. CABI Publishing, Wallingford, UK.

Vike, H. (2001). Norden. In: M. Melhuus and S. Howell (eds.) Fjærn og nær. Gyldendal Norsk Forlag, Oslo, Norway, pp. 534-556.

Vittersø, G. (1994) Ta vare på helheten. Økologisk landbruk i utvikling. Report No. 1. Alternativ Framtid, Oslo, Norway.

Vittersø, G., Lieblein, G., Torjusen, H., Jansen, B. and Østergaard, E. (2005). Local, organic food initiatives and their potentials for transforming the conventional food system. Anthropology of Food 4: 2-18.

Animal welfare in assurance schemes: benchmarking for progress

S. Aerts
Centre for Science, Technology and Ethics, K.U.Leuven, Kasteelpark Arenberg 1, 3001 Leuven, Belgium,
Stef.Aerts@biw.kuleuven.be

Abstract

Many, if not all, of the large retail companies run some sort of assurance scheme for their fresh meat and egg products. Other schemes exist, publicly and privately owned. This means that a large part of the animal products available to the consumer are produced under an assurance scheme. Most of these contain a section of animal welfare related requirements. Considerable differences between animal production schemes exist with respect to animal welfare requirements. But, a simple numerical comparison is dangerous, as qualitative differences between schemes and requirements are difficult to code. Assurance schemes are able to deliver an major added value to animal production. Through such a system, important sustainability elements can be enforced across the production chain, and other ethical concepts such as consumer choice and corporate social responsibility can play a more important role in animal production. A more open communication might result in push and pull effects towards better assurance schemes. Improving schemes is not sufficient. Appropriate legislation is indispensable in order to assure that all animals are produced according to decent standards, including those brought to the market as processed products.

Keywords: assurance scheme, animal welfare, sustainability, retail

Introduction

The retail sector is continuously concentrating and the market share of big distribution companies is therefore growing considerably. For example, in 2004, the four larges retail groups of Western Europe had a combined grocery market share (CR4) of 19.9% (Vander Stichele, 2005b) and in the US the CR5 reached 46% that year, up from 24% in 1997 and 38% in 2001 (Hendrickson and Heffernan, 2005). Within most European countries, CR5 values are even more extreme: Finland, 88.2%; Belgium, 78.6%; France, 70.4%, United Kingdom, 56.3%, to name just a few (Vander Stichele, 2005a). According to Grievink (2003), buying power within the agricultural sector is extremely concentrated.

Many, if not all, of the large retail companies run some sort of assurance scheme for their fresh meat and egg products. Many other schemes exist, publicly and privately owned. This means that a large part of the animal products available to the consumer are produced under an assurance scheme. Some of these are presented as an independent scheme; others are marketed as a (retail) brand.

Most of these schemes contain a section of animal welfare related requirements, but it is notoriously difficult to obtain a copy of the scheme documents (especially with the privately owned schemes). As communication about these schemes is often limited to a general overview, it is therefore unclear what level of animal welfare is reached under these schemes. This paper will discuss some results of a Belgian comparative study and comment on the ethical merits of animal production assurance schemes.

Benchmark study

Materials and methods

A benchmark study was conducted on 55 assurance schemes (including 17 foreign or international schemes) for beef (9), veal (3), pork (20), chicken (10), milk (4) and egg (9) production.

For each category, this included the assurance schemes of Belgium's three largest retail companies, the organic scheme, and the large public schemes. For some categories (most notably pork and poultry) small local schemes were included. When made available, other private schemes were included. Applicable laws were analysed using the same criteria as the schemes.

For each animal type, a different set of criteria was agreed upon, using a step-wise identification procedure: (1) welfare related elements mentioned in assurance schemes; (2) additional elements present in legislation; (3) elements scientifically known to be important, but not mentioned in schemes or legislation (to be identified by the working group).

For each additional step of the production chain (e.g. farm, transport, slaughter), a new set of criteria was devised. This resulted in a varying number of criteria applied to different schemes.

For each scheme, each criterion was scored '++' (supralegal requirement), '+' (explicit mentioning of legal requirement), 'w' (legal requirement not mentioned), or '-' (no requirements included). If none of these codes fitted, an alternative (description) was used.

In this paper, only the chicken, pork and beef assurance schemes will be used, as other categories do not include enough schemes or do not cover the entire production spectrum (e.g. no battery egg schemes).

Results

Table 1 summarises the average prevalence of each score type for chicken, pork and beef schemes, as well as its standard deviation (SD), minimum and maximum.

In Table 1 the different number of criteria to which a scheme is scored ('total') is an indication of how many parts of the animal production system are covered by the individual scheme. Especially in chicken production, which has a complex production chain, but also in pork production, a considerable variance exists between different schemes.

It can also be seen that a high number of legal requirements is not covered, throughout the different categories. Again, considerable variance exists between schemes. The higher number of 'w' and '+' for pork production, is easily explained by the existence of more detailed legislation for this production type.

Organic production consistently scored above average for each score type (data not shown), but only for egg production, it scored the highest number of '++'.

When referring to pork production alone, it can be seen that Belgian schemes tend to have lower coverage of legal requirements (less '+', more 'w'), but do carry more supralegal requirements than the foreign and international schemes analysed.

Table 1. Score type prevalence for chicken, pork and beef schemes.

	Chicken (B) n = 7			Beef (B) n = 8			Pork (B) n = 13			Pork (int.) n = 7		
	Av	Min	Max	Av	Min	Max	Av	Min	Max	Av	Min	Max
++	17 ± 8,58	2	26	14 ± 6,53	6	23	18 ± 4,53	8	24	12 ± 7,09	5	24
+	5 ± 3,55	0	10	4 ± 1,68	3	7	9 ± 3,05	4	16	16 ± 8,40	1	22
w	17 ± 5,73	9	23	14 ± 1,29	12	16	34 ± 7,16	19	45	24 ± 10,37	14	43
-	36 ± 5,76	26	43	27 ± 6,34	17	33	24 ± 3,23	18	28	32 ± 4,20	24	37
Other	9 ± 4,49	5	18	6 ± 2,48	5	8	4 ± 4,76	0	17	8 ± 5,86	0	18
Total	82 ± 10,74	76	107	65 ± 1,99	63	69	88 ± 3,55	79	95	91 ± 3,95	88	97

Legend: Av = average prevalence of code (± standard deviation), min = minimum prevalence, max = maximum prevalence; '++' = supralegal requirement, '+' = explicit mentioning of legal requirement, 'w' = legal requirement not mentioned, '-' = no requirements included, 'other' = none of the previous. B = Belgian schemes; int = foreign and international schemes (mainly Dutch).

Discussion

Production schemes analysis

A simple comparison of score prevalence, as done above, does not do right to the complexity of animal production schemes. Between supralegal requirements, for example, important differences might exist that are not reflected by the scoring system. Adding an '+++' category is difficult as it is nearly impossible to define across criteria and indicators. Additionally, it is clear that not all criteria are equally important for animal welfare. One should therefore always refer to the entire analysis matrix before drawing any firm conclusions.

Although it is subtle, there is an important difference between a '+' and a 'w' score for a certain criterion. Although both requirements are identical, the former holds a stronger assurance towards the consumer (and therefore, towards the animal). Although anyone involved in the animal production chain (these are not only stockholders) is obliged to follow all legal requirements, controls are scarce and superficial. A simple time and budget calculation is sufficient to explain why. When such a legal requirement is included in a production scheme, however, controls will be - or at least they should be - regular and thorough, which results in a higher degree of assurance. Even schemes that carry only a literal transcription of animal welfare legislation will therefore have positive effects for animals raised under those schemes.

Sustainability

Many assurance schemes date back to the late 1990's, when several big food scares hit the European food production (en distribution) sector. To re-establish trust, many assurance schemes were installed, by retailers, public authorities, (semi-)independent organisations and others. Although food safety was (and often is) the main objective of such schemes, soon other elements were included, such as price premiums for primary producers and ensured supply for processors and retailers.

It can be suggested that during the years between the creation of the assurance schemes and now, most schemes have installed a stable and reliable food safety system. This means that the mere existence of such a scheme does not longer suffice to enable consumers to differentiate between products. Indeed, there is a clear trend (within the Belgian assurance schemes) to supplement and update the existing requirements with more animal welfare related elements. Many of the schemes analysed have recently issued a new version or are in the process of doing so.

This means that, within the concept of sustainability, assurance schemes are moving towards a position that enables them to aid in delivering a more sustainable animal production. They already incorporated economic and social aspects, and they are enforcing their social relevance by taking a more active position on animal welfare.

What most schemes do not incorporate substantially, is the ecological aspect of sustainability. With the exception of the 'Milieukeur' scheme in The Netherlands and the organic scheme, most do not seem to incorporate many (supralegal) requirements on environmental matters. It seems possible that schemes will start to focus more on ecology after they have enforced the animal welfare elements. Considering the favourable socio-political climate, this may even become a selling proposition for some.

Food ethics

Assurance schemes are not value-free. They explicitly communicate trustworthiness towards the consumer and many other 'value' elements are used in communication about a scheme. This means that a discussion about rights, responsibilities and justice is appropriate in the context of assurance schemes. An in-depth analysis is necessary, but in the following paragraphs a few elements will be discussed.

An important question is who decides what is included (or not included) in a scheme. Owners may wish to include only what they know to be important for consumers to choose between products, or they might want to include what consumers say they consider important (for example in a survey). This of course leads to a new aspect of the classic consumer-citizen dilemma. Is a scheme to deliver what the consumer buys, or what he says he wants? It seems that scheme owners with a strong tradition of corporate social responsibility (CSR) will be more inclined to choose the latter option, particularly because that option is (usually) quite generally considered the morally better option. Additionally, hiding behind the consumer defines all that is known about the efficacy of commercials, product placement and other consumer-influencing techniques. Considering the strong position of some assurance schemes (especially, but not only retail-led schemes), scheme owners could therefore be expected to take responsibility and lead the way towards a more sustainable (animal friendly) agricultural production.

One could also ask to which extent consumers should have access to information about the assurance scheme. Although these schemes are ment to reassure the consumer about the quality of the product, they are also a competition instrument for many of the scheme owners (particularly in the case of distribution companies). Therefore, they might have important interests in keeping the requirements confidential. Puzzlingly, one of the schemes within the study was marketed as 'XXX, you know what you eat!', but the researchers were asked not to disclose the requirements in the scheme (except for the coded form in the study).

It seems there is a lot to be gained by a more open communication. It may be evident to expect schemes that receive any form of public funding to make all parts of the scheme public; but, even private schemes should make as much information available as possible. First, it would make the consumer more informed, and thus more able to make conscious choices; this would possibly create a pull effect towards better assurance schemes (on social and ecological aspects). Second, benchmarking between schemes - which is routinely done - would push schemes to incorporate more or better requirements. The results in table

1 confirm that a considerable room for manouvre exists. Personal experiences during the present study confirm the latter effect.

Finally, such an open communication about assurance schemes would deliver important information about the distribution of gains between the different parties concerned. Considering the important buying power of retail companies within the food chain, there is reason for concern that they will not distribute the price premiums of their assurance scheme across the chain. Again, companies with a strong CSR policy will benefit most from an open communication policy. This might create similar effects as with animal welfare requirements.

In short, assurance schemes are able to deliver important advances in animal production, and this would be even more so with a more open communication strategy. Unfortunately, in most cases only fresh products are marketed through assurance schemes (organic products are a marked exception), but a large part of agricultural production is destined for processed products. Although other types of schemes might apply to such production, most of the aforementioned effects will not be relevant in those cases. Legislation will therefore remain an important regulatory framework.

Conclusion

Considerable differences between animal production schemes exist with respect to animal welfare requirements. This small-scale study reveals the high variance in the number of legal and supralegal requirements incorporated by assurance schemes. A comparison between Belgian and foreign schemes (mainly Dutch) for pork production hints at a different culture towards such schemes, as Belgian schemes mention less requirements overall, but do have more supralegal elements. Importantly, a simple numerical comparison of schemes is dangerous, as qualitative differences between schemes and requirements are difficult to code.

In general, assurance schemes are able to deliver an major added value to animal production. Through such a system, important sustainability elements can be enforced across the production chain, and other ethical concepts such as consumer choice and corporate social responsibility can play a more important role in animal production. A more open communication might result in push and pull effects towards better assurance schemes.

Although assurance schemes are important in animal production, especially in the current economic constellation, an important part of the produce is not marketed under such a scheme. Improving schemes is therefore important and necessary, but is not sufficient. Appropriate legislation is indispensable in order to assure that all animals are produced according to decent standards, including those brought to the market as processed products.

Acknowledgements

The author would like to thank DP21 and Cera for funding this project, the members of the working group for their active cooperation, and the assurance scheme owners for making their schemes available to the working group.

References

Grievink, J.-W. (2003). The changing face of the flobal food industry. Presentation OECD Conference The Hague 6 February 2003, Cap Gemini Ernst and Young: 41p.

Hendrickson, M. and Heffernan, W. (2005). Concentration of agricultural markets. University of Missouri, Columbia, USA.

Vander Stichele, M. (2005a). EU members in Western Europe (15): share of modern grocery distribution of the top 5 retailers in each country (%) - 2004. [online]. http://www.marketsharematrix.org [Accessed May 15, 2007].

Vander Stichele, M. (2005b). Western Europe: Grocery Retailing 2004. [online]. http://www.marketsharematrix.org [Accessed May 15, 2007].

Best Practice in animal production: a grassroots evaluation framework

S. Aerts and D. Lips
Centre for Science, Technology and Ethics, K.U.Leuven, Kasteelpark Arenberg 1, 3001 Leuven, Belgium,
Stef.Aerts@biw.kuleuven.be

Abstract

In a stakeholder working group dialogue representatives from a farmers union, a welfare activist group, government and scientists, have drafted an evaluation framework for the qualitative comparison of animal production systems. The working group has devised a set of indicators directly connected to animal production and readily understandable to people working in the field. An animal production system (which ranges from selection to slaughter) cannot be judged in its entirety and must be divided into (at least) 6 phases. Each of these phases is to be assessed through a set of 28 indicators distributed over 7 areas of concern. These indicators are animal type and system phase independent. A qualitative code has been developed that allows a uniform assessment throughout all indicators. The system addresses a range of practical and ethical issues. Therefore, existing evaluations are easily integrated into the BP assessment, thus avoiding duplicating previous work. This also means that information will be unavailable for some (or many) of the indicators. Although this evidently complicates the evaluation, it reflects the situation in which an economic actor will be forced to choose between the alternatives. The framework has not yet been applied to specific animal production sectors and neither does it contain a formalised weighting procedure.

Keywords: best practice, animal production, practice-oriented ethics

Introduction

As a follow-up to the Societal Animal Welfare Indicator project (Aerts and Lips, 2006), a working group of five representatives of research, government, agriculture and animal welfare groups, set out to prepare a common framework for the identification of 'Best Practices' (BP) in animal production.

The project's goal was to deliver a framework capable of assisting in a qualitative assessment and comparison of animal production systems, without attempting to engage into concrete assessments and without being limited to a single production sector. The framework has been created through monthly workgroup meetings (October 2006 - January 2007) with interim electronic rapportage and feedback.

Results

The BP framework consists of a set of spreadsheet-like three-dimensional matrices with columns containing the assessment criteria (indicators) and rows describing production system types. Each animal type has its own matrix and within each matrix each phase of the animal production chain is evaluated independently. These phases are: (1) selection, (2) feed production, (3) housing, (4) transport, (5) gathering (markets, ...), (6) slaughter. As the project has originated from an animal welfare project, the framework is limited to those phases of the production chain that have direct effects on the animals concerned (although other concerns are addressed as well). This framework may or may not be applicable to production phases involving processed products.

In Figure 1 the general structure of the BP framework is illustrated for one animal type. This structure is to be repeated for each animal type. Table 1 summarises the indicators used, grouped in seven 'areas of concern'.

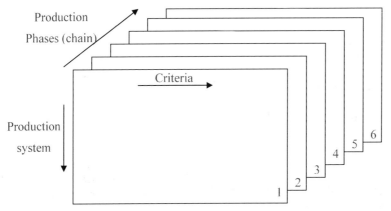

Figure 1. General structure of the Best Practice framework (for one animal type).

Table 1. Areas of interest and assessment criteria (indicators) in the Best Practice framework.

Food quality	Environment	Societal acceptance	Animal welfare
Technical product quality	Dust	Attitude towards system	Pain
Risk	NH3	Visual disturbance	Behaviour
	Greenhouse gasses	Odour	Mortality
	Water		Health
	Energy efficiency		Hunger
	Nutrient efficiency		Thirst

Labour circumstances	Economic sustainability	Management demands
Health	Income	Knowledge
Safety	Price (compared to	Attitude
Workload	standard)	Variable cost
Ergonomy	Support dependency	
	Income security	

The following codes should be used while coding an assessment:
++ This system scores very well for this indicator
+ This system scores well for this indicator
0 This system scores neutral for this indicator
- This system scores bad for this indicator
-- This system scores very bad for this indicator
? No data is available for this indicator
>< Contradictory data

Discussion

Concept

The major improvement of this system over existing 'best practice' excercises, is its broad nature. Where other documents, such as the IPPC reference documents (e.g. EC, 2003), focus on a 'single' issue (environment), this framework aims to integrate and expand all such research and assessments, and combine them with the assessment of previously unrecognised (or underrated) areas of concern.

Such an integrated approach corresponds well with the reality in which corporate decisions are made. Single issue assessments are relevant and valuable, but while making an investment decision, economic actors will often have to make a trade-off between different demands and expectations, including those typically recognised as ethical and societal demands. An concise overview of all existing information is therefore informative and necessary on a very practical level.

More generally, this practical decision situation is mirrored at the theoretical ethical level where competing claims exist. Irrespective of the general normative theory used, different applied ethical theories will introduce different claims about the same practical situations. In animal production, a typical example would be the disparity between environmental and animal welfare requirements. As soon as one wishes to step beyond the borders of a single applied ethical theory, an integration of these competing claims becomes necessary.

In this respect, the 'open-source' nature of the BP system assures an inclusion of different relevant claims. This open structure also facilitates dialogue about the areas of concern, the indicators and the assessment, ensuring a broader basis for the conclusions reached through the framework.

Remaining issues

Three important issues remain to be resolved: uncertainties, management influence, and integration.

There is a clear lack of sufficiently detailed scientific knowledge needed to perform a thorough assessment. In theory, an almost indefinite number of different production systems is possible for some of the production system phases. For example, housing systems for a single type of animal can vary substantially, and minor differences might have an important impact on some of the indicators. In theory, all these different systems would have to be described and included as a new 'row' in the matrix. There will never be enough scientific data to enable such a detailed comparison between systems. Therefore, a high degree of uncertainty will remain. One might be able to circumvent part of this problem by defining a series of 'reference systems' across the spectrum; this might also considerably decrease the workload associated with a BP framework application.

An important factor that is not included in the framework is the management itself. Although some estimate is made of the competence level needed to operate the production system, there is no direct assessment of the actual manager. As the framework is intended for an *a priori* assessment, this is unavoidable, but the outcome of such an assessment will always be a conditional result. Within animal welfare, for example, it is known (see e.g. Hemsworth and Coleman, 1998) that stockholders have a significant influence on animal welfare (positive or negative), independent of the production system used. It seems reasonable that this applies equally to other actors, indicators and phases within animal production.

The third issue, integration, is relevant at two different levels of the BP framework. First, it will be necessary to integrate knowledge about different measures that provide information about issues relating to a single indicator. For example, in order to investigate 'Labour circumstances - health' (at the farm) one might use scientific data on dust and ammonia levels. Even at this basic level in the framework, a trade-off will be necessary, even when these two 'measures' are not readily comparable. The broad coding definitions might facilitate this step. At the highest level of the framework, a similar situation exists. A final decision can only be made by integrating all 28 indicators into one final BP score. This can only be done by assigning weights to indicators and/or areas of concern.

Embedding within ethical theory

The BP framework aims to collect and summarise all ethical issues relevant to animal production. This means that it draws upon the conclusions of a range of applied ethical theories (animal ethics, environmental ethics, ...) which, in their turn, may rely on different general normative theories. As such, the BP framework is an addition to the ethical/societal deliberation toolbox at the level of practice-oriented ethics (Aerts, 2006; Aerts *et al.*, this volume).

As the framework aims to include all issues, it does include issues typically associated with animal production such as sustainability and animal welfare. The latter is addressed explicitly, the former through its major components: economy, ecology, and social aspects. It also explicitly includes 'food quality' and 'management demands', as these are areas of concern that, directly or indirectly, influence many of the other sets of indicators. Therefore, explicit attention to these areas seems justified.

The major question not confronted by the BP framework is whether animal production itself is ethically justified. It is clear that this is a fundamental issue, but it is not relevant within the practice-oriented ethical constellation in which the BP framework is to operate. The framework is intended for use in the day-to-day reality in which, at the present time, animal production is clearly not contested by a large majority of people. If, at a fundamental level, animal production is considered wrong, the framework will evidently be without ground; only when animal production is disappeared it will be without use.

Conclusion

The development of the BP framework is a new illustration of the strength and flexibility of the stakeholder dialogue model and the practice-oriented ethics concept (Aerts, 2006). Through a constructive cooperation between different stakeholder groups, common ground can be found and steps can be taken towards better communication.

The BP framework presented in this paper is a first step in the creation of a real Best Practice identification framework. It delivers a means to collect all existing information about animal production systems, and assess and compare those systems. The identification of *the* 'Best Practice' will only be possible when the weighting issue is resolved. This is evidently most relevant in a best practice discussion and will be necessary to be able to deal with contradictions. This was beyond the scope and possibilities of the current project, and therefore the current system needs further development. Even so, in its current form, the framework could be used in practice, although all users will have to assign their own priorities.

Acknowledgements

The authors would like to thank DP21 and Cera for funding this project, the members of the working group for the cooperation, and the Katholieke Hogeschool Sint-Lieven for the use of their infrastructure.

References

Aerts, S. (2006). Practice-oriented ethical models to bridge animal production, ethics and society. PhD. Katholieke Universiteit Leuven, Leuven, Belgium, 180p.

Aerts, S. and Lips, D. (2006). Stakeholder's attitudes to and engagement for animal welfare. In: M. Kaiser and M. Lien (Eds.) Ethics and politics of Food. Wageningen Academic Publishers, Wageningen, The Netherlands, pp. 500-505.

EC. (2003). Integrated Pollution Prevention and Control (IPPC). Reference document on Best Available Techniques for intensive rearing of pigs and poultry. European Commission, Brussels, Belgium. Available online at http://www.epa.ie/Licensing/IPPCLicensing/BREFDocuments/FileUpload,461,en.pdf [accessed May 15, 2007].

Hemsworth, P.H. and Coleman, G.J. (1998). Human-livestock interactions: the stockperson and the productivity and welfare of intensively farmed animals. CAB International, Wallingford, UK, 152p.

Part 8 - Sustainability and animal welfare: workshop 'Science and public expectations'

Can animal science meet the expectations in the 'animal welfare' debate?

Karel (K.H.) de Greef and Bram (A.P.) Bos
Animal Sciences Group of Wageningen UR, P.O. Box 65, 8200 AB Lelystad, The Netherlands

Abstract

In animal welfare projects, it is often expected that animal science can ultimately give reliable knowledge -and even the final answer- on the question what animal welfare is, and whether the welfare of a specific animal is good or not. There is reason to doubt that this expectation can be met. Firstly, although factual and truthful claims about physiological and ethological parameters can be made, nobody can claim to be sure about what animals feel and what these parameters mean for them. Secondly, not only animal feelings are relevant for the 'definition' of animal welfare, animal welfare is also an expression of how humans position themselves in relation to animals, like a duty to care, the intrinsic worthiness of living beings, or aesthetical revulsion against certain husbandry practices. Doing away with these convictions as 'irrational' or 'subjective' will not help: we might end up with reliable knowledge, but this will not be seen as the full story about animal welfare - i.e. it will not be 'socially robust'. This does not mean that scientists have no role in the ongoing debate on animal welfare, nor that science's role would be limited to the delivery of 'neutral facts'. On the contrary, in order to produce socially robust knowledge (mode II science), scientists additionally engage themselves in the societal debate, that should be explicit about norms and facts at the same time. One implication of this is that science should refrain from the ambition to monopolize knowledge (and *therefore*: judgement) about animal welfare.

Keywords: animal welfare, professional role, reflexive design

Animal welfare debate and the actual role and identity of animal welfare specialists

In Western Europe, the way we produce food by the use of (farm) animals is a societal issue, as can be seen by its occurrence and continued presence in the media and on the agendas of parliaments and governments. Generally, intensity of practices and suffering of animals are the key issues. Problems of the latter category are not only mistreatment of animals at incidental and separate occasions, but especially deal with rejection of systematic farming procedures that do not fit with societal demands and expectations. This political and social controversy on animal welfare is to be resolved somehow in the arena between the farming sector, science, society and governments. But in what way? Who will decide what animal welfare is? And what is the role of science, ideally and in practice?

Classically, governments set out research programmes to study animal welfare. Results went into three directions: extension into practice *(improve the system)*, scientific knowledge into science journals *(understand the animal)*, and directives into legislation *(force the farmer)*. For the latter, the translation from animal science to policy advice and measures was often carried out by (expert) working groups. Prominent examples of this are the Farm Animal Welfare Committee in the United Kingdom (FAWC), the Raad voor Dieraangelegenheden in the Netherlands (RDA) and the European Food Safety Authority of the European Union (EFSA).

In this way, the public authorities determined the welfare-requirements, being predominantly fuelled by research and science. This position, in combination with societies trust in 'objective science', has paved a way for animal scientists to be seen as the major representative of animals in the welfare debate. This professional group is thought to be able to judge or assess the animals' quality of life in specific situations. Or to advise what is needed to improve 'animal welfare'. Animal science positively responds to this and shows a high level of positivity in its activities and attitudes around this theme: *'the more we measure or model, the more we will know about the animal and the better statements we can make about it'*. This attitude is happily accepted by the environment: *'who could make better good judgements on the welfare of animals than those who study them?'* Thus, a neat division of roles is established that corresponds to the classical roles of state, science and market. A division of roles that at first glance even seems to satisfy anyone in involved.

Governance in animal welfare

At the same time, two general trends contradict this neat and classical division of roles. Firstly, science is increasingly challenged to produce not only reliable, but also socially robust knowledge (Gibbons, 1999), known after Gibbons as Mode II Science. An example is the debate on GMOs in Europe. Mode I science may produce lots of reliable knowledge on the health implications of GMOs, but without taking into account a range of other perspectives this reliable knowledge will have little effect. More generally: knowledge that may be relevant and reliable within its own disciplinary domain, is not necessarily applicable and accepted in other contexts.

An important driver for this trend towards, and the perceived need for mode II science is the increasing complexity of the problems we face. In animal production, the animal welfare issue is closely connected to economical and environmental issues, that are difficult to be disentangled and solved in isolation. This implies closer collaboration and integration, not only between disciplines but also between science, industry and society.

Secondly, governments increasingly refrain from solving societal issues by itself, and take a more facilitating role in which other societal actors (including science) may come to a resolution in cooperation with each other (governance).

These two trends imply that the clear cut division of roles sketched above, in which science delivers 'the facts', politics attaches the norms, and practice obeys to the rules set accordingly, is not self-evident, and that sole trust on science it might be counterproductive to a resolution of the animal welfare debate in the end - just like science was not able to be conclusive on its own on the European perception of health implications of GMOs.

Thus, facts cannot be produced without problems outside their perceived context of application, and (political) norms are established increasingly by the interaction of a diversity of societal actors. These general trends cast doubt on the possibility that animal welfare science can give the final answer on animal welfare in mode I, that is: autonomously and on its own.

Two cases illustrating high expectations

The role of science in the animal welfare debate thus isn't as unproblematic as current expectations of stakeholders and scientists themselves might suggest. Two short examples of animal welfare projects will be given in which expectations from classical, objective science are high. The first example deals with a seemingly straightforward scientific effort to 'objectively' bring together al possible welfare indicators into one scientifically sound index. The second example describes a multi-stakeholder project, in which animal scientists are thought to bring the answer to the welfare claim of an improved husbandry system.

Case 1: Pure and objective science to measure animal welfare

The Dutch government finances a considerable portfolio of farm animal welfare research. One of the two central goals of the main Dutch Animal welfare research programme '*Animal Welfare - room for natural behaviour and transparency)* is phrased as '*to develop objective criteria and methods to assess animal welfare*'. In this program, a central position is reserved for the development of a 'welfare index', with which '*welfare can be appraised in an objective, reliable and scientifically sound way*'. Interviews with the program manager and with a project manager revealed that the idea is that science delivers a list of on farm assessable parameters that are relevant for animal welfare, and makes an algorithm or model that integrates them into the best practical appraisal of farm animal welfare. The latter would then be expressed in a limited number of scores. In a next phase, it should serve as the base of a consumer directed clarification of the animal welfare history (background) of the product of choice. The process is seen in chronological order. First science-dominated development of an objective instrument. Then application of it for the public.

The scientists involved thus put themselves in the role of scientific objectifier, whose actions precede and exclude the influence of other interests. In this situation, the scientists themselves are the key players that force animal science into the role of deliverer of the 'truth'. Despite the context the result will have to fit in.

Case 2. multi-stakeholder welfare improvement

Early this age, a scientific organisation and an animal protection organisation jointly developed the ComfortClass principle in animal husbandry (de Greef *et al.*, 2003). This concept comprises a husbandry design such that all needs of the animals are met. From the premise that in that way, quality of life as perceived by the animals (Bracke, 2001) is good and thus infringement on animal welfare (from the animal's interest) is not at stake. The group had made a deliberate choice for the animal as the premium object for optimisation (de Greef *et al.*, 2006), and base this solely on animal science expertise. Soon after presentation of the results in 2003, the major Dutch farmers' organisation and the major Dutch animal protection organisation adopted this concept together. They raised funds to organise a test facility and a research program. This program aimed at testing the claim that animal welfare is indeed good in a facility designed and managed according to the Comfort Class principle. In the so called 'Proof of Principle study', scientists are hired to observe behaviour and facility use and to monitor health problems and skin damages.

For the present contribution, the noteworthy observation from this latest phase is that the parties involved fully rely on the research scientists to be able to produce the required answer. As far as the information of the authors reaches not a single remark or worry has been put down about the authority of science and the degree of reality of bringing that important and final answer to light. And also the environment (not only opposing farmers groups and welfare organisations, but also scientists and media) refrain from expressing worries about these expectations. Strongly stated: the responsible actors rather point into a moment in the near future 'when we'll know whether the stable is as welfare friendly as we expect it to be'.

The actors thus put animal science in the role of experimenters that will give the answer to the burning question: is animal welfare good in this facility? In essence, however, both parties postpone their quarrels on how pigs should be kept, until science says 'something'. Something with authority, because 'it is scientifically correct'.

Reasons why animal welfare science cannot meet the expectations

There are two fundamental reasons to doubt that the expectations can be met. In short: the technical problem that animal feelings can not be measured; and an arena problem: animal welfare comprises considerably more than animal feelings.

On the first reason: although we can make factual and truthful claims about physiological and ethological parameters of animals, we cannot claim to be sure about what animals feel and what these parameters tell about that. Animal scientists primarily study and discuss behavioural and physiological states and discern abnormalities in especially health and behaviour. Measuring animal feelings or emotions is beyond our abilities. Thus, there is a gap in what we are expected to evaluate ('the quality of life experienced by an animal') and what we can measure ('physiology'). Even the fanciest index of all thinkable physiological measurements will lack the golden standard for its relation to animal feelings. And this also holds for the behaviour-related parameters, which are thought to be closest to 'understanding the animals' and thus approaching their feelings.

As for the second reason: the societal concept of '*animal welfare*' comprises more than the actual animal status. Animal welfare is *also* an expression of how humans position themselves in relation to animals, like a duty to care, the intrinsic worthiness of individual living beings, or an aesthetical revulsion against certain practices in animal husbandry. The public is not as 'reductive' in approaching problems. They rather reason inclusive.

Thus, not only animal feelings are relevant for the 'definition' of animal welfare, but also (explicitly human) cultural judgements and ethical convictions. Doing away with these convictions as 'irrational' or 'subjective' will not help: we might end up with reliable knowledge, but this knowledge will not be seen as the full story about animal welfare - i.e. it will not be 'socially robust'. Animal science will not be able to resolve the issue of animal welfare *alone*.

Animal sciences new contract with society (mode II)

Especially the second fundamental problem shows the necessity of science contributing to the resolution of the animal welfare debate in close connection with other voices. If animal science would continue to try to solve it autonomously, in their own terms, insensitive to the 'non-scientific' meanings and values, it will probably produce reliable knowledge, but will diminish its contribution to the debate as a whole. Science has the challenge to maintain a meaningful connection between its own framing of the problem (in language and concepts), and the diversity of other perspectives on animal welfare. This does not mean that scientific expertise is irrelevant for the ongoing debate on animal welfare, nor does it mean that science's role is limited to the delivery of 'neutral facts'. On the contrary. The problem is that the position of animal experts *tend to take* is typical for mode I science, in which science itself autonomously defines what are the relevant questions, and produces reliable answers to these. In order to produce socially robust knowledge (mode II science), scientists should keep doing this, but *additionally* actively engage themselves in the societal debate, that is fundamentally about norms and facts at the same time. Paraphrasing Gibbons (1999), animal science should sign a new contract with society. The first implication of this is that science should refrain from the ambition to monopolize knowledge (and *therefore*: judgement) about animal welfare, another is that it should make its work more permeable to perceptions and convictions from outside, and do this in an explicit way. Surely, there is a need for science-driven knowledge production as well, but in that case scientists should be even more modest in their claims about their knowledge covering animal welfare.

Mode II science challenges the professional identity of scientists. A legitimate question would be how can science keep being reliable (and, as you wish: 'objective'), while engaging in the slippery fields of political and social debate at the same time? Instead of giving an argument detached from ourselves, we first give a personal reflection.

On being involved in mode II science

Both of the authors of the present paper engage in mode-II like R&D projects in animal husbandry, but their background and orientation differs considerably.

The first author is trained as an animal scientist, and considers himself a relativist in no respect. He finds himself acting in two rather discrete roles. When approached as the classical mode I scientist, the role is to deliver information about animals as independent and objective as possible, by putting it in an experimental and scientific perspective. However, being active in multi-stakeholder networks that collaborate on new welfare arrangements, he has to take a principally different role. Then, choices have to be made what aspects of housing of animals deserve attention. Then, results of studies have to be interpreted in a wider context. Then, un-researchable questions have to be answered on basis of inspired guesses without experimental verification. All in an environment where it is not usual and not easy to be explicit in indicating one's normative position. In this role, the author is consciously constructing facts and norms in reality, in a specific context. The author thus expresses both a constructivist and a positivist approach to science. His ideal is to bring both roles in consonant, but it requires daily effort to do so. A preliminary answer is, to try to be extremely explicit on behalf of what one is acting or speaking, but there is no natural peergroup to corroborate the soundness of this, like the referee process provides in mode I. One can imagine that being an animal scientist of this type in case 2 feels like roller-skating on ice.

The other author is trained in biology, science studies and philosophy, but happens to be in the animal production domain. He could be said to be a constructivist in theory and in practice. He is convicted that scientific facts do not possess truth by themselves, but *become* true in interaction with reality. He is not a relativist, however, if we mean by this that any claim about reality has equal value. Claims about reality become stronger, as the number of connections with objects and other claims increases (Latour, 1987). This can be done in the laboratory, or in other experimental settings, but can be done as well in more complex and real-life contexts. Thus, producing socially robust knowledge does in no way mean that the claims we make about reality become softer by definition. On the contrary, they might become harder (more robust), but it requires a lot more effort, because more actors and more objects have to be aligned. A non-animal science constructivist as presented can happily flourish in current post-normal science initiatives, in our post modern time frame.

This leads to two slightly different solutions to the perceived challenge of operating as a scientist in mode II in the animal welfare arena. Both are based on a specific relativation of their claims for authority. Author I may tap on his expertise from the animal science field, but has to specify in a much more precise way what relevance this knowledge has in the context at hand, and has to be aware of the assumptions that are taken-for-granted in his own field of expertise, but might frame the issue in other contexts in an undesired way. He may maintain the claim for truth of his knowledge, *within its specific context of origin*. Author II will not claim any specific authority on the subject, but is trained to use a conceptual apparatus that facilitates the connection between different knowledge domains (both scientific and practical) and between different conceptual languages. Both authors will not do claims about any situation being 'good' or 'bad', but will rather work on common definitions of 'good' or 'bad' (animal welfare) in specific instances.

Which role to take?

Ideally, the societal search for 'good welfare' is a balanced process between and within a diverse group of actors and stakeholders. Too high raised reliance on, or expectation from 'hard science' may hinder finding a robust balance. Our role as classical animal scientists is considerably smaller than thought at first sight: a position in delivery of verifiable information on animal behaviour and physiology. Is this limited role problematic for animal scientists? By no means! But we do have to make choices, our role has to be clear. We have the choice for two perspectives.

Role 1. Remain the classical mode I scientist. In that role, we happily take up our role as 'intelligent measuring devices'. Learn about animals by studying them, make them speak so to say. Publish this in our own media and make this expertise available to those who need it. In that role, it is of joint benefit to share the awareness of this limited position: we don't deliver 'the final truth' but have the ambition to be a humble and trusted provider of animal information (be it limited and incomplete); we are not a stakeholder in the welfare debate, rather experts on a limited domain.

Role 2. Be a hybrid: remain the scientist and take the mode II role. Play the role of science-based information deliverer *and* participate in constructing societal robust balances between the many interests related to our dealing with food producing animals. Possibly, most animal scientists already are more involved in (and affected by) mode II contexts than they are aware of. The scientists introduced in case 1 were explicitly driven by a search for balances between interests, which is more than being in the autonomous scientific objectifier role. A principal aspect of the real hybrid role would be to step into the themes of the other actors. In stead of rejecting anthropomorphic projections of people on animals, seek ways how the animal science instruments can help solve those issues (taken in relation to the other issues). For example, when constructing a welfare friendly system, animal scientists are well equipped to deal with human non-animal-needs requirements such as naturalness, respect, and public welfare icons (see Bos, 2007). Also in this role, we have an interest in making and keeping our position clear. We have to beware that scientifically true facts do not dominate over other views, interests or values. Our training programmes might need some philosophy of science in order to make us more comfortable in that co-constructor role: being a scientist in a process where science has to share its authority with others.

Conclusion

No, animal scientists indeed cannot meet the expectations. But that is as much more a problem of societies' overestimation of science as it is our defect. But it is our responsibility to be clear in our position. In the role of ultimate objectifier, animal science cannot meet the expectations, as knowledge and technology are not developed enough to conclude on the key point: assessment of how animals experience their living conditions. And because the scientific approach towards animal welfare is considerably narrower than that of the public, thus giving incomplete answers and solutions. But, regarding the way science and society develop, it is not a big problem that animal science cannot answer all questions. That is typical for science. Animal scientists are best equipped to (try to) speak for the animals about how they experience their living conditions, based on the state of knowledge they have. They can proudly take that position within the arena of actors. Others can bring in other perspectives. The key challenge is to find and clarify the role and position taken, and not to monopolise the term animal welfare and thereby risking a reduction of the social robustness of hard work contributions. In order to prevent confusion, maybe we should temporarily abolish the oxymoron term 'animal welfare science'. Animal science is hard science, animal welfare is different.

References

Bos, B. (2007). Instrumentalization theory and reflexive design in animal husbandry. Social Epistemology, in press.

Bracke, M.B.M. (2001). Modelling of animal welfare: the development of a decision support system to assess the welfare status of pregnant sows. PhD thesis Wageningen.

Gibbons, M. (1999). Science's new social contract with society. Nature 402: c81-c84.

Greef, K.H. de, W.G.P. Schouten, W.G.P., Groenestein, C.M., Hoope, R.G. ten and Jong, M. De (2003). Husbandry systems from the animal's point of view: Pigs. In: Y. van der Honing (ed.) Book of Abstracts of the 54th annual meeting of the European Association for Animal Production., EAAP/Wageningen Academic Publishers, 566pp.

Greef, K.H. de, Stafleu, F.R. and Lauwere, C.C. de (2006). A simple value-distinction approach aids transparency in farm animal welfare debate. Journal of Agricultural and Environmental Ethics 19:57-66.

Latour, B. (1987). Science In Action: How to Follow Scientists and Engineers Through Society. Harvard University Press, Cambridge Mass., USA.

Empirical facts in farm-animal welfare discourses

Herwig Grimm
Institute TTN (Institute Technology, Theology and Natural Sciences), Marsstraße19, 80335 Munich,
Germany, herwig.grimm@elkb.de

Abstract

Anybody dealing with farm animal welfare issues should be familiar with technical aspects of farm practices. Ethical reasoning by itself will not suffice in solving farm animal welfare issues. However, the issue has seen an empirical turn. It has become increasingly common to look solely to natural science and empirically based research for advice in order to solve ethical issues in farm animal welfare. This approach - to base the ethical discourse on *objective* empirical facts - promises to facilitate moral decision making and to settle moral controversies in animal production at first sight. However, further reflection reveals that the narrow view on 'animal welfare' incorporates implicit normative statements (as seen by many authors) and therefore can not be considered a neutral basis. The aim of this paper is to focus on how the empirical approach hinders successful discourse. This will be shown by considering five implications of the narrow scientific approach: (1) comparative analyses and its limits, as a commonly used procedure to deal with established data; (2) priority of short and midterm experiences, as opposed to lifetime; (3) difficulties in dealing with probabilities as typical scientific findings; (4) reduction of irreducible complexity; (5) the danger of losing touch with people involved in animal welfare discourse. In conclusion it will be argued that we need scientists and ethicists to combine forces to provide concepts which lead to more serviceable and helpful evidence. This challenge is essential to sustainable agricultural development since public discourse and debates on animal welfare increasingly influence and guide policy making in Europe.

Keywords: scientific reductionism, epistemology, value-laden facts

Introduction - Animal welfare as a matter of facts

Whenever controversial ethical questions are raised, reliable scientific data is called for. This seems to be especially true when we deal with animal welfare issues. Animal welfare science informs ethical decision making and provides empirical evidence for moral judgements. However, this approach has a significant problem which is rooted in the very beginning of animal welfare science. When farm animal welfare was established as a subject for research at universities and agricultural research institutes, it was important to establish it as a genuine subject of scientific study and not something tainted by political motives or subjective emotions. In order to live up to scientific criteria, the conventional scientific approach was adopted, with experiments focusing on effects of single factors under controlled circumstances (Sandøe *et al.*, 2003). Animal welfare scientists have become aware of the various methodological problems of this approach (Tannenbaum, 1991; Verhoog, 2003; Webster, 2005). Despite their insights, the mentioned approach still appears to be highly influential.

The aim of this paper is to examine and focus on how characteristics of empirical findings - stemming from a narrow scientific approach - influence farm-animal welfare discourse, and to what extent they can provide guidance to solve ethical questions of farm-animal welfare. The the increasing awareness of the narrow scientific approach's limitations and related alternatives are not included in the discussion of this paper.

The narrow scientific approach

Most importantly, the crucial feature of the narrow scientific approach is that farm-animal welfare assessments have to be based on observed, measured effects on animals under controlled circumstances. Whether a housing system, e.g. farrowing crates, fattening bulls on slatted floors, or a certain farming practice, such as dehorning calves, are morally acceptable, depends on their influence on parameters, perceived as indicators of good or bad welfare.[8] Usually, such parameters encompass physical, physiological and behavioural aspects, which - provided an adequate background theory - give insights in the animal's welfare. An altered welfare state of the animal would cause observable effects on animal based parameters (Broom, 1991). Put in other words: Animals are used in order to check whether a housing system or farming practise has impact on identified parameters, as, for instance, formulated by Unshelm (2002):

- Anomalous behaviour (behaves in repetitive stereotypical ways such as bar-biting etc.).
- Performance of individual animals (irregular gain in weight, anomalous milk quantity etc.).
- Physiological parameters (heart rate, temperature, fitful respiration, blood count, etc.).
- Medical evidence (abrasion, infections, illnesses, intestinal parasites etc.).
- Mortality rate and causes.

If a certain farming practice or housing system has no measurable effect, the practice or housing system is thought not to be problematic from an animal welfare point of view. Certainly, the big advantage of such lists is that animal welfare can be based on observable effects. Sometimes effects can be measured directly as levels of stress hormones (e.g. cortisol) in blood, or other bodily secretions, or by post-mortem pathological changes in organs or tissues. Behavioural measures of welfare are more complex, and therefore usually applied to identify behaviours which are clearly abnormal and indicative of pathological mental state (e.g. stereotypes). However, under this perspective animal welfare can be described objectively by scientists. Tannenbaum (1991) alleges a number of reasons for ethical limitations and implications of the 'pure scientific model.' With reference to Tannenbaum, Verhoog (2007) develops this approach and shows its severe methodological limitations and its bias against positive welfare. As experimenters try to avoid difficult philosophical discussions about analogies and homologies between animals and humans, they usually adopt a minimalist strategy, relating welfare to the avoidance of states of pain or frustration (Verhoog, 2007). Accordingly, to reduce animal welfare to measurable effects on parameters - as the narrow scientific approach holds - can not be considered neutral or value free.[9] This illustrates the vital and essential question of how to identify the relevant parameters and what we can learn from them, about the actual welfare of farm animals.

Comparative analyses

The most commonly used procedure to evaluate farm practices and housing systems is to compare them on the basis of measurable effects. Given comparable scientific findings from different housing system e.g. the behavioural parameter 'constraints on natural behaviour' in battery cages and enriched cages for laying hens, we can easily state that enriched cages are better in terms of animal welfare. But this settles the question only half way. Comparing measured effects on particular parameters in different housing systems informs us whether one system is better or worse than others, but does not show whether the measured effects themselves are relevant to animal welfare. However, empirical research can indeed inform us about the measurable impact on the animal, but not how to deal with them except in a comparative way. To set up ethical benchmarks on reasonable grounds is not in scope of this approach.

---

[8] This is a parallel to the tendency in bioethics where evidence for moral judgements is thought to be found in empirical data (Borry et al., 2005).

[9] For more salient remarks on the 'value-ladenness' of objective, empirical data in natural science cf. Putnam (2004).

Furthermore, the question arises which parameter outweighs others. The science-driven approach tends to direct discussions towards the weight of particular welfare indicators and parameters. E.g. talking about the farrowing crate is likely to be reduced to the discussion about the mortality rate of piglets compared to the sow's restricted movement for several weeks. Usually, inhibited social contact with the piglets, increase of MMA (mastitis, metritis, agalactica) in sows and so forth do not come into focus. The essential part, to select relevant factors and rank them according to their importance, can not be deduced from empirical findings.

When farmers take part in animal welfare discourses they find themselves unsupported by scientific opinion. Confronted with questions, such as whether enriched cages for laying hens meet animal welfare requirements, they compare different housing systems but do not relate to the animal's welfare state itself. A narrow scientific approach only helps to compare different practices or housing systems on the basis of identified parameters. Although this is of crucial importance in case of decision making, this approach tends to reduce discussions to few relevant aspects and deal with implicit superior normative concepts of animal welfare. For instance, to answer the question why we should treat behavioural aspects prior to e.g. physiological parameters implies a normative decision. To step back from measurement to judgement about the welfare state of the animal involves interpretations that are never value-free. Despite the fact that any empirical data has to be interpreted in the light of some animal welfare concept in order to assess farming practices and housing systems, experimenters often stick to the method to provide data only. This leaves participants in animal welfare discussions without guidance. Consequently, empirical results of animal welfare experiments on their own are of limited relevance, or worse, reduce animal welfare discussions to single factors only.

Short time versus lifetime experiences

Any scientific experiment is designed to answer a specific question under controlled circumstances guided by theories (Schäfer, 2006). Stable and controlled circumstances are crucial, since they allow the experimenter to reproduce findings and consequently draw conclusions from the experiments. This is of importance to the question of animal welfare for the reason that the farm animal's lifetime is divided according to specialised production cycles. Most farm animals are kept in various housing systems during their life. Since relevant data can only be gained under more or less stable circumstances, the empirical research focuses on particular periods in specific production systems (stable circumstances). Therefore, results relate rather to certain periods in the animal's lifetime than to the whole lifespan. Concerning farming practices, the scientific approach tends to concentrate on the actual short time procedures, such as the actual castration of pigs or the act of dehorning calves. Considering housing systems, maternal behaviour in farrowing crates, behaviour in pregnancy stalls, etc. come into focus.

To highlight short time experiences and investigate periods of the animal's lifetime tends to neglect the importance to consider lifetime welfare or good life. For instance, dehorning calves to prevent injuries within the adult cow herd is considered to be an adequate procedure since the calves suffer pain only a short time (if done without anaesthesia). The reduction of the behavioural repertoire throughout lifetime does not get into focus. Another example is again the sow's fixation in the farrowing crate to control them better and lower the risk of loss in piglets per litter. Considering that the sow is fixed again straight after the piglets are weaned, animal welfare considerations would differ considerably, since it doubles the time of fixation and prolongs negative effects during lifetime. It is highly contradictory to talk about animal welfare in terms of short and mid term experiences. Due to the incommensurability of parameters used in the narrow scientific approach, there is no easy way or sophisticated method to express overall lifetime animal welfare. Therefore, an ethical discussion about what is a good animal life and how it can be achieved should be an integral part of discussions surrounding the assessment of farm animal welfare (Sandøe *et al.*, 2003). Again, this question can not be answered within a narrow scientific approach but is evidently ethical in nature.

Laboratory findings versus farming experience

Another problem is whether scientific results gained under laboratory conditions are transferable to field conditions. This problem is not specific to animal welfare science since it applies to all experimental research. Relevant epistemological criteria have been formulated which allow us to transfers results from the laboratory to the field (Köchy, 2006). However, looking at farm animal welfare, results gained under laboratory conditions, when transferred to practical circumstances, develop severe problems. The huge effect - which is known to vary considerably between farms and laboratories - namely the factor defined by management and stockmanship, can not be reflected by epistemological criteria. Even if the experiment is realistically designed, the predictability of the gained results is only given, if the farmer runs the production system in a similar way, which is not very likely. Given these factors, what do these scientific results then tell us about the farming practice or housing system? Since management can improve things or make them worse, probabilities gained in experiments can not easily be transferred to farms.

Taking this aspect into account, in animal welfare discourses we face the problem of how to deal with probabilities. For instance, what does the occurrence of an illness in 15 out of 100 animals under laboratory conditions mean to the individual animal on the farm? Such numbers are very difficult to understand in terms of welfare since they present an average, meaning that not every animal was in bad condition, but only some. Defining where the threshold lies appears to be a problem. Solely focusing on such numbers, we could infer that in a certain respect a specific housing system is only animal welfare unfriendly for some animals and does not say a lot about the housing system itself. What if negative effects described and published in scientific bodies just do not appear on a specific farm? Why should the farmer take these findings into consideration? Despite such difficulties, percentages and probabilities can certainly provide reasonable orientation of farm animal welfare.

Inconvenient irreducible complexity

Animal welfare - as an ethical issue - should not be reduced solely to the animal's welfare for moral reasons. There is a tendency to neglect practical aspects of animal husbandry which makes it exceedingly difficult for farmers to implement animal welfare recommendations. When we look at realistic, practical settings and circumstances, the approach to equate the ethical issue of animal welfare with scientific animal welfare, appears helplessly unsatisfying. The problem is that we are always dealing within complex networks of interdependent aspects. Besides animal welfare, the farmer's area of responsibility incorporates morally relevant objectives that are necessarily bound to livestock husbandry. This plurality of areas of responsibility related to livestock husbandry is a potential source of moral conflict, namely:
- Using animals and their products as means to economic ends (income of farmer).
- Production of safe and healthy foods (hygiene standards, residues, etc.).
- Protecting the environment (taking care of the countryside, reduce emissions, etc.).
- Legal requirements concerning farms (environment, hygiene, etc).

Empirical data provided by welfare scientists is only one aspect of the ethical issue. It just can not be separated from its related areas of moral responsibility. To introduce best practice in the sense of environmental pollution or hygiene may endanger animal welfare in specific cases. The crucial moral question is, however, how to reach optimal animal welfare practice within a plurality of ethical values and demands? The narrow scientific approach is not open to such considerations. Therefore, ethical decision making has to be supported by frameworks which consider the issue's complexity. Herein we find a reason why animal welfare discourses often fall short of practical relevance. In this field a lot of work is to be done, because discourses on farm animal welfare will not be able to provide sensible solutions without taking its irreducible complexity into account (Grimm, 2006).

Common sense versus scientific perception

Undoubtedly, the pure scientific concept of animal welfare differs significantly from the common sense perception. The missing interrelation between scientific and common sense perception was highlighted early on from Rollin (1990) with regard to D. Hume. Science, he puts forward, has become increasingly remote from common sense and ordinary experience. Becoming educated as scientists, we often abandon common sense and the categories which govern our interpretation of ordinary experience (Rollin, 1990). These sociological aspects of scientific practice and training were seen already by Fleck (1980[1935]). Presuppositions are passed on from teachers to pupils, scholars to students etc. and determine what counts as real, as facts, and as valid data or explanations. Such implicit knowledge orientates research. For instance, as Verhoog criticises, data about conscious experience or the animal's mind are not considered legitimate data, for ideological and philosophical presuppositions (Verhoog, 2007). This leads to the absence of experiential data about animal consciousness and consequently the moral irrelevance within the laboratory setting. Strictly speaking, natural science cannot say anything about the subjective experiences of animals, because the use of the value-free method excludes it. But exactly these subjective experiences are at the central core of the lay-public's definition of animal welfare. This seems not to be taken into account by welfare-scientists who apply minimalist strategies to avoid philosophical discussions about similarities between animals and humans and try to relate welfare to the avoidance of states of pain or frustration only (Verhoog, 2007). Whereas lay people want to know whether the animal is well off in the sense of 'happy', data and figures are provided, which are very remote from their original idea about animal welfare. As Rollin puts it, concepts of science are often not even interpretable or explicable in the language of common sense, to rich in details irrelevant to common sense, or even inadequate or irrelevant for common sense (Rollin, 1990).

This aspect is of crucial importance when farmers and lay-people participate in discussions about animal welfare. Whenever farmers want to explain how they take responsibility, trying to base their opinion on scientific findings, they have to bridge the gap between the scientific and common perception of animal welfare. It is not enough to tell people that data and statistics exist which support their viewpoint. Of greater importance is to interpret and explain them. Farmers are not usually familiar with interpreting scientific studies, so they easily fall back into their practical background. This reflects the point, that if there is no reflection on how results of animal welfare research fits into the societal context and public perception, there is a danger that these results will not be properly utilised (Sandøe *et al.*, 2003). To express scientific results within common language is indispensable for successful discourse. Accordingly, the challenge is not only to produce data, but to present it in an understandable manner. Otherwise science ends in counterintuitive solutions. Verhoog (2007) mentions, for instance Rollin's suggestion (1990), to change the telos of chickens through genetic engineering so that they no longer have an urge to nest. In a scientific approach this is perfectly consistent with animal welfare since it means to remove a source of suffering for animals held in battery cages. They are obviously better off than before. Of course, Rollin agrees that it may be better to change the housing conditions, but as long as this is not expected to occur in our present societies, it is better to decrease the suffering, even if this has to be achieved by means of genetic engineering. From a narrow scientific point of view this is perfectly consistent. However, most lay-people would find this position unacceptable and therefore not contribute.

Conclusion

The aim of this paper was to inquire on the narrow scientific approach's problematic implications on animal welfare discourse. It has been shown that in certain cases the focus on empirical facts leads to new ethical questions, which are difficult to solve. Science often leaves us to empirical data which only gain relevance in the light of interpretation and normative animal welfare concepts. This applies especially to results of comparative analyses, data from short or midterm time investigations and probabilities of good or bad animal welfare. For ethical discourse it is of importance to put empirical data in context and establish data relevant to the context. Further, to develop adequate solutions it is of essential importance to broaden the scope of animal welfare science and open it to the conflicting facets and complexity of the societal embedded issue. Lastly, the point was made that the narrow scientific approach provides information which is remote from common perception of animal welfare, which appears to be a problem in animal welfare discussions.

If the narrow scientific approach is broadened and aspects of the aforementioned kind are considered, there is a higher possibility to improve and develop discourses on sustainable farm animal welfare among lay-people, stakeholder, ethicists and scientists.

References

Broom, D.M. (1991). Animal welfare: Concepts and measurement. Journal of animal science 69: 4167-4175.

Fleck, L. (1980). Entstehung und Entwicklung einer wissenschaftlichen Tatsache: Einführung in die Lehre vom Denkstil und Denkkollektiv (1935). eds. L. Schäfer/Th. Schnelle, Frankfurt /M.

Grimm, H. (2006): Animal welfare in animal husbandry - How to put moral responsibility for livestock into practice, in: M. Kaiser and M. Lien (eds.) Ethics and the politics of food, Wageningen.

Köchy, K. (2006). Lebewesen im Labor. Das Experiment in der Biologie. Philosophia naturalis 43: 74-110.

Rollin, B.E. (1990). The unheeded cry. Animal consciousness: animal pain and science. Oxford University Press, New York. (expanded ed. 1998)

Sandøe, P., Christiansen, S.B. and Appleby, M.C. (2003). Farm animal welfare: The interaction of ethical questions and animal welfare science. Animal welfare 12: 469-478.

Schäfer, L. (2006). Die Erscheinung der Natur unter Laborbedingungen. Philosophia naturalis 43: 10-30.

Tannenbaum, J. (1991). Ethics and animal welfare: The inextricable connection. Journal of the American Veterinary Medical Association 198, 1360-1376.

Unshelm, J. (2002). Indikatoren für die Tiergerechtheit der Nutztierhaltung. In: W. Methling and J. Unshelm (eds.) Umwelt- und tiergerechte Haltung von Nutz-, Heim- und Begleittieren. Parey , Berlin, p. 242-248.

Verhoog, H. (2003). Biotechnologie und die Integrität des Lebens. Scheidewege 32: 119-141.

Verhoog, H. (2007). The tension between common sense and scientific perception of animals: Recent developments in research on animal integrity. NJAS Wageningen Journal of life Sciences 54: 361-373.

Webster, J. (2005). Animal welfare. Limping towards Eden. Oxford. Blackwell.

Part 9 - Ethics of organic farming

On the ethical dimension of organic agriculture

Tatjana Kochetkova
Louis Bolk Institute, Hoofdstraat 24, 3972 LA Driebergen, The Netherlands, t.meira@louisbolk.nl

Abstract

This paper analyses the concept of organic agriculture (OA) from the viewpoint of philosopher Ken Wilber's Integral model. This analysis determines that for OA it is essential to include the interior perspective (e.g., specific worldviews, values and consciousness) and discusses its content. The ignorance of the interiority by some of the definitions of OA makes them merely partial. After a brief introduction to OA and its history in section 1, and discussion of the integral model in section 2, interiority will be described in terms of Ken Wilber's four-quadrant model in section 3. Finally, in section 4, cases are shown of recent research on OA that actually includes interiority and has thus an integral character.

Keywords: organic agriculture, ethical assumptions, sustainability, integral

Introduction

Organic agriculture (OA) emerged in response to the unsustainable nature of the industrial agriculture as it emerged during the 19th and 20th century. The most well-known among the problems of industrial agriculture are the hazards associated with the use of pesticides and chemical fertilisers (OECD, 2003).

From its very start, OA has shared goals with environmental protection movements, along with the goal of producing healthy food in an environment-friendly way. For this purpose, OA chose to avoid synthetic pesticides and artificial fertilisers, as well as to use both scientific and traditional knowledge in order to achieve a safe and environmentally responsible agriculture based on a holistic worldview. OA has evolved from various roots, among which are Rudolf Steiner's bio-dynamic agriculture, Müller and Rusch's organic-biological agriculture, the American school of Sir Albert Howard, Lady Eve Balfour, and Rodale, as well as Rachel Carson's book 'Silent Spring' and Schumacher's 'Small is Beautiful' (Padel *et al.*, 2007; 28). Carson`s 'Silent Spring' described how DDT entered the food chain and accumulated in the tissues of animals, including humans, causing cancer and genetic damage. She concluded that DDT and other pesticides had irreversibly harmed birds and animals and had contaminated the entire world food supply.

However, what exactly is meant under OA - what is the nature of the alternative that it offers to conventional agriculture - remains a subject to an ongoing discussion and various definitions. For instance, the United States Department of Agriculture`s (USDA) definition of the word 'organic' is: 'An ecological production management system that promotes and enhances biodiversity, biological cycles and soil biological activity. It is based on the minimal use of off-farm inputs and on management practices that restore, maintain and enhance ecological harmony' (OECD, 2003: 31).

This definition includes only objective factors of production, without reference to world-view or subjectivity. Basically of the same character is the National Standard of Canada for Organic Agriculture (NSCOA) with its six principles of organic production (Prpich, 2005):
1. environmental protection;
2. maintaining long-term soil fertility;
3. maintaining biodiversity;
4. recycling materials and resources;

5. promoting health and behavioural needs of livestock;
6. maintaining the integrity of organic food.

Opposed to these two definitions, which both come from governmental agencies and define OA in objective terms, there is also a definition that came from the organic farmers themselves, and which stresses the values and specific view of OA. To quote the Nordic Platform of Ecological Associations:

'Organic farming describes a self-sustaining and persistent agro-system in good balance. As far as possible the system is based on local and renewable resources. It builds on a holistic view that incorporates the ecological, economic and social aspects of agricultural production in both the local and the global perspectives. In organic farming Nature is considered as a whole with its own innate value and Man has a moral obligation to farm in such a way that cultivated landscape constitutes a positive aspect of nature' (DARCOF, 2000).

As we have seen, some of the definitions of OA emphasise only its objective characteristics, while others consider values and world views. Are values and world-views indeed so essential? I here argue that values and world-views indeed are essential, based on an integral analysis of organic agriculture. Based on these and other definitions, the major principles of OA, according to IFOAM, include behavioural or exterior (ecology and health) and cultural, or interior principles (care and fairness) (IFOAM, 2005).

These principles go beyond behavioural/ positivist approaches by showing the subjective, interior side of OA practice. A different possibility, which sheds more light on the interior side of OA, is the integral approach proposed by Ken Wilber, an American philosopher whose name many organic researchers may not even know. In the next section, his integral approach will be briefly outlined, in order to be used to interpret the concept of OA. After that, we will discuss the compatibility between recent OA research and Wilber's integral approach.

The AQAL model

Ken Wilber claims that every entity, natural as well as social, has four major characteristics - individual and collective, interior and exterior (Wilber, 2000). These four characteristics give us four dimensions or quadrants, as on the following schema (Figure 1).

The integral approach claims that the four quadrants cannot be reduced to each other, and they are manifestations of every entity. The difference between the individual behavioural quadrant, and the collective behavioural quadrant, in my view, is that the holons of the individual quadrant are composed of parts which are more heterogeneous, and belong to different levels of complexity. In this sense, atoms or an individual animal are individual holons. As for collective holons, like 'heterotrophic ecosystems' (level 4), or tribes, they are largely composed of big amounts of holons of approximately the same degree of complexity. Whatever relative is the difference between individuals and groups - both are composed of parts (i.e. organisms are composed of cells, and ecosystems are composed of different types of organisms) - it is still not negligible.

Further, to understand an entity, one needs to trace interrelations between all four quadrants, because they are different dimensions of one and the same entity. Understanding must thus be AQAL, i.e. 'all-quadrants, all-levels'. The entire right half of Figure 1, the exterior half, can be described in objective terms and can be studied empirically (in terms of behaviourism or positivism). Depending on which part of the scheme is considered, there are differences between the Right hand and the Left hand approaches. The limited character of reductionism is that is sees only the Right hand approach, and further reduces all the quadrants to the Upper Right quadrant ('materialism'), which produces purely mechanistic

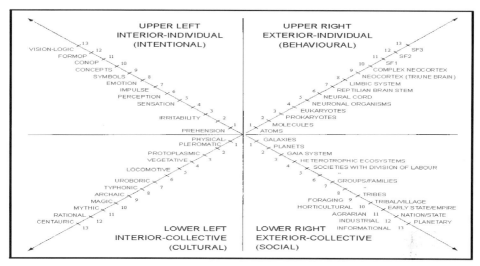

Figure 1. Ken Wilber's four quadrants scheme

results. This is the problem of 'flatland', i.e. the collapsing of everything onto the Right-hand dimension, recognising only the sensory perceivable empirical data, thus ignoring, for instance, the direct farmer's experiences of animals.

The integral model is also 'all-level' in the sense that it accounts for different levels of development in every quadrant. But the important point about AQAL for the present discussion is its requirement to integrate interiority, or the Left-hand of the model, considered as autonomous dimension, irreducible to the Right-hand. With respect to OA, this means including more than the administration and functional fit of the proper physical management system in relation of the ecosystem and the production methods that take into account the physiological and behavioural treatment of the livestock (the Right-hand dimension). There are also values and specific world-views characteristic of OA, together with the corresponding levels of interior development (the Left-hand dimension). Let us now consider why these elements are so essential.

The interior dimension of OA: a heuristic potential and its exploration

Philosophers have always battled over the issue of an *interior* side to humans (materialism - e.g., modern-day philosopher Daniel Dennett - claiming that the interior side was actually external, and idealism - e.g., Hegel, Schopenhauer - claiming that the external world was actually part of the internal side; there were also many intermediate positions). *Stricto sensu*, the existence of an interior side cannot be *externally* (or *objectively*) demonstrated: people are either aware of their internal dimensions - in which case demonstration is not necessary - or then they are not (or maybe they are but deny that this 'internal dimension' is really separate from its external counterpart - which is what Wilber calls 'subtle reductionism'). This philosophical debate, of course, goes far beyond the scope of the present paper. Here, we assume that human beings have an internal dimension, which is crucial for their happiness and well-being.

Wilber's AQAL model further claims that the interior dimension is not a feature specific to humans, but is actually an inherent part of every entity (which he calls *holons*: four-quadrant entities which are always made of smaller four-quadrant entities and are always part of larger four-quadrant entities). This claim, easier to accept for higher animals (e.g., dogs, cats, cows), is perhaps harder to extend to inanimate entities, molecules, and atoms. For our purposes, however, it suffices to say that certain entities - e.g., living beings - also have an interior side, albeit less rich or developed than the human interior side.

If it is assumed that the interior (intentional/cultural) side of the participants plays a fundamental role, then OA also has to take it into account. To neglect, e.g., the world-views and mutual perceptions of the participants (farmers, animals) would mean running the danger of causing internal harm (unhappiness, decrease in well-being) of a kind that cannot be healed by mere food production. In fact, in a sense, the essence of OA rests on concern for things other than mere food production (sustainability, ecosystem preservation, etc.); to take into account the interior side of the involved entities is simply to take the premises of OA one step further: towards seeing the 'organic farm' as a well-balanced unit which maximizes its own well-being and is capable of adapting to and learning from new circumstances.

For instance, Bloksma and Stuik (2007: 422) propose a metaphor which uses a human being to describe a farm. According to this metaphor, intentionality in a healthy farm corresponds to the mind in a healthy human. The intentional dimension in an organic farm includes its mission, its clear identity, and the capacity of the enterprise to learn from what is being produced. Intentioniality in agriculture also includes its meaning of existence, the people's awareness of the uniqueness of the farm, current values providing directions for the future, and the farm reflecting the special choices that are being made.

Note that taking the interior dimension into account in OA implies the same also for the concept of sustainability. The popular definition of sustainable development as the one which meets the needs of present generations without compromising future is too limited, and purely anthropocentric. An ecocentric view of sustainability is better suited for an OA which takes the interior dimension seriously: sustainability, in this perspective, is the preservation of the life-giving capacity of the Earth, which presupposes concern for earthly beings, and presupposes interior development as the factor that makes this concern unconditional (Prpich, 2005: 143). The ecocentric concept of sustainability and the concurring OA become thus intertwined: indeed, if the sustainable orientation was not in the values that form the very identity of the organic farm but is pursued simply because it is currently profitable or fashionable, then this orientation might change depending on circumstances (Prpich, 2005: 143).

Interiority is not limited to ethical values alone, as we have already seen - it includes identity, motivation, aims, aesthetics. An important aspect of interiority is the aesthetic experience, which can be shown to exist, for instance, in case of appreciation of the landscape. This experience is aesthetic and spiritual, forming a part of the OA identity, as put by the Nordic Platform of Ecological Associations: 'the cultivated landscape constitutes a positive aspect of nature.' The aesthetic side of interiority is based on gestalt perceptions. The perception of landscape is a gestalt-based: 'a range of forms and colours, structures and smells, its dynamics over time and its links to the observer's reminiscences and spiritual meanings' (Pedroli et al., 2007: 435). The gestalt experience of nature is one of the strongest motivational factors for its protection, and this is what corresponds more clearly both to the individual intentionality and cultural-collective dimension of OA.

There also is nothing wrong with considering ethical values as some of the identifying features of OA. This is actually necessary and socially efficient. One must simply note that ethical values are not the *only* motivating factor. OA definitely has an aesthetic dimension: an organic farm can fit naturally and beautifully to the landscape; it can be cultivated positively, like gardens, or individual plants and animals, with the consequent desire to protect their naturalness. Ethics, albeit important, needs to be complemented by aesthetics.

The aesthetic perception comes very close to being tacit knowledge, difficult to verbalise but very important in a farmer's experience. Moreover, the aesthetic perception, essential to the Goetheanistic-phenomenological method of research, is practiced in some kinds of biodynamic and organic agriculture, as Verhoog mentions (2007: 362).

Cases of research on organic agriculture which actually includes interior levels

There is much evidence of the organic approach's tendency towards integral holism. Thus, the four principles of organic agriculture, as presented by IFOAM, are integral and reflect both the Right-hand and the Left-hand dimensions. So is, for instance, the suggestion of Henk Verhoog that scientific perception of animals should be complemented by the common sense perception, as well as the process-focus of organic agriculture versus the product-orientation of biotechnology (Verhoog, 2007: 364).

Looking at the integral nature of the organic approach, it is logical to start with the principles of organic agriculture by IFOAM (Luttikholt, 2007: 358), which cover three out of four Wilber's quadrants. Thus, I relate the principle of Health relates to the individual exterior (physical), the principle of Ecology to the collective exterior (social quadrant), the principles of Fairness and Care to the collective interior (cultural). The principal of Health, the criteria of which are immunity, resilience and regeneration, is considered here as belonging primarily to the individual dimension because health can be measured in the individual organisms, and health of groups is derived from the health of the individuals. As for the principle of Ecology, it was placed here in the collective exterior because the referent of ecology - ecosystems - were placed by Wilber into the collective exterior (see level 4 of Wilber's scheme, 'heterotrophic ecosystems') as primarily collective entities. Also, it is important to remember that quadrants are no separate entities but merely different expressions of each one and the same entity, so their division is relative.

When Henk Verhoog shows that the holistic perception of animals includes, along with the scientific perception of animals, their aesthetic perception, characteristic of daily life, this can be interpreted as a plea for inclusion of interiority. His argument is that a merely objective perception of animals is one-sided, and it nullifies the moral side of treating animals. On the other hand, the inclusion of the aesthetic perspective on animals, in which we abstract from their instrumental value, provides the genuine vision of the animals and their ethical treatment (Verhoog, 2007: 362). This perspective has heuristic potential because combining aesthetic and intuitive with an objectivistic perspective on animals may lead to new findings, as in the case of Maslow's research on the social behaviour of chimpanzees (Maslow, 1940).

Table 1. Values of Organic Agriculture on the 4-quadrant scheme.

	Interior	Exterior
Individual	Standard: *Truthfulness* (Intentional)	Standard: *Truth* (Physical) Organic principle: *Health* (Immunity, resilience and regeneration)
Collective	Standard: *Justness* (Cultural) Organic principle: *Care* (Precautionary principle) Organic principle: *Fairness* (Equity, respect, justice and stewardship)	Standard: *Functional fit* (Social) Organic principle: *Ecology* (Production based on ecological process and recycling)

The organic rehabilitation of the interior (aesthetic, intuitive, ethical) perspective on animals resulted in the research program of 'Calf with the cow' carried out at Louis Bolk Institute (the Netherlands), in the research and practice of implementation of a suckling system in calf rearing. Despite some negative implications of the suckling system (stress after weaning, loss of marketable milk), some organic farmers tend to adopt this system because it better reflects the natural living and better serves animal welfare (Wagenaar and Langhout, 2007). Again, we encounter here a practice based on the integral view of animals.

An interesting case of a vertical (all-level) approach can be found in the use of 'healthy human' as a metaphor for farm health (Bloksma and Stuik, 2007). The health of both human and organic farm is considered at three levels: physical, social-cultural, and mental-spiritual. Applying this metaphor led to interesting insights on what constitutes a healthy farm: being 'efficient, direct and self-conscious', and able to fulfill one's own potential as a reason for one's existence (Bloksma and Stuik, 2007: 415) The conclusion is that being economically remunerative is only one side of a healthy farm, while the other is its socio-cultural functioning, and the mental health as the farm's capacity to provide inspiration, and having a mission. An application of a metaphor of healthy human to a think about farm, in fact, had the meaning of applying 'all-quadrant, all level approach', because the three levels that Bloksma and Struik have identified in the human being are essential parts of the integral approach.

Conclusion

The earlier analysis of the definitions of OA has contrasted the one-dimensional policies, restricted to behaviours and systems (USDA and NSCOA definitions) with that of the integral, two-dimensional definitions that come from the organic movement itself (e.g. Nordic Platform of Ecological Associations and four principles of IFOAM). The article has attempted to demonstrate the need to include the interior (cultural and intentional) dimension of OA for its all-encompassing analysis. A number of examples were given of actually integral research on organic agriculture, which benefited from the inclusion of interiority. Interiority in OA was considered as a line of development that encompasses world-views, identity, ethical and aesthetical values, motivation, and aims. As much as can be seen now, this seems necessary for the organic movement to pay attention to interiority, if it wants to remain true to its essence. Otherwise it is at risk of becoming a playing ball, only busy with including new trends and external objectives into standards and guidelines due to the focus on trade, consumer trust and legal aspects.

References

Bloksma, J. and Struik, P. (2007). Coaching the process of designing a farm. Netherlands Journal of Agricultural Sciences 54: 413-429.

DARCOF (2000). Principles of organic farming. Discussion document prepared for the DARCOF Users Committee, Danisch Research Centre for Organic Farming, Foulum.

IFOAM (2005). The Principles of Organic Agriculture, Bonn.

Kaltoft, P. (1999). Values about Nature in Organic Farming. Sociologia Ruralis 39: 39-53.

Luttikholt, L. (2007). Principles of organic agriculture as formulated by IFOAM. Netherlands Journal of Agricultural Sciences 54: 347-360.

Maslow (1940) Dominance-quality and Social Behaviour in Infra-human Primates, Journal of Social Psychology 11: 255-270.

OECD (2003). Organic Agriculture. Sustainability, Markets and Policies. CABI, 406p.

Padel, S. *et al.* (2007). Balancing and integrating basic values in the development of organic regulation and standards. Organic Revision Project, at www.organic-revision.org.

Pedroli, B.G.M., Van Elsen, T. and Mansvelt, J.D. (2007). Values of rural landscapes in Europe. Netherlands Journal of Agricultural Sciences 54: 431-444.

Prpich, W. (2005). An Integral analysis of the national standard of Canada for organic agriculture. World Futures: The Journal of General Evolution 61: 138-150.

Verhoog, H. (2007). The tension between common sense and scientific perception of animals. Netherlands Journal of Agricultural Sciences 54: 361-373.

Wagenaar, J and Langhout, J. (2007) Practical implications of increasing 'natural living' through suckling systems in organic dairy calf rearing, Netherlands Journal of Agricultural Sciences 54: 375-386.

Wilber, K., (2000) *Sex, Ecology, Spirituality*, Shambhala http://wilber.shambhala.com/html/books/cowokev8_intro.cfm/xid,5431661/yid,21177236

Organic farming: for the sake of nature?

Karsten Klint Jensen
Danish Centre for Bioethics and Risk Assessment, Institute of Food and resource Economics, Faculty of Life Science, University of Copenhagen, Rolighedsvej 25, DK-1958 Frederiksberg C., Denmark, kkj@foi.dk

Abstract

This paper attempts to discuss the value of organic farming in the light of environmental ethical positions. It appears to be a widespread belief that organic farming is benevolent to nature more than conventional farming is. However, the relevant sense in which it is 'benevolent to nature' is contested within environmental ethics. I shall argue that it is very difficult to cash out intuitions about the benefit of organic farming to nature in terms of the organic approach being in some sense 'more natural'. To make room for a position within environmental ethics that will allow us to value the contribution of organic farming as regards nature, I sketch an alternative view. This is a unified theory of the intrinsic value of nature, interpreted by analogy with the value of art and history, according to which there may be reason to value both certain cultivated and certain uncultivated areas. Within this theory, there is room to claim that organic farming, in certain respects, is better for nature than conventional farming.

Keywords: cultivated areas, ecosystem health, intrinsic value, wild nature

Introduction

It seems widely accepted that organic farming is more 'natural' than conventional farming, where this 'naturalness' is taken to imply an acknowledgement of the intrinsic value and integrity of living organisms, species and ecosystems (e.g. Verhoog *et al.*, 2003). By contrast conventional farming is largely seen as driven by anthropocentric concerns only. In other words, from the point of view of environmental ethics, in the emphatic sense of an ethic claiming that nature deserves moral consideration for its own sake, we ought to favour organic farming over conventional farming.

In this paper, I shall challenge this conventional wisdom. It seems clear that within environmental ethics there are conflicting views about how best to understand what is valuable about nature. These conflicts have implications for the evaluation of farming practices.

The aim of this discussion is twofold: one aim is to uncover complexities in the relationship between organic farming and nature, and to identify conflicting perspectives which organic farming will have to deal with in its future development. The other aim is to sketch a unified theory of the intrinsic value of nature.

Preliminary remarks

Most debates about organic versus conventional farming are probably concerned either with the effects of farming, and of its products, on human health and well-being; or with the significance for mankind, in the short- and long-term, of the effect of farming on the environment. In what follows I shall leave aside traditional anthropocentric concerns about the environment and human health. I am exclusively concerned with the effect of organic farming on nature when it is evaluated in a non-anthropocentric environmental ethics. More specifically, I explore the evaluation of organic farming (as compared with conventional farming) from positions that, in one sense or other, attach intrinsic value to nature.

First-generation holistic environmental ethics: the ecosystem itself sets the norm

Perhaps the widespread belief that organic farming respects the intrinsic value of nature is best understood in the light of the first-generation, non-anthropocentric, *holistic* environmental ethics inspired by Aldo Leopold (1949) and the science of ecology.

Early environmental ethicists were preoccupied with the question of what kinds of entity are entitled to moral consideration for their own sake, as that question was raised by Feinberg (1974). Perhaps encouraged by Singer (1976), who argued that the circle of morally considerable individuals should be expanded to include sentient animals, some environmental ethicists (e.g. Goodpaster, 1978, Stone, 1972) asked if the circle should be widened even further to also include plants.

However, in a seminal publication Callicott (1980) argued that this kind of extended individualism was unable to accommodate some of our central intuitions about the value of nature - intuitions roughly captured in Leopold's (1949) suggestion that we reappraise 'things unnatural, tame and confined in terms of things natural, wild and free'. The shift of values implied by this reappraisal involves concern for the ecosystem as such (see also Goodpaster, 1979) - for the 'land' or the 'biotic community', as Leopold calls it - at the cost of equal treatment of the interests of individuals.

As Rolston III (1975) observed, ethics can be 'ecological' in a primary and a secondary sense. Ethics is ecological in the second sense if it has a predefined set of (anthropocentric) values and then uses insights from the science of ecology to learn how these values are affected in environmental developments. It is ecological in the primary sense if insight into the ecosystem's holistic character prompts moral concern for the ecosystem itself. For Rolston III and Callicott, this means a concern for the good of the ecosystem, which Leopold identifies in its 'integrity, stability and beauty'.

According to this position, ecosystems posses intrinsic value, and there is a notion of the optimal functioning of ecosystems from which norms for human behaviour can be derived, prescribing a life-form in accordance with the cycles that sustains the optimum of the system. Precisely such a view is apparent in the Principles of Organic Farming (IFOAM, 2005). Here it is stated that 'The role of organic agriculture, whether in farming, processing, distribution, or consumption, is to sustain and enhance the health of *ecosystems* and organisms from the smallest in the soil to human beings' (my italics).

The idea is that human beings, just like any other living organism, live by and in ecosystems, and that human well-being is sustainable in the long run only if these ecosystems thrive. Therefore, humans should acknowledge this interdependence and live in a way that promotes the health of the ecosystems. In particular, farming should promote ecosystem health in a way that is sustainable over time.

However, the idea of an ecosystem health from which normative guidelines can be derived has been met with the criticism that no optimal state of ecosystems that generally coincides with what we intuitively value intrinsically about nature exists. The prime candidate for an optimal state of an ecosystem is likely to be the end-state of a succession - a state often assumed to be mature, fairly diverse and relatively stable. But what is valuable about such a state? Is it the stability as such or the diversity? Many ecosystems are not stable, either as a result of human intervention or as a result of some natural disruption such as a fire or attack by a predator; still, they can be very diverse. Rain forests, as we presently understand the matter, are not very stable ecosystems, but they are extremely diverse. Some very stable ecosystems, such as desert or tundra, on the other hand, are not very diverse.

It seems hard to avoid the conclusion that the judgement that one state of an ecosystem is intrinsically better than another cannot be read off from the ecosystem itself; we shall have to base it on a human value judgement. As we shall demonstrate later, this does not reduce the value of nature to instrumental status. But it does mean that when we call one form of agriculture more valuable, as far as nature is concerned, than another, we need to specify the values this judgement is based upon. It is not enough to refer to 'ecosystem health'.

Second-generation holistic environmental ethics: Leave (some) nature alone

Probably in response to this problem, later writers (Taylor, 1986; Rolston III, 1988) have located the prime environmental value of nature in the fact that natural processes unfold without human intervention, be they stable or not, diverse or not. This value was already more or less implicit in Leopold's and Callicott's favouring of 'things natural, wild and free' over 'things unnatural, tame and confined'.

Thus, according to this position, areas of *wild* nature, i.e. areas not cultivated by humans, should be left uncultivated. Here I shall not discuss how this is to be understood in detail. The important thing, in the present context, is that, on this view, cultivated areas have no value as nature from the point of view of environmental ethics. From this perspective, they can only be valued in terms of their effects on the valuable uncultivated areas.

In one way, organic farming is likely to fare better than conventional farming here, because it has no impact, via pesticides and fertilizers, on the natural areas surrounding it. On the other hand, to the extent that organic farming gives lower yields, it fares worse, because, other things being equal, it will require a larger area for cultivation. Hence, there seems to be a genuine conflict over land-use: the value of low impact on surrounding natural areas has to be weighed against the value of there being more uncultivated areas.

However, whereas the first-generation position on environmental ethics did not distinguish between cultivated and uncultivated areas, and consequently evaluated all ecosystems from the same environmental ethics perspective, the second-generation position gives only a partial evaluation of ecosystems. It leaves the evaluation of farming and cultivated natural areas unanswered from the point of view of environmental ethics. Resources for evaluating these areas ethically will have to come from somewhere else.

The problem is not only to find criteria for evaluating cultivated areas as landscape. It is also that the concern for the living individual organisms involved in agriculture (animals and plants) will have to be based on norms different from those involved in the concern for organisms living under wild conditions. The latter should be allowed to live freely, under natural conditions, without human interference. Taylor (1986) implies that the life of farm animals and plants should be evaluated according to a different standard, but he is silent on what the ethical constraints on the treatment of farm animals and plants should be. However, the implication that there are different standards seems unsatisfactory from a theoretical perspective.

Rolston III (1988) appears to wish to claim that naturalness, in the sense of processes unfolding without human intervention, makes up a norm that can guide farming - so that, for example, the conditions in wilderness can serve as a norm for farm animal welfare. More generally, less human intervention could be considered better as regards nature.

In my view, this approach is unsatisfactory, because although organic farming could be understood as involving less human intervention, and could therefore be considered better than conventional farming, the implication seems to be that no farming at all would be far better still, which I believe is absurd as a general view.

A unified perspective

I approve of the value judgement that there should be wild nature for its own sake, i.e. that some uncultivated areas should be left uncultivated and, perhaps, that some areas should be taken out of cultivation and returned to nature. However, I suggest that this holistic concern for nature is better understood as a human value, rather than a concern for ecosystems as organisms with a good, or health, of their own (Callicott, 1980) or as a concern that is derived from the concern for the good of individual organisms living in uncultivated ecosystems (Taylor, 1986).

Although the holistic concern for nature is conceived by me as a human interest, I do not consider it instrumental or subservient to other concerns; it is a concern for nature for its own sake (see O'Neill, 1993 for a similar view). To me, this position is comparable to the view that certain works of art are valuable for their own sake, or the view that we should preserve certain historic monuments for their own sake.

I also suggest that the welfare, or good, of individual animals (I want to put questions about plants aside) is judged by the *same standards* where farm animals and wildlife are concerned. In many cases, there would probably be worse long-term consequences, overall, if humans were to attempt to 'correct' the course of nature in uncultivated areas. But obviously there could be conflicts between the interests of individual wild animals and the holistic concern for nature - e.g. in cases where suffering could be avoided by human intervention. For Taylor and Rolston III, such cases involve an inherent conflict in our concern for nature. Taylor and Rolston III tend to conceal their existence. Within my perspective, however, such conflicts are like those in farming, where concern for animal welfare often conflicts in a similar way with human interests.

Hence, I suggest that my sketch of a theory provides a theoretically more satisfying, unified perspective.

Conclusion

I shall conclude by outlining what this unified perspective implies as regards the effect of organic farming on nature. The first thing to note is that, from my perspective, cultivated areas may also possess characteristics that are valuable for their own sake. To mention some examples, this could be landscapes of historical (natural or human), physical, biological, geographical or some other kind of interest. Again, cultivation often results in characteristic types of nature, and representative exemplars of these familiar types might possess value in themselves.

The second thing to note is that, in many cases, organic farming may be better equipped to protect such values than conventional farming is, not least because it involves a commitment to being low-input and to adapting to, and respecting, local natural and social conditions.

The third thing to note is, however, that organic farming does not automatically possess this advantage. For one thing, organic farming, just like conventional farming, is challenged by the general trend towards increased productivity, and this might lead to more intensified production in which organic farming could lose some of its advantages as regards valuable nature. Secondly, conventional farming practices might in some cases be better rooted in local conditions than organic farming. For instance, sheep farming in mountain areas, which could be considered an activity rightly valued for its own sake, is often dependent on the use of feed additives during the winter. Suppose the organic ban on feed additives ensures that organic sheep farming would not be economically feasible. We would then have a case where the principles of organic farming entail a loss of value as regards nature.

From these observations I draw the conclusion that, given a more detailed understanding of nature's value, those who advocate organic farming will have to ask whether that value requires organic practices to be revised in the future.

References

Callicott, J. B. (1980). Animal Liberation: A Triangular Affair. Environmental Ethics 2: 311-338.

Feinberg, J. (1974). The Rights of Animals and Unborn Generations. In: W.T. Blackstone (ed.) Philosophy and Environmental Crisis. Athens: University of Georgia Press, 43-68.

Goodpaster, K. E. (1978). On Being Morally Considerable, Journal of Philosophy 78: 308-325.

Goodpaster, K. E. (1979). From Egoism to Environmentalism. In: K.E. Goodpaster and K.M. Sayre (eds.) Ethics and problems of the 21st Century. University of Notre Dame Press.

IFOAM (2005). The Principles of Organic Agriculture. http://www.ifoam.org/about_ifoam/principles/index.html

Leopold, A. (1949). A Sand County Almanac and Sketches Here and There. New York: Oxford University Press.

O'Neill, J. (1993). Ecology, Policy and Politics. Human Well-being and the Natural World. London, Routledge.

Rolston III, H. (1975). Is There An Ecological Ethic. Ethics 85: 93-109.

Rolston III, H. (1988). Environmental Ethics. Philadelphia: Temple University Press.

Singer, P. (1976). Animal Liberation. A New Ethics for Our Treatment of Animals, New York: New York Review and Random House. (Later reprinted as: Animal Liberation. Towards an End to Man's Inhumanity to Animals, Wellington, Northamptonshire: Thorsons Publishers).

Stone, C. F. (1972). Should Trees Have Standing? Toward Legal Rights for Natural Objects. Los Altos: William Kaufmann.

Taylor, P. (1986). Respect for Nature. Princeton University Press.

Verhoog, H., Matze, M., Lammerts van Bueren, E.T. and Baars, T. (2003). The role of the concept of the natural (naturalness) in organic farming. Journal of Agricultural and Environmental Ethics 16: 29-49.

Mutilations in organic animal husbandry: ethical dilemmas?

Susanne Waiblinger[1,2], Christoph Menke[2] and Knut Niebuhr[1]
[1]*Institute of Animal Husbandry and Welfare, Department of Veterinary Public Health and Food Science, University of Veterinary Medicine, Veterinärplatz 1, 1210 Wien, Austria*
[2]*Association of Animal Welfare - VEAT, 93055 Regensburg, Germany*

Abstract

Organic standards (e.g. IFOAM) as well as the EU-regulation on organic production (2092/91) restrict the use of mutilations but do not prohibit them completely. The EU-regulation allows mutilations (called 'operations'), if not carried out systematically, for human safety reasons (e.g. dehorning) or to improve animal health, welfare or hygiene. Consequently, dehorning of cows and goats is still very common also in organic farming, beak trimming in laying hens can be found.

Farmers performing mutilations argue, in a first line, that animal welfare considerations or human safety issues make mutilations necessary. However, research has shown that those problems are avoidable using optimized care, management, housing and handling, which are as well basic principles of organic husbandry. Economic interests that truly do or are perceived to conflict with implementation of those optimum conditions thus often are an important motivation for performing mutilations. For the animals, mutilations entail short-term (anxiety, suffering, pain) and partly long-term effects (e.g. sensory deprivation). Also, mutilations obviously violate the integrity of the animal, an ethical principle which is claimed to be part of organic values. Further, consumers hold the view that good animal welfare standards come as close as possible to nature.

We therefore argue in this paper that if the values and principles of organic husbandry are taken seriously, mutilations should not be performed nor are they really necessary. For sustainable organic farming, the aim therefore should be to renounce, and possibly ban, such mutilations in the future.

Keywords: dehorning, tail-docking, beak-trimming, animal welfare, organic values

Introduction

Mutilations are very common in animal husbandry all over Europe and worldwide. Besides castration of male animals, dehorning of cattle and goats, beak trimming in laying hens or tail docking in pigs and sheep are performed widely, whereas other mutilations are used regularly only in few countries, for example tail docking in dairy cows. Those painful procedures are justified by arguing with (1) welfare benefits for the whole herd on the long term by reducing risk of injuries, stress or disease outweighing largely the short term pain of the mutilation (all mutilations mentioned above), (2) reduced risk of accidents for humans working with the animals (horned cattle and goats), (3) improved working conditions for humans (tail docking in sheep and dairy cows), and at last (4) economic benefits including for example a reduced value of horned animals on markets, lower working time or decreased costs for building.

Organic animal husbandry restricts the performance of mutilations, but does not ban them completely. According to Council regulation (EEC) No. 2092/91 'operations such as .....tail docking, trimming of beaks and dehorning must not be carried out systematically in organic farming', but some 'may be authorized by the inspection ...body for reasons of safety (for example dehorning in young animals) or if they are intended to improve the health, welfare or hygiene of the livestock.' IFOAM basic standards (2006) state: 'standards shall require that: mutilations are prohibited; the following exceptions may be used: castration, tail-docking of lambs, dehorning, ringing, mulesing...'.

Therefore not surprisingly, several mutilations are common also in organic animal husbandry, e.g. dehorning in dairy cows and goats, and - at least in some countries - tail docking in sheep or beak-trimming in laying hens. However, all those mutilations entail anxiety, suffering, pain and, e.g. in laying hens, sensory deprivation. They may change the species-specific behaviour, and they obviously violate the integrity of the animals, an ethical principle which is claimed to be part of organic values (Verhoog *et al.,* 2004). Systems should be adapted to the animals instead of adapting the animals to the system. Further, the importance of providing the animals with possibilities to perform their natural behaviour in organic systems is underlined by several authors (for review Verhoog *et al.,* 2004). This also is in line with consumers' view that good animal welfare standards come as close as possible to nature (Ouédraogo, 2002).

Do organic farmers thus face ethical dilemmas regarding differing interests of different individual animals, or of the animals and themselves? Or are solutions existing which meet the interests of all parties involved? We will examine and discuss these questions by taking dehorning of dairy cows and goats and beak trimming in laying hens in European countries as an example.

Dehorning

Situation

Since centuries horned cattle are living with humans, except some few hornless breeds. Only in the middle of the last century parallel to the intensification of agriculture, an increase in herd size and the development of loose housing, dehorning of cattle became more widespread. In mountainous countries or regions where tie stalls are still very common this development was retarded, but now dehorning of cattle, even if kept in tie stalls, is widespread. With respect to organic farming, more farmers keep horned dairy cows in loose housing compared to conventional ones, but dehorning is still found frequently, except on biodynamic farms where it is not allowed. In recent studies in Germany, the proportion of farms dehorning their cattle in organic farms ranged from 46% (both tie stalls and loose housing) to 80% (loose housing).

In contrast to dairy cows, genetically hornless dairy goats are common. However, because the gene for hornlessness is associated with the gene for inter-sexuality and thus infertility, it is common to have genetically horned and hornless goats and still many goats are dehorned. In Austria and Germany, dehorning is performed by the majority of goat farmers, also when producing organically.

Functions of horns and effects of dehorning

Dehorning or disbudding, i.e. the removal of the horn bud, is a very painful procedure (for cattle see Graf and Senn, 1999; Faulkner and Weary, 2000). In goats, the risk for severe injuries caused by the procedure is higher due to different anatomical features (relatively larger horn buds and thin skullcap).

Further, horns do have functions in social behaviour and dehorning affects social behaviour (for review in cattle Menke *et al.,* 2004): The presentation of horns influences the rank of individual cows and can reduce the frequency of agonistic behaviour. While in horned herds social rank is mainly influenced by age and size of horns, in herds without horns weight is highly associated to rank. This is important with regard to social stability when older animals loose weight and their rank is contested more easily by younger animals.

Reasons for dehorning

Two main reasons are mentioned when justifying dehorning: (1) keeping horned dairy cows comprises an unacceptable risk for (severe) injuries and stress for the animals, especially subdominant ones; (2) the risk of accidents for the farmer is higher. Due to the common use of dehorning nowadays another reason has emerged: farmers that principally are reluctant to dehorn their animals face the problem that it gets more and more difficult to sell horned animals. Breeding organisations sometimes even require dehorning in cattle or auction markets only allow dehorned animals to be presented on the market. Even insurance companies sometimes (try to) impose dehorning in cattle.

Discussing the arguments: are horns a risk for welfare of animals and humans?

Two on farm studies regarding horned dairy cows in loose housing in Switzerland, Germany and the Netherlands demonstrate that on most farms keeping horned dairy cows in loose housing is possible without unacceptable levels of injuries or social stress, but also, that some farms had considerable problems with high levels of agonistic behaviour and injuries caused by horns and poor welfare for at least some of the animals (Menke, 1996; Menke *et al.*, 1999; Baars and Brands, 2000). These studies identified appropriate management guided by a high quality and intense relationship of the farmers with their cows being the main factor for success; but also housing factors showed some relevance (Menke *et al.*, 2004). Thus, negative consequences of keeping horned animals are avoidable. In dairy goats no study has been conducted so far, which explicitly investigated injuries due to agonistic behaviour. Studies on horned as well as mixed and hornless goat herds suggest that social stress and thus risk of injuries depends again very much on management and also on housing factors (e.g. Loretz *et al.*, 2004; Noack and Hauser, 2004). Further, social stress affecting negatively the welfare of individual animals or the whole herd also occurs in hornless (polled or dehorned) cow or goat herds (Bøe and Færevik, 2003; Noack and Hauser, 2004). Even injuries are reported as well, though the likelihood of severe injuries is certainly lower.

In sum, keeping horned dairy cows and goats is possible without an unacceptable risk of social stress or injuries, if the farmer adopts appropriate management strategies (and offers appropriate housing) which seem to depend largely on farmers' attitudes and the relationship with their cows (Menke *et al.*, 2004).

The human-animal relationship is also the key factor for reducing the risk of accidents for humans working with animals (Waiblinger *et al.*, 2006; Boivin *et al.*, this issue) regardless whether animals are horned or without horns. Reports of insurance companies (e.g. Switzerland, France) document that horns do hold a risk of injury, but injuries caused by horns of cattle only constitute a smaller amount (<20%) of accidents with animals and mainly happen in tie stalls. Dehorning thus cannot prevent a large majority of injuries/accidents with animals. For general prevention of accidents with cows improving or maintaining a good human-animal relationship is most effective.

Dehorning, horns and economics

An economic evaluation of keeping of horned or dehorned dairy animals was not yet performed. Regarding housing costs recommended space and design of housing does not differ between horned and dehorned herds in dairy cows, but requirements often are not yet implemented in practice. In dairy goats probably there is a need for more space in horned and mixed herds compared to dehorned herds especially in small groups (Loretz *et al.*, 2004), but experiences on existing farms suggest that the space requirements of organic agriculture in the EC are sufficient. Again, these are not always fully implemented (e.g. with respect to access to an outside yard). Management of horned herds may cause some more work, but no quantitative figures exist. Selling of horned animals could be a problem, as described above, and this may imply an economic loss especially for successful breeders. Farmers may change their practice to dehorning to be able to continue participating in the auction markets, mainly for economic reasons, not for welfare or security considerations.

Beak trimming in laying hens

Situation

Beak-trimming in laying hens has been performed in all European countries and in all husbandry systems. Only recently there has been a move in some countries to ban it completely, in some at least in organic layers beak-trimming is not performed anymore or is almost absent (e.g. Austria, < 1% in 2006). Due to European legislation (Directive 1999/74) beak-trimming may only be performed in chicks of an age of less than ten days, which means that only preventive beak trimming is allowed. Beak-trimming consists of the removal of the tip of the upper and lower beak. It is only effective, if a sufficient part of the beak, including not only the horn layer but as well bone structures and connective tissue with sensory receptors are removed. This leads to a change in the mechanical properties of the beak and a sensory deprivation. As a preventive measure beak-trimming is always performed on the whole flock, which could as well prima facie be considered as fulfilling the term 'systematically'. Nevertheless in organic agriculture beak-trimming still seems to be performed in different European countries. For instance, in Germany in a survey on 70 organic laying hen farms 14 (20%) had flocks with beak-trimmed birds (Hörning *et al.*, 2004). In a recent study in the Netherlands Kijlstra and van der Werf (2006) found beak-trimmed birds in all of 32 organic laying hen farms visited. Nonetheless, there seems to be a general move towards avoiding it.

Reasons for beak trimming

In the view of many farmers beak-trimming is performed mainly to avoid damage to the hens inflicted by other hens through feather-pecking (FP) or injurious-pecking (IP, also called vent pecking). FP and IP are thought to be closely linked but may as well develop independently. During FP feathers are often grasped and pulled out, which is painful, leads to impaired insulation and sometimes to wounds. As a consequence of these wounds IP may be triggered. Pecking in IP is targeted directly at the skin and can, if wounds are severe, lead to the death of the victim. Especially as a consequence of IP mortality increases, but as well FP is often accompanied by a higher number of deaths. FP and IP have a multi-factorial origin and causes are often difficult to detect. A second argument commonly encountered is therefore the easier management of flocks. Thirdly potential psychological problems of caretakers in the case of wounded or dead birds are put forward, as well as negative economic consequences.

Solutions in intact hens

Due to the multi-factorial and complex origin of FP and IP single measures are often unsuccessful. Influencing factors known today can be categorized belonging to four different groups, being genetic predisposition, composition of feed and husbandry conditions during rearing and during lay. Breeding companies have tried to select hens, which are less likely to show FP and IP and practical experience from Austria suggests a decline of the propensity to peck in brown hybrids. Adequate feed composition poses a considerable problem especially in organic laying hen husbandry, but at least in Austria efforts to optimize rations during rearing and laying and closer monitoring of feed intake and weight gain during the start of the laying phase seemingly had a positive effect. During rearing lower stocking densities especially in the first weeks, good air quality, good hygiene management and an even distribution of light in the house could further contribute to reduce the risk. Especially optimized husbandry conditions during the laying phase including again an even distribution of light and good air quality, smaller herd sizes and the avoidance of concurrent infections through optimized hygiene and disease prophylaxis seem to be a means to avoid FP and IP. Experience of farmers seems to be an additional important factor. In conclusion, in order to tackle the problem a more holistic approach and optimisation of the whole system is needed including breed, feeding and husbandry conditions together with better management and an improved human-animal relationship (Niebuhr *et al.*, 2005; 2006). FP and IP in non beak-trimmed laying hens are avoidable, as the Austrian situation shows, resulting not only in a decline in the incidence (from 18% in 2003, n=254 flocks to 7% in 2006, n=284 flocks) but as well in severity of single outbreaks.

Discussion: dehorning, beak trimming and organic values

Council regulation (EEC) 2092/1991states that 'housing conditions for livestock must meet the livestock's biological and ethological needs (e.g. behavioural needs as regards appropriate freedom of movement and comfort)' and 'disease prevention in organic livestock production shall be based on....the application of animal husbandry practices appropriate to the requirements of each species, encouraging strong resistance to disease and the prevention of infections'. These standards are in line with organic principles, e.g. in the IFOAM basic standards (2006), stating that 'organic livestock husbandry is based on...respect for the physiological and behavioural needs of livestock'. From the results shown above it can be argued that rather than offering the animals an adequate environment to perform their species-specific behaviour, as stipulated in the standards, dehorning or beak trimming is merely a means of adapting the animals to the husbandry system. This also contradicts the expectations of consumers regarding animal welfare and naturalness of the production system, an issue that can even be seen in the advertisement of organic dairy products using very often pictures of horned cows. Yet, promoting dairy products with pictures of horned cows, which have in fact been produced widely by dehorned ones, poses another ethical problem.

IFOAM basic standards state a further general principle: 'Organic farming respects the animal's distinctive characteristics'. Verhoog *et al.* (2004) call for an inclusion of the concept of integrity into organic values, which comprises 'respect for the wholeness, harmony or identity of a living entity'. Horns are characteristic for all cattle, except some few polled breeds, as well as for wild and many domesticated goats. Dehorning can be considered as taking away part of what makes a cow a cow, or a goat a goat. It takes away a characteristic of the animals as part of the species that has evolved during phylogeny (despite the fact that the gene for hornlessness is dominant). Also beak trimming takes away species-typical sensory and behavioural capacities of hens. Dehorning as well as beak trimming thus contradicts the integrity approach and the respect for the animal's characteristics.

While beak trimming actually is already comparatively restricted in organic farming in many countries, dehorning is (still?) widely accepted. The case of beak trimming clearly shows that the negative impact on welfare when keeping intact hens due to cannibalism or feather pecking decreases largely by improving (farmers') knowledge and forcing them to change their management (due to the lack of the alternative of beak trimming). Why isn't dehorning banned as well more strictly? This may have two reasons. First, the necessity for beak trimming may have been linked more directly to poor husbandry conditions and stress, thus being perceived more easily as a way to adapt the animal to poor, not appropriate, and thus not acceptable conditions instead of adapting the conditions to the animal. Secondly, a strong argument for dehorning besides animal welfare is human safety. This is supported by the fact that the EEC directive explicitly mentions security reasons in the first place together with dehorning. Thus, it is harder for the inspection body to refuse a claim for dehorning when justified by security; who takes the responsibility if an accident will happen? The more important role of the human-animal relationship and the way of handling animals in prevention of accidents is generally less acknowledged.

According to organic standards, economic aspects (such as difficulties in selling of animals) are no reason for acceptance of dehorning. This aspect however should not be forgotten. Farmers maintaining horns in their animals should be supported.

Conclusion

In an ethical framework based on organic values mutilations do not fit in. The ethical dilemmas brought into discussion - violating integrity and causing short-term pain to the animal or accepting more stress and (severe) injuries in some animals as well as a higher risk of accidents - loose their ground when husbandry follows the organic principles and optimized care, management, housing and handling are applied, by this minimizing problems due to intact animals.

If, however, economic reasons, e.g. higher costs due to more labour or a stricter selection for 'temperament' traits instead of just high milk yield or exclusion from the auction markets, are the real reasons covered by other arguments, these have to be made explicit to find solutions.

References

Baars, T. and L. Brands (2000). En koppel koeien is nog geen kudde. Driebergen, NL:Louis Bolk Instituut, 67p.

Bøe, K.E. and Færevik, G. (2003). Grouping and social preferences in calves, heifers and cows. Applied Animal Behaviour Science, 80: 175-190.

Faulkner, P.M. and Weary, D.M. (2000). Reducing pain after dehorning in dairy calves. Journal of Dairy Science 83: 2037-2041.

Graf, B. and Senn, M. (1999). Behavioural and physiological responses of calves to dehorning by heat cauterization with or without local anaesthesia. Applied Animal Behaviour Science 62: 153-171.

Hörning, B., Trei, G. and Simantke, C. (2004). Ökologische Geflügelproduktion - Struktur, Entwicklung, Probleme, politischer Handlungsbedarf. Bericht, Geschäftsstelle Bundesprogramm Ökologischer Landbau, Bundesanstalt für Landwirtschaft und Ernährung (BLE), Bonn. http://orgprints.org/8215/01/8215-02OE343-ble-unikassel-2004-sq-gefluegel.pdf.

Kijlstra, A. and van der Werf, J. (2006). Effects of the compulsory indoor confinement of organic layer poultry: a dust storm! ASG Report 06/I00502, Animal Production Division, Animal Sciences Group, Wageningen University and Research Centres, Wageningen http://orgprints.org/9822/01/Effects_of_confinement_of_poultryak.pdf.

Loretz, C., Wechsler, B., Hauser, R. and Rüsch, P. (2004). A comparison of space requirements of horned and hornless goats at the feed barrier and in the lying area. Applied Animal Behaviour Science 87: 275-283.

Menke, C. (1996). Laufstallhaltung mit behornten Milchkühen. Dissertation, Institut für Nutztierwissenschaften, ETH, Zürich, Switzerland, 162p.

Menke, C., Waiblinger, S., Fölsch, D.W. and Wiepkema, P.R. (1999). Social behaviour and injuries of horned cows in loose housing systems. Animal Welfare 8: 243-258.

Menke, C., Waiblinger, S., Studnitz, M. and Bestman, M. (2004). Mutilations in organic animal husbandry: dilemmas involving animal welfare, humans, and environmental protection. In: Vaarst, M., Roderick, S., Lund, V. and Lockeretz, W. (eds.). Animal health and welfare in Organic Agriculture. CABI Publishing Wallingford, UK, pp. 163-188.

Niebuhr, K., Zaludik, K., Baumung, R., Lugmair, A. and Troxler, J. (2005). Injurious pecking in free-range and organic laying hen flocks in Austria. Animal Science Papers and Reports 23 (Suppl. 1): 195-201.

Niebuhr, K., Zaludik, K., Gruber, B., Thenmaier, B., Baumung, R., Lugmair, A. and Troxler, J. (2006). Epidemiologische Untersuchungen zum Auftreten von Kannibalismus und Federpicken in alternativen Legehennenhaltungen in Österreich. Final report BMLFUW, Wien, 109p.

Noack, E. and Hauser, R. (2004). Der ziegengerechte Fressplatz im Laufstall. FAT Bericht Nr. 622, Agroscope FAT Tänikon, Etternhausen, Schweiz, 12p.

Ouédraogo, A.P. (2002). Consumers' concern about animal welfare and the impact on food choice: social and ethical conflicts. INPL-Nancy, 84-86.

Verhoog, H., Lund, V. and Alrøe, H.F. (2004). Animal welfare, ethics and organic farming. In: Vaarst, M., Roderick, S., Lund, V. and Lockeretz, W. (eds.). Animal health and welfare in Organic Agriculture. CABI Publishing Wallingford, UK, pp. 73-94.

Waiblinger, S., Boivin, X., Pedersen, V., Tosi, M.-V., Janczak, A.M., Visser, E.K., and Jones, R.B. (2006). Assessing the human-animal relationship in farmed species: A critical review. Applied Animal Behaviour Science 101: 185-242.

The development of organic aquaculture as a sustainable food production system: exploring issues and challenges

S. Tomkins and K.M. Millar
Centre for Applied Bioethics, School of Biosciences, University of Nottingham, Sutton Bonington Campus, Loughborough, Leics, LE12 5RD, United Kingdom, sandy.tomkins@nottingham.ac.uk

Abstract

Aquaculture production has increased dramatically in the EU since the 1970s, but recently concerns have been raised regarding the sustainability of the systems employed, particularly in terms of environmental and animal welfare impacts. A number of systems are being developed and applied in order to improve the overall sustainability of aquaculture production. Organic aquaculture has been claimed to be one of the most sustainable systems currently applied by the industry. However, a number of commentators are still concerned about the sustainability of the industry as a whole and a number of these concerns also relate to organic production systems. The UK Soil Association (SA) has only recently granted full status to their organic aquaculture standards (July 2006) and this potentially exemplifies the challenges that face not only the organic industry, but also the aquaculture industry as a whole, as they seek to develop sustainable methods. This paper explores the SA's work to develop organic standards for aquaculture and to manage the concerns raised about the sustainability of the current systems. For organic salmon and trout production the greatest challenge appears to relate to the use of sustainable sources of fish feeds and the animal welfare standards of the systems. Examining the recent development of UK SA organic aquaculture standards is an interesting case that may inform other aspects of the aquaculture sustainability debate. It is therefore hoped that this paper highlights a far more extensive empirically-led ethical analysis of these new standards, and the certified systems that result, is needed.

Keywords: organic aquaculture, organic standards, sustainability, soil association

Aquaculture production

The major source of fish is capture fishing, however, current practice is proving increasingly unsustainable. Global marine stocks are diminishing at alarming rates due to over exploitation which is threatening marine biodiversity. Recent figures from the FAO indicate that fish stocks for a number of major species are at the point of collapse, with 75% of the world's fish stocks fully-exploited, over-exploited or depleted (FAO, 2007). These significant concerns regarding marine overfishing that have resulted from an ever-increasing global market for fish, has stimulated increased production from aquaculture[10]. Levels have increased rapidly over the last twenty years, partly due to technological innovation, with the industry providing over 32% of the overall marine product market (fish, molluscs, crustaceans, etc) (FAO, 2007). In addition, globally aquaculture has been characterised as a notable force for social development, playing a highly significant role in providing local food security, alleviating poverty, improving rural livelihoods, creating employment and generating income in some of the poorest regions. Increased aquaculture production has also increased the fish protein supply for communities, both rural and urban, in developing countries.

---

[10] As defined by FAO, aquaculture is the 'farming of aquatic organisms including fish, molluscs, crustaceans and aquatic plants. Farming implies some sort of intervention in the rearing process to enhance production, such as regular stocking, feeding, protection from predators, etc. Farming also implies individual or corporate ownership of the stock being cultivated'

Within the EU aquaculture production has increased dramatically since the 1970s, and although this has reduced prices, increased product availability, alleviated some pressure on capture fishing and contributed to food security, questions have been raised about the sustainability of the current production systems. With the rapid expansion of production, there are also concerns that production efficiency drivers may dominant, so that the potential benefits of these systems, such as wider environmental protection, economic development and increased social justice, will not be delivered. At its most significant, unsustainable aquaculture practices could intensify some of the existing problems facing the industry and create new ones.

Organic aquaculture

The aquaculture industry has a number of developmental trajectories (Millar and Tomkins, 2006; Muir, 2005) with increased production of organic fish products as one possible market option. As well as responding to the economic opportunities of a growing market, the application of organic principles may help to improve the sustainability of existing systems and combat some of the criticisms levelled against current methods. A number of systems are being developed and applied that aim to improve the overall sustainability of aquaculture production (CEC, 2002; Naylor *et al.*, 2001). However, questions have been raised as to whether any current aquaculture systems can be considered as 'sustainable', with a number of organisations highlighting some of the challenges that the industry faces (EU, 2004; Greenpeace, 2005). Boehmer *et al.* (2005) highlights some of these challenges as follows:

'Organic' in the context of food production connotes standards and certification ... as well as more elusive characteristics such as consumer expectation for food quality and safety and general environmental, social, and economic benefits for farmers and for society. The variety of species produced in aquacultural systems and vast differences in cultural requirements for finfish, shellfish, mollusks, and aquatic plants add to the complexity of defining this sector. Some species and some production systems may prove quite difficult to adapt to a traditional 'organic' system'.

There are a number of definitions of sustainability that could, and have been, applied to aquaculture production e.g. by the FAO Committee on Fisheries in 1991 (cited by Singh-Renton, 2002). In this context it maybe useful to consider three different concepts of sustainable production (Ziman, 2006), these are sustainability as: (1) a production system requiring a balance of inputs and outputs with the ultimate goal of ensuring the long-term supply of a product (e.g. food); (2) as a form of stewardship that guides production methods that respect nature and the integrity of ecological systems; (3) the application of a principled or value-based approach that underpins production practices, e.g. surmounting to elements of communitarianism, agrarian values, etc. What appears to be underpinning aspects of the current discourse surrounding the nature of sustainable production for aquaculture and the role of organic systems, is the issue of whether the further development of these systems is driven by an adherence to a fundamental set of principles (valued-based approach) that guide organic farming, or a production-oriented form of assessment of agricultural sustainability. Pelletier highlights the potential tension that may arise between sustainability drivers and organic principles, '*These overarching general principles must be connected to the specific conditions of aquaculture systems while remaining fully consistent with the basic philosophy of organic production*' (Pelletier, 2003). This paper briefly explores the development of organic standards in the UK, focusing on the Soil Association's (SA) recent work, and the way in which the organisation has responded to a number of specific concerns. Exploring the issues surrounding the development of UK organic aquaculture standards may help to inform the overall debate on the sustainability of aquaculture production systems.

Development of UK organic aquaculture standards

Standards for organic practices have existed for many years[11] and organic fish producers must comply with the same regulations as other organic certified producers, however interpreting standards developed for terrestrial species into practices and standards relevant to aquatic species is a major challenge, as it is claimed that most cannot be directly applied to aquaculture systems (Pelletier, 2004). The SA has only recently granted full status to their organic aquaculture standards (July 2006), this potentially exemplifies the challenges that face not only the organic industry, but also the aquaculture industry as a whole, as they seek to develop sustainable methods. Although industry representatives have noted the benefits of organic production, a number of commentators are concerned about the sustainability of the aquaculture industry as a whole and these concerns also relate to organic production systems.

The SA developed its first draft organic standards for aquaculture in the late 1990s, in response to requests from fish farmers who wanted 'better practice' to be recognised. In 1998, these Standards were given interim approval by the Soil Association Council (SAC) and full organic status by the UK Government's Advisory Committee on Organic Standards (ACOS). Organic trout and salmon went on sale in the UK in 1999 (Greenpeace, 2005). In July 2000, the SA, Food Certification Scotland and the Organic Food Federation published their joint UK Organic Aquaculture Standards, which were recognised by the UK Department for Environment, Food and Rural Affairs (DEFRA). However the SA's own governing body wanted greater clarity on the potential impacts of fish farming, so they held back for eight years from granting salmon-farming and other forms of aquaculture full organic status restricting it to interim status until July 2006. A programme of research and development (Aquaculture Development Programme) set up in 2003 led to the publication in 2006 of this new, significantly amended set of standards to which the SAC gave its full backing (Soil Association, 2006a).

The range of species for which organic standards have been developed now extends to salmon, trout, Arctic charr, shrimp, carp and shellfish (mussels and oysters) (Soil Association, 2006c). Raven, SA Scotland Director, (2006) hailed this granting of full organic status as a crucial step forward in organic aquaculture:

'*As with land-based organic farming, the Soil Association's aim is to achieve the most sustainable production for aquaculture. Our new standards represent carefully targeted key improvements on their 'interim' predecessors...... We now embark on a major programme of continuing work to develop the standards further - focusing on priorities such as sustainable fish feeds, moving away from potentially polluting veterinary treatments, and farming multiple species of fish, sea-weed and crustaceans to minimise nutrient losses - replicating the diversity of cropping and species found on land-based organic farms.*' (Soil Association, 2006a).

Key issues for the organic industry

The SA have thus claimed that their aquaculture standards address the main areas of concern such as: potential risks associated with environmental impacts from the increased use of pesticides and antibiotics; discharge of production wastes and pollutants; ecological impacts from potential escapees; frequency and diversity of disease outbreaks etc. However, a number of challenges still remain, such as the development of sustainable sources of fish feeds, pollution and animal welfare issues. Although focusing on the issues of feed and animal welfare, the following sections highlight how the SA has attempted to address some of these issues:

---

[11] SA's first organic standards were published in 1967, www.soilassociation.org/web/sa/saweb.nsf/Aboutus/Timeline.html

Siting of salmon pens and pollution - Fish waste such as faeces and uneaten food are major environmental hazards. In general 15-20% of feed used at salmon farms enters the surrounding environment uneaten (Weber, 2003). According to the SA, some studies (e.g. Kutti *et al.*, 2007; Kempf *et al.*, 2002) show that in well sited non-organic salmon farms, apart from short-term build up of waste (fish faeces and waste food) on the seabed below the pens, there is little impact on the wider environment (Soil Association, 2006c). However this is disputed (FAO, 2007; EU, 2004). The SA aquaculture standards seek to address this by specifying that pens can only be sited in areas with strong tidal flushing (Soil Association, 2006c). In order to break parasite life-cycles and aid recovery of the sea bed, SA aquaculture standards specify a mandatory 6-week fallowing period after each harvesting of fish (Soil Association, 2006c).

Use of chemicals and antibiotics - Organic farming restricts the use of pesticides and antibiotics. For salmon farmers, the main pest challenging the health of fish are parasitic sea-lice which when incubated in fish farms can attach themselves to and kill young wild salmon and sea trout as they migrate. The SA permits the control of sea-lice using sea-wrasse and two licensed chemicals (that require long withdrawal periods to ensure residue-free fish at harvest). A strict set of standards to control sea lice are embedded in the new Standards (Soil Association, 2006b). In addition, SA organic farmers are only allowed to use physical methods for keeping nets clear from colonisation by seaweeds, small shellfish, etc, antifoulant chemicals are not permitted (Soil Association, 2006b).

Escapes and biodiversity - Escapes of fish from organic systems are acknowledged. Impacts of escaped farmed fish include interbreeding with wild fish thereby diluting their unique genetic make up and escapees have the capacity to spread infectious diseases, such as sea lice to wild fish populations (Staniford, 2002). SA salmon farmers are required to avoid siting pens in areas of importance for wild salmon and sea trout populations. The SA states the problem of escaping fish is being addressed by the industry as a whole through better engineering and improved codes of practice, but refers to this issue as a '*serious Achilles heel*' and states '*it is hard to see how organic farmers can do more*' (Soil Association, 2006c).

Animal welfare - Fish welfare is a complex and highly debated area. Concerns have been voiced regarding whether the use of net pens is acceptable in an organic system, particularly if the species is migratory (e.g. Pelletier, 2003). The SA approach is '*based on respecting the life stages as the fish grow through different phases, and ensuring their physiological needs are met to avoid adaptive responses to adversity, such as aggression and marine migration in search of food*'. The SA states that at low stocking densities there should be no evidence of poor welfare, and claims there is no evidence that increasingly domesticated farmed salmon suffer in any way from not being able to migrate to feeding grounds (Soil Association, 2006c). Maximum stocking rates are specified at half the density of the average non-organic farm. The SA believes that their specified stocking rate would have to be more than doubled to compromise welfare (Soil Association, 2007).

Sourcing of organic fish feed - A major concern with the organic production of salmon and trout (carnivorous fish species) is that fish meal and fish oil are used in organic feeds. Specifically, questions centre on whether a product derived from wild-caught animals can be certified organic. If organic principles are to be upheld these products should be '*obtained from sustainably managed fisheries; ...derived from locally available fishery products not suitable for direct human consumption; ... free from synthetic additives and unwanted contaminants; and ... only fed to farmed organic aquatic species with naturally piscivorous feeding habits*' (Tacon and Brister, 2002). SA standards state '*At least 50% of the fish-based components must come from the by-products of fish that have been wild-caught for human consumption (waste from filleting, etc), and the rest must be from sustainably managed sources*' (Soil Association, 2006b). The SA claim that recycling such wastes into high quality food is a justifiable use of this resource, and reduces pressure on wild fish stocks typically used for fishmeal, therefore what would have been wasted becomes part of a sustainable food chain.

Conclusion

This initial characterisation has indicated that both the SA and external commentators identify the greatest challenges to sustainable organic aquaculture systems to be the issues of sources of fish feed, waste management and animal welfare. The SA attempts to address these concerns, however the organisation highlights that '*While we have made progress, we are aware that we have further to go before our aquaculture standards meet our highest principles. We anticipate further positive changes over the next few years - at least as radical as those we've made so far*' (Soil Association, 2007). An important part of the discourse surrounding the development of organic standards appears to be the tension between whether the further development of these systems should be driven by an adherence to a fundamental set of principles that guide organic farming, or a more production-oriented form of assessment of sustainability.

It could be argued that applying a holistic systems approach, might lessen the environmental impacts associated with facility siting; provide an integrated approach that safeguards and enhances biodiversity; ensure that feed requirements are sustainable; and may improve the safety and quality of food products through banning the use of antibiotics and pesticides. However as recently noted, '*In order to realize these benefits ... the resulting organic aquaculture standards must be stringent and reflect a primary commitment to developing low impact and sustainable production methods*' (Center for Food Safety, 2006).

Examining the recent development of UK SA organic aquaculture standards and the interplay between guiding principles and concepts of sustainability is a valuable case that may inform other aspects of the aquaculture sustainability debate. It is therefore hoped that this paper demonstrates that a far more extensive ethical analysis of these new standards, and the certified systems that result, is needed.

References

Boehmer, S., Gold, M., Hauser, S., Thomas, B. and Young, A. (2005). Organic aquaculture AFSIC notes 5. National Agricultural Library, US Department of Agriculture, USA.

CEC (2002). A strategy for the sustainable development of European aquaculture, COM(2002)511. Commission of the European Communities, Brussels, 26p.

Center for Food Safety (2006). Comments on the interim final report of the Aquaculture Working Group (Winter 2006) http://www.centerforfoodsafety.org/pubs/CommentsOrgAquaculture 4.05.06.pdf accessed 15 May 2007.

EU (2004). Sustainable EU fisheries: facing the environmental challenges. Conference report. European Parliament, 8 - 9 November 2004, Brussels, 96p.

FAO (2007). The state of world fisheries and aquaculture 2006. FAO, Rome, Italy, 180p.

Greenpeace (2005). A recipe for disaster. Supermarkets' insatiable appetite for food. Greenpeace, London, 96p.

Kempf, M., Merceron, M.,Cadour, G., Jeanneret, H., Méar, Y. and Miramand, P. (2002). Environmental impact of a salmonid farm on a well flushed marine site: II. Biosedimentology. Journal of Applied Ichthyology 18: 51-60.

Kutti, T., Ervik, A. and Kupka Hansen, P. (2007). Effects of organic effluents from a salmon farm on a fjord system. I. Vertical export and dispersal processes. Aquaculture 262: 367-381.

Millar, K. and Tomkins, S. (2006). The implications of the use of GM in aquaculture: issues for international development and trade. In: M. Kaiser and M. Lien (eds.) Ethics and politics of food. Wageningen, Academic Pers, p 442-445.

Muir, J. (2005). Managing to harvest? Perspectives on the potential of aquaculture. Phil Trans. R. Soc. B 360: 191-218.

Naylor, R.L. Goldberg, R.J., Primavera, J.H., Kautsky, N., Beveridge, M.C., Clay, J., Folke, C., Lubchenco, J., Mooney, H. and Troell, M. (2001). Effects of aquaculture on world fish supplies. Issues in Ecology 8, 14p.

Pelletier, P. (2003). Organic certification for farmed salmon. BC Organic Grower 6: 22-26. Pelletier, P. (2004). Going organic - a new wave in aquaculture. AAC Spec Publ No 8: 77-79.

Singh-Renton, S. (2002). Introduction to the sustainable development concept in fisheries. FAO Fisheries Report No 683. FAO, Rome, Italy, p229-234.

Soil Association (2006a). Soil Association embraces organic aquaculture. Press Release 16/08/2006 (version 1). SA, London.

Soil Association (2006b). Organic salmon - myths and realities. Information Sheet 10/11/2006 (version 1). SA, London.

Soil Association (2006c). Fish farming and organic standards. Information Sheet 25/11/06. SA, London, 4p.

Soil Association (2007). Sea Change. Information Sheet 08/02.2007 (version 1). SA, London.

Staniford, D. (2002). Sea cage fish farming: an evaluation of environmental and public health aspects. Paper presented at Aquaculture in the European Union: Present Situation and Future Prospects, 1st October 2002, 39p.

Tacon, A.G.J. and Brister, D.J. (2002). Organic aquaculture: current status and future prospects. In: N.E.H. Scialabba and C. Hattam (eds.) Organic agriculture, environment and food security. FAO, Rome, 20p.

Weber, M.L. (2003). What price farmed fish: a review of the environmental and social costs of farming carnivorous fish. SeaWeb, Rhode Island, USA, 53p.

Ziman, R.L. (2006). Agriculture's ethical horizon. Elsevier, Academic Press, 229p.

Part 10 - Sustainable disease control

Keeping backyard animals as a way of life

Nina E. Cohen[1], Elsbeth N. Stassen[1] and Frans W.A. Brom[2]
[1]Animals and Society, Animal Production Systems Group, WIAS, Wageningen University, P.O. Box 338, 6700 AH Wageningen, the Netherlands, nina.cohen@wur.nl
[2]Ethics Institute, Utrecht University, Utrecht, the Netherlands

Abstract

Many people keep backyard animals for non-commercial reasons, such as for company, or to breed a rare or fancy species. Keeping backyard animals is an emerging animal practice which is distinct from commercial farm animal keeping. This distinction became clear during recent epidemic outbreaks of animal diseases. To eradicate the diseases, many infected and healthy animals, not only from farmers, but also from non-commercial animal keepers, were culled. This stamping-out strategy was met with much resistance from backyard animal keepers, who felt that culling their animals to protect the Dutch export position, was unjust, since their animals were not commercially sold or exported. The Dutch authorities then realised that future prevention and control strategies need support from all animal practices involved, which in turn requires sufficient knowledge of the nature of the various animal practices. In this paper the practice of backyard animal keeping is described in terms of the nature of the practice and the nature of the human-animal relationship.

Keywords: animal practice, human-animal relationship, status, typology, animal disease epidemics, control policy

Some fifty years ago the Dutch rural area was dominated by agricultural businesses, and most animals that were seen grazing the fields were farm animals kept for commercial purposes by farmers. This countryside has changed now, due to a decrease of agricultural activities and a migration to the rural area of people who keep animals for other than strictly commercial reasons, such as for company, for riding, in deer parks, children's playing zoos, green care farms, shelters, and nature management.

To describe these different animal keepers and their activities, the term animal practice is used. An animal practice is a set of activities involving animals, with a certain aim, and with its own internal human-animal relationship (Waelbers *et al*., 2004, Velde *et al*., 2002). All these emerging animal practices have their specific interests, depending on the reason for keeping these animals and the nature of the human-animal relationship.

One emerging animal practice is that of keepers of backyard animals. Backyard animals are kept for non-commercial reasons and include species that are usually kept for production purposes, such as cows, sheep, goats, poultry and pigs, but, since there is no comprehensive definition of backyard animals, species such as pigeons and other birds, rabbits, bees, deer, and even camels, llamas and wallabies can be defined as backyard animals (Treep *et al*., 2004).

The practice can further be sub-divided into breeders and non-breeders. Non-breeders keep their animals in the first place for the company. Breeders breed animals for sport (e.g. pigeon racing, showing their animals at shows), for aesthetic reasons (fancy breeds), breed rare breeds, or aim to return an Old-Dutch breed to its original region.

It was not until recently, when the Netherlands was confronted with three major outbreak of animal disease epidemics, that it became clear that this animal practice has interests other than those of commercial farmers.

In line with the European non-vaccination policy, the Dutch livestock had not been vaccinated against very contagious animal diseases. This non-vaccination policy was based on economic motives: routine vaccination of livestock was more expensive than eradicating a disease epidemic by culling. Furthermore, the policy stimulated the free market trade of animal products between countries who had adopted this policy (Koninklijke Nederlandse Akademie van Wetenschappen, 2002).

During the three recent outbreaks of classical swine fever, foot and mouth disease and avian influenza, animals were culled to stop the disease from spreading and to safeguard the export of animals and animal products. This stamping-out strategy included not only infected animals, but also healthy animals. In Europe, more than 58 millions animals were culled to stop these epidemics (www.OIE.int/Handistatus II).

This control policy was met with resistance by those who had to have their animals culled, which included not only farmers, but also non-commercial animal keepers of backyard animals, rare animals, zoo animals and animals in nature reserves. These animal keepers felt that they were entitled to a status separate from commercial animal keepers, since their animals are usually not sold or exported, and are often of a rare or special breed. Furthermore, many were convinced that the contribution of backyard animals to the spread of the disease, as compared to commercially kept animals, was negligible.

The Dutch government then realised that a control strategy which was based on the massive culling of healthy animals, was no longer acceptable and would be met with even more resistance in the future. New prevention and control strategies were developed, acknowledging that strategies can only been successful when they are supported by those animal keepers who will become subject of a control policy. Therefore, more knowledge about these emerging animal practices, such as the nature of the practices, their human-animal relationship and their specific interests, is imperative.

In this paper the results are given of a study performed in 2005 and 2006 among backyard animal keepers, which consisted of 24 interviews and a survey among 214 respondents. The goal of the study was to describe this animal practice in terms of the nature of the practice and the nature of the human-animal relationship. The results may be helpful in the discussion and development of new prevention and control policies.

Nature of the animal practice

The nature of the animal practice of keepers of backyard animals is described by means of the demographic characteristics of the participants - such as gender, age, and education - animal species, and the reason to keep backyard animals. The nature of the human-animal relationship is described by the core element defining the relationship, the relative status of the animals, and typology.

The backyard animal keepers, who had participated in the study were predominantly male, middle-aged, and with a high education. Poultry was the species kept by most. Treep *et al.* (2004) in their study found that pigeons were the species kept by most. As compared to cloven-hoofed animals, poultry and pigeons need less space, and in keeping are not as expensive. This in part can account for their popularity.

Most of the people interviewed were born and brought up in the country and were familiar with backyard and companion animals from childhood, which was an important reason to keep animals as an adult. This is in line with Miura *et al.* (2002), and Endenburg *et al.* (1994, 1995), who also found that childhood experience with animals is a motivation to keep animals later in life. The reason to keep backyard animals was first of all for their company. Breeding rated second place. The usefulness of the species (grazing, animal products), and the beauty of the animals, shared a third place.

Living in the countryside to these animal keepers is a way of life, and is appreciated for the experiencing of nature, the quiet, the clean air and the space to keep animals. Keeping animals in this respect is an important element of this way of life.

The human-animal relationship

Nature of the relation

The nature of the human-animal relationship in an animal practice is the core element which forms the basis of this relationship, and can be either relational or functional. In other words, one can keep animals either for company, or to utilise them. In most practices though, one can find a combination of relational and functional elements.

The nature of the human-backyard animal relationship is predominantly relational, with functional elements. Most people keep backyard animals to enjoy their company, their beauty and their behaviour, and to care for and interact with them.

Breeding backyard animals has a functional foundation, because these animals are functional in a cultural-historical, sporting or aesthetic context. Breeding is relational as well, because successful breeding requires contact, knowledge and care of the animals and sometimes, the animals are merely bred for the pleasure of having young animals around the house. So for a breeder, keeping animals may be an end in itself, and a means to an end at the same time.

Status

A second element of the human-animal relationship is the status of backyard animals. We propose to define status as composed of intrinsic, relational, functional and situational elements.

The intrinsic properties of an animal are those that are intrinsic to its species or to the individual. Backyard animal are appreciated for their character, intelligence, behaviour, naturalness, beauty and shape.

Relational elements describe the relation between an animal and its surroundings, such as with a person, in a human community, with other animals, or the spiritual or religious role of an animal. The relational qualities of backyard animals which are valued are contact, vicinity, familiarity with the species, activities performed with the animal, care, the ability to reciprocate the attachment, and trust in the person.

The functionality of an animal is defined by its utility to people or its function in an ecosystem. The functional properties of backyard animals valued are grazing, animal products, their therapeutic use in green care farms, and their contribution to the biodiversity.

Situational elements define the status of an animal in a particular place or time. An animal may be rare in a certain region or time, but considered a pest in another area, or may be kept as a pet in one moment but may be eaten in the next, etc. Some situational properties of backyard animals valued are rareness, their cultural-historical value and their visible presence in the countryside. Furthermore, in each situation, a weighing of interests between humans and animals takes place. These four elements and the weighing of interests ultimately define the status of an animal or animal species.

In this paper, the situational focus is on animal disease epidemics. In a situation in which an epidemic may jeopardise the livestock sector, the interests of the sector may be favoured over the lives of farm animals or individual farmers, therefore, the culling of healthy farm animals in a stamping-out strategy, may be considered justified to protect the trade position of a country. During an epidemic, the status of backyard animals is higher than that of farm animals. This is due to the fact that the situational status is higher, because backyard animal are not part of the export market, are considered to hardly contribute to the spread of the disease, and are highly valued as special or rare breeds. Furthermore, the relational status of backyard animals is higher, and is closer to the status of companion animals. Therefore, from the point of view of the backyard animal keeper, in the weighing of the interests between those of the livestock sector and those of backyard animals an their keepers, the interests of the latter are favoured.

Typology

A typology describes people's basic views on the role and position of animals, with respect to humans and the natural world. The typology of De Cock Buning (2005) was used, who proposed four types, which we described as functional, relational, biocentric and holistic. From the functional viewpoint, animals are subservient to humans and are mainly valued for their utility to humans. From the relational viewpoint, the core element is the interpersonal relation between a person and an animal. In the biocentric view, the emphasis is not so much on a relation between two individuals, but considers animals and humans to be interconnected and dependent on each other in the natural order of things. The holistic viewpoint takes this one step further and holds that animals are our teachers and show us that man and animal are both elements of a greater unity.

The biocentric viewpoint scored the highest among keepers of backyard animals, followed by the relational, the functional, and the holistic view respectively. The biocentric view describes an interdependent relationship between humans, other animals and the natural world. In this respect, this animal practice is a way of life, in which keeping backyard animals is seen as a way to interact with the living nature.

Animal practices compared

When comparing the backyard animal practice to farming practices and to the companion animal practice, it shows that the backyard animal practice integrates certain elements of the hobby farming practice (Holloway, 2000, 2001) and of the companion animal practice (Endenburg *et al.*, 1994; Wilkie, 2005).

Non-breeders are motivated by the personal experiencing of the presence of animals, as in the practice of companion animals, but backyard animals have a lower status than companion animals. This can be explained by the higher relational value of companion animals, because companion animals live in the house, and are therefore included into the innermost circle of the family. This role of physical proximity is also described by Lookabaugh Triebenbacher (1999) and Shore *et al.* (2006), who find that close proximity of an animal intensifies the attachment. Treep *et al.* (2004) furthermore suggest that backyard animals are less capable of reciprocating the affection, which renders the attachment less intense. Due to their higher status, companion animals are not usually kept for a functional purpose, or killed for food. Backyard animals may be or may not be killed and eaten. In our study, 20% of the respondents consumed their own poultry products, and 12% consumed products of their cloven-hoofed animals, but this does not necessarily mean that the animals are killed. Sijtsema *et al.* (2005), in a study among 682 respondents, found that only 25% consumed their own animals.

Breeders are closer to hobby farmers than non-breeders. Hobby farmers breed and sell their animals - which may be of a rare breed - and animal products, but this farming practice does not constitute the basic income of the hobby farmer. Backyard animal breeding is motivated by other reasons as well, such as for sport and showings. In both practices, the killing of animals is acceptable. In the breeding practice, animals are selected on the basis of their gender and appearance, and may be killed after selection. The similarity between breeders, non-breeders and hobby farmers is that keeping animals is most of all *a way of life*, which involves the interaction with and experiencing of animals in a rural environment.

Conclusion

Backyard animals are mainly kept for company and breeding. The practice of backyard animal keeping is mainly relational and based on mutual attachment. The animals are not kept for commercial purposes, but are part of a way of life: to interact with animals in a rural environment. During an epidemic, the relational and situational status of backyard animals is higher than that of farm animals. Furthermore, the stamping-out policy infringes upon the way of life of these animal keepers. Therefore, the culling of healthy backyard animal for economic or veterinary reasons is considered unjust.

References

Cock Buning, Tj. de, Kupper, F., Krijgsman, L., Bout, H. and Bunders, J. (2005). Denken over de eigen waarde van dieren in Nederland [Thinking about the intrinsic value of animals in the Netherlands], Athena Instituut, Vrije Universiteit, Amsterdam.

Endenburg, N., 't Hart, H. and Bouw, J. (1994). Motives for acquiring companion animals, Journal of Economic Psychology 15: 191-206.

Endenburg, N. (1995). The attachment of people to companion animals, Anthrozoös 8:

Holloway, L. (2000). Hobby-farming in the UK: producing pleasure in the post-productivist countryside. Anglo Spanish Symposium on Rural Geography, University of Valladolid, Spain, July, 2000.

Holloway, L. (2001). Pets and protein: placing domestic livestock on hobby farms in England and Wales. Journal of Rural Studies 17: 293-307.

Koninklijke Nederlandse Akademie van Wetenschappen (2002). Bestrijding van mond- en klauwzeer. 'Stamping out' of gebruik maken van wetenschappelijk onderzoek? [Controlling foot and mouth disease. 'Stamping out' or using scientific research?] Koninklijke Nederlandse Akademie van Wetenschappen, Amsterdam.

Lookabough Triebenbacher, S. (1999). Re-evaluation of the companion animal bonding scale. Anthrozoös 12: 169-173.

Miura, A., Bradshaw, S. and Tanida, H. (2002). Childhood experiences and attitudes towards animal issues: a comparison of young adults in Japan and the UK. Animal Welfare 11: 437-448.

Shore, E.R., Riley, M. L. and Douglas, D.K. (2006). Pet owner behaviours and attachment to yard versus house dogs. Anthrozoös 19: 325-334.

Sijtsema, S., van der Kroon, S., van Wijk-Jansen, E., van Dijk, M., Tacken, G. and de Vos, B. (2005). De kloof tussen de hobbydierhouders en overheid; over passie voor dieren en perceptie van wet- en regelgeving die voor landbouwhuisdieren gelden. [The gap between keepers of backyard animals and the government. On passion for animals and perception of policy for farm animals]. The Hague, Landbouw Economisch Instituut.

Treep L., Brandwijk, T. and Olink, J. (2004). Verkenning Hobbydierhouderij, The Hague.

Velde, H. te, Aarts, N. and van Woerkum, C. (2002). Dealing with ambivalence: farmers' and consumers' perceptions of animal welfare in livestock breeding. Journal of Agricultural and Environmental Ethics 15: 203-219.

Waelbers, K., Stafleu, F. and Brom, F.W.A. (2004). Not all animals are equal. Differences in moral foundations for the Dutch veterinary policy on livestock and animals in nature reservations. Journal of Agricultural and Environmental Ethics 17: 497-514.

Wilkie, R. (2005). Sentient commodities and productive paradoxes: the ambiguous nature of human-livestock relations in Northeast Scotland. Journal of Rural Studies 21: 213-230.

Animal disease policy as a moral question with respect to risks of harm

Franck L.B. Meijboom[1, 3], Nina Cohen[2] and Frans W.A. Brom[1]
[1]*Animal Breeding and Genomics Centre, Wageningen University*
[2]*Animals and Society, Animal Production Systems Group, WIAS, Wageningen University*
[3]*Ethics institute, Utrecht University, Heidelberglaan 2, 3584 CS Utrecht, The Netherlands, F.L.B.Meijboom@uu.nl*

Abstract

European animal disease policy seems to find its justification in a 'harm to others' principle. Limiting the freedom of animal keepers - by culling their animals - is justified by the aim to prevent harm, i.e., the spreading of the disease. The picture, however, is more complicated. Governmental policy is not limited to controlling outbreaks, but includes the prevention of notifiable animal diseases as well. Consequently, the government does not merely wish to limit further harm, but also to limit a risk of harm. In this paper we argue that when the policy focus shifts from 'limiting harm' to 'limiting the risk of harm', the classical harm-principle is no longer sufficient justification for governmental intervention. A policy that aims to 'limit the risk of harm' could also entail new risks of harm. Therefore, it is not primarily a matter of limiting harm, but one of dealing with and distributing conflicting risks of harm. This requires additional value assumptions that guide the process of assessing and distributing risks of harm. The current policies are based on assumptions, which are mainly based on economic considerations. In order to show the limitations of this policy, we use the interests and position of keepers of backyard animals as an example. Based on the problems they face during and after the recent outbreaks we defend that in order to develop a sustainable animal disease policy other than economic assumptions need to be taken into account.

Keywords: animal disease, harm principle, risk, keepers of backyard animals

The justification of animal disease policy as a method to prevent harm

As in many parts of Europe, The Netherlands were recently confronted with several outbreaks of notifiable animal diseases: classical swine fever in 1997-98, foot and mouth disease in 2001 and avian influenza (bird flu) in 2003. The disease control policy of the Netherlands is in line with the stringent EU regulations for the control of notifiable animal diseases: a non-vaccination policy, and in case of an outbreak, a stamping-out strategy. Animals that are either infected, possibly infected or are a potential carrier of the disease, are culled in order to prevent further spread of the disease. By eradicating the disease the interests of individual animal keepers are harmed. However, the culling of the animals and the harm for the keepers of the animals are justified from the benefits for the animal production sector, especially the export position on the global market.

It seems that the current policy is based on a 'harm to others' principle. The classical harm principle, as was introduced by Mill (1859/1979) and further developed by many other authors (cf. Hart, 1961; Feinberg, 1984; 1994) roughly states that governmental intervention is justified when it is aimed to prevent harm to others. Harming is in this context defined as a wrongful setback of interests (Feinberg, 1994: 4, 34). This seems applicable in the case of government intervention during an outbreak of an animal disease. The government intervenes in the freedom of choice of individual actors in order to limit economic and veterinary harm. Animals are culled, transportation is restricted, resulting in welfare problems for the animals, and sometimes even the freedom of movement of people is restricted. Nonetheless, these measures are considered justified in that they aim to prevent society from further harm, i.e., the spreading of the disease and the subsequent consequences for food production, transport and trade.

Governments however do not only intervene during an outbreak, but also formulate strategies for the prevention of notifiable animal diseases. In this paper we argue that in the latter situation the policy focus shifts from 'limiting harm' to 'limiting the risk of harm'. Consequently, the classical harm-principle is no longer sufficient justification for governmental intervention. A policy that aims to 'limit the risk of harm' should acknowledge and anticipate conflicting risks of harm and therefore should aim at a just distribution of risk of harm. This implies that in distributing the risks, the values at stake for those who run the risks should be taken into account. This leads to a critical assessment of current policy, which is mainly based on economic considerations and does not include the values and interests of non-commercial animal keepers - such as keepers of backyard animals[12]. We defend that in order to develop a sustainable animal disease policy, other than economic assumptions need to be taken into account.

Policy as a method to deal with risks of harm

In the case of an outbreak the situation is relatively clear: in order to prevent the spreading of the disease serious interventions in the freedom of the keepers of animals is justified. However, animal disease policy is not just the script for action during an outbreak, but also aims at preventing that a disease will occur. To achieve this aim interventions that infringe on the freedom of individual and collective agents are necessary. For instance, the introduction of specific housing conditions or new criteria for animal feed that contribute to the reduction of the risk of an outbreak entail an infringement of the freedom of the individual animal keeper. However, the justification with harm prevention is less clear. The reason for the intervention is not the fact that these animal keepers directly harm others, but since their actions may impose a risk of harm on others. The BSE feed restrictions that were designed to reduce the spread of BSE are a good example. The so-called ruminant-to-ruminant feed ban implies the prohibition of the feeding of ruminant animals, e.g. cattle, sheep, and goats, with animal proteins of mammalian origin. The introduction of this prohibition entails an intervention in the freedom of several actors in the animal food sector. Nevertheless, it is considered to be an issue that has to be subject of governmental law and policy. Not because these actors directly harm others when they feed their animals with animal proteins of mammalian origin, however since it involves a risk of harm. This illustrates that the policy aim is not restricted to limiting harm, but is extended to limiting a *risk* of harm.

This complicates justification. The government is confronted with a web of different, overlapping, but also conflicting risks of harm. Even though all stakeholders agree on the importance of the prevention of an outbreak of notifiable animal diseases, the question is at what cost this aim should be achieved. The different positions of commercial actors in the livestock sector and of keepers of backyard animals can highlight this point. Both groups sincerely aim to prevent an outbreak of an animal disease. However, they differ with respect to the view on what risk of harm are acceptable. For instance, commercial keepers may consider the presence of actors that keep animals for reasons of hobby as a risk factor, since they are less organised and operate in a less standardised way and thus the traceability of possibly infected

---

[12] Keepers of backyard animals are a diverse group. Due to their diverse nature, no satisfactory definition has yet been formulated. It is agreed that backyard animals are not kept for commercial purposes and usually include poultry, cloven-hoofed (cattle, sheep, goats, pigs), horses and ponies, but also species such as lama's, wallabies, racing pigeons and exotic birds. In contrast to pet animals, these animals are not kept in the house. Keepers of backyard animals are a substantial group in The Netherlands. An estimated 400.000 citizens keep a few million animals and spend at least a few billion euros to their hobby each year (Treep *et a.,l* 2004, Den Boer *et a.,l* 2004, Sijtsema *et al.,* 2005). In a quantitative study performed by Cohen *et al.,* (in press 2007) the main reasons for keeping animals were human-animal contact and breeding of fancy or rare breeds.

animals is low. Consequently, this way of animal keeping may jeopardise the business security of the former. On the other hand, keepers of backyard animals may consider a risk of harm of the interests of the commercial sector as not sufficient justification for a restriction in their freedom of lifestyle. They even may claim that these commercial keepers and their focus on trade and export are a risk to their idea of the good life and their way of caring for their animals. Furthermore, keepers of backyard animals are convinced that their animals are not the cause of disease outbreaks, and do not contribute to a further spread of the disease.

Such conflicting positions cannot simply be addressed by the harm principle only. The issue is not that one partner is harmed by the other if the government does not intervene, but that that the partners anticipate running a risk to be harmed by each other. The harm principle, however, does not provide much guidance in this process, since it only shows that doing harm is morally not permissible and should be prevented, yet it does not provide the tool to compare conflicting risk of harm. To cope with this situation the government need additional, value-laden assumptions and decisions in order to justify the intervention that result in limiting one risk of harm at the costs of leaving another unchanged.

This illustrates that the debate on animal disease prevention policy should focus on the value assumptions that underlie the evaluation of the risks of harm, rather than the harm principle itself. The criticism on governmental intervention is not so much focused on the (risk of) harm done, but more on people's feelings of being wronged by this intervention, since they do not agree on the underlying moral assumptions and evaluations of the policy. One can be compensated for the harm done by means of financial compensation. However, when one believes to be wronged, one is not so easily compensated through financial means. This is the position in which many keepers of backyard animals find themselves in. They believe that the then current EU animal disease control policy was not justified in causing such an infringement on their personal liberty and on their moral duty and responsibility towards their animals.

Distributing risks: a value discussion

We have argued that a policy that aims to 'limit the risk of harm', can at the same time be the cause of risk. Therefore, a policy needs to incorporate a fair weighing and distributing of conflicting risks of harm. From this argument it follows that governmental policy requires additional value assumptions in addition to the harm principle. However, it is economic and veterinary considerations that play a crucial role in the current value assumptions.

Since animal husbandry in Europe has been developed to ensure 'food security', i.e., to secure the production of safe food for all, policy, market structures and technologies aim to increase food production and facilitate free trade. This is the focus of animal disease policies as well. There is a clear tendency to assess animal diseases and its consequences in economic terms. For instance, the decision of the European Union to adopt a non-vaccination policy was informed by veterinarian arguments, but mainly justified on economic grounds. It was calculated that the economic costs of preventive vaccination were higher than the costs involved with controlling an epidemic. Furthermore, a non-vaccination policy stimulates free market trade of animal products between countries, who have adopted this policy (cf. KNAW, 2002). This economy-based approach, however, has become subject to serious criticism. Economy-based animal disease policy is criticised for its too limited view on its responsibility, while neglecting other interests that are considered valuable in society. An animal disease policy, which is strictly focused on economic trade, evaluates the value of goods merely in terms of its economic benefits. Empirical research, however, has shown that arguments based on economic benefit and export are not longer considered sufficient to justify the culling of healthy animals or to cause very severe animal welfare problems (Treep *et al.*, 2004, 55ff; Noordhuizen-Stassen *et al.,* 2003: 43-44). Furthermore, economic motives are no longer justified to restrict individual liberty.

Furthermore, the criticism of keepers of backyard animals highlights another problematic aspect. The clear emphasis on trade and export entails that stakeholder interests are only taken into account when they can be assessed within an economic framework. Consequently, the interests of stakeholders that do not easily fit within this framework are often left out of the equation. For instance, the role of non-commercial animal keepers is mainly perceived from a veterinary and economic perspective, i.e. the contribution of backyard animals to the spread of the disease, thus jeopardising the trade position of the livestock sector. Since these animal keepers play only a very modest role in this trade, their interests and claims are often subservient to those of the livestock sector that trade and export animals and animal products. This is problematic for two reasons.

First, it neglects inherent worth of hobby as a way of life. Keepers of backyard animals consider their hobby as a predominantly private affair (Sijtsema, 2005), with its own internal values. The reason to keep animals is part of their idea of the good life. It is not just a hobby; it is part of their lifestyle and an essential element of who they are and what they consider worthwhile in life. Therefore, policy measures that entail the risk of the culling of animals do not only jeopardise the lives of the animals themselves, or the interests of a certain animal practice, but are considered a serious infringement of one's way of life. This lifestyle element, however, is difficult to translate into economic sound values or in terms of free trade. Consequently, an economy-based approach to conflicting 'risk of harm'-based claims, cannot deal with the importance of this value. This is problematic since common sense opinion in society now recognises these individual values as highly important and worth respecting. Hence, the use of economic arguments to overrule the moral claims of these animal keepers is no longer acceptable.

Second, keepers of backyard animals believe that in keeping these animals, they fulfil public interests that need to be protected and cannot be so easily discarded by economic arguments. For instance, the breeding of rare breeds or endangered species is considered a public good in itself for its contribution to the biodiversity. Furthermore, keeping non-commercial livestock to be seen grazing the fields, represents a type of rural life that has almost disappeared as a result of processes such as the intensification in agriculture, and of urbanisation and industrialisation. The visible presence of these animals in the countryside is now highly valued for nostalgic reasons, because it is a reminder of former days when farm animals were still seen outside. All these goods, even though they may not contribute to food production or the export, have public value.

The need to go beyond economy-based arguments

The emphasis on economic considerations in dealing with conflicting risks of harm may be effective, but it disregards stakeholders as end in themselves and denies that private and public goods can have inherent value other than an instrumental value to the agro-food market.

The current approach tends to treat stakeholders as mere means to economic ends, instead of as ends in themselves. It neglects keepers of backyard animals as deserving a voice within public policy even when they have no substantial role in the export and trade market. The conflicting claims on risk of harm cannot be settled by translating all claims into economic or even monetary terms, when not all stakeholders agree on the importance and value of this overall claim.

Moreover, weighing and coordinating conflicting risks of harm merely within an economy-based frame, only recognises the importance of individual lifestyle and the public goods related to keeping backyard animals, in as far as they are economically beneficial. However, for keepers of backyard animals this lifestyle aspect with its public goods has an inherent value independent of any economic aspect. This inherent value, however, is often disregarded on economic arguments. Consequently, the infringement of their lifestyle is justified on grounds they do not agree with. This results in a situation in which they believe to be harmed and wronged by governmental policy. Moreover, they consider the culling of perfectly healthy animals to be senseless, and feel they have betrayed their animals, because they could not fulfil their moral duty to protect and not to harm their animals (Cohen, 2007, forthcoming).

This illustrates the need to include other than strict economic arguments, by recognising that other public goods are of equally high value. For instance, the inherent value of the hobby (way of life) of keepers of backyard animals and its public value need to be taken seriously.

Acknowledgement

This article is part of the research project entitled 'New foundations for prevention and control of notifiable animal diseases' that is funded by The Netherlands Organisation for Scientific Research (NWO).

References

Boer, M. den, Cante, L., Dekker, A., Dyvesteyn, B., Geveke, H., Jansen, R., Kort, M. and van der Mark, R. (2004). De crisis tussen mens en dier, evaluatie bestrijding AI crisis. Utrecht, The Netherlands.

Cohen, N.E., van Asseldonk, M.A.P.M. and Noordhuizen-Stassen, E.N. (2007). Social-ethical issues concerning the control strategy of animal diseases in the European Union: an inventory, Agriculture and Human Values, in press.

Feinberg, J. (1984). The Moral Limits of the Criminal Law, Volume 1: Harm to Others, Oxford UP, Oxford, UK.

Feinberg, J. (1994). Freedom and Fulfilment, Princeton University Press, Princeton, USA.

Hart, H.L.A. (1961). The concept of Law, Clarendon Oxford UP, Oxford, UK.

KNAW/ Royal Academy of Sciences (2002). Bestrijding van mond- en klauwzeer,'stamping out' of gebruik maken van wetenschappelijk onderzoek. Koninklijke Nederlandse Akademie van Wetenschappen, Amsterdam, The Netherlands.

Mill, J.S. (1859/1979). On liberty. Edited with an introduction by Gertrude Himmelfarb. Harmondsworth, Penguin Books, London, UK.

Noordhuizen-Stassen, E.N., Rutgers, L.J.E. and Swabe, J.M. (2003). Het doden van gehouden dieren, ja mits...of nee tenzij. Utrecht University, Utrecht, The Netherlands.

Sijtsema, S., van der Kroon, S., van Wijk-Jansen, E., van Dijk, M., Tacken, G. and de Vos, B. (2005). De kloof tussen de hobbydierhouders en overheid; over passie voor dieren en perceptie van wet- en regelgeving die voor landbouwhuisdieren gelden. LEI, The Hague, The Netherlands.

Treep, L., Brandwijk, T., Olink, J., Tillie, F., Veer, M. and Verhoek, A. (2004). Verkenning hobbydierhouderij. Expertisecentrum LNV, nr. 255, EC-LNV, The Hague, The Netherlands.

Avian influenza and media coverage: a qualitative study in four European countries

Asterios Tsioumanis
Aristotle University of Thessaloniki, 127 Wordsworth Road, LE2 6ED, Leicester, UK

Abstract

During the last decades, the food industry has been dealing with a variety of safety issues: beef hormone scares, salmonella scandals, *E. coli* outbreaks, BSE, and avian flu all took place during the last 20 years. Research was subsequently undertaken, not only to limit these incidents but also towards the channels that convey information to consumers and build consumer trust. While relevant literature identifies numerous channels of information towards the general public, the media is undoubtedly also an important agent. The exact extent and intensity of the capacity of the media in forming attitudes is open to debate, however, it is generally accepted that dramatic news sequences can have stronger effects on risk perception. In order to understand how the available information is likely to impact on public attitudes, content should also be examined.

This paper focuses on avian influenza and newspaper coverage and analyses newspaper coverage of bird flu in four different European countries (UK, France, Spain and Greece). Two newspapers from each country were selected and all written material relating to bird flu, over a time-span of five months, was analysed. Comparisons of newspaper reporting in the four selected countries prove useful in examining potential differences in their focus and interpretation of the risks.

The media has been seen as the 'distorting lens' between scientists and the public and has been blamed for utilising rhetoric of fear, creating uncertainty and thriving in blame attribution. Addressing the written material on bird flu of the eight newspapers in question and comparing these to public quotes by experts on the field, interesting conclusions are drawn and the 'distorting lens' argument collapses.

Keywords: bird flu, risk information, media coverage, food safety

Avian influenza: myths and realities

Avian influenza, which is also called avian flu or bird flu, refers to flu from viruses adapted to birds, but is sometimes mistakenly used to refer to both other flu subsets (such as H5N1 flu) or the viruses that cause them (such as H5N1). Citations of avian flu usually refer to one specific type, called avian influenza A, which has caused infections in both birds and humans. Even then, it has to be noted that there are many subtypes of type A influenza viruses, which vary according to changes in hemagglutinin [HA] and neuraminidase [NA] proteins, found on the surface of the virus. There are 16 known HA subtypes and 9 NA subtypes but many different combinations are possible (CDC, 2006).

Domestic poultry (chicken, turkeys, ducks, geese) are susceptible to the virus in two forms, which in lay terms may be distinguished as 'low' and 'highly' pathogenic. In the first case, symptoms are mild including ruffled feathers or a decrease in egg production. The highly pathogenic form spreads rapidly, affects multiple internal organs and has a mortality rate that can reach 90-100% often within 48 hours (CDC, 2006).

There are only three known A subtypes of influenza viruses (H1N1, H1N2, and H3N2) currently circulating among humans. Although the risk from avian influenza is generally low, cases of infection have been reported since 1997 when a three-year-old boy died in Hong Kong. Until April 2007, multiple cases have been reported mainly in Asia (18 in Hong Kong, 81 in Indonesia, 24 in China, 25 in Thailand). However confirmed infections of humans have also been reported in Nigeria, Turkey, and Egypt. In Western Europe the virus has appeared only in birds in Spain, Germany, UK, Denmark, Sweden, Austria, Switzerland, France and Greece (WHO, 2007).

Egypt, Indonesia and Nigeria have not been able to contain the disease yet, effectively making them reservoirs of the virus for possible introduction to other countries. The spread of avian influenza viruses among humans has been reported very rarely, and has been limited and unsustained. The most serious threat is that each human case contracted from poultry offers a new possibility for the virus to mutate into a form that can spread rapidly from person to person (FAO, 2007).

Laboratory research indicates that some of the prescription medicines approved for human influenza viruses should work in treating avian influenza infection in humans. However, influenza viruses can become resistant to these drugs. It is difficult to provide accurate estimations on the number of sick people, but H5N1 has killed at least 171 people worldwide; 66 in Indonesia, the highest human death toll of any country (FAO, 2007).

Risk communication and media coverage

During the last decades, the food industry has been dealing with a variety of safety issues: beef hormone scares, salmonella scandals, E. coli outbreaks, BSE, avian flu of which all took place during the last 20 years. Research was subsequently undertaken, not only on limiting these incidents but also towards the channels that convey information to consumers and build consumer trust. Risk communication can be approached as the process of providing the public with information that serves to reduce anxiety and fear as well as provide suggestions for planning, which will assist the public in responding appropriately to some crisis situation.

While relevant literature identifies numerous channels of information towards the general public, the media is undoubtedly also an important agent. The quality press has been considered as one of the most trusted sources of food-related risk information (Frewer and Shepherd, 1994).

The media has been seen in the past as 'the distorting lens obscuring communication between scientists and the public' (Hargreaves, 2001). Although this view is often regarded as over-simplistic and simultaneously dismissive towards questions that are unable to answer, the recent debate over genetically modified organisms has refuelled the same argument. Public distrust over modern biotechnology, especially its applications in agriculture has been attributed, by defenders of genetically modified crops, to mere ignorance. The 'distorting lens' view of the media ignores work by social scientists on the complex meanings of media texts and their social, historical, political and economic context (Hargreaves, 2001).

Anderson (2000) notes that the relationship between the media, the food industry and the consumer is probably at its lowest point, underlining that the gulf between the food industry, food safety experts and the public has grown. He also argues that this rift has been fuelled by the media.

The question that arises, however, is whether the media can be blamed for utilising rhetoric of fear, creating uncertainty and thriving in blame attribution. Studying newspaper coverage on avian influenza in a multi-cultural context on the one hand and scientific discourse on the same issue on the other, may provide useful insight.

Method

Analysing newspaper articles is no simple task. Frewer and Shepherd (1994) demonstrated that the type of newspaper is of crucial importance, as the public trust that surrounds them may vary. The quality press has been shown to be one of the most trusted sources of risk information in the UK (Frewer *et al.*, 1993). Tabloids are often regarded as prone to risk amplification and their headlines are often exacerbating in an effort to engage their reader sentimentally.

A total of eight newspapers have been sampled, two in each country under study. In the UK, The Times and The Guardian were selected; in Spain El Pais and El Mundo; in France Le Monde and La Liberation and finally in Greece Eleftherotipia and Kathimerini. The choice was based on popularity with a prerequisite being that their circulation is at national level, and there was also an effort to include different political orientations.

These newspapers were sampled over a period of 5 months, between October 1st 2005 and February 28th 2006. The chosen period studies the arrival of the virus in Europe, although before that there were many reported incidents in birds and humans in Asia. October 2005 was a benchmark for Europe as far as the influenza virus is concerned. In that specific month, Turkey, Croatia and Romania announced their first confirmed cases of infected poultry, and consequently the attention of European newspapers towards avian flu increased considerably, providing the data for this study.

The actual number of articles each newspaper published as well as their chronological appearance allows an interesting comparison in a cross-cultural context. In addition, headlines were drawn into the analysis in terms of content. 'The nature of an article's headline is liable to have a disproportionate influence on reader perceptions compared to the rest of the article' (Rowe *et al.*, 2000). The nature of each headline was categorized either as alarming or reassuring in a way that linguistically is approached as pragmatic meaning, namely the meaning or force of utterances or speech acts used in a situation. 'From a pragmatic point of view one can distinguish between three factors in verbal communication: locution (what utterances say), illocution (what utterances do - perform), and perlocution (what utterances achieve)' (Nerlich and Halliday, 2007*)*. Based on Austin's work, the study of the way words are used in order to elucidate meaning is central to this paper. Illocutionary as well as perlocutionary force markers (e.g. warn, alert, alarm, scare, threaten) were used to make the distinctions, as the force of speech acts can be conveyed directly through them.

Finally, locations in the headlines have been studied separately. As avian influenza was approaching Western Europe it was interesting to see whether the focus of the newspaper reports were at the national level or from an international context, as the border-transcending nature of the virus would probably suggest. In the same context, special attention has been drawn upon references on vaccines or other pharmaceuticals, such as antiviral medicines and articles concerning the market, poultry business and other related sectors.

Results and discussion

The number of articles as well as their chronological sequence are portrayed in Figure 1. Four small charts, one for each country in question, reveals the number of articles for each of the five months under study.

The pattern is evident. During the five months under study, there is an obvious decline in the number of articles towards December followed by a subsequent increase. The question that arises is whether this pattern can be explained by the actual number of avian influenza cases in the respective months.

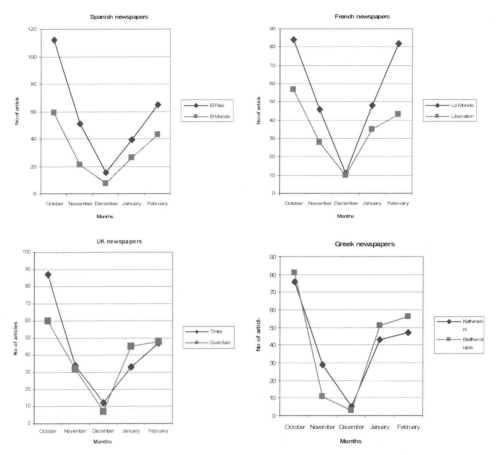

Figure 1. Number of articles on avian flu in eight European newspapers over a period of five months.

One would expect that the number of articles would be proportionate to the actual number of cases or their importance.

While December 2005 did not constitute a 'hot' month as far as the spread of the virus is concerned, such a decrease is rather disproportionate to the actual evidence. In early December, Ukraine reported its first outbreak of avian influenza in domestic birds in the Crimea, and continued to issue similar reports thereafter in the following months. Towards the end of the same month, Turkey reported a new outbreak in poultry in the eastern province of Igdir. Control measures in that case included the culling of poultry in Iraqi Kurdistan and all backyard poultry in Iran within ten kilometers of the Turkish border.

A couple of telephone interviews with senior journalists of the Greek newspapers under study did not provide any insight on this issue. The suggestion that this trend is interrelated with the festive Christmas atmosphere can not be proven but may not be disregarded either. It is a fact however that all the newspapers under study reported on the problems that the poultry business faced after the outbreak of the virus as consumers avoided chicken and related products. One must also not forget that the traditional Christmas dinner includes turkey or chicken.

In Table 1 the headlines have been categorized as alarming or reassuring using the trigger words already mentioned. It also provides an insight into the use of geographical locations, in order to approach whether the focus has been on the national level, as well as it providing information on the number of articles referring to the market for vaccines and other relevant antiviral medicines.

As seen in Table 1, a lager percentage of the newspaper articles' headlines included mainly foreign locations in their headlines, which was expected as it portrays the spread of the virus. Greece had the greatest number of references on market problems, mainly associated with the poultry business. A substantial percentage of articles on vaccines and antivirals were related to Tamiflu. It is also obvious that the percentage of alarming headlines is relatively low. That does not mean that such type of articles did not exist as The Times' 'Bird flu could kill 750,000' or Le Monde's 'Scenarios catastrophe pour une grippe fatale' (catastrophic scenarios for a fatal flu) headlines would clearly suggest. However, the argument that the media are to be blamed for creating rhetoric of fear can not be sustained based on this evidence. It can be argued, on the other hand, that if the analysis had included tabloid newspapers, which are much more prone to risk amplification, results could be qualitatively different. As an example one might use The Daily Mail's headline stating 'Bird flu will hit Britain and kill 50,000'. Similar headlines may be found in the tabloid papers from the other countries in question.

However, even the tabloids' headlines failed to compete with quotes from the so called experts on avian influenza. David Nabarro, one of the leading WHO public health experts notes that 'the range of (human) deaths could be anything between five and 150 million' (BBC, 2005). Robert Webster, a world-known virologist goes one step further pointing out that 'society just can't accept the idea that 50% of the population could die. And I think we have to face that possibility' (ABC News, 2006). The Secretary General of the European Public Health Alliance, which represents 115 European health groups says: 'Millions and millions would die, and a pandemic would change society as we know it. And no-one seems prepared' (The Sunday Herald, 2005). Dr Julie Gerberding, head of the Centre for Disease Control and Prevention in the US identifies avian influenza as 'the single biggest threat to the world right now.' She is also quoted addressing avian influenza as 'a very ominous threat to human beings'; a quote that was reported in the British medical journal The Lancet in its editorial on March 5th 2005 under the title 'Avian influenza: perfect storm now gathering?'.

Table 1. Features on the headlines reporting on avian influenza in eight European newspapers (% of the total number of articles on avian flu).

	Monde	Liberation	Times	Guardian	Pais	Mundo	Kathim.	Elefther.
Headlines								
Alarming	21	17	18	13	17	14	19	17
Reassuring	6	4	3	4	4	5	3	4
Both	2	1	2	2	2	2	2	2
Neither	71	78	77	81	77	79	76	77
Location								
Domestic	12	9	7	7	8	7	9	8
Foreign	30	28	14	30	20	43	21	25
Market	7	4	1	3	6	3	8	7
Vaccines/antivirals	14	12	14	13	10	7	7	7

Michael Osterholm, Director at the Centre for Infectious Disease Research and Policy at the University of Minnesota sounds alarming enough when stating that 'an influenza pandemic of even moderate impact will result in the biggest single human disaster ever - far greater than AIDS, 9/11, all wars in the 20th century and the recent tsunami combined. It has the potential to redirect world history as the Black Death redirected European history in the 14th century' (The Gazette, 2005). Dr Lvov, director at the Institute of virology at the Russian Academy of Medical Sciences is not doing any favours either: 'Up to one billion people could die around the whole world in six months.... We are half a step away from a worldwide pandemic catastrophe' (MosNews, 2004).

Shigeru Omi, WHO Western Pacific Regional Director, said the world is in 'the gravest possible danger of a pandemic'. Nerlich and Halliday examining avian flu in the UK mass media recollect this quote six times in the two-month corpus (February -March 2005) they studied (Nerlich and Halliday, 2007).

On February 5, 2005 an editorial for *New Scientist* warned readers that a bird flu outbreak, in which the virus would be transmitted not from poultry to people but between people, could kill 1.5 billion and that science and society were not prepared (Editorial/*New Scientist* 2005). Professor Hugh Pennington, retired head of the Institute of Animal Health, fears that 'this is the biggest threat to the human race. It far outweighs bioterrorism, this is natural bio-terrorism. It won't spare anybody' (The Express, 2005). The list is not exhaustive but the above are simply illustrative.

Politicians have been much more reassuring regarding the spread of avian influenza although Liam Donaldson caused unrest when he announced several times that 'at least 50,000 people might die if a flu pandemic struck the UK' (Nerlich and Halliday, 2007). The Spanish President, Jose Luis Rodriguez Zapatero states that avian influenza requires prevention and prudence reassuring that adequate measures have been taken (El Pais, 2005). The mayor of Athens Nikitas Kaklamanis notes that there is neither space for panic not for complacency (Kathimerini, 2005). French President Jacques Chirac sounded very reassuring also (Le Monde, 2006), while Dominic de Villepin moved one step further eating chicken in public in order to reduce anxiety (Liberation, 2005).

Concluding remarks

The case of the 'distorting lens' may be a convenient way for scientists to transfer part of the blame towards the media, but is not necessarily reflecting, in all cases, reality. As the analysis in this paper shows, uncertainty and fear has also been created via government officials and scientific researchers. The press may amplify the message under certain circumstances and intensify an emergent rhetoric of fear. However, in the case of avian influenza the conclusion reached by Nerlich and Halliday, which states that it is not possible to blame the media entirely for hyping up fears about a global influenza pandemic (Nerlich and Halliday, 2007) is valid taking into consideration the press coverage in the four European countries under study.

References

Anderson, W.A. (2000). The future relationship between the media, the food industry and the consumer. British Medical Bulletin 56: 254-268.

CDC (Centers for Disease Control and Prevention) (2006). Key facts about avian influenza and avian influenza virus, June 2006.

FAO (2007). Fewer bird flu outbreaks this year. FAO report, 2007.

Frewer, L.J., Raats, M.M. and Shepherd, R. (1993). Modelling the media: the transmission of risk information in the British quality press. Journal of Mathematics Applied in Business and Industry 5: 235-247.

Frewer, L.J and Shepherd, R. (1994). Attributing information to different sources: effects on the perceived qualities of the information, on the perceived relevance of the information and effects on attitude formation. Public Understanding of Science 3: 385-401.

Hargreaves, I. (2000). Science, Society and the Media. Presented at the British Academy, Sept. 12th, 2000.

Nerlich, B. and Halliday, C. (2007). Avian flu: the creation of expectations in the interplay between science and the media. Sociology of Health and Illness 29: 46-65.

Rowe, G., Frewer, L. and Sjoberg, L. (2000). Newspaper reporting of hazards in the UK and Sweden. Public understanding of science 9: 59-78.

WHO (2007). H5N1 avian influenza: timeline of major events. May 2007.

Websites consulted

http://abcnews.go.com/WNT/AvianFlu/story?id=1724801, ABC, 14 March 2006 (consulted 4 May 2007).

http://news.bbc.co.uk/1/hi/world/asia-pacific/4292426.stm, BBC, 30 September 2005 (consulted 9 May 2007).

http://findarticles.com/p/articles/mi_qn4156/is_20050828/ai_n14909675/pg_3, The Sunday Herald, 28 Aug. 2005 (consulted 2 May 2007).

http://www.mosnews.com/news/2004/10/28/pandemic.shtml, MosNews, 28 Oct 2004 (consulted 17 May 2007).

http://www.articleco.com/Article/The-Bird-Flu-Boost-Your-Immune-to-Increase-Your-Chance-of-Survival/9907, The Gazette (Montreal), 9 March 2005 (consulted 15 May 2007).

Using participatory methods to explore the social and ethical issues raised by bioscience research programmes: the case of animal genomics research

K.M. Millar[1], C. Gamborg[2] and P. Sandøe[2]
[1]*Centre for Applied Bioethics, School of Biosciences, University of Nottingham, Sutton Bonington Campus, LE12 5RD, United Kingdom, Kate.Millar@nottingham.ac.uk*
[2]*Danish Centre for Bioethics and Risk Assessment, Faculty of Life Sciences, University of Copenhagen, Denmark*

Keywords: animal genomics, Ethical Matrix, participatory methods, ethical reflection

Abstract

One of the possible strategies for including sustainability objectives in animal production systems of the future is to embed these objectives in the planning phase of research programmes through a process of ethical reflection. This paper presents an attempt, within a European Network (project) on genomics of host-pathogen interactions, to engage involved scientists in reflections about ethical issues they should address. The challenge is on the one hand to clarify key ethical values at stake and ensure priorities are set, whilst also raising awareness and discussion capacity among scientists. A workshop was conducted with a sub-group of members of the EADGENE Network. By applying an adapted version of the Ethical Matrix (EM), Network members considered whether current research trajectories and technology innovations might infringe upon or respect broadly defined ethical principles for a number of interest groups. Participants mapped out what they saw to be the key ethical challenges that the Network would have to respond to as a research group. These issues related to the research field and more broadly to animal production systems.

There are a number of methods that can be applied to facilitate ethical review, with each method having strengths and weaknesses. However, it appears that a modified EM method may be helpful when structuring ethical reflection discourses. Overall, participants were positive about the use of an ethical framework to structure wider consideration of these issues, as a starting point for further ethical reflection. Applying a modified Matrix Method: (1) allowed participants to structure the ethical discourse; (2) facilitated discussion of a broad range of issues, including IPR and welfare implications; and (3) produced informative outcomes.

Introduction

Scientists, particularly those working in genomics-related research areas, are being increasingly encouraged and challenged to include some form of ethical engagement in their research programmes. Moreover, this type of reflection is no longer perceived as an optional extra, funding programmes as well as research institutions require explicit ethical analysis (e.g. EC Seventh Framework Research Programme) either as part of the application stage (project proposal submission) or as part of the implementation phase (research implementation and execution).

For researchers who work in the agri-food sector, alongside the more established societal discussion of risk and responsibility, this ethical engagement includes a reflection on and discussion of 'sustainability principles' and how these can be taken into account within research and technology development (RTD) programmes. A key strategy for including sustainability objectives in animal production systems, is to embed both objectives and ethical reflection in the early research planning and implementation phases.

As a result of this explicit and perceived need for ethical review, there is growing interest in methods that can facilitate ethical reflection and a wider consideration of sustainability objectives. There are a number of emerging methods that appear to have the potential to facilitate this form of ethical 'upstream' engagement / reflection. This paper presents an attempt, within a European project on genomics of host-pathogen interactions, to engage the involved scientists in reflections about ethical issues that they should consider and potentially address.

The details of a workshop that was conducted with partners from the European Animal Disease Genomics Network of Excellence for Animal Health and Food Safety (EADGENE) are presented here as a case study highlighting a participatory approach for structuring this form of ethical reflection. An adapted version of the Ethical Matrix (EM) was used to structure the discussion and process of engaging in ethical reflection. The outcomes of the workshop conducted with a sub-group of Network members and the outcomes of the process and the perceived contribution of an adapted version of the adapted Matrix are discussed. More specifically the paper will explore the potential for this form of ethical reflection within bioscience research areas (such as animal agriculture, environmental management) to facilitate a wider consideration of a broad range of ethical issues, including the inclusion of sustainability objectives. Animal genomics and breeding research is examined here as a case study.

The role of ethical reflection in research planning and implementation

Before examining details of the workshop and discussing outcomes, it is valuable to explore the role of ethical reflection strategies in research planning and implementation and in particular the role these approaches play in bioscience research programmes, using animal genomics and breeding research as a case study. Ethical review strategies and participatory technology assessment are increasingly becoming embedded methods within research and technology development (RTD) programmes. As well as the pre-funding evaluations seen at both a national and EU level (such as the EU Framework ethical review (http://cordis.europa.eu/fp7/ethics_en.html)), participatory technology assessment [pTA] strategies are observed either as complementary Ethical, Legal and Social Aspects (ELSA) programmes (as seen in Genome Canada Programme (www.genomecanada.ca/xresearchers/researchPrograms/projects/index.asp?o=d&d=6&l=e)) or as components of research projects and research training networks, such as specific workpackages with EU funded projects (For example, 6th Framework Research Projects such as EADGENE and CLONET (www.eadgene.info/ and www.abc.hu/dinnyes/clonetenglish.htm)).

There is currently a significant drive to stimulate wider consideration and reflection on research trajectories and how these impact on a number of stakeholders, both immediately and in the long-term. Ethical reflection can help researchers challenge assumptions and clarify their goals, as well as informing interested parties of the potential outcomes (positive and negative), increase trust through transparency and help to initiate dialogue (by reporting on and extending this ethical reflection process). Embedding ethical reflection during the early research planning phase may also encourage a wider reflection on sustainability objectives in agri-food production systems. For the case considered here, animal genomics and breeding, the concepts of sustainability relating to farm animal breeding have been previously discussed and defined (Gamborg and Sandøe, 2005; Liianamo and Neeteson-van-Nieuwenhoven, 2002).

Alongside more established discussion of risk assessment and responsibility, ethical review of a research programme can take several forms, such as: (1) *ethical regulation* - e.g. formally conducted by an external (ethical) committee review resulting in judgement; (2) *ethical engagement* - e.g. involving a discourse with a wider range of external stakeholders that may be directly affected by the RTD process; (3) *ethical analysis* - e.g. conducted externally by an individual, committee or consultative group; (4) *ethical reflection* - e.g. facilitated through internal ethical analysis and value-based discourse.

Although some commentators have contested the extent to which ethical regulation and some forms of engagement should be included in the RTD process, there appears to be general acknowledgment of the utility of some forms of ethical analysis and reflection. However, exploring the important question of the extent and usefulness of different forms of ethical review is not within the remit of this paper. This paper will explore the inclusion of ethical analysis and reflection, and not ethical regulation. Therefore, it is assumed here, from positions of both substantive and procedural reasoning, that encouraging ethical analysis and reflection is an important component of research planning and implementation. One of the questions that this paper will therefore address is: how should ethical reflection be initiated and managed within a research programme?

Participatory approaches that facilitate the inclusion of ethical reflection may be deemed as particularly important for animal genomics and breeding research programmes. These programmes once initiated may set in motion unsustainable production paradigms that could impact on the industry and other affected groups for years to come. The challenge for any ethical reflection (participatory) approach is on the one hand to identify and clarify key ethical values at stake and ensure priorities are set, whilst also raising awareness and discussion capacity among scientists.

Ethical Matrix (EM) as a participatory tool: facilitating ethical reflection

A number of participatory methods have been developed to facilitate stakeholder engagement, ethical reflection and decision-support for the agri-food sector (Beekman *et al.*, 2006; Millar *et al.*, 2007). However, there appear to be only a limited number of tools that can be applied to facilitate ethical reflection within research projects. One of these methods, a form of ethical framework that has been used to map potential ethical impacts raised by research trajectories in the biosciences, is the Ethical Matrix (EM) (Mepham, 2000).

This method is used to map issues for the defined interest groups and is based on the application of *prima facie* ethical principles that encapsulate traditional ethical theories. The method has been previously applied to a number of biotechnology cases (e.g. Kaiser and Forsberg, 2001; Mepham *et al.*, 2006). The application of the EM facilitates the assessment of a proposed strategy (i.e. the use of biotechnology, development of experimental animal models, etc) in terms of respect (or lack of respect) for three ethical principles, viz wellbeing, autonomy and fairness, as applied to a defined set of interest groups (a generic version of the EM is set out in Table 1). The 'weight' or significance assigned to each ethical impact is determined by the evaluation of forms of evidence. Evidence is defined as '*anything that provides material or information on which a conclusion or proof is based*' (Mepham *et al.*, 2006).

Table 1. Generic Ethical Matrix (Translation of the ethical principles for the corresponding interest group).

	Wellbeing	Autonomy	Fairness
Treated animal	Animal welfare	Behavioural freedom	Intrinsic value
Producers	Satisfactory income, and working conditions	Managerial freedom	Equitable IPR conditions, trading and market systems
Consumers (including affected citizens)	Food safety and quality of life	Informed democratic choice	Equitable access to food
Environment (Biota)	Protection and conservation	Biodiversity	Sustainability

The classical use of the EM, as applied in this case, does not result in a definitive outcome. However, it is valuable to note that a variation of the method has been developed which focuses on producing a final judgement (Forsberg, 2007). The method is also not prescriptive, yet the value of the approach is that it requires the analysis of ethical principles as applied to defined interest groups, this results in the approach being much more than merely a descriptive methodology. The method is intended to fulfil a number of functions which include: raising awareness; clarifying value conflicts, making explicit the evidence used to justify a position; encouraging ethical reflection; and acting as a starting point for ethical deliberation.

Applying the EM to a 'genomics of host-pathogen' research programme

In order to apply the EM and facilitate reflection on the key ethical issues raised by the project, a two day workshop was convened in November 2006, with thirteen participants from the EADGENE Network (representing eleven partner institutions). Prior to the workshop, a first phase ethical analysis was conducted through a series of interviews with project participants. This process allowed the researchers to map what was deemed to be the prominent ethical issues (positive and negative) raised by the project and the key principles that should guide the Network's researchers (Meyer, 2005). This work informed the workshop's ethical discourse (structured by use of the EM) and helped to focus some of the key discussion points.

The overall aim of the workshop was firstly to enable Network members to explore ideas and opinions relating to ethical and social issues raised by the EADGENE research agenda and secondly to build capacity within the full Network so that these issues could be mapped and addressed as appropriate. The process of ethical reflection was initiated with a preamble of the goals of the workshop, followed by an introduction to bioethics and the EM. The EADGENE workshop participants applied an adapted version of the EM. The modified version was applied in order to structure the discussion, and at the same time facilitate ethical reflection rather than impose a process of ethical regulation. For the analysis of the case, application of animal genomic knowledge and technologies, six interest groups were defined (Table 1). The interest groups were defined as: (1) Producers / Industry (including farmers and commercial breeders), (2) Consumers / Citizens; (3) Scientists (e.g. those involved in animal genomics research); (4) Animals in used in research; (5) Production Animals; and (6) the Environment (e.g. biota, water quality).

The workshop was conducted with two distinct sessions (afternoon session on Day 1 and a morning session on Day 2) and participants were divided into two sub-groups in order to facilitate more in depth discussion and reflection. Each group was supplied with a generic matrix and asked: (1) if they wished to extend the specification, i.e. re-specify or elaborate on the original specification of the principles and then; (2) to fill in the cells considering whether innovations and current research trajectories might infringe upon or respect defined ethical principles for each of the interest groups. The aim was not to arrive at some form of consensus on any of the identified issues and potential ethical impacts, but to clarify, record perspectives and increase understanding. As part of the second session, participants were asked to finish mapping the issues for each of the individual cells of the matrix, and then to prioritise the issues / ethical impacts identified. The outcomes from the two groups were reviewed in plenum and further discussed in an attempt to draw out some of the more salient issues. Again the aim of this plenum session was not to prescribe any particular decisions or positions, but to raise awareness of embedded ethical values, clarify issues, and assist in the identification of contrasting perspectives and concerns. In particular, the outcomes of the ethical reflection were intended to serve as a starting point for further ethical analysis and internal deliberation within the Network.

Workshop outcomes and participant feedback

The process of mapping the issues and elaborating on the affected interest groups was one of the outcomes of the ethical discussion. In terms of specifics, the sub-groups identified 'key issues to focus on in the network' and 'the main responsibilities of EADGENE researchers'. In specific terms, three core issues were identified by the first Group: (1) Intellectual Property Rights (IPR) and openness; (2) use of animals in research; (3) impacts on the public at large. In addition to these, Group Two identified: (1) the need to be realistic about the benefits of research and open about uncertainty; the value of increased collaboration and availability of biological resources. When considering the main responsibilities of Network researchers, both groups identified the need to consider the wider impacts of IPR strategies, the need to apply high animal ethics standards (e.g. the 3Rs), and to engage with with the public at large and wider societal groups about the benefits and limitations of the research (Gamborg and Sandøe, 2007). The issues and responsibilities identified demonstrate the groups' reflection on a number of sustainability-related objectives particularly in terms of longer-term societal impacts.

With regard to the EM, most participants identified the benefit / value of applying this method to structure the ethical reflection of their work and responsibilities, especially when a more specific 'genomics of host-pathogen' case was used to prompt discussion. In addition, several participants also noted that they did not see this method as exhaustive, but rather as a starting point for more specific discussion of the ethical questions that arise from animal disease genomics research and technology developments.

Conclusions and further work

The overall positive response of participants to the use of an ethical framework to structure wider consideration of these issues, as a starting point for further ethical reflection, indicates that the method could also be applied successfully in other areas. Using a modified 'top down' Matrix Method approach allowed participants to discuss key issues, such as environmental and welfare implications of the research programme. Identification of these issues at an early stage of the research and technology development process, for areas such as proposed breeding strategies, should enhance integrated management. One potential limitation of this form of internal ethical reflection is the absence of contentious or challenging views. One way of combating this would be to include external stakeholders who may challenge the project group. Using a specified case instead of analysing the wider field of animal genomics may also stimulate further, although potentially case-specific, ethical reflection, In addition, specific sustainability objectives that relate to this field of research were not directly discussed by the participants, however this may be facilitated by introducing sustainability concepts (again exposure to different views) as part of the introduction to the methodology.

If ethical reflection programmes, facilitated through participatory methods, are to be routinely embedded within RTD programmes then the methods applied need to facilitate effective reflection and be resource efficient. Researchers involved in RTD programmes should also be able to identify the wider value of these processes and the merits of the outcomes. This requires structured approaches that lead to notable end points. Rather than being a restrictive method that 'pre-frames' the discourse and the 'included' issues, it appears that the workshop participants found the EM method useful and the outcomes informative.

Acknowledgements

The contributions of the project partners and the financial support from the NoE (FOOD-CT-2004-506416) European Animal Disease Genomics Network of Excellence for Animal Health and Food Safety' (EADGENE) are kindly acknowledged.

References

Beekman, V., de Bakker, E., Baranzke, H., Deblonde, M., de Graaff, R., Ingensiep, H., Lassen, J., Mepham B., Thorstensen E., Tomkins S., Nielsen, A., Skorupinski, B., Brom, F., Kaiser, M., Millar K. and Sandøe, P. (2006). Guide to an ethical bio-technology assessment toolbox for agriculture and food production. Ethical Bio-TA Tools final report (QLG6-CT-2002-02594). Agricultural Economics Research Institute (LEI), The Hague, The Netherlands, 36pp.

Forsberg, E.M. (2007). A Deliberative Ethical Matrix Method - Justification of Moral Advice on Genetic Engineering in Food Production. Faculty of Humanities, Oslo.

Gamborg, C. and Sandøe, P. (2007). Conclusions from the full internal EADGENE seminar. EADGENE Project Report . Danish Centre for Bioethics and Risk Assessment, Copenhagen. 13 pp.

Gamborg, C. and Sandøe, P. (2005). Sustainability in farm animal breeding: a review. Livestock Production Science 92: 221-231.

Kaiser, M. and Forsberg, E.M. (2001). Assessing fisheries - using an ethical matrix in a participatory process. Journal of Agricultural and Environmental Ethics 14: 192-200.

Liinamo, A.-E. and Neeteson-van-Nieuwenhoven, A.-M., (eds.) (2002). Inventory and options for sustainable farm animal breeding and reproduction. SEFABAR First Annual Report. AnNe Publishers 62 pp. http://www.sefabar. org/public/2001/First_annual_report_2.pdf. (Accessed 09/05/07).

Mepham, B. (2000). A framework for the ethical analysis of novel foods: the ethical matrix. Journal of Agricultural and Environmental Ethics 12: 165-76.

Mepham B., Kaiser, M., Thorstensen, E., Tomkins, S. and Millar K. (2006). Ethical matrix manual. Agricultural Economics Research Institute (LEI), The Netherlands, 45pp.

Meyer, G. (2005). A Study of Ethical and Societal Issues. EADGENE Project Report. Danish Centre for Bioethics and Risk Assessment, Copenhagen. 34pp.

Millar, K., Thorstensen, E., Tomkins, S., Mepham, B. and Kaiser, M. (2007) Developing the ethical delphi. Journal of Agricultural and Environmental Ethics 20: 53-63.

System approach to improve animal health

Albert Sundrum, Klaas Dietze and Christina Werner
Department of Animal Nutrition and Animal Health, Faculty of Organic Agricultural Sciences, University of Kassel, Kassel, Germany, sundrum@wiz.uni-kassel.de

Abstract

A high level of animal health is demanded by an increasing number of consumers and expected to be realised especially from organic livestock farming. Furthermore, animal health can be used as an indicator for sustainability within the farm system. In a current study on 20 organic sow herds, conducted to assess the effects of the implementation of animal health plans, constraints which limit the possibilities of appropriate diagnosis, treatment and preventive measures are evaluated. Previous results show a high variability with regard to the prevalence of production diseases, the availability of relevant resources such as labour time or investments, the perception of diseases by farmers, and the expertise and management skills to deal with diseases. In order to think through the complexity of production diseases it is reasonable to grasp the farm as an ecosystem and to define animal health as an emergent property of a farm system. While the inductive approach has failed to provide valid information to identify the main causes of multi-factorial diseases, a deductive approach is recommended. Within the system approach, animal health precaution plans can be developed as a suitable frame for feedback mechanisms within the farm system. The use of feedback mechanisms, however, requires a clear guideline concerning the expected output of the system. Consequently, there is a need for a change in the paradigm from a standard oriented to an output oriented approach.

Keywords: organic pig production, production diseases, aetiology, diagnostic

Introduction

Consumers are becoming increasingly interested in foods produced according to ethical aspects of animal health and welfare principles (Verbeke and Viaene, 2000). A high level of animal health in the herd indicates that the farm animals are able to cope with their living conditions (Broom, 1991). These are characterised by what is offered by the farm management in terms of housing, nutrition and animal care. Hence, a high level of animal health is an indicator for the status of longevity and sustainability of the herd and the farm system.

Organic farming is often directly associated with an enhanced level of animal health and welfare, and many consumers conflate organic and animal-friendly products (Harper and Makatouni, 2002; McEachern and Willock, 2004). This is reasonable as the minimum standards in the EU Regulation of organic livestock production (EEC 2092/91) clearly exceed the legal minimum requirements of conventional livestock production in many areas.

However, analysis of the literature reveals that the average prevalence of production diseases in conventional and organic livestock production is comparably high and at the same time varies widely between farms and between countries (Hovi *et al.*, 2003; Sundrum *et al.*, 2004). Thus, the current approach of organic livestock production to ensure animal health by upgraded minimum standards lacks efficiency in the implementation of organic principles (Sundrum *et al.*, 2006). Reasons for the variation within and between countries are multi-factorial and can not be generalised. Due to the contrast between the claim of organic agriculture and the reality, the question arises on how to improve animal health on organic farms. Difficulties and possibilities in realizing a high level of animal health are discussed exemplary in the case of organic pig production in Germany.

Organic pig production

In an ongoing study, focusing on the implementation of animal health plans in 20 organic sow herds selected for a representative study, the status quo of animal health showed distinct variation. Livestock data and disease prevalence varied highly and often did not meet the standards obtained in conventional production (Dietze *et al.*, 2007). This corresponds with results published by Leeb and Baumgartner (2000) and Vaarst *et al.* (2000). Detailed on-farm assessment brought up weak points in hygienic, nutritional and animal health management. Measures considered to be standard in conventional farming such as electronic livestock data acquisition, barn disinfection, regular feedstuff analysis or effective disease prevention measures (vaccination-, de-worming protocols, quarantine etc.) were implemented consequently only on few organic pig farms, leading to an overall unsatisfactory situation with regard to health precaution.

Balanced diets for sows, known to be a key factor for animal health in general, are accomplished by less than 50% of the farms due to the absence of a regular analysis of diet components. The emphasised parasite burden in organic pig farming (Leeb and Baumgartner, 2000) is antagonised by de-worming regimes whose effectiveness is not monitored by over 60% of the farms, leading to positive findings in almost every faecal analysis. These factors come along with a high variability regarding the perception of diseases by farmers and the expertise and management skills to deal with health disorders, as indicated by a questionnaire that has been carried out on these farms. In general, farmers were not able to address a clear goal with regard to the animal health status on their farms or a reference point marking the difference between prevalence rates that are seen as still tolerable or already non-acceptable.

The farms showed huge differences with regard to the availability of relevant resources such as labour time, feedstuffs of high quality or investments. Hence, farmers have to deal with corresponding conflicts of aims when they have to decide where to set priorities. To do so, they need to assess which measures can be expected to provide a high effectiveness and which can be implemented on a low cost level. Further constraints seem to be evident in relation to the ability of the persons involved to think through the complexity with regard to the background and the development of production diseases. Due to the complex interactions among farm animals, microbes and living conditions, the identification of the specific causal factors responsible for the corresponding production disease on each farm is a big challenge. However, without being aware of the specific aetiology of the predominant production diseases, it is not possible to develop appropriate strategies for improvements.

Identification of problem areas

In the current study, the on-farm assessment showed that on many farms it is very difficult to identify the aetiology of production diseases. This is due to the fact that many farmers, when asked at the beginning of the project, were not able to provide concrete data about the current or previous prevalence rates of the predominant diseases on their farms. Information about individual diagnosis, the pathogens possibly involved or the specific nutrient supply was often missing. This information, however, is essential to narrow down the possible causes of production diseases on each farm.

In general, the approach to diagnose production diseases in herds or groups is inductive, and goes from specifics in a few individuals to generalities about the group (Slenning, 2001). While this procedure is adequate in the case of infectious diseases, where the case of illness in the herd is dominated by specific pathogens, the inductive approach lacks validity in relation to the aetiology of multi-factorial diseases. The inductive approach does not take into account the interactions between the diverse factors involved, the dynamic of the process, and the heterogeneity in the reaction on these factors between single farm animals. As the signs of multi-factorial production diseases are rarely precise in relation to the possible aetiology, farm animals are often treated according to their symptoms while the possible causal factors remain unclear and therefore unchanged.

Several precautions and hygienic measures are known from the textbooks for their efforts to reduce the incidence of production diseases. To implement those measures into practice would lead to an increase in the production costs. On the other hand, the farmer tries to reduce costs and efforts to a minimum, and strives for an optimum between the implementation costs and the value of intervention. Precaution measures, however, vary to a high degree among farms with regard to the effectiveness and the costs of implementation. As the optimum can only be estimated in the specific farm situation, it is not possible to conclude from single factors, which have been proven under *ceteris paribus*-assumptions to generalities about possibilities to reduce production diseases. Thus, generalised statements about the effectiveness of measures are not scientifically valid. Due to the limitations in the availability of resources (labour time, nutrients, investments etc.) farmers are often very uncertain about what to do and where to set priorities in conflicting areas. Consequently, appropriate measures often fail to appear. There is a need for a more comprehensive approach that takes into account the diverse interactions and the resulting complexity on the farm level and the difficulties to provide valid information about the aetiology of the predominant production diseases.

Deductive diagnosis within a system approach

There is a growing understanding within the scientific community that it is necessary to develop more comprehensive concepts which simultaneously consider a larger number of causal relationships. Thus, the isolated view under *ceteris paribus*-assumptions should be replaced by a holistic or system approach (DFG, 2005). In order to deal with interactions and dynamic processes on a farm it is reasonable to grasp the farm as an ecosystem. The system approach and the theory of 'open systems' provide a general concept to deal with complexity (Von Bertalanffy, 1968). A theoretical framework for analysing farms as an entity has been reviewed and outlined by Noe and Alroe (2003). The authors pursue the idea of a farm as a self-organizing system in a complex of heterogeneous socio-technical networks of food, supply, knowledge, technology, etc. that must produce and reproduce itself through demarcation form the surrounding world. Within the system, single parts are related to each other and generate emergent properties. The system approach can be applied also in the case of livestock production and the production goal of animal health as the result of complex interactions within a farm system. In this context, animal health can be defined as an emergent property of a farm system which does not belong to any of its constituent parts, but emerge from the relationships and interactions within the farm system.

Due to the interaction between various factors within a farm system, the causes of production diseases vary considerably between farms. At the same time, a comparable level of production diseases on different farms can be based on very different causes while a high animal health status can be obtained from different initial conditions and in different ways.

Hence, generalized conclusions from single signs and factors with regard to the development and maintenance of production diseases in the herd are inadmissible. A causal and encompassing diagnosis of the predominant production diseases requires rather a deductive approach reaching from general factors and indicators of the farm system to specific aetiologies. While the deductive diagnosis is standard in individuals, it has to be transformed for the use in herds by making a list of comparisons where each deviation from a reference has to be considered as a clue to the cause of the production diseases.

The meaningfulness and validity of the deductive diagnosis result from the comprehensiveness and plausibility, by which the various pieces of circumstantial evidence are used for a coherent statement about the aetiology of the production diseases on each farm. At the same time the diagnosis functions as the starting point for the development of corresponding strategies to improve animal health status.

Regulation of herd health is based upon dynamic interactions and feedback mechanisms, monitoring back information on deviations from the state to be maintained or the goal to be reached. Thus, animal health monitoring systems and herd health precaution plans have to be developed for feedback within the farm system. They could be used as a tool to assess and follow the current state, to support an appropriate diagnosis and to assess the effectiveness and the cost-benefit relationship of interventions and preventive measures.

Conclusion

The current animal health status on many organic farms reveals severe discrepancies between the claim and reality with regard to organic products. As the implementation of minimal standards obviously is not an appropriate tool to predict the level of animal health, there is a need to develop on-farm control measures and assessment tools with regard to the animal health status to meet the expectations of consumers.

In the face of the high prevalence of production diseases in organic farming, it can be concluded that the lack of clear goals with regard to animal health, and the lack of feedback and control mechanism within the farm system is one of the main reasons for the huge variation in the animal health status between farms. There is reason to assume that how a problem is viewed influences how the tools that affect that problem are viewed.

The use of feedback mechanisms is expected to provide a very helpful tool for improvements. They require, however, a clear guideline concerning the level of production diseases expected as the output of the farm system. Currently, there is no common agreement about a threshold acceptable for organic livestock farming or about categories that allow to group different animal health levels in a range between very good and bad.

References

Bertalanffy, L. von (1968). General System Theory - Foundations, Development, Application. George Braziller, New York, 295 pp.

Broom, D.M. (1991). Animal welfare: concepts and measurement. Journal of Animal Science 69, 4167-4175.

DFG (Deutsche Forschungsgemeinschaft) (2005). Future perspectives of agricultural science and research. Wiley-VCH.

Dietze, K., Werner C. and Sundrum, A. (2007). Status quo of animal health of sows and piglets in organic farming. In: U. Niggli, C. Leiffert, T. Alföldi, L. Lück andH. Weller (eds.) Proceedings of the 3rd International Congress of the European Integrated Project Quality Low Input Food (QLIF), 20.-23.032007, University of Hohenheim, Germany, p. 366-369.

Harper, G.C. and Makatouni, A. (2002). Consumer perception of organic food production and farm animal welfare. British Food Journal 104, 287-299.

Hovi, M., Sundrum, A. and Thamsborg, S.M. (2003). Animal health and welfare in organic livestock production in Europe - current state and future challenges. Livestock Production Science 80, 41-53.

Leeb, T. and Baumgartner, J. (2000). Husbandry and health of sows and piglets on organic farms in Austria. Proc. from the 13th IFOAM Scientific conference, p. 361.

McEachern, M.G. and Schröder, M.J. (2002). The role of livestock production ethics in consumer values towards meat. Journal of Agricultural and Environmental Ethics 15, 221-237.

McEachern, M.G. and Willock, J. (2004). Producers and consumers of organic meat: A focus on attitudes and motivations. British Food Journal 106, 534-552.

Noe, E. and Alroe, H.F. (2003). Farm enterprises as self-organizing systems: a new trans-disciplinary framework for studying farm enterprises? International Journal of Sociology of Agriculture and Food 11, 3-14.

Offermann, F. and Nieberg, H. (2000). Economic performance of organic farms in Europe. Organic farming in Europe: Economics and policy. Vol 5. University of Hohenheim.

Slenning, B.D. (2001). Quantitative tools for production-oriented veterinarians. In: O.M. Radostits (ed.) Herd health: Food Animal Production Medicine. 3rd edition, W.B. Saunders Company, p. 47-95.

Sundrum, A., Benninger, T. and Richter, U. (2004). Statusbericht zum Stand der Tiergesundheit in der ökologischen Tierhaltung - Schlussfolgerungen und Handlungsoptionen für die Agrarpolitik. http://orgprints.org/5232/.

Sundrum, A., Padel, S., Arsenos, G., Kuzniar, A., Henriksen, B.I.F., Walkenhorst, M. and Vaarst, M. (2006). Current and proposed EU legislation on organic livestock production, with a focus on animal health, welfare and food safety: a review. Proceedings of the 5th SAFO Workshop, 01.06.2006, Odense/Denmark, 75-90.

Vaarst, M., Roepsdorff, A., Fenestra, A., Hogedal, P., Larsen, A., Lauridsen, H.B. and Hermanden, J. (2000). Animal health and welfare aspects of organic pig production. Proc. from the 13th International IFOAM Scientific Conference, Basel, p. 373.

Verbeke, W.A. and Viaene, J. (2000). Ethical challenges for livestock production: meeting consumer concerns about meat safety and animal welfare. Journal of Agricultural And Environmental Ethics 12, 141-151.

Will nano-enabled diagnostics make animal disease control more sustainable?

Johan Evers, Stef Aerts and Johan De Tavernier
Centre for Science, Technology and Ethics, Kasteelpark Arenberg, 1 bus 2456, B-3001 Leuven, Belgium,
johan.evers@biw.kuleuven.be

Abstract

The ethical principles 'right to know', 'right not to know' and 'duty to know' originate from the moral decision making process in complex biomedical ethical issues such as predictive genetic testing. Applied to intensive livestock production, these principles can help to clarify present broad public uneasiness related to animal disease control strategies such as stamping out with massive animal suffering. Implanted nano-enabled biosensors integrated in an autonomous sensor network are precision techniques in advanced sensing and information technology that offer promising opportunities for better animal disease surveillance and management strategies in future livestock production systems. Especially the 'duty to know' principle from the viewpoint of producers, citizens and official authorities will have a strong pull effect towards the implementation of advanced diagnostics in future livestock production. This article puts the promising technological advances in nanobiotechnology (the convergence of nanotechnology and biotechnology) into a perspective of more sustainable animal disease control, which seeks to satisfy the needs of producers and authorities as well as the aspirations of broader society.

Keywords: livestock production, nanobiosensors, sustainability

Introduction

The last decade, the European Union (EU) has repeatedly witnessed outbreaks of animal diseases like Foot and Mouth Disease (FMD) in 2001 and Avian Influenza (AI) in poultry stocks in 2003 and 2006 causing economic losses and animal suffering. During the AI outbreak of 2003 in Belgium and the Netherlands a total of respectively 4.1 million and 30.3 million birds were culled (FAVV, 2004; Stegeman *et al.*, 2004). Of that number, in the Netherlands, 24 million healthy birds were killed in order to prevent the further spreading of the disease (i.e. pre-emptive culling) and for welfare reasons (LNV, 2004). FMD in 2001 in the United Kingdom lasted for over 6 months and led to the destruction of four million animals including 3 million sheep, 600,000 cattle and 138,000 pigs. The epidemic cost the national treasury £2.7 billion including £1.2 billion compensation paid to farmers for animals slaughtered under control measures, £701 million spent on eradication measures and £471 million compensation for animals killed for welfare reasons (Davies, 2002). Although disease control is necessary to overcome natural diseases in our highly productive livestock systems, the public seems to be reluctant to accept these massive slaughter practices of commercial and 'backyard' animals and the images of overcrowded production plants in the recent disease outbreaks.

A common disease control strategy is stamping out which may include culling of infected animals, setting up a protection zone around infected farms with pre-emptive culling and a surveillance zone with a regional or national stand still of transport, and an export stop. A significant part of the animals affected by such stamping out is not a carrier of the disease and hence are killed in an attempt to prevent the disease spreading. If the standstill lasts to long, many animals have to be killed for welfare reasons, i.e. the highly productive animals suffer because they are not transported or slaughtered in due time. Preventive vaccination for highly contagious viral infections like Avian Influenza and FMD is not a legal EU control policy option since the early 1990s through EU Directives respectively 92/40/EEC and 90/423/EEC. This political choice for stamping out strategies instead of preventive vaccination has been contested many times (EFSA, 2005). Therefore, scientists and legislative authorities are looking for alternatives for present disease diagnostics and control strategies.

The array of classic diagnostic tools currently at our disposal is mainly deployed as soon as farmers or veterinary surgeons recognise disease symptoms such as lower food or water intake, abnormal behaviour, higher morbidity or mortality. This implies a huge delay in disease management because then it is often already too late; infectious agents of most (highly) contagious diseases are spread before symptoms of the infection occur. Furthermore, conventional 'off-site' analysis requires samples for disease diagnostics to be sent to a laboratory for testing. These methods allow the highest accuracy of quantification and the lowest detection limits, but are expensive, time consuming and require the use of highly trained personnel. Immunoassays for the diagnosis of animal diseases are generally based on the detection of antibodies to the pathogen of interest (Tothill, 2001; Schmitt and Henderson, 2005). Not only immunological reactions, but also physiological, behavioural and performance parameters (such as stress indicators in blood, saliva or excreta, loss of weight, lack of mobility, lower milk yield or egg output...) can be assessed on individual or on group basis. Failure of stockpersons and/or veterinary surgeons to notify and adequately interpret disease symptoms, variation in clinical signs and inappropriate specimen collection from suspect clinical cases are the main sources of diagnostic uncertainty (AusVet, 2005).

Biosensors for advanced diagnostics from a technological point of view

Nano-enabled biosensor technology has great potential for improving the monitoring of animal health and welfare in live production systems. Nanobiotechnology (1 nm = 10^{-9} meter) is the convergence of engineering and molecular biology, two existing but distant worlds, and is leading to a new class of multifunctional devices for biological and chemical analysis with better sensitivity, specificity and a higher rate of recognition (Fortina, 2005). Hence, nanotechnology is an enabling technology that is believed to offer far more reliable, better-built, fast and inexpensive diagnostic tools.

A biosensor can be defined as a device that consists of a biological recognition system, often called a bioreceptor (enzyme, a DNA strand, an antibody, a cell or part of a cell) and a transducer (Vo-Dinh *et al.*, 2006). The interaction of the analyte with the bioreceptor is designed to produce an effect measured by the transducer, which converts the information into a measurable effect, such as an electrical signal. Biosensors and biochips (i.e. an set of individual biosensors) for implantable applications should exhibit minimal fouling and should be able to function under a wide range of biological conditions such as temperature, pH, humidity and enzymatic activity (Desai *et al.*, 2000; Vo-Dinh and Cullum, 2000).

Biosensors are up to now mainly developed for human needs (e.g. biosensors for glucose monitoring, Wilkins and Atanasov, 1995). Few applications exist for veterinary ends: in vitro detection of milk progesterone (Xu *et al.*, 2005), Bovine Viral Diarrhoea (Muhammad Tahir *et al.*, 2005) and mastitis (Mottram *et al.*, 2006). The latest generation of biochip technologies combines performance and analytical features not yet available in any bioanalytical system. For instance, recently biochips have shown promising progress towards recognition of multiple biological agents (Yacoub-George *et al.*, 2007).

Currently, research on implantable nano enabled biosensors for pathogen and contaminant detection has not yet resulted in applications. But the need for fast, on-line and accurate early diagnostics for infectious diseases in livestock has a strong pull effect. On a longer term, such biosensors will be integrated into on-farm sensing systems where the immunological activity of animals is constantly monitored through a network of sensors connected with a central information device. As soon as any disturbance is detected, appropriate measures can be taken by stockholders, veterinary surgeons or competent authorities.

Besides the advantages for better and earlier disease control, sensors could also give interesting feedback information about the animal's physiology. Both for intensive and extensive farming systems (especially dairy cattle), this online diagnostic system could be supplemented with existing systems for tracking animal movement, identification and registration of animals such as electronic tag readers and injectable transponders and collars (Rossing, 1999; Sikka *et al.*, 2006).

On the conceptual level it might be clear what the demands are; on the operational level there are still many scientific and technological uncertainties. Many reasons could be given for the slow biosensor technology transfer from research laboratories to the marketplace: cost, instability of the new technologies, sensitivity issues, quality assurance and instrumentation design (Velasco-Garcia and Mottram, 2003). In order to become a commercial and efficient early on-farm diagnostic tool, the sensor system should be easy-to-use in daily farm conditions, (relatively) inexpensive, able to carry out multiple analyses and provide fast and accurate information. Furthermore, sensitivity, reliability and selectivity towards one or more analytes, the solution to possible interference of multiple analyses and sensor signals, the organic or inorganic characteristics of the chip material, and calibration and resetting after detection are important technological challenges. Implanted sensors have to function in living organisms - an environment known to be unkind to non-biological components. When working with living organisms, low power consumption and wireless communication are recommended as animals may bite or pick the wires. Finally, data interpretation models have to be developed that can provide the stockholder or veterinary surgeon with accurate and transparent information.

Nano-enabled diagnostics in livestock production from an ethical point of view

Aside from technological feasibility, socio-economic desirability aspects have to be taken into account if new disease diagnostics technologies are to contribute to a more sustainable animal farming in the 21st century. Current intensive livestock production is generally not considered sustainable (Dahlberg, 1991). Sustainability can be defined as the overlap between what people collectively want, reflecting social values and economic concerns, and what is ecologically (biologically and physically) possible in the longer term. This overlap is dynamic because both societal values and ecological capacity continuously change (Bormann *et al.*, 2004). Sustainability assessment often results in conflicting economic, ecological and societal and cultural interests of the relevant actors such as affordability for producer and consumer, animal welfare, environmental pollution food safety and public health.

Additionally, new diagnostic technologies will bring new ethical deliberations to sustainability ethics in future livestock production such as those currently raised in modern medicine about predictive gene testing. Decisions in biomedical ethics are based on a process of deliberation that involves the impacts of proposed actions in the light of four prima facie principles: beneficence, non-maleficence, justice and autonomy (Beauchamp and Childress, 2001; Mepham, 2005). These principles are very general guides which need to be specified in each case and for each relevant actor. For instance, in the case of a doctor treating a patient, autonomy can be specified into the '*right to know*', the '*right not to know*' and the '*duty to know*'. First, the 'right to know' principle is a fundamental ethical and legal principle in human healthcare as the patient has the right to be informed about a certain medical intervention or treatment. Secondly, recent advances in predictive genetic testing make an increasing number of people aware they are at risk of a serious disease without any real chance of reducing that risk, or of obtaining an effective treatment. Hence, such people can have a desire not to be informed and this desire is called the '*right not to know*' principle. Both the '*right to know*' and the '*right not to know*' are principles explicitly recognised by recent ethical and legal instruments such as the Belgian Patient's Right Act of 2002 (Article 6) and the European Convention on Human Rights and Biomedicine (Article 10.2) (Adorno, 2004). Finally, the *right to know* and the *right not to know* (as most rights) are not absolute; in some - often very delicate

- cases such parental decision-making pro or contra selective abortion following prenatal screening, the right not to know can be turned into a *duty to know* with obliged disclosure of information.

Daily reality urges that innovative technological applications have economic benefits for producers. Permanent investments in sensing networks have to pay off for livestock producers. At first sight, producers can have the *right to know* or the *right not to know*, depending on the price of such sensor systems and the individual farmer's financial situation. In other words, they can deny the right to be continuously informed about the health status of their animals. After all, the commercial implementation of implanted sensor networks will depend on the early adoption of such disease management system by single stockmen entrepreneurs, who are clearly convinced of the benefits (cost reducing) of such system in the long term. In this initial stage, the legal framework may not yet oblige producers to implement advanced sensing networks for disease control because they still have the *right not to know*. However, as technology reliability and accuracy increase and production price decreases this *right not to know* can be overruled by a national and supranational legislative force into a *duty to know*. The present high number of farm animals per production unit or per nation, is a clear indicator that there is a vast potential market for sensors in livestock production. Hence market prices will probably become affordable for producers and authorities. Thus, considering the decrease in cost price and the need for additional disease control measures by the authorities, it is to be expected that farmers will be pushed from the *right to know* or the *right not to know* to the *duty to know*.

As previously stated, current stamping out strategies and the rejection of alternative strategies such as preventive vaccination provoke uneasiness by the broader public because they do not overlap with broad societal aspirations. However, a consumer's attitude (purchasing habits) and citizen's aspiration do not have to converge in the same individual (Aerts, 2005). Hence, different ethical principles can apply for both. First, from a citizen's viewpoint, enforcing a *duty to know* about the health status of animals is a strong societal push for the development and implementation of nanobiotechnology enabled diagnostics, However, it can also be stated that citizens have a *right to know*, i.e. transparent and up-to-date information about health safety issues about the technology itself such as bio-accumulation and biodegradability of nano enabled biosensors used in the food chain. Secondly, from a consumers' viewpoint, each technology investment in disease control leading to an increase in product price, will be carefully weighted against human health risk probability and risk perception, trust in information sources and other consumer interests such as taste and visual quality characteristics. Finally, a strong push (*duty to know*) from such technology can be expected for veterinary surgeons and authorities as they have to deal with immense responsibilities, time pressure and high cost expenses in the case of a disease outbreak.

Conclusions

Depending on the outcome of technological feasibility, nano enabled biosensors integrated in autonomous remote on-farm sensor networks may sooner or later serve as powerful alternatives to the conventional diagnostic technologies and disease control strategies. They will provide us with faster, more reliable and more accurate information about the health status of animals. This new continuous stream of information will partly influence the autonomy of the relevant actors (producers, citizens, consumers and authorities). It is to be expected that nano enabled diagnostics in future livestock production, will be favoured by the *duty to know* principle pull effect from the viewpoint of citizens, disease management authorities and producers. A consumer's viewpoint will heavily rely on risk, risk perception, and trust in information sources. He could have the right to know and the right not to know. These ethical principles need to be taken into account in order to promote sustainability criteria in nano-enabled future livestock production systems.

References

Adorno, R. (2004). The right not to know: An autonomy based approach. Journal of Medical Ethics 30: 435-439.

Aerts, S. (2005). Dierenwelzijn en consumptie. In: J. De Tavernier, D. Lips and S. Aerts (eds.) Dier en Welzijn. Lannoo Campus, Leuven, Belgium. p.84.

AusVet (2005). Review of the potential impacts of new technologies on Australia's Foot and Mouth (FMD) planning and policies. AusVet Animal health Services.

Beauchamp, T. and Childress, J. (2001). Principles of Biomedical Ethics, 5th Ed, Oxford University Press, Oxford, United Kingdom.

Bormann, B.T., Brookes, M.H., Ford, E.D., Kiester, A.R., Oliver, C.D. and Weigand, J.F. (1994). Volume V: a framework for sustainable-ecosystem management. Gen. Tech. Rep. PNW-GTR-331. Portland, OR: U.S. Department of Agriculture, Forest Service, Pacific Northwest Research Station. p. 6 Available online at http://www.fs.fed.us/pnw/publications/gtr331/pnw_gtr331a.pdf [Accessed on May 16, 2007].

Dahlberg, K.A. (1991). Sustainable Agriculture - fad or harbinger. Bioscience 41: 337-340.

Davies, G. (2002). The Foot and Mouth Disease (FMD) epidemic in the United Kingdom in 2001. Comparative Immunology, Microbiology and Infectious Diseases 25: 331-343.

Desai, T.A., Hansford, D.J.; Leoni, L., Essenpreis, M. and Ferrari, M. (2000). Biosensor and Bioelectronics 15: 453-462.

EFSA (2005). Annex to the EFSA Journal (2005) 266, 1-21; Animal Health and welfare aspects of Avian Influenza. Available online at http://www.efsa.europa.eu/etc/medialib/efsa/science/ahaw/ahaw_opinions/1145.Par.0003.File.dat/ahaw_op_ej266_avianinfluenza_en2.pdf [Accessed May 8, 2007].

FAVV (2004). Activiteitenverslag 2003, Brussel, Federaal Agentschap voor de Voedselveiligheid van de Voedselketen. Available online at http://www.favv-afsca.fgov.be/home/pub/doc/rapport/AV_2003_s.pdf [Accessed on May 15, 2007].

Fortina, P., Kricka, L.J., Surrey, S. and Grodzinski, P. (2005). Nanobiotechnology: the promise and reality of new approaches to molecular recognition. Trends in Biotechnology 23: 168-173.

LNV (2004). Avian Influenza in the Netherlands. Outbreak 2003. Presentation at the SPS enhanced meeting on Article 6 (regionalization). January 30-31, 2004, Geneva. Available online at http://www.wto.org/english/tratop_e/sps_e/meet_jan06_e/netherlands_e.ppt#286,4,Facts%and%figures(2) [Accessed May 15, 2007].

Mepham, B. (2005). Bioethics: An introduction for the biosciences. Oxford University Press, Oxford, United Kingdom, p 47.

Mottram, T., Rudnitskaya, A. Legin, A., Fitzpatrick, J. and Eckersall, P. (2007) Evaluation of a novel chemical sensor system to detect clinical mastitis in bovine milk. Biosensors and Bioelectronics 22: 2689 - 2693.

Muhammad Tahir, Z., Alocilja, E.C. and Grooms, D. (2005). Polyaniline synthesis and itsbiosensor application. Biosensors and Bioelectronics 20: 1690-1695.

Rossing, W. (1999). Animal identification: Introduction and history. Computers and Electronics in Agriculture 24: 1-4.

Schmitt, B. and Henderson, L. (2005). Diagnostic tools for animal diseases. Revue scientifique et technique de l'Office international des épizooties 24, 243-250.

Sikka, P. Corke, P., Valencia, P., Crossman, C., Swain, D. and Bishop-Hurley, G. (2006). Wireless Adhoc Sensor and Actuator Networks on the farm. IPSN'06, April 19-21, Nashville, Tennessee, United States of America.

Stegeman, A., Bouma, A., Elbers, A.R.W., de Jong, M.C.M, Nodelijk, G., de Klerk, F., Koch, G. and van Boven, M. (2004). Avian Influenza A Virus (H7N7) Epidemic in The Netherlands in 2003: Course of the Epidemic and Effectiveness of Control Measures. The Journal of Infectious Diseases 190: 2088-2095.

Tothill, I.E. (2001). Biosensors developments and potential applications in the agricultural diagnosis sector. Computers and Electronics in Agriculture 30: 205-218.

Velasco-Garcia, M.N. and Mottram, T. (2003). Biosensor Technology addressing Agricultural Problems. Biosystems Engineering 84: 1-12.

Vo-Dinh, T. and Cullum, B. (2000). Biosensors and biochips: advances in biological and medical diagnostics. Fresenius' Journal of Analytical Chemistry 366: 540-551.

Vo-Dinh, T. Kasili, P. and Wabuyele, M. (2006). Nanoprobes and nanobiosensors for monitoring and imaging individual living cells. Nanomedicine: Nanotechnology, Biology and Medicine 2: 22-30.

Wilkins, E. and Atanasov, P. (1995). Integrated implantable device for long-term glucose monitoring. Biosensor and Bioelectronics 10: 485-494.

Xu Y. F., Velasco-Garcia, M. and Mottram, T.T. (2005). Quantitative analysis of the response of an electrochemical biosensor for progesterone in milk. Biosensors and Bioelectronics 20: 2061-2070.

Yacoub-George, E., Hell, W., Meixner, L., Wenninger, F., Bock, K., Lindner, P., Wolf, H., Kloth, T. and Feller, K.A. (2007). Automated 10-channel capillary chip immunodetector for biological agents detection. Biosensor and Bioelectronics 22: 1368-1375.

Is it possible to make risk-reduction strategies socially sustainable?

Sara Korzen-Bohr[1] and Jesper Lassen[2]
Danish Centre for Bioethics and Risk Assessment, Department of Human Nutrition, Faculty of Life Sciences, University of Copenhagen, Rolighedsvej 30, 1958 Frederiksberg C, Denmark, skb@life.ku.dk, jlas@life.ku.dk

Abstract

The public is involved in the assessment of different strategies for reducing food-related risks through perception studies that examine the social and cultural sustainability of these strategies. In this paper, we argue that this public involvement is based on the false assumption that ordinary people have an active perception of risk-reduction strategies. Thus, such studies run the risk of being futile or, in the worst case, of providing a misleading image of public perception. We outline some theoretical and methodological issues that need to be addressed when members of the public are invited to take part in qualitative and quantitative perception studies.

Keywords: meat safety, perception studies, public perception, social sustainability

Introduction

Food safety issues involving meat have developed into an important international political issue. The political prominence of these zoonotic risks is due to the occurrence of outbreaks of illnesses such as Creutzfeldt-Jakob disease (commonly known as mad-cow disease), or contamination of meat with bacteria such as *Escherichia coli*, *Listeria* or *Salmonella*. Considering the public and media concern following such scandals, it is not surprising that political attention has resulted in a demand for action that can eliminate, or at least reduce, the risks. Besides this conflict-driven demand for action, public policies are also motivated by the high economic costs to society of hospitalization and lost workdays that are caused by the outbreaks. Thus there is political and societal interest in the development and implementation of strategies that increase the safety of our meat.

The desire to handle these zoonotic risks has, among other things, resulted in national and international action plans involving surveillance of outbreaks and systems of (self) control. Parallel to this, research projects have been launched to develop and investigate new risk-reduction strategies: such projects often include assessments of economical effectiveness as well as of the ability of strategies to reduce the zoonotic problems. Such risk-reduction strategies, however, may be met with public scepticism and, consequently, poor implementation or controversy. Food irradiation as a means to reduce bacterial zoonosis is one example of a strategy that was rejected by consumers, despite widespread positive appraisals of this technology by experts. The lesson learned from food irradiation, and from other new technologies such as genetic engineering, is that a public acceptance does not necessarily follow support for a technology by experts. Hence there is a need to involve the public in the assessment of new strategies to reduce zoonotic risks in meat. One widespread method of involving the public in such issues is by means of qualitative or quantitative studies of public perception. The question, however, is the extent to which sociological studies of perception are able to capture the public's perception of, for example, risk-reduction strategies and thus serves as a basis for choosing strategies that are socially sustainable. Based on the results from an ongoing cross-disciplinary project about risk-reduction strategies (http://www.bioethics.kvl.dk/), the remainder of this paper will discuss such methodological difficulties.

Safety: not a problem to the public

There is a striking contrast between the importance of meat safety on the political agenda and public concerns as expressed through studies of perceptions and everyday practices. A number of studies have shown that, to lay people, safety is not a primary concern when buying, cooking and eating meat. Rather than safety, consumers seem to be preoccupied with taste, texture and appearance, as well as issues related to the practical organization of food-related activities in a busy everyday life. Shopping, cooking and eating has to be easy, practical and ensure the social cohesion of the family (e.g. see Holm and Kildevang, 1996; Holm and Mohl, 2000; Lassen *et al.*, 2006). This low level of attention towards food safety in studies of everyday practices is, however, contrasted by claims made when people are asked directly about contaminated meat; when asked, of course, nobody wants contaminated meat or to become ill. This stresses the importance of context for the kinds of answers you get in sociological studies: in an everyday context of consumption, people are concerned about issues such as eating quality and managing the many practicalities involved. For reasons we will discuss in the following sections, zoonotic risks are not a concern in this context.

As a part of our study we organized six focus-group interviews about people's perceptions of meat quality, meat safety and strategies to reduce microbial risk in meat. Among other things, the study was designed to uncover the impact of different contexts on public perception. To do this, the interviewees were placed in three different contexts during the interviews. In each context, the moderators guided the participants through discussions aimed at revealing their perceptions of quality, safety and strategies for reducing risk. The three contexts were as follows. (1) Everyday food practices and meals. This context was introduced by asking the interviewees what they had for dinner yesterday and why. (2) Meat as a product. This context was introduced by reminding the participants that meat is the result of a series of processes in the food sector, from agriculture to food-production/-packaging industry. (3) The final context was management of meat safety.

In accordance with the results of other studies (see above), the discussions within the everyday context did not include safety as an important issue; similarly, risk-reduction strategies were an almost non-existent theme. When changing the context and introducing meat production as the framework for discussion, safety got a central position in line with a group of immaterial qualities like animal welfare and environmental issues. Within these first two contexts it was easy for the participants to express themselves and participate in a discussion of the issues put forward.

In the last part of the interviews we introduced the safety context, first by asking what risk-reduction strategies the participants were aware of and subsequently by finding out how they assessed a number of different strategies presented to them by the moderator. The strategies presented to the interviewees represented different technologies, interventions at different stages in the sequence from production to distribution to consumption, and different levels of impact on the safety problem. In the first section, where participants discussed the strategies they listed themselves, the suggested strategies were limited to individual ones such as strategic shopping (e.g. buying meat in shops with a high level of credibility) and careful preparation (e.g. making sure the meat is well cooked or that vegetables are kept apart from raw meat). In addition, improvements to the public food-control system were suggested as a means to avoid safety problems in the food sector. This theme of control was most likely found to be important in all groups because of a pending large-scale scandal in Denmark concerning defective public control of meat sold for consumption. This was also illustrated by the fact that even though most interviewees agreed that there should be more control, they had a hard time explaining in detail how more control could secure safe meat. In the final section, when introduced to a set of concrete risk-reduction strategies, the interviewees were able to rank them meaningfully according to acceptability, but they found it difficult to defend their ranking and to discuss the strategies in relation to each other.

This study demonstrates that, despite an ongoing and intense public debate about meat safety in Denmark, the public does not have active perceptions of risk-reduction strategies. On the one hand, the participants were aware of and able to relate to and discuss strategies that were closely related to their everyday practices, but they did not do so by themselves when placed in either an everyday context as a consumer or in what we have termed the production context. On the other hand, when forced to consider risk-reduction strategies they were generally unaware of strategies at other levels of the food chain, and when asked to discuss these they were short of arguments. Nevertheless, when faced with a list of possible risk-reduction strategies, almost all the interviewees participated in the ranking of the strategies according to their acceptability as a means to solve zoonotic problems. The discussions taking place when deciding the ranking demonstrated that the arguments were limited to more-or-less general statements such as prevention being better than treatment, or chemicals not having a place in food production.

In the study we found that the interviewees did have perceptions of technologies or strategies, even though they had neither knowledge about nor technical understanding of the issue. We interpret this kind of perception as passive perception. Using Michael's concept of comprehension, apprehension and prehension (Michael, 2002) we might shed more light on the concept of passive perceptions and the consequences for studies of perception. Michael (2002) discusses the meaningfulness of the public understanding of science in different traditions: first, a tradition understanding the individual as a cognitive subject; second, a tradition seeing the individual as a social subject; and third, a tradition where the subject is seen in interaction with the physical world. According to this division comprehension is the intellectual capacity to perceive and understand the world around us. Apprehension is the process of assessing the status of the source of knowledge and of social (re)production. Finally, prehension is the process of grasping the world in a corporal sense, seen as an interaction between the subject and the object being grasped; in this way the individual is linked to the physical world in a corporal process (Michael, 2002).

In sociological studies comprehension is easily studied by means of a variety of methods, whereas apprehension and prehension call for more sophisticated methods. In the present study we addressed the problem of studying apprehension by introducing different contexts and thereby focusing on the social and cultural aspects of perceptions. Below we will address the question of how to include prehension in perception studies.

Explaining lay perceptions in light of expert perceptions

Most perceptions studies start with problems or questions raised by experts. Thus many studies have focused traditionally on either trying to reduce the gap between expert and lay perceptions by transmitting the expert understanding of science to lay people, or on illuminating the different cultural and social contexts of the understanding of science and technology (Michael, 2002). In both cases the starting point is that the public have an understanding of science and technology that differs from that of the experts. As a consequence lay perceptions are studied within the framework of expert perceptions. The fact that the questions are raised on the basis of expert perceptions and agendas set by experts means that researchers run the risk of studying lay perception, which is passive perception. Before we return to the problem of studying passive perception we will discuss three levels of explanation for the relationship between expert and lay perceptions.

1. Different sets of values
At the first level of explanation, differences in expert and lay perception are based on different sets of values. These different values determine the classification of safety and the scale used when assessing a safety problem. The classification of safety compared to other quality parameters is based on values

and interests (Bowker and Star, 2000). The different interests among experts as well as lay people reflect different contexts. For the expert, safety is a professional issue reflecting their professional identity. By contrast, for lay people safety is, in line with other issues, seen in a perspective reflecting their everyday identities and practices. When lay people are confronted with the political setting of food production, safety becomes an issue of concern similar to the way in which the political system reacts to food scandals (Jensen *et al.*, 2005). At this level of explanation we can talk about both comprehension and apprehension, because it allows for intellectual and contextual understanding of perceptions of safety.

2. Coping with risk in everyday life
In the eyes of the expert the lay management of risk is often viewed with scepticism, both because of the selection of strategies by lay people and because of the risks that are considered relevant. Lay management strategies of risk are about developing routines and practical strategies that eliminate the continuous assessment of the problem on one hand, and establishing an understanding of the risk as unproblematic on the other. At this level of explanation we can talk about apprehension because of the importance of social and cultural contexts.

3. Understanding based on corporal experiences
At this level of explanation understanding of safety and strategies is seen as more than a cognitive and social process. Lay people and experts have corporal experiences of safety and safety strategies. Lay people, for example, have a bodily experience of food telling them that they can normally eat it without becoming ill. The everyday handling of meat thus becomes a practical process developed from the interaction between the subject (the individual) and the object (the meat). At this level of explanation we can talk about prehension because of the awareness of corporal interactions.

Conclusion

The discussions above raise three issues that have to be addressed to answer the question of how to make risk-reduction strategies socially sustainable. (1) What is the context of the public's perceptions? (2) How are passive perceptions detected and interpreted? (3) How is prehension identified and understood? In our present work we have addressed all three issues and found that sociological methodology alone cannot solve the problems.

First we have, in our ongoing work, addressed the issues on contextuality by defining contexts of perceptions and integrating them into our study design. In this way we have suggested a methodological means with which to capture apprehension as well as comprehension. This is done by placing people in the contexts of everyday life, food production and risk management. This strategy was implemented successfully in our focus-group study. The next step in our study is a quantitative survey of the Danish population. During the design of the questionnaire we experienced difficulties separating the contexts and interpreting the data unambiguously. So far, our best suggestion for overcoming this obstacle has been to pose relatively long and precise questions and to use a mix of questions about the public's attitudes and practice.

Second, we have addressed the concern that study data might result from the expression of passive perceptions by designing the focus groups so that the contexts closest to people's everyday lives are met first. Thereby we can analyse the level of people's active perception of issues surrounding successive contexts (relating to food production and the management of meat safety) with reference to the previous context relating to their own lives. The problem of passive perceptions arises from the need to study perceptions of issues arising from other actors' agendas. In our study we compare the perceptions of lay people with those of experts and actors in the retail sector. For the quantitative survey the passive perceptions pose the question of what dimensions we can include meaningfully. This question can only be answered by sociological consideration based on knowledge of the field.

Third, we have tackled the matter of prehension as a problem of the ability of the methods used to study corporal processes. Prehension is not directly observable in the transcripts of focus groups but needs to be extracted from the material as a complex whole (Michael, 2002). In quantitative surveys this problem seems even more present as the method is not sensitive to corporal and nonverbal information. The issues of prehension call for development of methods across disciplinary boundaries. Sociological methods, both quantitative and qualitative, are limited to studying data that can be formulated and communicated between people. The discussion in this paper points to the need to realize the limits of sociological methods, and furthermore initiate future cooperation between sociological science and other disciplines, such as psychology and ethnography, to develop studies of lay perceptions of risk-reduction strategies that are socially sustainable.

References

Bowker, G.C. and Star, S.L. (2000). Sorting Things Out. Classification and Its Consequences. Cambridge: The MIT Press.

Holm, L. and Kildevang, H. (1996). Consumers' views on food quality. A qualitative interview study. Appetite 27: 1-14.

Holm, L. and Mohl, M. (2000). The role of meat in everyday food culture: an analysis of an interview study in Copenhagen. Appetite 34: 277-283.

Jensen, K.K., Lassen, J., Robinson, P. and Sandoe, P. (2005). Lay and expert perceptions of zoonotic risks: understanding conflicting perspectives in the light of moral theory. International Journal of Food Microbiology 99: 245-255.

Lassen, J., Sandoe, P. and Forkman, B. (2006). Happy pigs are dirty! Conflicting perspectives on animal welfare. Livestock Science 103: 221-230.

Michael, M. (2002). Comprehension, Apprehension, Prehension: Heterogeneity and the Public Understanding of Science. Science Technology Human and Values 27: 357-378.

New understanding of epigenetics and consequences for environmental health and sustainability

Veronika Sagl¹, Roman Thaler¹, Astrid H. Gesche² and Alexander G. Haslberger¹
¹Dept. for Nutritional Sciences, Center for Ecology, Univ. of Vienna, Althanstrasse 14, A-1090 Vienna, Austria
²Applied Ethics and Human Rights Program, SHHS, Queensland University of Technology, GP Campus, X209, 2 George Street, Brisbane, Qld. 4001, Australia, Alexander.haslberger@univie.ac.at

Abstract

Genetic variation is controlled by two different mechanisms: genetic and epigenetic. Genetic variations are based on differences in DNA-sequences due to mutation and recombination events. Epigenetic variations, on the other hand, are not encoded through the nucleotide sequences of DNA, but rather through the chemical modification of either DNA or its associated proteins which results in certain genes being turned on or off. It appears that methylation, but also acetylation or ubiquitylation, lead to different molecular outcomes, resulting in phenomena such as the inactivation of the X-chromosome, genomic imprinting, or different types of cancer. Epigenetics will have profound effects on our understanding of human and environmental health by forcing us to look afresh on interactions between (wo)men with their natural and social environment and by adding a transgenerational, even evolutionary, aspect to the debate. These findings could strengthen emerging thoughts about sustainable and responsible care taking of our environment and consequently of our health through it.

Keywords: epigenetics, nutrition, environmental health, sustainability, ethics, impact

Introduction

Even though the human genome is completely sequenced, we still cannot read the code. Until recently Darwinists looked in a depreciatory way at Lamarcks' theories. Charles Darwin (1809-1882) postulated that evolution is based on natural selection and inheritance. Today, we can add gene drift and gene flow to his original postulate. Jean-Baptiste Lamarck (1744-1829), on the other hand, suggested that every creature passes on new acquired characteristics. Lamarck believed that evolution was not based on random selection, but, instead, was goal-orientated. Today, latest research indicates that inheritance is indeed composed of two different mechanisms: genetic and epigenetic. Genetic variation is based on differences in DNA-sequences, spawned by mutation and recombination. Epigenetic is a dynamic, stable and partially hereditary modification of gene expression, which does not result from a change in DNA-sequence. Epigenetics does not focus on the sequence or the organisation of genes, but rather how, when and why they are turned off and on. It also investigates which interactions of genes lead to novel or special functions and products, and, ultimately, to a particular phenotype (Watters, 2006).

Nutritional, environmental, chemical and physical factors have the potential to modify in many different ways the epigenome. Especially three genomic regions are likely to be involved in epigenetic changes: the promoter regions of some housekeeping genes; transposable elements that lie close to genes with metastable epiallele; and regulatory elements of imprinted genes. Epigenetic regulation arises from a number of molecular modifications like DNA methylation, which often takes in CpG islands. These regions of the DNA-strand have many cytosine-guanine base pairs. The molecular mechanisms lead to changes in the chromatin-packaging of DNA; to changes in regulation by non-coding RNAs, such as microRNAs; or to various packaging of the DNA-strand around histones. As methylation is stable, the attachment can even be passed on to the next generation. Chemical mechanisms are also possible, such as acetylation, ubiquitylation and phosphorylation.

Epigenetic impacts are many. Some of the impacts reported have been the inactivation of the X-chromosome, genomic imprinting (one of a pair of alleles are silenced during early development), the development of various types of cancer, an increase of disorders after IVF treatment, inefficacy in cloning, and, possibly, a shortening of cellular and organismal viability (for review: Bjornsson *et al.*, 2004).

The impact on epigenetic factors on inheritance and disease occurrence and progression

In 2000, Randy Jirtle, professor of radiation oncology at Duke University Medical Center, began experimenting on Agouti-mice. These rodents have the agouti-gene modification, which makes the mice highly susceptible to diseases such as obesity and diabetes. The mice also have an increased susceptibility to tumours. By changing their food intake to one that is high in methyl groups, the animals were found to have considerably fewer diseases. In fact, the offspring still had the agouti-gene, but in an innoxious version (Dolinoy *et al.*, 2006).

Latest research points to a high correlation between epigenetic inheritance and environmental, nutritional, lifestyle and other socio-economic factors, such as social stress. Several scientists see an imprinted component in behavioural traits, as seems to be the case for the Turner- or the Prader-Willi-syndrome. Other authors assert that autism, Alzheimer's disease or schizophrenia may also result from imprinting errors during early brain development. We also know that many chronic diseases can be impacted by different factors, such as the influences exerted by several other genes, or by environmental conditions, age or nutritional status. It is possible that in many diseases, in particular in many types of cancers affecting seniors, the increasing number of methylation of 'tumour suppressor genes' and the dis-methylation of 'oncogenes' play an important role (for review: Feinberg *et al.*, 2004).

Another example pertains to Vincozolin, an insect repellent frequently used in winegrowing. It is known to have toxic adverse effects on human health by inactivating hormones. Lately, a direct inheritable damage to fertility has been discovered, which is caused by a modification of DNA methylation. Other chemicals, like BPA (Bisphenol A), have been found to act by a hormone pathway, resulting in toxic and epigenetic consequences by slowing down normal, age related gene expression, because important enzymes were missing. These effects were thought to link to tumours in seniors, adiposity and other, hormonally influenced diseases (Anway *et al.*, 2005, Vom Saal, 2007, Ho *et al.*, 2006).

Which sites are under epigenetic influence?

Several new studies indicate that behaviour, such as pup licking and grooming in nurturing, is not inherited via genes, but handed down to the offspring directly from the mother during the first days of postnatal life by epigenetic means. Latest research results indicate that this 'maternal programming' is correlated to DNA methylation and histone modification of the NGFIA (nerve growth factor inducible protein A) transcription factor-binding motif. Epigenetic factors appear to be associated with changes in gene expression affecting the hypothalamic-pituitary-adrenal-axis and might be caused by behavioural responses to stress. Furthermore, epigenetic factors may influence TSA (trochostain A, a histone deacetylase inhibitor), which influences maternal adroitness. Early postnatal experiences may lead to changes in behaviour via the epigenome. Recognising these factors could result in new therapeutic strategies in complex diseases.

A better understanding of epigenetic mechanisms could also cast a cloud over in vitro fertilisation, particularly in cases where the embryos are deep-frozen, defrosted at a later date and placed in the uterus for induction of pregnancy. Several studies report, that the Beckwith Widemann Syndrom (BWS), responsible for disturbances in organs and for an increased risk of tumours, affects test-tube babies more frequently than babies conceived naturally. BWS is caused by a mutation in a growth regulation gene on chromosome 11. Test tube babies are also more likely to be suffering from the Angelman syndrome. The risk of ectopic pregnancies also increases. It is possible that the defrosting-procedure interrupts early developmental stages, which could have their origin in epigenetics.

Why did epigenetic regulation evolve? It is feasible that epigenetic regulation has some advantages in evolutionary terms. One theory suggests that imprinting mimics somewhat parthenogenesis, which results in an unfertilised egg to develop into an embryo. In these instances, the risk of imprinting of a few genes is smaller than the genetic vantages of sexual reproduction. Yet, most scientists support the conflict hypothesis, which puts the idea of imprinting forward as the result of a reproductive contest between the sexes, based on polyandry (a female has sexual contact with more than one male). The true reason behind the phenomena is probably a combination of evolutionary and environmental mechanisms.

Research regarding the first knowable imprinted genes supports the conflict hypothesis. IGF2 (insulin-like growth factor 2) and its receptor, IGF2R, regulate mechanisms from the cell growth to the growth of the whole organism. When we analyse molecular biological data, imprinting first appeared only 180 million years ago. Imprinting of these genes does not occur in e. g. birds or monotremes, but in mammals which do not lay eggs, indicating that imprinting and being able to make epigenetic modifications to the genome may play an important role in speciation (Guerrero-Bosagna *et al.*, 2005).

Recent studies in mammals show that telomere-loss raises epigenetic modifications at telomeric and subtelomeric chromatin domains. Generally, the shortening of telomeres seems to play a role in the aging process, including in human aging and in tumour development. Too much telomere reduction leads to a loss of vitality. Chromatin is conjoined in nucleosomes, which are composed of histones and DNA. Histones, which are often under epigenetic influence, control chromatin structure and gene expression. Admittedly, DNA is also influenced by epigenetic mechanisms and directs chromatin structure and gene expression as well. So, when telomeres reduce, chromatin compaction is also lost. New studies indicate that as histone acetylation increases, telomere shortening through DNA methylation decreases. As a result, the activity of telomerase (the enzyme is responsible for re-completing the ends of the DNA stand) declines. All these modifications could lead to drastic effects on gene expression having an effect on the aging process and on tumour growth (Benetti *et al.*, 2007).

Possible consequences for human health, environment and nutrition

Epigenetics represents a huge opportunity to study an alternative pathway that explains why individuals respond differently to environmental signals e.g. why do certain disease genes seem to affect some people more than others. The still fragmentary knowledge regarding the epigenetic regulatory mechanisms can form the basis for interdisciplinary collaborations with regards to healthcare and environmental policy. The Environmental Genome Project, for example, is currently working on a three phase programme. The first phase attempts to establish a technology for identification in order to increase the number of data sets. From this information, allele-disease links can be created and functional studies of allelic variants conducted. Animal models for disease susceptibility and dose-response relationships as well as risk assessment studies are also necessary. In the third phase, a diseases screening program of high-risk populations is scheduled. All phases are conducted by also considering possible ethical, legal and social impacts.

When screening-tests become available, epigenetic datasets may lead to individualized prevention strategies, or novel ways of diagnosis and curative treatment. It is possible that personal predispositions to certain diseases can be tackled more successfully and in a manner that reduces hazards and paves the way for pharmaceuticals tailored to a particular individual. This knowledge could lead to a new type of biomedical industry.

Nutrition is one of the most important players in the epigenetic repertoire. For example, maternal diet during pregnancy is very important in fetal development. The maternal reproductive tract, arguably, is the environment most critical to the developing mammalian embryo. Gene-nutrient interactions can produce visible as well as stealth changes in embryonic or fetal development, but they set the stage for an adult's susceptibility to a host of diseases and behavioral responses. Unlike defective genes, which are damaged for life, methylated genes can be demethylated. And, methyl tags that are knocked off can be regained via nutrients, drugs, and enriching experiences. No longer are mutant genes sought as the sole cause of disease. The dramatic rise in obesity, heart disease, diabetes and other conditions of prosperous nations are increasingly pegged as epigenetic in nature, and may well claim their origins in faulty embryonic development. We are, quite literally, what we eat as well as what our parents and even grandparents ate. In South Asia, undernutrition in one generation is followed by fat-laden fast foods the next. Children are set up in utero to experience an environment of low nutrition and find themselves in the land of plenty (Duttaroy, 2006)

This understanding could lead to novel interventions with respect to nutrition. One option is to perhaps direct human diets towards a more vegetarian diet (Helms, 2004), which could be supplemented or enhanced in an equally targeted way, based on our increased knowledge in epigenetics and how to modulate its effects. We know already that an abnormal caloric restriction by a prospective mother during pregnancy can have an observable effect on the adult phenotype of her offspring by inducing physiological changes which may lead to diabetic and uterine abnormalities or growth defects for at least three generations. Indications are that epigenetic processes may also impinge on behavior and the psychological make-up of an individual, such as a person's physical, social, and cognitive competencies (Bjorklund, 2006). Furthermore, there is evidence that toxins can lead to epigenetic disturbances of hormones with detrimental affects, which points towards the need for an increase in toxicological assessments of foods in the future.

Possible consequences for society

Many potential effects of epigenetic are complex and not yet well-understood. As further investigations are being conducted, a better understanding of the mechanisms involved could provide the key to improved human health. However, what is needed is cautionary, sensible and rational approach to scientific and normative issues that may arise. Questions might arise that could be similar in scope than those that were raised regarding the use of hereditary information connected to the Human Genome Project. How should society treat people who are at high risk to become ill? Will society compel individuals to take greater responsibility for their own well-being - especially when it becomes apparent that a simple change in diet could avert negative outcomes to an individual's health, or that a healthy, nutritious diet can result in a long and healthy life? If individuals are afflicted with avoidable illnesses, should society pay the cost for medication or hospitalization?

When it becomes evident that toxins and other environmental agents have a damaging effect on the environment, and, through its food chain, a damaging effect on human health, who will be liable for that damage? And how can society rectify that damage? (Olden, 2000). As epigenetic factors impact not only on one generation, but on subsequent generations as well, what impact might this have on family relationships? Furthermore, if we can influence certain qualities through epigenetic means, will we one day decide which qualities we intend to improve, and which ones we want to suppress? What will the impact and outcome of such control and interference be? How important will society regard autonomy and respect for person?

There is a related, but different aspect to epigenetics, which links the environment and sustainability to human flourishing and the common good. The common good pertains to those goods that serve all of us, not just a particular individual, institution, or society. Contributions to the common good often do not offer immediate and tangible benefits, but we regard them as 'the right thing to do'. It provides a rationale for parents looking after their children and other family members; it provides the basis for sacrificing present pleasures in exchange for making provisions for the future; it provides nourishment in times of conflict; and, applied to our topic, it provides a vision for environmental health and sustainability. Lastly, it provides the interconnectedness between humanity and life in general. It is built on relationships, on respect for person, integrity, solidarity, justice and an ethics of care.

It is feasible, that a better experience of epigenetic mechanisms will have a profound effect on society and the understanding of evolutionary mechanisms. Interactions between adaptative and selective processes have been explained in the model of recursive causality as defined in Rupert Riedl's systems theory of evolution. One of the main features of this theory also termed as theory of evolving complexity is the centrality of the notion of 'recursive' or 'feedback' causality, 'the idea that every biological effect in living systems, in some way, feeds back to its own cause'. 'Recursive" or 'feedback' causality may provide a model for explaining the consequences of interacting genetic and epigenetic mechanisms (Haslberger *et al.*, 2006).

A bottom line is that all interferences effecting epigenetics are able to alter gene expression and change the phenotype by modifying the epigenome. In order to avoid detrimental outcomes for human and environmental health, many factors have to be considered. Rather than pursuing reparatory concepts, a change in our behaviour (by simply eating a better diet and exposing us and our environment to less stress) could lead to marked improvements in well-being for every individual, for society and sustainability.

References

Anway, M.D., Cupp, A.S., Uzumcu, M. and Skinner, M.K. (2005). Epigenetic transgenerational actions of endocrine disruptors and male fertility. Science 308:1466-1469.

Benetti, R., Garcia-Cao, M. and Blasco, M. A. (2007). Telomere lenth regulates the epigenetic status of mammalian telomeres and subtelomeres. Nature genetics 39: 243-250.

Bjorklund, D.F., (2006).Mother knows best: Epigenetic inheritance, maternal effects and the evolution of human intelligence. Developmental Review 26: 213-242.

Bjornsson, H.T., Fallin, M.D. and Feinberg, A.P. (2004). An integrated epigenetic and genetic approach to common human disease. Trends in Genetics 20: 350-358.

Dolinoy, D.C., Weidman, J.R. and Jirtle, R.L. (2007). Epigenetic gene regulation: Linking early developmental environment to adult disease. Reprod Toxicol 23: 297-307.

Duttaroy, A.K. (2006). Evolution, Epigenetics, and Maternal Nutrition. Darwins Day Celebration.

Feinberg, A.P. and Tycko, B. (2004). The history of cancer epigenetics. Nature Reviews 4: 1-9.

Guerrero-Bosagna, C., Sabat, P. and Valladares, L. (2005). Environmental signalling and evolutionary change: can exposure of pregnant mammals to environmental estrogens lead to epigenetically induced evolutionary changes in embryos? Evolution and Development 7: 341-350.

Haslberger, A., Varga, F. and Karlic, H. (2006). Recursive causality in evolution: a model for epigenetic mechanisms in cancer development. Med Hypotheses 67: 1448-1454.

Helms, M. (2004). Food sustainability, food security and the environment. British Food Journal 106: 380-387.

　　　　　　　　　　　　　　　　　　　　　Sustainable food production and ethics

Ho, S.M., Tang, W.Y., Belmonte de Frausto, J. and Prins, G.S. (2006). Developmental exposure to estradiol and bisphenol A increases susceptibility to prostate carcinogenesis and epigenetically regulates phosphodiesterase type 4 variant 4. Cancer Reserach 66: 5624-5632.

Jirtle, R.L. and Skinner, M. K. (2007). Environmental epigenomics and disease susceptibility. Nat Rev Genet 8: 253-262.

Jirtle, R.L. and Weidman, J.R. (2007). Imprinted and More Equal. American Scientist: 143-149.

Olden, K. and Wilson, S. (2000). Environmental health and genomics: vision and implications. Nature Reviews 1: 149-153.

Vom Saal, F. (2007). Environmental Health News. 'Fetal exposure to Common Chemicals Can Activate Obesity', 02.2007. http://www.ens-newswire.com/ens/feb2007/2007-02-16-02.asp

Watters, E. (2006). Discovermagazine. 'DNA is not destiny', 11.2006. http://www.geneimprint.com/media/pdfs/1162334912_fulltext.pdf

Local knowledge and ethnoveterinary medicine of farmers in Eastern Tyrol about wild plant species: a potential basis for disease control according to EC Council Reg. 2092/91 in organic farming

Brigitte Vogl-Lukasser, Christian R. Vogl, Susanne Grasser and Martina Bizaj
Working Group for Knowledge Systems and Innovations, Institute of Organic Farming, Department for Sustainable Agricultural Systems, Univ. of Natural Resources and Applied Life Sciences, Vienna, Gregor-Mendel Straße 33, A-1180 Wien, Austria

Abstract

Due to the restrictions for the use of allopathic medicine in organic farming, and an explicit legal statement to favour phytotherapy (EC-Regulation 2092/91), the local knowledge of farmers about phytotherapy is of importance for the organic farming movement. In this project in Eastern Tyrol the local knowledge of 144 organic and non organic farmers about gathered wild plant species and their use as home made remedy in animal husbandry was studied in 2004 and 2005 with tools and methods of ethnobotany and ethnoveterinary medicine. 109 respondents mention 51 plant species that they know explicitly as to prevent diseases and maintain animal health and 144 respondents reported 98 plant species (52 out of them were gathered) and other ingredients, used alone or in combination in home made remedies. The most frequently mentioned plant species are *Achillea millefolium*, *Allium cepa* and *Arnica montana*. In total 1,328 home made remedies were reported. Most home made remedies are known from 'earlier times' only and are not used any more. Local knowledge about the use of plant species in folk veterinary medicine is disappearing rapidly and urgent activities have to be undertaken to safeguard this knowledge as a basis for further experimental trials.

Keywords: home made remedies, ethno-veterinary medicine, organic farming, phytotherapy

Introduction

Gathering of wild plant species is a typical and important activity of many people in rural communities worldwide. Gathering of wild plant species is also an activity done frequently by farmers, including organic farmers. The species gathered by farmers are used usually as food, teas, ornamentals, but also as fodder or for human or veterinary medicinal purposes.

Due to the restrictions for the use of allopathic medicine in organic farming, and an explicit legal statement to favour phytotherapy (EC-Regulation 2092/91), the local knowledge of farmers about phytotherapy is of importance for the organic farming movement. The authors hypothesize, that local knowledge of (organic) farmers about plant based home made remedies to maintain animals health or to cure animals' diseases include time tested experiences of these farmers about effective species, recipies, ways of application and combination with other remedies. This local knowledge might be a starting point for the further development of sustainable animal health care programmes worldwide. Local knowledge about folk veterinary practices is usually studied under the disciplinary context of ethnobotany (Martin, 1995, Alexiades *et al.*, 1996) and ethnoveterinary medicine. Ethnoveterinary medicine is the discipline that puts its focus on local knowledge about prevention and cure of animal diseases (Mc Corkle *et al.*, 1996, Martin *et al.*, 2001). Ethnoveterinary medicine is also called veterinary anthropology (Mc Corkle, 1989).

In this project in Eastern Tyrol the local knowledge of farmers about gathered wild plant species and their use as home made remedy in animal husbandry was studied with tools and methods of ethnobotany and ethnoveterinary medicine. In the study area Eastern Tyrol gathering is a practice widespread among farmers (Christanell *et al.*, 2007). The project (Vogl-Lukasser *et al.*, 2006) is part of an ongoing initiative to analyze local knowledge of (organic) farmers (Vogl and Vogl-Lukasser, 2003).

Study area

The Alpine landscape of Eastern Tyrol is characterized by spruce forests up to 1,700 m above sea level and alpine pastures up to 2,500 m. Annual precipitation in the region is 826 - 1,354 mm and the mean annual temperature is 2.8 - 6.9 °C. Local values depend on exposure to the sun and altitude. This broad range of natural conditions within a small area has lead to a highly diverse pattern of human- environment relationships (Staller, 2001). In Eastern Tyrol, 2,313 farms are managed by families. An additional 445 farms are managed by associations of varying legal status. Of the total of 2,758 farms, 20.3% of them are run on a full time basis; the rest combine farming with off-farm labour. 83 farms are not accessible the entire year by car, 11 farms are accessible only by foot (Brugger, 2001). In Eastern Tyrol, the historical form of agriculture in this region can be described as mountain cereal grazing (Netting, 1981), where arable farming (cereal cultivation, field vegetables, fibre crops etc.) and dairy cattle were the main components of the subsistence system until the 1970s. Large parts of today's meadowlands were once tilled up to 1,700 m altitude. Farming systems in Eastern Tyrol have been in a process of change in the last few decades. Cultivation of cereals, fibre crops and field vegetables (e.g., Pisum sativum, Vicia faba, Brassica rapa ssp. rapa) has been declining during the last three decades, due to unfavourable economic circumstances and their need for high inputs of labour. The economy is dominated by meadowland in lower zones, where hay is produced for winter fodder, and by pastureland in the higher alpine zones, where cattle remain throughout the summer. Nowadays the majority of mountain farms in Eastern Tyrol are based on cattle breeding, milk production and timber harvesting for cash income. Some farmers offer beds for tourists and/or process milk, meat and other products from the farm. For their own consumption, farmers diversify their basic activities by adding activities such as keeping sheep, goats, pigs, hens or bees, and/or growing fruit, herbs or vegetables (e.g., potatoes). Farming is combined with different kinds of off-farm labour and federal subsidies play an important role as income (Vogl-Lukasser, 1999).

Data collection

In the years 2004 and 2005 144 respondents were interviewed about the use of gathered plant species in folk veterinary medicine. Sampling was done as snowball sampling (Bernard 2002), based on local people recommended by the respondents as being 'knowledgable about' home made remedies used in animal husbandry'. The first respondents sampled were those known from previous studies (Vogl-Lukasser, 1999, Vogl and Vogl-Lukasser, 2003, Christanell *et al.*, 2007) as being knowledgeable about this topic. Methods used for data collection included e.g.:
- Freelists (i.e., answering spontaneously to a request, mentioning all items that come to the mind of a respondent when thinking about a given topic or domain; Bernard, 2002) about plant species known (1) to maintain the health of kept animals; (2) to cure diseases of kept animals; (3) home made remedies known; (4) lokal experts for this topic known.
- Semistructured interviews (Bernard, 2002) about a diverse spectrum of topics about harvest, processing and application of gathered plant species in animal husbandry.
- Structured interviews (Bernard, 2002) on socio-demographic data of the respondents.
- Informal interviews (Bernard, 2002) with older people about changes in the gathering of plants and their application in the study area.

In addition, non-participant and participant observation was carried out during gathering, processing, preparation and application of home made remedies.

Collected data was stored in an MS ACCESS database and then analyzed according to ranks and frequencies. Data presented here were analysed according to frequency. In addition (not shown here) freelists were analyzed according to average rank, salience, and consensus (Weller and Romney, 1988; Bernard, 2002) with ANTHROPAC 4. The network data (recommendations of local experts) was visualised and analyzed with UCINET.

Selected results

109 respondents could mention per person between one and 22 ingredients of fodder from 51 plant species that they know explicitly as to prevent diseases and maintain animal health. 18 plant species were mentioned as to be gathered in the wild for this purpose.

All 144 respondents reported 98 plant species (52 out of them were gathered) and other ingredients, used alone or in combination in home made remedies. The 20 most frequently mentioned plant species (*=gathered) were: *Achillea millefolium**, *Allium cepa*, *Arnica montana**, *Artemisia absinthium**, *Avena sativa*, *Brassica rapa* ssp. *rapa*, *Calendula officinalis*, *Camellia sinensis*, *Cetraria islandica**, *Cinnamomum camphora*, *Coffea arabica*, *Gentiana lutea**, *Hordeum vulgare*, *Juniperus communis**, *Larix decidua**, *Linum usitatissimum*, *Matricaria chamomilla**, *Picea abies**, *Sambucus nigra** and *Secale cereale*.

In total 1,328 home made remedies (multiple answers possible) were reported. 1,085 home made remedies have at least one plant based ingredient. Most home made remedies are known from 'earlier times' only and are not used any more.

The social network of the respondents and the persons recommended by the respondents as to be an 'expert' can be characterized as (1) highly fragmented (e.g. only few recommendations of persons in other villages and other valleys; no recommendations of persons outside the district), (2) uncertain about experts on the topic (e.g. many persons with high *outdegree* but no persons with high *indigree*) (3) and missing of persons that might be described from an outside perspective as knowledgeable about the topic. Respondents also report that veterinary doctors of the study area do not recommend plant based remedies.

Conclusion

A remarkable high amount of knowledge about plant based remedies is held not by single folk experts, but dispersed in ideosyncratic units over a large diversity of respondents, who mostly do not practice this knowledge any more. Local knowledge about the use of plant species in folk veterinary medicine is disappearing rapidly and urgent activities have to be undertaken to safeguard this knowledge.

The diversity of home made remedies used by farmers can be a basis for sustainable approaches to disease control, but further efforts have to be made by veterinary science to test them, by veterinary doctors to be trained in their use and by advisory agents to make explicit their importance in organic farming.

The involvement of (organic) farmers' knowledge into the development of sustainable animal husbandry and veterinary programs is not only a potentially rich source of time tested experience. It is of ethical importance: (Organic) Farmers must be understood not as targets, but as key actors in discussions and steps towards animal welfare and organic farming.

References

Alexiades, M.N. and Sheldon, J.W. (1996). Selected Guidelines for Ethnobotanical Research: A Field Manual. The New York Botanical Garden Press; New York, U.S.A.

Bernard, H.R. (2002). Research Methods in Anthropology - Qualitative and Quantitative Approaches. Altamira Press. Walnut Creek, USA.

Brugger, R. (2001). Landwirtschaft. In: Katholischer Tiroler Lehrerverein (ed.) Bezirkskunde Osttirol. Löwenzahn, Innsbruck, Austria, pp. 132-136.

Christanell, A., Vogl-Lukasser, B., Vogl, C.R. and Guetler, M. (2007) The cultural significance of wild gathered plant species in Kartitsch (Eastern Tyrol, Austria) and the influence of socio-economic changes on local gathering practices. In: M. Pardo de Santayana, A. Pieroni and R. Puri (eds.) Ethnobotany in Europe. Berghahn Books, Oxford. In Press.

Martin, M., Mathias, E. and Mc Corkle, C. (2001). Ethnoveterinary Medicine. An Annotated Bibliography of Community Animal Healthcare. Indigenous Knowledge and Developing Series. ITDG Publishing, London, UK.

Martin, G. (1995). Ethnobotany. Chapman and Hall, London, UK.

Mc Corkle, C.M., Mathias, E. and Schillhorn Van Veen, T.W. (eds.) (1996). Ethnoveterinary Research and Development. Intermediate Technology Publications, London, UK.

Mc Corkle, C.M. (1989). Brief Communications - Veterinary Anthropology. Human Organizaiton 48, 156-162.

Netting, R.M. (1981). Balancing on an Alp - Ecological Change and Continuity in a Swiss Mountain Comunity, Cambridge University Press, Cambridge,UK.

Staller, M. (2001). Das Klima. In: Katholischer Tiroler Lehrerverein (ed.) Bezirkskunde Osttirol. Löwenzahn, Innsbruck, Austria, 107-109.

Vogl, C.R. and Vogl-Lukasser, B. (2003). Lokales Wissen von Biobauern über ausgewählte Elemente der Agrarbiodiversität im Bezirk Lienz (Österreich): Zur Bedeutung, Anwendung und Weiterentwicklung ethnobiologischer Forschungsfragen und Methoden in der Forschung im Ökologischen Landbau. In: Freyer B. (ed.) Proceedings: 7. Wissenschaftstagung zum Ökologischen Landbau 'Ökologischer Landbau der Zukunft'. 403-406. Division for Organic Farming. University for Natural Ressources and Apoplied Life Sciences, Vienna, Austria.

Vogl-Lukasser, B. (1999). Studien zur funktionalen Bedeutung bäuerlicher Hausgärten in Osttirol basierend auf Artenzusammensetzung und ethnobotanischen Analysen. PhD Dissertation, University of Vienna, Austria.

Vogl-Lukasser, B., Vogl, C.R., Bizaj, M., Grasser, S. and Bertsch, C. (2006). Lokales Erfahrungswissen über Pflanzenarten aus Wildsammlung mit Verwendung in der Fütterung und als Hausmittel in der Volksheilkunde bei landwirtschaftlichen Nutztieren in Osttirol. Final report Nr. 1272, GZ 21.210/41-II1/03 (Part 1). Available at: http://www.nas.boku. ac.at/fileadmin/_/H93/H933/Personen/Vogl/PDF_1272_VOGL_Wildsammlung03042006.pdf.

Weller, S.C. and Romney, K.A. (1988). Systematic data collection. Sage Publications, Newbury Park, California, USA.

Part 11 - Sustainable food production and consumption

Sustainable value chain analysis

David Simons
Food Process Innovation Unit, Cardiff University, Aberconway Building, Colum Drive, Cardiff, CF10 3EU, United Kingdom, simonsdw@cardiff.ac.uk

Abstract

The Stern report has been a catalyst to raise the profile of carbon emissions in the food sector with several retailers responding with strategic plans. Tesco have embarked on a course of labelling all their products so that customers can compare their carbon footprint (Leahy, 2007). This paper contributes a measure of a product's supply chain footprint, which includes one social value as well as environmental cost. A specialist dairy cheese supply case is used to test the measure, which indicates that the sustainable development footprint of a product will differ from the carbon footprint. This raises the possibility that the reasoning for some categories of local products in the UK is more convincing from a social value standpoint rather than an environmental cost. Therefore it is concluded that there is potential benefit in a robust Sustainable Development measure to identify the trade-offs between carbon footprint and other important social and environmental components of stakeholder value.

Keywords: supply, environment, social, assessment, local

Introduction

Following the Stern report it is clear that carbon emissions are and will remain a key business issue going forward. The report says that 'climate change will eventually damage economic growth', that 'tackling climate change is the pro-growth strategy', and that 'the earlier effective action is taken the less costly it will be'. In a recent speech, Sir Terry Leahy (Tesco CEO) stated that such early action was a major component of the company's future strategy. In particular on understanding the environmental value chain he said that 'We will therefore begin the search for a universally accepted and commonly understood measure of the carbon footprint of every product we sell - looking at its complete lifecycle from production, through distribution to consumption. It will enable us to label all our products so that customers can compare their carbon footprint as easily as they can currently compare their price or their nutritional profile.' (Leahy, 2007).

Such an approach is argued to have limitations as it does not address the underlying issue of consumption patterns and consumer preferences (Lang and Heasman, 2004). Specifically in relation to the Tesco initiative above, (Macmillan, 2007) writes 'communicating food ethics effectively and fairly means dialogue within the food chain about values as well as facts'. To inform such dialogue on values, this paper applies the premise that a supply chain measurement system should be holistic in two respects. Firstly, as suggested by Leahy, it needs to cover the whole lifecycle, and secondly it should be multi-dimensional expressing amongst other things consumer value(nutrition, preferences), social value (employment, communities), environmental cost (carbon footprint) and biodiversity. This paper contributes an incremental step by building on existing environmental supply chain analysis through the contribution of a measure which includes one aspect of social value as well as environmental cost.

The measurement system is applied to a specialist cheese supply chain. A dairy company based in the South West of England supply a distinctive semi-hard cheese to multiple retailers and specialist outlets. The channel analysed is through to the deli counter of one of the major retailers.

Sustainable cost/benefit measure

This paper is based on operations management and supply chain thinking that has moved from a functional to a process approach in recent years (Stalk and Hout, 1990; Womack *et al.*, 1990; Hammer and Champy, 1993; Hines, 1994; Christopher, 2005). The measurement system in this paper is based 'Lean' business movement (Womack *et al.*, 1990; Womack and Jones, 1996), which has advocated process based supply chains and developed techniques to support this goal (Hines and Rich, 1997;Rother and Shook, 1998; Jones and Womack, 2002). These techniques assess the economic effectiveness of each supply chain step that the product traverses from raw materials to the consumer. Also, synergies between economic and environmental focus of lean processes have been identified (Romm, 1994), and environmental analysis has been undertaken (Mason *et al.*, 2002; Sheu and Lo, 2005). This paper extends lean mapping to include social value and proposes that Sustainable Cost/Benefit of the supply chain is expressed as:

$$\text{Sustainable Cost/Benefit} = \frac{\text{Environmental Cost CO}_2 \text{ per kg of product}}{\text{Social Value Minutes of Employment per kg of product}} \tag{1}$$

Where:
- Environmental Cost is CO_2 equivalent of all emissions including methane and nitrous oxide.
- Social Value Minutes of employment per kg of product assesses the quality of jobs through employment multipliers, job quality and job risk.

The whole chain was mapped by a cross-company team of managers; collecting data on energy inputs and employment at each stage of the chain. The main steps in the chain are the dairy farm, cheese factory, consolidation centre, retailer distribution centre and retail store. For the environmental assessment, where primary data was not available estimates were made based on secondary data. Findings are presented on environmental costs and social value between farm and retail point of sale.

Environmental cost

Figure 1 shows each stage of the supply chain along with corresponding cumulative CO_2. The annual fertiliser usage on farm was used to estimate the CO_2 equivalence, principally from nitrous oxide. Taking account of the proportion of milk entering the cheese factory and the yield of the milk to manufacture, this represents over 2kg of CO_2 per kg of finished product, which is about a quarter of whole chain CO_2 (Figure 1). More generally this value is likely to be higher as the farm mapped had made considerable progress in reducing fertiliser input whilst maintaining milk production.

The soil sink in the UK has generally been deteriorating by 0.5% per annum over the last twenty five years (Bellamy *et al.*, 2005). However, there is evidence to suggest that farming practices and deep rooted grasses can reverse this trend. On the farm investigated no longitudinal measurement was available to indicate on the carbon sink performance. However, the pasture had been improved with secondary measurements indicating that the humus content of the soil had improved in recent years. Therefore the production of grass in this instance is assumed from secondary sources to have a negative carbon value.

The growth of cattle and production of milk was the key source of CO_2 equivalence through ruminant methane emissions. This is expected since 'methane production from cattle and other ruminants also contributes about 20% of atmospheric greenhouse gas (GHG) emissions associated with global warming.' (Boadi and Wittenber, 2004). A modelling system developed by University of Melbourne(Crawford and Eckard, 2006) was used to estimate farm emissions, and a conservative value was assumed.

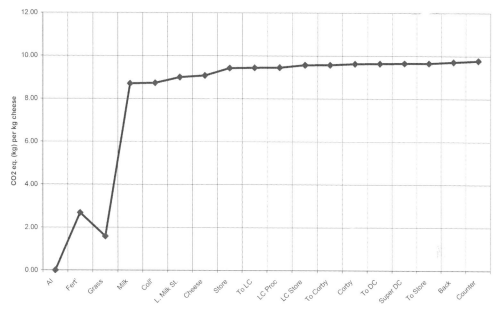

Figure 1. Environmental cost.

Factory emissions were calculated using actual energy use and established power generation CO_2 rates. Transport, chilled distribution and retail were also calculated using standard published CO_2 rates. The cumulative effect was around 10% of the whole chain emissions.

Social value

The first bar on Figure 2 shows the minutes of actual employment required to produce one kilogram of finished cheese. The second bar on the chart takes account of the contribution of the job to the community through factors of full-time, pay rate, locality employment, employment multiplier effect and job vulnerability. The actual employment time in the chain is 31 minutes, which generates 28 minutes of social employment value when these qualitative social values of jobs are considered. The farm and cheese manufacturing processes contribute jobs that have a high qualitative impact as they have a significant multiplier effect in the local economy. There are a significant number of retail jobs as this product is served on the Deli counter but by contrast these jobs have a lower qualitative value due their part-time nature.

Sustainable cost/benefit

Figure 3 depicts cost (kg of CO^2) per benefit (minutes of weighted employment). This indicates that the biggest single cost with little societal employment benefit in the chain is fertiliser due to its high carbon equivalence and largely automated production process. The transportation of product reacts in a similar way with relatively high cost per unit of employment. By contrast the highest environmental cost, ruminant methane, requires significant employment to support the activity giving it a relatively low sustainable cost/benefit.

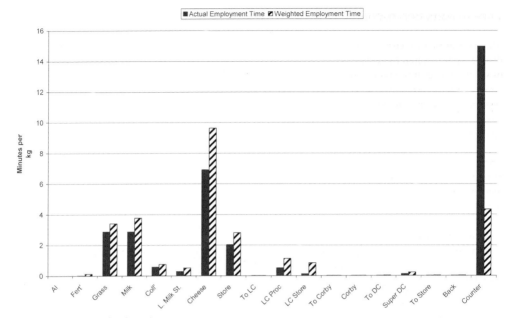

Figure 2. Social benefit of employment.

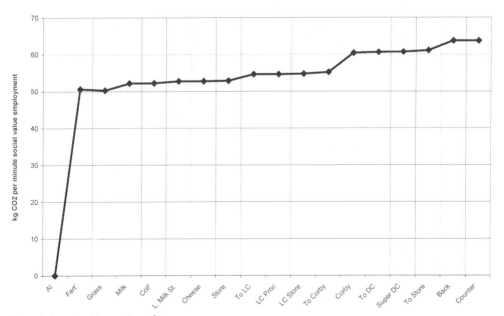

Figure 3. Sustainable cost/benefit.

Sustainable food production and ethics

Conclusions and further research

The environmental cost of this chain is almost entirely based in farm inputs and production, with distribution through food miles (Lang and Heasman, 2004) as a much smaller issue. From observation of similar farms in the same research programme, the farm studied is likely to have lower than average emissions and therefore tend to understate this position. Further study and measurement of these farm systems are required to confirm this and to pin-point farm improvement strategies. In respect of these environmental costs there are significant local benefits in terms of high quality employment in support of rural society.

The CO_2 Sustainable Cost/Benefit assessment provides a different set of priorities than environmental cost alone. It indicates that the biggest issue in the chain is management of fertiliser, and that transport food miles are of significance. Ruminant methane has a relatively small value due to farm employment benefit.

Establishing the carbon footprint of products (Leahy, 2007) is a major step forward in assessing their environmental cost. This covers a major part of sustainable development but does not include the social pillar. This introduction of an overall sustainable development supply chain measure in this paper is one step towards an overall metric. Early results indicate that the sustainable development footprint of a product will differ from the carbon footprint. Such an analysis can question the reasoning for local products in the UK for which there is significant support and a high profile. For the product analysed the local argument is supported principally by increased social value rather than environmental cost. This is unlikely to be an isolated incidence with other scenarios such as airfreighting from disadvantaged developing countries likely to have very high social value. This indicates that there is value in further research to develop a robust Sustainable Development measure to build on assessment of the global warming aspect of the environmental pillar.

References

Bellamy, P.H., Loveland, P.J., Bradley, R.I., Lark, R.M.and Kirk, G.J.D. (2005). Carbon losses from all soils across England and Wales 1978-2003. Nature 437: 245-248.

Boadi, D. and Wittenber, D. (2004). Feeding practices can reduce methane production from cattle operations! Faculty of Agricultural and Food Sciences University of Manitoba.

Christopher, M. (2005). Logistics and Supply Chain Management. London, Pitman.

Crawford, A. and Eckard, R. (2006). Tools for Cutting the Hot Air. University of Melbourne.

Hammer, M. and Champy, J. (1993). Reengineering the corporation. Nicholous Brearly Publishing.

Hines, P. (1994). Creating World Class Suppliers. London, Financial Times/Pitman.

Hines, P. and Rich, N. (1997). The Seven Value Stream Mapping Tools. International Journal of Operations and Production Management 17: 46-64.

Jones, D.T. and Womack, J. (2002). Seeing the Whole - Mapping the Extended Value Stream. Massachusetts, The Lean Enterprise Institute: 1-100.

Lang, T. and Heasman, M. (2004). Food Wars. Earthscan.

Leahy, T. (2007). Green Grocer? Tesco, carbon and the consumer. Tesco, London.

Macmillan, T. (2007). Communicating Food Ethics: whether it likes it or not, Tesco can teach us a lesson. Eursafe News 9: 5-8.

Mason, R., Simons, D., Peckham, C. and Wakeman, T. (2002). Wise Moves Modelling Report, Transport 2000.

Romm, J. (1994). Lean and Clean Management: How to Boost Profits and Productivity by Reducing Pollution. Kodansha Amer. Inc.

Rother, M. and Shook, J. (1998). Learning to See. Massachusetts, The Lean Enterprise Institute: 1-100.

Sheu, H.J. and Lo, S.F. (2005). A New Conceptual Framework Integrating Environment into Corporate Performance Evaluation. Sustainable Development 13: 79-90.

Stalk, G. and Hout, T. (1990). Competing Against Time. New York, Free Press.

Womack, J. and Jones, D.T. (1996). Lean Thinking: Banish Waste and Create Wealth in your Corporation. New York, Simon and Schuster.

Womack, J., Jones D.T. and Roos, D. (1990). The Machine that Changed the World. New York, Rawson Associates.

Sustainable green marketing of SMEs' ecoproducts

Maarit Pallari
MTT, Tutkijantie 2, 96900 Saarenkylä, Finland, maarit.pallari@mtt.fi

Abstract

Sustainable development and entrepreneurship has been studied in Finland from the point of view of small and medium-sized enterprises (SMEs) and their ecoproducts. The aim of the study was to develop means and opportunities for SMEs to communicate the ecological and sustainable features of their products and overall business and to develop tools to assist the entrepreneur in decision-making and marketing. The study was conducted by interviewing four entrepreneurs and two organic farmers. Organic farming and marketing of organic food is regulated in the EU and therefore organic farmers and their marketing communication was used as an example in the study. Sustainable marketing communication focuses on all dimensions of sustainable development, cultural, social, environmental and economic, instead of focusing purely on environmental features of a product. It leads to richer marketing argumentation and informs consumers better about the product's qualities. Most often the environmental features that are used in marketing argumentation originate from life cycle assessment (LCA) analysis and communication is production process oriented. In addition to this, mostly ecological and environmental aspects are communicated which leads to a one-sided marketing communication. The green marketing and environmental features appeal mostly only to the so-called green consumer group and it could be assumed that bringing other dimensions of sustainable development to marketing would attract wider audiences. Informing SMEs better about the possibilities of sustainable entrepreneurship would help them to include sustainable dimensions as part of business strategy and therefore enhance sustainable entrepreneurship.

Keywords: sustainable green marketing, marketing communications, SMEs, eco-positive thinking

Ecopositive positioning

Welford (1995) and Doyle (2001) underline value-based sustainable environment strategy as a part of development and research field for enterprises. It is evident that values have a strong position in sustainable green marketing. Sustainable green marketing uses Peattie's (2001) concept of radical, new sustainable marketing and joins Charter and Polonsky's (1999) concept of green marketing to it. Thompson's (1979) Rubbish theory and Hofstetter's (1998) ideas of LCA analysis are also considered. The results presented in this text are included in my yet unpublished doctoral dissertation. SME's value thinking has an influence on how environmentally friendly action is valued in an enterprise and what is considered a environmentally friendly action. Images can also be negative, when ecoproductization is perceived as a threat for a SME and not as a resource that increases economical and social well-being. On the other hand, a SME is rarely alone with its values; also the immediate surroundings have an influence on the decisions. Ecopositive positioning is a starting point for an environmental marketing management system for SMEs. Ecopositive thinking makes it possible to target the creation process of sustainable marketing correctly. Goal of the value-based marketing strategy is functional integrity where added value is more than just one part of the whole.

Ecopositive thinking comes up when we choose the criteria that are used to value the environmental friendliness. The question is what are criteria of environmental friendliness and what is the starting point of the thinking process. At this moment one has to accept the fact that the criteria of an environmentally friendly product are defined from different perspectives and there does not exist any coherent way of defining the criteria. The classification related to an environmentally friendly product has directed our thinking to questions of green product concepts, enterprises' production processes and strategy decisions (Charter and Polonsky, 1999, Doyle, 2000).

In the ecopositive positioning (Figure 1) it is important to consider the goal of communication and how changing values and stable values can be raised to create strong profile and entrepreneur identity in sustainable green marketing. The positioning clarifies the entrepreneur's identity and the enterprise's profile. Theoretical and practical knowledge and know-how are needed in the positioning to create differentiated ecoproducts. The theory and praxis form possible interlinked ways to develop differentiated ecoproducts.

There are two schools in ecoproduction thinking in between of which we should find a harmonizing connection and they should follow the same line. Other school suggests that product is environmental damaging already when it is born whereas the other school suggests that already before the product is born it can be environmentally protective. Even though the product would already be in markets it can still become a target of eco-production through the theory of New Product Development (NPD, Crawford, 1996) and production process and marketing communication can be improved benchmarking (Barber and Pellow, 2006; Merino New Zealand Inc., 2005) and sustainable green branding (Hartmann *et al.*, 2005; Wong and Merrilees, 2005; Krake, 2005).

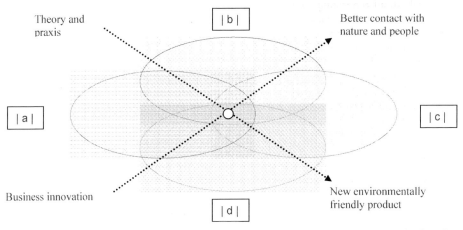

Figure 1. Ecopositive positioning, [Value, |x| = a,environmental +b,economical +c,social +d,cultural]

Production-oriented ecological communication

The entrepreneur has a crucial role in eco-production and in the decision-making process. The content the entrepreneur gives to the product has to be taken into consideration and the concepts the entrepreneur uses have to be evaluated. The weighting of the concepts depends on the entrepreneur's cultural environment, working environment and value-basis. The concepts concerning the production process and the entrepreneur have to be observed and evaluated separately so that marketing arguments can be analysed systematically and become verified. The following chapter shows how this process is realized in practice by interviewing two eco-producers about eco-production and marketing of their products in Austria and Finland. Having interviewees from two different countries gives a wider approach to the European eco-producers' sustainable green marketing opportunities and also difficulties associated with it.

Two eco-producers, one Austrian and one Finnish, have given their expertise concerning eco-production for this research. Interviews showed pilot entrepreneurs' strong commitment to an open and transparent sustainable eco-production and entrepreneurship that improves people's well-being. Eco-producers shared the same kind of ethical, value-based way to see eco-production, they both had an official label for their eco-product and they used the same kind of language. Yet, because of coming from different countries and from different working environments, the pilot entrepreneurs naturally stressed different kinds of expressions. This is why Austrian and Finnish pilot entrepreneur's eco-criteria are presented separately in the following figures (Figure 2 and Figure 3).

Both eco-producers described eco-products and production with the expressions like clean, safe, eco-efficient and natural. Neutral expressions they used were expensive, ideological, standardized, local and traditional. Both eco-producers claimed that the expression 'luxurious' goes really far from eco-product and gives a negative sound to it. Interviews showed that it was important for both entrepreneurs to produce natural, healthy, ecological products for normal families and especially for children. The ideology of producing something luxurious that only few people can afford was strongly negative to them (Figure 2)

Most interesting, when asked about marketing of their ecoproducts, the entrepreneurs did not know what to say. The whole time during the interview (three hours each) the entrepreneurs were talking enthusiastically about ecoproducts, but when it came to the matter of their product's marketing, the entrepreneurs did not have anything to answer. It became obvious that these pilot entrepreneurs did not actually know how to market their ecoproducts, what are the real and functioning marketing channels or what kind of marketing argumentation was included in their products. This was due to their own expertise in primary production which then had effect on the marketing communications and on the ideas of what was considered to be marketing communications. Their marketing relied heavily on the information about the production process. In the discussions with the pilot entrepreneurs it also became clear that the ecoproduct markets are very sensitive to price fluctuations and single failures affect the

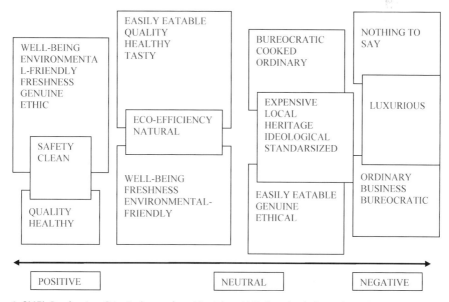

Figure 2. SME's Production Criteria (upper box: Finnish, middle box: both, lower box: Austrian).

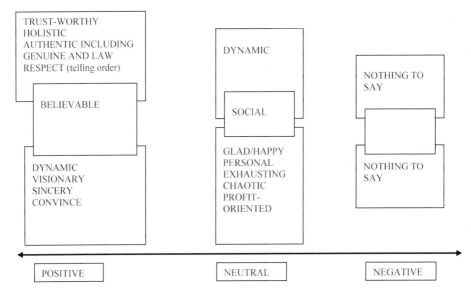

Figure 3. SME Marketing Criteria (upper box: Finnish; medium box; both, lower box: Austrian).

entire eco-production field. Only a few eco-producers seem to have a functioning marketing whereas the majority has to survive with poor marketing opportunities because they lack the central system with functioning networks. The interview also showed that they need clearer visions and personal effort creates trust among customers. Normal supermarkets need a sufficient selection of ecoproducts and they need to be easily accessible to customers.

Marketing-oriented ecological communication

As already mentioned the interviewed pilot enterprises had difficulties to define the marketing of the ecoproducts in detail (Figure 3). As a conclusion both the Austrian and the Finnish entrepreneur emphasized reliability as the strongest criterion for successful and efficient marketing of ecoproducts. The Finnish pilot highlighted the importance of a holistic and genuine ecoproduct whereas the Austrian pilot considered the convincing and dynamic approach to be the most important factor of an ecoproduct's marketing. Both Finnish and Austrian entrepreneur described eco-marketing at the same time chaotic and difficult to do in practice, but also satisfying and rewarding. They also found eco-marketing as a social event. They did not use any analyzing methods to define their ecoproducts' ecocriteria.

Case enterprises did not describe the use of natural resources in own business and in products nor how they have been used in marketing of the enterprise's products. For example, issues like by which energy the products are grown or manufactured or how other natural resources are used did not come up in the interview even though wind power was used in the production. I have approached the issue at first through four different SMEs. None of these is an eco-entrepreneur per se, but each emphasizes some aspect of sustainability in their business idea.

During the interviews the production-oriented way to describe eco-production was emphasized, as respondents were able to describe specific details of the production process, regulations and other official standards connected with eco-products. The pilot entrepreneurs emphasized their contribution to eco-production development and they were worried about the current situation. They found bureaucracy really important but still said that it should not be highlighted too much because it affects the picture that people have of eco-products. The entrepreneurs claimed that arguments arising from the eco-production are not sufficient because they can also have a negative echo. The important part is to find own tools and ways to realize sustainable research and development. Marketing planning cannot be left only to the SMEs' responsibility since they have neither the required know-how nor monetary resources for this. SMEs need analysis, which helps the decision-making processes, for example utility value analysis.

The interviewed pilot entrepreneurs found their work as something that promotes social well-being and cherishes the traditions. Communal activity is the power of eco-production. The SME needs social networks to aid it in marketing. The entrepreneur should build marketing networks by choosing suitable partners from the actors operating outside the enterprise. To strengthen competitive advantage, the entrepreneur should also network with enterprises that have similar values and ways of thinking. My study indicates that eco-producers are strictly committed to the different aspects of sustainable development. If sustainable development is seen as a holistic policy that highlights social well being one can find eco-products' marketing as a way to reach functional integrity of marketing.

Conclusions

Sustainable communications should originate from eco-positive thinking and not focus on the damage-thinking and the language that approach produces. The starting point of eco-positive thinking is environmental protection. The language it produces focuses on protection instead of damaging or harmful aspects of products and production. Eco-positive argumentation should avoid moral tone in communication. The idea is to promote the positive aspects of ecoproducts and environmental protection.

The communication that focuses on LCA analysis (Hofstetter, 1998) and damage-thinking does not realize the full potential of communicating the dimensions of sustainable development. In addition to ecological information, the SMEs should have possibilities to communicate also the cultural, economic and social dimensions of their sustainable entrepreneurship. This way it is possible to promote the entrepreneur's sustainability of business and move the emphasis on environmental issues towards a holistic view of sustainable development's dimensions. The emphasis in marketing communications should therefore move towards sustainable green communications.

References

Barber, A. and Pellow, G. (2006). Life Cycle Assessment: New Zealand Merino Industry. Merino Wool Total Energy Use and Carbon Dioxide Emissions. The AgriBusiness Group: New Zealand: Auckland.

Crawford, C. M. (1996). New products Management. 5 Ed. Irwin series in marketing. Chicago. P. 508. ISBN 0-256-18778-9.

Charter, M. and Polonsky, M.J. (eds.) (1999). Greener Marketing. A Global Perspective on Greening Marketing Practice. UK: Greenleaf Publishing Limited, 432p.

Doyle, P. (2006). Value-based marketing (3rd edition). John Wiley and Sons. England. 370p.

Hartmann, P., Ibáñez, V.A. and Sainz, F.J.F. (2005). Green branding effects on attitude: functional versus emotional positioning strategies. Marketing Intelligence and Planning 23, 9 - 29.

Hofstetter, P. (1998). Perspectives in Life Cycle Impact Assessment. A Structured Approach to Combine Models of the Technosphere, Ecosphere and Valuesphere. Kluwer Academic Publishers. 484p.

Krake, F.B.G.F.M. (2005). Successful brand management in SMEs: a new theory and practical hints. Journal of Product and Brand Management 14, 228 - 238.

Laszlo, C. (2003). The Sustainable Company: how to create lasting value through social and environmental performance. USA, 215p.

Merino New Zealand Inc. (2005). Eco Benchmarking of New Zealand Merino Production Systems. June 2005.

Peattie, K. (2001). Towards Sustainability: The Third Age of Green Marketing. The Marketing Review 2: 129-146.

Thompson, M. (1979). Rubbish Theory. The Creation and Destruction of Value. Oxford University Press, 228p.

Welford, R. (1995). Environmental strategy and Sustainable development. The corporate challenge for the 21st century. Routledge, USA and Canada, 217p.

Wong, H. Y. and Merrilees, B. (2005). A brand orientation typology for SMEs: a case research approach. Journal of Product and Brand Management 14: 155 - 162.

The role of women in promoting ethical consumerism: an historical perspective

Laura Terragni
Norwegian National Institute of Consumer Research (SIFO), P.O. Box 4682, Nydalen, N - 0405 Oslo, Norway, laura.terragni@sifo.no

Abstract

This paper explores the roots of ethical consumerism, focusing on the role of women. Through an analysis of relevant literature, the paper indicates that consumerism has a long tradition dating back to the 19th century, when a modern definition of *consumption* and *consumer* developed. Middle class women in particular played an important role. Excluded from citizenship rights, they turned to the opportunity to use their power as consumers. Consumption became an arena for exerting expertise and for weaving social relations. Boycotting and *buycotting* campaigns developed in Europe and in the USA at the end of the 19th century. Issues such as child labour, workers' rights and the exploitation of nature and of animals were often at the core of these initiatives. One of the most prominent examples is the funding of consumers organisations such as for instance the Consumers League. This organisation, in which women were at the forefront both in the USA and in Europe, urged its members to choose merchandise from a *white list*, as well as to reform their shopping habits. By looking at these first - and gendered - steps of consumer activism, the paper contributes to an understanding of how ethical consumerism is framed in the current debate.

Keywords: ethical consumption, consumer's role, women activism

Introduction

The past decade has seen increasing research into political consumerism, defined as forms of political participation expressed through consumption (boycotts and buycotts or alternative ways of acquiring goods, or the choice of alternative life styles). Ethical issues, in particular environmental concerns, social justice, child labour or discrimination, are often at the core of political consumerism (Micheletti, 2003). In that respect consumers are regarded as political subjects, and consumption is perceived as an arena for expressing and attaining citizenship. Studying forms of contemporary mobilisation expressed through consumption indirectly raises the question of when consumers started to be framed as a specific social actor.

Consumption has always been part of the social life of people, in the sense that consumption has always existed; but is it possible to recognise a turning point in which consumption and consumers began to been defined as such?

This search for the origin of consumers as social actors and of political consumerism has been the subject of numerous historical studies of consumerism (Strasser *et al.*, 1998; D'Aunton and Hilton, 2001; Hilton, 2003; Trentman, 2006; De Grazia, 1996). According to these studies, the rise of consumers as social actors is related to the development of capitalistic society. As Trentmann observes, '*to put it simply, all human societies have been engaged in consumption and have purchased, exchanged, gifted or used object or services, but it has only been in the specific context of the 19th and 20th century that some (not all) practices of consumption have become connected to a sense of being a consumer, as an identity, audience or category of analysis*' (Trentmann, 2006).

Particularly relevant in this process of the formation of a new social actor are a sharper distinction between the spheres of production and consumption; the rise of a middle class that had money to spend; the emergence of advertising; and government interest in consumption as a way of enhancing public policies.

A closer look at the literature of the history of consumption reveals that women were particularly significant in promoting a specific consumer role. In fact, women not only became the subject of special attention as consumers; they also served to give a political relevance to consumption. Lacking political citizenship - as was the case in most nations at least until the first decades of the 20th century - women found that consumption afforded them with an opportunity to express expertise and identity and to establish relations that were both social and political.

This was particularly true for upper and middle class women. Contrary to working class women, who could count the experience of participation from trade unionism or the socialist movement, upper and middle class women (typically non-workers) struggled more in defining autonomous identities and forms of social participation. For them consumption represented an entrance into the public sphere as a recognisable collective subject.

The construction of an identity based both on gender and on the consumer role was the result of several, often interrelated, factors. One of them was the physical creation of spaces specifically devoted to consumption (i.e. shopping arcades); a second was the increasing relevance of the 'housewife' as a partner for the accomplishment of social policies; a third factor was the involvement of women in associations and boycott campaigns focusing on consumption. In the next section, we will briefly consider these three aspects.

The growth of a (gendered) consumer identity

As mentioned above, the creation of physical spaces dedicated to consumption was an important step towards the creation of consumers as specific subjects. Shopping arcades were one of most visible sign of the rise of a consumer society in the late 19th century. Often perceived as a symbol of decadence and of feminine coquetry (Brenna, 2002), shopping arcades have also been regarded - by feminists as well - as venues where women could obtain a public space and where they could move freely and meet each other undisturbed by a masculine presence. As pointed out by Michelle Perrot (1995), the conquest of public spaces in the city represented an important symbolic discontinuity with the past. Previously, women's presence in the cities' landscape had often been related to specific activities and social class (as in the case of servants or working class women) or regarded with suspicion (as in the case of women visiting *cafés*). Respectable women were seldom out alone. Shopping arcades offered these women the opportunity to cross the household boundaries without losing their reputation. Meeting other women at the *tearoom*, an indispensable part of the experience of visiting the arcade, helped middle class women to develop autonomous friendships. Moreover, in order to get to these arcades, which were usually placed in the city centre, women had to make a journey through the streets or take public transport, increasing their visibility as autonomous subjects in the life of the modern city. The rise of a consumer society (here identified by shopping arcades) contributed to the institutionalisation of an identifiable group of women, for whom being identified as 'consumers' was an important factor for the creation of a personal as well as a collective identity.

The definition of consumers as a specific group in which women played a predominant role is even more relevant in the context of policy making. This is particularly true with regard to public policies which explicitly referred to and were based on the presence of a particular category of women, who made consumption their main activity (not only by purchasing goods but also transforming and organising them and making them available for others): the housewives. The supremacy of a *male breadwinner regime* heightened the distinction between the places of production and the places of re-production and consumption, fundamentally characterised by a gender segregation. Taking care of the house, cooking proper meals for the family and being an efficient administrator of the household finances became essential values in the definition of a gender identity and, in particular, in the construction of a gendered subject - the housewife - that was both private and public. Many policies aimed at protecting or enhancing national wealth or nationalisic values appealed to housewives, as for instance described by Reagin in her study of interwar Germany (1998), De Grazia's analysis of women under fascism in Italy (De Grazia, 1996) or Thein's description of the mobilisation of Norwegian housewives after the Second World War (Thein, 2005). Housewives' organisations, which were founded from the 1920s onwards, represented an arena for social participation and involvement that widened the opportunities for non-working women to become involved in activities outside the home.

Finally, consumption became relevant as a direct form of mobilisation and activism. Following the pattern of the food riots in the Ancient Regime, women were often at the forefront of food protests, demonstrating against the shortage of specific goods or opposing price hikes. Furthermore, the women affiliated with the socialist movement had an important role in establishing and leading the consumers' co-operative movement. Among the increasing mobilisation of people as 'consumers' that took place in the late the 19th and early 20th century, of particular interest is the creation of consumers' associations. These organisations, which tended to distance themselves from the associationism inspired by socialist ideology, mobilised participation from a more heterogeneous background, ranging from the Catholic activism inspired by Pope Leon XIII to the involvement of an upper or middle class that developed criticism towards some of the aspects of the capitalistic society.

Starting in 1890, consumers' organisations were established both in Europe and in the USA. The history of these organisations, at least until the 1920's, is most of all a history of women's activism: women were in fact at the very core of these organisations, both as members and as leaders.

Among the initiatives promoted by these consumers' organisations, of particular relevance was the 'white list campaign' (which later became the 'white label campaign' in the USA) aimed both at expressing protest about goods produced under questionable working conditions and at promoting products which, instead, were manufactured in accordance with legislation. The next section examines this particular campaign in more detail.

The white label campaign

'The responsibility for some of the worst evils from which producers suffer rests with the consumers, who seek the cheapest markets regardless of how cheapness is brought about. It is, therefore, the duty of consumers to find out under what conditions the articles they purchase are produced and distributed and to insist that these conditions shall be wholesome and consistent with a respectable existence on the part of the workers' (Josephine Shaw Lowell, head of the Consumers League of New York - quoted in Storrs, 2000: 21).

These words characterise the essence of the initiative known as the white list or white label campaign: the assumption of responsibility by consumers for the conditions in which production takes place. Kish Sklar, in her study of the white label campaign in the USA, defines this as an *informed morality*, which implies a threefold responsibility on the part of consumers: to recognise the direct relationship with the producer, to learn about producers' working class conditions and to limit purchases to goods produced under ethical conditions. (Kish Sklar, 1988: 27). The consumers' organisations, in the USA as well as in Europe, encouraged their members to inform themselves about the working conditions of dressmakers and shop girls in department stores. Members were asked to choose 'correctly' from a 'white list' of 'good dressmakers' who were thought to treat their workers well and who promised not to make them work at night or on Sundays.

The origins of the white list (or white label) campaign can be traced across the Atlantic as a result of cooperation between reformers in Europe and the USA. Experience with the white list campaign can in fact be traced in some European countries (such as France, Italy or France -Chessel, 2006), although it is probably in the USA that this initiative achieved the greatest relevance - at least if we consider the number of studies focusing on the American experience.

According to Storrs, the first white label was affixed to cigar boxes in 1874 to identify the product of white labour, as opposed to the labour of Asians receiving substandard wages. However, the textile industry and department stores were more often the main target of the campaign (Storrs, 2000). In 1891, for instance, a 'white list of department stores' was compiled in New York by Josephine Shaw Lowell and other members of the New York City Consumers League.

The white label campaign replaced the white list initiative when Florence Kelly was elected general secretary of the National Consumers League in 1899 (Kish Sklar, 1998: 19). As one of her first initiatives, she designed the Consumer White Label and launched the campaign on a national scale. The campaign focused mainly on *'women's and children's [machine] stitched white cotton underwear'*, including *'corsets and corset substitutes, skirt and stocking supporters, wrappers, petticoats, and flannelette garments'* (quoted in Kish Skal, 1988: 24).

Members of the League were granted the status of inspectors by the government. During their visits to factories and department stores, they controlled that federal legislation governing labour conditions and child labour was in place. At the height of its success, in 1904, the League had licensed sixty factories.

Although the White Label Campaign was officially concluded during the First World War, it had already begun to slow down in 1906. Difficulties in keeping pace with inspection activities, combined with an unwillingness to compete with trade unions, made the League turn its efforts towards other goals, such as minimum wage legislation (Kish Sklar, 1998).

It is difficult to evaluate the impact of the white list or white label campaigns in terms of improved working conditions. What these initiatives undoubtedly achieved was to provide middle class women with an arena for activism and increased social awareness. According to Kish Skal, 'the campaign deserves a place as one of the most extensive expressions of women's political activism before the passing of the nineteenth amendment in 1920. The campaign drew women into public life in ways that validated what might be called their social citizenship almost twenty years before the passage of the women's suffrage amendment to the constitution' (Kish Sklar, 1998: 33-34).

Conclusions

This article has described in brief the role played by consumption in framing the growth of a new political subject: the consumer. As argued by the literature that has investigated the origin of the consumption society, this new political subject was - at least at its origin - deeply characterised by a gender dimension. Consumption appealed to women as users of new facilities as well as a main target of public policies. Furthermore, as we have seen, consumption gave women - particularly middle class women - new opportunities for social and political participation. The 'white label' movement, which developed both in Europe and in the USA in the late 1800s and early 1900s, is an example of this activism. This experience can be regarded as one of the first examples of 'ethical consumption', as it asked consumers to make choices of products and producers based on the way these products were manufactured. Violations of work regulations, use of child labour, and unhealthy working conditions were the core concerns of the white label activists, showing a remarkable parallelism with the current 'sweatshop' campaigns (e.g., the Clean Clothes Campaign). The similarities do not end here: now, as then, women tend to be most active in practising ethical consumerism (Micheletti, 2003).

Moreover, as pointed out by Chessel, the white label campaign was and remained a middle class phenomenon. At the core, it was fundamentally paternalistic (or materialistic), and the gap between consumers (non-workers) and workers was a fundamental dimension of that experience. *The consumers' league did not seek to help workers but to educate men and women of the middle-class in to help workers* (Chessel, 2006: 85). A similar divide tends to characterise contemporary campaigns as well. They make a distinction between consumers in the richer part of the world and producers in the poorest countries; these campaigns are addressed particularly towards wealthy consumers. The concept of 'informed morality' used by Kish Sklar in describing the white label campaign may, to a large extent, aptly describe the current emphasis on ethical consumption. Finally, we may argue that contemporary forms of mobilisation related to consumption generate arenas of political participation for subjects (in this case both women and men) who have lost stronger forms of affiliation and are lacking defined collective identities.

These considerations suggest the presence of relevant links between past and present forms of ethical consumption and between forms of social inclusion -and exclusion- and political consumerism. Further investigation on these topics is however needed.

References

Brenna, B. (2002). De kjøper, altså er de. Skisser til historien om forbrukerforskningssamfunnets framvekst. Tidsskrift for kulturforsking 1,2:5-20.

Chessel, M. (2006). Women and Ethics of Consumption in France at the Turn of the Twentieth Century: the Ligue Sociale d'Acheteurs, in F. Trentmann (ed.). The Making of the Consumer. Knoweledge, Power and Identity in the Modern World. Berg, Oxford, United Kingdom, pp.81-97.

D'Aunton M. and Hilton, M. (2001). The Politics of Consumption. Material Culture and Citizenship in Europe and America. Berg, Oxford, United Kingdom, 310p.

De Grazia, V. (1996). Nationalizing Women: the competition between Fascist and Commercial Cultural Models in Mussolini's Italy. In V. de Grazia (ed.) Sex of Things. Gendered Consumption in Historical Perspective. University of California Press, Berkeley, California. pp.337-357.

Hilton, M. (2003). Consumerism in the 20th Century Britain. Cambridge University Press, Cambridge, United Kingdom. 382p.

Kish Sklar, K. (1998). The Consumers' White Label Campaign of the National Consumers' League, 1989-1918. In S. Strasser, Charles McGovern and M. Judt (eds.) Getting and Spending. Cambridge University Press, Cambridge, United Kingdom, pp.17-35.

Micheletti, M. (2003). Political Virtue and Shopping. Palgrave, New York, 247p.

Perrot, Michelle (1991). Uscire. In G. Duby and M. Perrot (eds.) Storia delle Donne: L'ottocento. Laterza, Bari pp.256-281.

Reagin, N. (1998). Comparing Apples and Oranges: Housewives and the Politics of Consumption in Interwar Germany. In S. Strasser, Charles McGovern and M. Judt (eds.) Getting and Spending. Cambridge University Press, Cambridge, United Kingdom, pp.241-261.

Storrs, L. R.Y. (2000). Civilizing Capitalism. The National Consumer's League, Women Activism, and Labour Standards in the New Deal Era. The University of North Carolina Press, Chappel hill and London, 392p.

Strasser, S., McGovern Charles, Judt M., (1998). Getting and Spending. Cambridge University Press, Cambridge, United Kingdom, 478p.

Theien, I. (2005). Campaigning for milk. Housewives as consumer activist in post-war Norway, in H. Roll-Hansen and G. Hagemann (eds.) Twentieth-century Housewives: Meaning and Implications of Unpaid Work, Oslo Academic Press, Oslo.

Trentmann, F. (2006). The Making of the Consumer. Knoweledge, Power and Identity in the Modern World. Berg, Oxford, United Kingdom, 318p.

Waste not want not - the ethics of food waste

André Gazsó[1] and Sabine Greßler[2]
[1]*University of Vienna, Faculty for Geosciences, Dept. of Risk Research, Turkenschanzstrasse 17/8, 1180 Vienna, Austria*
[2]*Forum Österreichischer Wissenschaftler für Umweltschutz, Mariahilfer Strasse 77-79, 1060 Vienna, Austria*

Abstract

Several contradictory trends are currently going on in industrialised countries: while a growing amount of persons are able to spend an increasing portion of their income for healthier and higher-quality food the share of housholds which are suffering permanently or temporarily from food insecurity is still growing. On the one hand intensified industrial food processing and dense supply networks are able to provide an ever growing variety of products and brands and on the other hand this surplus contributes to the solid waste of municipalities. Food waste includes uneaten, unsold or unserved portions of meals and food preparation remainders of kitchens and restaurants. Although food waste is only the third-largest component of generated waste by netweight it is the largest component of discarded waste by weight because of its low composting rate.

Introduction - facts and figures

Every US citizen is producing app. 250 to 600 g of food waste per day. The EPA concludes that increased consumption of packaged foods was a key factor causing food waste's share of the solid waste stream to decrease by one-sixth in the period from 1960 to 2000. The USDA estimates that higher percentages of fresh fruits and vegetables, dairy and grain products are thrown away, while lower percentages of meat, dried beans, nuts and processed foods are disposed of. Every year normal households lose up to 600 USD on food (Jones, 2005).

While food losses occur at all levels of the whole food system (farm, farm-to-retail, retail, consumer and food services), losses and waste production on most of these levels cannot be controlled by the individual consumer. At retail level for example 5 billion pounds of food were lost in the US in 1995 (less than 2% of the edible food supplies), but over 90 billion pounds at consumer level (26% of edible food supplies). Fresh fruits and vegetables accounted for nearly 20 percent of consumer and foodservice losses. This shows that at this level there is a considerable capacity to simultaneously reduce several food related risks.

In Austria every citizen causes between 110 and 404 kg of residual waste per year. Up to 60% are food discards and food packing materials. Between 6 and 12% of food waste in Vienna and Lower Austria are orginally packed products or food that has never been eaten (Wassermann, 2006). That means that Austrian households throw away up to 79 kg per person and year and spend up to about 390 Euros for food going to waste (Schneider und Wassermann, 2005).

UK households discard 6.7 million tons of food every year, accounting for around 30-40% of all the food bought. About half of this is edible food (WRAP 16.3.2007). Britains spend about 400 Pounds per year for food to end up in the garbage bin just because of sell-by dates (BBC News 14.4.05). Although there is limited information available on the amount of waste generated by the agricultural sector and waste generated by the retail food and drinks sector, several studies show that the food and drinks sector accounts for a substantial amount of the UK's annual commercial waste. By the end of 1990s the food manufactoring sector was estimated to generate over 8 million tonnes of waste annually (Food Processing Faraday, 2003).

But not only the consumers throw away food they have bought but never eaten, a large amount of products is thrown away before selling by supermarkets, bakeries and food companies. In Vienna per year approximately 1300 tons of edible food remain unused by discounters (3 companies, 93 branches). Every day 12,800 kg of bread and other bakery products are discarded by Viennese bakeries because they remain unsold at the end of the day (and the employees are not allowed to take unsold products home). This sums up to an incredible amount of 3,600 tons of perfectly fresh bread and pastries every year - the annual need of about 50,000 people (Schneider und Wassermann, 2004).

Food discards recovery

According to the U.S.Department of Agriculture Economic Research Service, if 5% of consumer, retail, and food service food discards from 1995 were recovered, savings from landfill costs alone would be about $50 million dollars annually (Kantor *et al.*, 1997). Food discards comprise 6.7% by weight of the total U.S. municipal solid waste stream. In 1995, 14,000,000 tons of food discards were generated. Of this, only 4.1% - 600,000 tons - were diverted, or recovered, from the traditional disposal destinations of landfills and incinerators (US EPA, 1997).

Depending on the quantity and type of food discards there are several methods of food recovery and deverting food discards from landfills. These methods involve all actors of the food production and consumption chain (US EPA, 1998).

- Food Donations: Food can be donated to local food banks, soup kitchens, and shelters provided that it is non-perishable or unspoiled.
- Animal Feed: Recovering food discards as animal feed is a well-established method in areas of pig farming. Some types of food, such as foods with high salt content can be harmful to livestock. Some food discards can be converted into a high-quality, dry, pelleted animal feed. Food discards are also used to make pet food.
- Rendering: Liquid fats and solid meat products can be used as raw materials in the rendering industry, which converts them into animal food, cosmetics, soap, and other products.
- Composting: Composting can be done both on and off-site depending on available landscape. On-site composting will need to assess carbon/nitrogen ratios. Temperature and aeration are other important factors that will determine how long it takes materials to compost. There are several forms of composting in use:
  - Unaerated Static Pile Composting: Organic discards are piled and mixed with a bulking material. This method cannot accommodate meat or grease.
  - Aerated Windrow/Pile Composting: Organics are formed into rows or long piles and aerated either passively or mechanically. This method can accommodate large quantities of organics as long as there is a careful temperature and moisture control available. Large amounts of meat or grease have to be frequently turned.
  - In-vessel Composting: Composting vessels are enclosed, temperature and moisture controlled systems including a mechanical mixing or aerating system. In-vessel composting can process larger quantities in a relatively small area more quickly than windrow composting and can accommodate animal products.
  - Vermicomposting: Worms (usually red worms) break down organic materials into a high-value compost (worm castings). This method is faster than windrow or in-vessel composting and produces high-quality compost. Animal products or grease cannot be composted using this method.

Organised food donation

In many Western countries great efforts are undertaken by non-profit organisations to fight food waste and to get food to the people who need it. For redistributing surpluses of agricultural products and of retail trade as well as overproduction in the food processing industry in 1967 the first 'Food Bank' was established in Phoenix, USA. Meanwhile there are more than 200 member Food Banks and food-rescue organizations serving all 50 states of the USA, the District of Columbia and Puerto Rico as well as 196 Food Banks in 17 national federations all around Europe. The Food Banks operate as a kind of wholesaler, picking up food provided free of charge by farmers, processing plants, wholesalers, importers, retailers, public authorities and civic organizations. They organize transport of the food collected, sorting and storage, quality control, monitoring of the storage conditions and inventory management. The distribution of food is carried out by voluntary associations which fight hunger on the ground. In 2005 almost 220,000 tons of food could be collected and distributed by the European Food Banks (Fedération européenne des Banques Alimentaires, http://www.eurofoodbank.org).

In Germany and Austria the 'Tafel' initiative redistributes food directly to the poor and people in need. And there are a lot of people even in the industrialized countries who are in need of emergency food. In Vienna, Austria, every day 2 tons of food is collected and dispensed. Food worth about 1.5 million Euros is distributed per year to 5,500 people in 55 social institutions (Wiener Tafel; http://www.wienertafel. at/). Even in the USA hunger is an every day problem for many. 'America's Second Harvest - The Nation's Food Bank Network' (A2H) serves an estimated 24 to 27 million people annually. Approximately 4.5 million different people receive emergency food assistance from the A2H system in any given week (Hunger in America Study, 2006; http://hungerinamerica.org/key_findings/).

Overproduction, waste and environmental impacts

There is a general agreement that the production, processing, transport and consumption of food accounts for a significant portion of the environmental burden imposed by any Western European country (Foster *et al.*, 2006). Industrialized agriculture causes soil erosion, contamination of water bodies with farm chemicals such as fertilizers and pesticides, loss of biodiversity and contributes considerably to global warming by producing greenhouse gases such as CO_2 and CH_4. Especially meat and meat products have the greatest environmental impact with estimated contributions in the range of 4-12% for Global Warming Potential (Foster *et al.*, 2006). In Austria an average release into the atmosphere of 2600 kg of CO_2 equivalents per year and capita is caused by human nutrition. This is equivalent to about 1000 liters of fuel (Salmhofer *et al.*, 2001).

For the irrigation of farm land an enormous amount of water is necessary; e.g. 40 liters of water are needed for growing 1 kg of tomatoes (Foster *et al.*, 2006). Many countries in Europe such as Spain and Italy face already severe problems in water supply caused by excessive use of water ressources and the effects of increasing drought due to global warming.

Transportation of food procucts has also severe environmental impacts, such as air pollution by exhausted gases, noise and the destruction of natural habitats and living space to build transnational transportation routes. The demand of the consumers for fresh fruits and vegetables all year round at low prices and the globalisation of the world economy result in long-distance transportation of all kinds of foodstuff with all its negative environmental effects.

Agriculture, food production and transport also have a high energy consumption. It is estimated that in the developed countries about 20% of the total energy is used for the food sector (Salmhofer *et al.*, 2001).

The industrialized overproduction of food and the behaviour of the consumers, who buy more food they can eat as cheap as possible albeit how and where the food has been produced, therefore contribute considerably to several environmental problems.

Sustainability, risk and waste

In general the term sustainability refers to the ability of a society, ecosystem, or any such on-going system to continue functioning into the indefinite future without being forced into decline through the exhaustion or overloading of key resources on which that system depends. Risk management and sustainable development are two strategic frameworks utilised for studying and managing the environmental consequences of human actions. Understanding sustainable development as risk mitigation strategies emphasises the conception that risk assessment and sustainability research complete each other insofar as sustainability research continues where risk assessment comes to an end. In other words goals agreed upon as sustainable can serve as management options within the risk regulation process.

Given that producing waste is not intrinsically immoral because waste occurs necessarily as byproduct of life sustaining processes, the ethical quality appears not until the aspect of uneven distribution of resources is considered. In this case the management of waste contains also questions of equity and justice, normally expressed as either imposing threats to the health of unprivileged social groups by hazardous material or by additionally deteriorating future chances of these groups by the destruction of environmental resources they are dependant to live on.

Food waste constitutes the largest single proportion accounting for an average of 21% of total commercial waste. Additionally most of the food has been processed, packaged and transported extensively before it reaches the end consumer. These activities themselves are often associated with the generation of large amounts of wastes. According to the four elements of the waste hierarchy (reduction - reuse - recovery - disposal; DETR 2000), the best way to eliminate food waste is not to create it. This can be accomplished by streamlining processes and using resources in a more efficient way.

The British organisation 'Sustain - Alliance for Better Food and Farming' uses certain criteria to indicate what sustainable food could and should be like (http://www.sustainweb.org):
- Proximity, meaning that the food originates from the closest practicable source or the minimisation of energy use in production, storage and transport.
- Healthy, which means that the product does not contain harmful biological or chemical contaminants and is part of a balanced diet.
- Fairly traded between producers, processors, retailers and consumers.
- Non-exploiting of employees in the food sector in terms of pay and conditions.
- Environmentally beneficial or benign in its production (e.g. organic).
- Accessible both in terms of geographic access and affordability.
- Applying high animal welfare standards in both production and transport.
- Socially inclusive of all people in society.
- Encouraging knowledge and understanding of food and food culture.

References

BBC News, 14.4.05. Britons throw away a third of food. http://news.bbc.co.uk/1/hi/uk/4443111.stm
Center for Food Safety (2007). Cloned Food: Coming to a Supermarket Near You? Food Safety Fact Sheet, January 2007 (http://www.centerforfoodsafety.org).
DETR - Department of the Environment, Transport and the Regions (2000). Waste Strategy 2000 for England and Wales. HMSO, London.

Food Processing Faraday (2003). Waste minimisation, reuse and recycling strategies for the food processor. Technical Report 27 pp.

Foster, C., Green, K., Bleda, M., Dewick, P., Evans, B., Flynn, A. and Mylan, J. (2006): Environmental Impacts of Food Production and Consumption: A report to the Department for Environment, Food and Rural Affairs. Manchester Business School. Defra, London.

Jones, T.W. (2005). Clean your plate already. Video, Discovery Channel, 12 Jan; (http://www.exn.ca/dailyplanet/view. asp?date=1/25/2005)

Kantor, L.S., Lipton, K., Manchester, A. and Oliveira, V. (1997). Estimating and Addressing America's Food Losses. USDA, 1997, page 8: Advance release of same article in FoodReview,Vol. 20, No. 1, Jan.-Apr.

Salmhofer, C., Strasser, A. and Sopper, M. (2001). Ausgewählte ökologische Auswirkungen unseres Ernährungssystems am Beispiel Klimaschutz. Natur und Kultur: Transdisziplinäre Zeitschrift für ökologische Nachhaltigkeit 2(2): 60-81.

Schneider, F. and Wassermann, G. (2004). SoWie. Sozialer Wertstofftransfer im Einzelhandel. Endbericht Initiative 'Abfallvermeidung in Wien', Institut für Abfallwirtschaft, Universität für Bodenkultur, Wien.

Schneider, F. and Wassermann, G. (2005). Original verpackte Lebensmittel im Müll. Forschungsbericht ABF-BOKU, Institut für Abfallwirtschaft, Universität für Bodenkultur, Wien.

U.S. Environmental Protection Agency (1997). Characterization of Municipal Solid Waste in the United States, pp. 5-6.

U.S. Environmental Protection Agency (1998). Don't Throw Away That Food. Strategies for Record-Setting Waste Reduction.

Wassermann, G. (2006). ABF-Newsletter, Institut für Abfallwirtschaft, Universität für Bodenkultur, Wien. März 2006.

WRAP (Waste and Resources Action Programme), 16 March 2007. New WRAP Research Reveals Extent of Food Waste in the UK. http://www.wrap.org.uk/wrap_corporate/news/new_wrap_2.html.

Ruminant feeding in sustainable animal agriculture

Wilhelm Knaus
Division of Livestock Sciences, Department of Sustainable Agricultural Systems, University of Natural Resources and Applied Life Sciences, Gregor Mendel-Strasse 33, A-1180 Vienna, Austria, wilhelm.knaus@boku.ac.at

Abstract

Animal agriculture is becoming a growing area of conflict between economists and ecologists, especially in countries where agricultural development has reached a high level of industrialisation and specialization. Environmental concerns have come up because animal farms are growing in size and increasingly dependant on purchased feeds, fertilizers and pesticides in order to maintain or increase the level of animal production. Sustainability ultimately depends upon our use of energy, because anything that is useful in sustaining life on earth ultimately relies on energy. The second issue of major concern regards the mass nutrient balance on farms. Nutrients normally concentrate on livestock farms because more are imported than exported in products sold. As more livestock farms have become highly dependent on exogenous inputs like feedstuffs, fertilizers and pesticides that have been produced by the use of non-renewable energy and resources, the promotion of and search for animal production systems that can be maintained by solar energy-driven, locally available resources is becoming urgent. Ruminant husbandry has the potential to play an enormously important role in sustainable agricultural systems. This is primarily due to the fact that ruminants are capable of feeding on plant substrate (grassland) that cannot be directly consumed and digested by humans. It is no coincidence that cattle, sheep and goats have been domesticated. These animals have the potential to transform energy and nutrients exclusively from feedstuff into foods (milk and meat) that play an essential role in human nutrition. The grazing of grassland is an example of one of the most sustainable forms of agriculture available. Domestic ruminants therefore already play a crucial role in food provision, and will continue to do so in the time to come.

Keywords: feeding, food, ruminant, sustainability, agriculture

Introduction

In affluent countries where food is produced in excess, animal agriculture is increasingly becoming an area of societal concern, as external inputs on farms have been elevated in order to reach and maintain a high level of animal performance. Animal nutrition plays a central role in animal production. The aim is the production of organoleptically and nutritiously high-quality foods such as meat, milk and eggs by feeding less valuable or otherwise not consumable feedstuffs to farm animals (Stangassinger, 1993).

A growing number of people are becoming ecologically aware and see the maintenance of an intact environment as the basis for the production of wholesome foods. According to Pfeffer (1992), intact ecosystems are characterized by the fact that nutrient cycles are complete and that no long-term depletion or accumulation of nutrients occurs. The transportation of nutrients in opposite directions over great distances must ultimately balance out to avoid nutrient imbalances.

Over the last decades the increasing industrialisation and specialisation of agriculture has moved away from this ideal at an ever growing pace, and is increasingly endangering its own production basis. Regions with an unnatural high density of farm animals have developed, because the shipping of huge amounts of feedstuffs over long distances has become technically and economically feasible. These farm areas have been the first to face serious nutrient imbalance problems.

In this article, we will look at the role of ruminant feeding in sustainable animal agriculture, a form of agriculture in which fibrous plant foodstuffs are transformed into milk and meat in the most efficient way, while minimizing possibly negative effects on the environment at the same time.

Defining sustainability

The term sustainability has become omnipresent: It obviously can no longer be excluded from any statement or discussion that is in any way oriented towards development, progress and our common future. The term originated in the forestry industry at the beginning of the 18th century, and plausibly describes a logical form of resource management, i.e., harvesting only as much wood as can be expected to grow back. Almost 200 years later, the Brundtland-Report (WCED, 1987) once again made it unmistakably clear what is meant by sustainable development: 'Sustainable development meets the needs of the present without compromising the ability of future generations to meet their own needs'.

This statement must be held up as a benchmark to current agricultural production and our use of natural resources. Mankind began to develop sustainable forms of agriculture millennia ago, but most scientists concur that current systems and their rates of resource extraction will lead us to a vastly depleted earth in the future (Vavra, 1996).

Agriculture and sustainability

A close look at the basic structure and function of ecosystems shows that agriculture can be defined as the handling of natural resources to capture solar energy by plants that can directly or indirectly (i.e. through animal products) serve as food for humans. An understanding of how ecosystems work helps us understand and define sustainable agriculture. According to Heitschmidt *et al.* (1996), 'sustainable agriculture may be broadly defined as ecologically sound agriculture and narrowly defined as eternal agriculture, that is, agriculture that can be practiced continually for eternity. It is those forms of agriculture that do not necessarily require exogenous energy subsidies to function. For example, grazing of indigenous grasslands is one of the most sustainable forms of agriculture known. This is because no other form of agriculture is less dependent on external finite resources, such as fossil fuels, and (or) external, potentially environmentally sensitive resources such as fertilizers, pesticides, and so on, than grazing of native grasslands'.

In livestock production, sustainability can be understood as the harvest of the same quantity of meat, milk or fibre from a given land area indefinitely. In other words, the removal of agricultural products (meat, milk, fibre) should not reduce the ability of the land-area to continue providing substrate (e.g. forage) for consecutive removals (Vavra, 1996).

Evolutionary adaptation of domestic ruminants

According to Engelhardt at al. (1985), there can be no doubt that mankind deliberately domesticated those ruminant species that were able to easily digest high-fibre feeds as a result of their evolutionary development. The feeds these animals are intended by nature to consume are inedible to humans. It seems therefore unnatural, perhaps even immoral, that in the wealthy countries of the world, ruminant production is based on grain feeding (Hofmann, 1989). As we all know, this trend has increased steadily until now, a development that can only serve to exacerbate the growing conflict between the affluent and the poor.

Van Soest (1994) describes very clearly the reason why ruminant nutrition has taken this course: 'The feasibility of feeding all-concentrate rations to ruminants was in doubt before 1950, but the fact that the cost per unit of net energy was less for corn grain than for forage pushed ruminant nutrition research to solve the problems of digestive disturbances that frequently resulted from concentrate feeding. (Most feedlot animals do not live long enough to experience the full toll of rumen acidosis, parakeratosis, and abscessed livers that are the result of overfeeding grains with too little dietary fibre)'.

Thus, the common practice of feeding ruminants the same diets as monogastric animals, i.e. ignoring the evolutionary purpose these animals are intended by nature to fulfil, is problematic not only from a societal and ecological point of view, but from the animal welfare perspective as well.

Competition with humans for food

Oltjen and Beckett (1996) state that cereal grains are not necessary for ruminant production, meaning that ruminants need not compete for human foods. These researchers at the University of California, Davis, USA, were the first to compare and document the edible returns from dairy cattle production. They investigated two types of rations fed to dairy cattle (lactation yield approx. 8,600 kg) with regard to plant products that could otherwise be used as food for humans (see Table 1).

The energy contained in the milk and meat produced, resulting from a ration consisting of 50% forage and 50% concentrates (Ration I), amounted to only 57% of the energy contained in that portion of the cattle ration which would otherwise have been directly consumable for humans. Not only was the energy balance negative, but also the comparison of the protein content of the ration investigated with the resulting food output was negative (meat and milk amounted to only 96% of the protein in the ration that was consumable for humans).

A positive food output balance was achieved when the ration consisted of approx. 70% forage and 30% concentrates, and the concentrate portion of the diet consisted of approx. 70% industrial by-products (e.g. cotton seed meal, soybean oil meal, etc.). This ration (Ration II) resulted in a 28% gain in energy consumable for humans, and an impressive 176% increase in protein available for human nutrition.

Table 1. Returns on food input from two different types of rations used in dairy production (lactation yield: 8,600 kg). Source: Modified from Oltjen und Beckett (1996).

Feedstuffs	Ration I	Ration II
	% of ration dry matter	
Corn silage	20	35
Alfalfa silage	30	34
Corn	37	-
Soybean oil meal	10	-
Barley	-	9
By-products (from mills, cotton production)	-	22
	Yield (milk + meat), % of food-quality feed input	
Energy consumable by humans	57	128
Protein consumable by humans	96	276

Conclusions

The practice of feeding ever-growing amounts of edible foodstuffs to livestock with the capacity for transforming grassland into human foods goes against common sense. Pushing dairy cattle to lactation yields of 8,500 kg and more through the use of grain-intensive rations means wasting millions of tonnes of food-quality grain, and is contrary to all ecological, social and ethical considerations. The use of up to 25% concentrates (based on the dry matter content of the ration) to balance out possible nutrient deficiencies is recommended to help achieve a better assimilation of the nutrients contained in forage-based feeds, thus contributing towards improving the quantity and quality of our food supplies.

References

Engelhardt, W. von, Dellow, D.W. and Hoeller, H. (1985). The potential of ruminants for the utilization of fibrous low-quality diets. Proceedings of the Nutrition Society 44: 37-43.

Heitschmidt, R.K., Short, R.E. and Grings, E.E. (1996). Ecosystems, sustainability, and animal agriculture. Journal of Animal Science 74: 1395-1405.

Hofmann, R.R. (1989). Evolutionary steps of ecophysiological adaptation and diversification of ruminants: a comparative view of their digestive system. Oecologia 78: 443-457.

Pfeffer, E. (1992). Umweltbelastung durch die Tierhaltung und Möglichkeiten zu ihrer Minderung - Ernährungsphysiologische Aspekte. Züchtungskunde 64: 254-261.

Stangassinger, M. (1993). Tierernährung und Umwelt - ein Konfliktbereich mit Lösungsansätzen? Vorträge zur Hochschultagung 1993. Schriftenreihe der Agrarwissenschaftlichen Fakultät der Universität Kiel, Germany.

Van Soest, P.J. (1994). Nutritional ecology of the ruminant. Second Edition. Cornell University Press, Ithaca, New York, USA, pp. 4-5.

Varva, M. (1996). Sustainability of animal production systems: An ecological perspective. Journal of Animal Science 74: 1418-1423.

WCED (World Commission on Environment and Development) (1987). Our Common Future. G.H. Brundtland (ed.) Oxford University Press, Oxford, England, United Kingdom.

What is 'regional food'? The process of developing criteria for regional food in Eastern Tyrol, Austria

Julia Kaliwoda, Heidrun Leitner and Christian Vogl
Working Group Knowledge Systems and Innovations, Institute of Organic Farming, Department for Sustainable Agricultural Systems, University of Natural Resources and Applied Life Sciences (BOKU), Gregor Mendelstrasse 33, 1180 Vienna, Austria

Abstract

Recently, there is a growing interest in regional food systems from a scientific point of view as well as from food retailing and regional development institutions. However, little attention has been paid to the development and application of criteria to define or evaluate regional food. Criteria to define and measure the regional character of food are needed to provide not only economic but also ecological and social benefits for the regional food system. In a literature review, over twenty regional marketing organizations were analyzed to obtain a comprehensive sample of existing criteria used to define the regionality of food products. A list of criteria was drafted and adopted to the specific economical, ecological and social conditions in Eastern Tyrol in a participative research process. The list of criteria was finalized in explorative interviews with producers, manufacturers and providers of regional lamb meat, bread and fruit products in Eastern Tyrol was concretized. The criteria used in the analysed regional marketing initiatives can be assigned to three dimensions of quality which define the regionality of food: provenance of ingredients, quality in the production process (process quality), and product quality. The participative research process in Eastern Tyrol has resulted in a list of criteria containing all three dimensions, although less detailed than expected mostly due to lack of time and expertise in the participatory process.

Keywords: regional food system, criteria, Eastern Tyrol

Introduction

Regional food systems are widely expected to increase chances and develop strategies for rural areas regarding sustainable development. Governmental as well as non-governmental organizations have launched awareness-raising campaigns to interest consumers in buying regional products. Super market chains and regional marketing organisations are attempting to establish the regionality of products as a unique selling proposition.

The marketing of regional products becomes a political and ethical issue, as not only are these products associated with a high product quality, but the production and consumption of regional products are supposed to be a morally desirable thing to do (Ermann, 2005).

Regional food systems face high expectations: from a regional economic development perspective, they are seen as a means to produce regional value added - an expectation which induces the creation of regional brands for food and other products in almost every region with active regional management. Additionally, regional production and consumption are assumed to reduce transport and therewith CO_2-emissions. Regional food systems are further claimed to support sustainable local small-scale farming associated with positive impacts on the environment and the cultural landscape.

Some scientific discussion but few concrete conclusions have so far dealt with the definition of regional food systems. Unlike successfully constituted criteria for 'organic' or 'fair-trade' products, there is no consensual definition of what makes a product 'regional'. As every product comes from a certain region, what is it that evokes and justifies such high interest and expectations?

Scientific literature does not provide a concrete definition, but offers some characteristics of regional food, named: the 'geographical' and 'emotional' proximity between producer and consumer, a small-scale, sustainable and artisan production, as well as natural, healthy and fresh products (Schade and Reuter, 2001, Ermann, 2002, Dorandt, 2005). So 'regionality' must be more than a geographical indication, as Ermann (2005) states.

Verifiable criteria for regional food concerning the product´s provenance (Where do the steps of production take place?), process quality (Under which conditions and regulations is it produced?) and product quality (How good is the product?) are needed to (1) guarantee sustainability, i.e. economical, ecological and social benefits and (2) gain transparency and trust between producers and consumers and besides justify a higher price.

Criteria for regional food are best set up in the specific region of provenance of the product, as this task has a normative component and depends on the aims, values and needs of the regional inhabitants (Ermann, 2005).

A research study[13] carried out in the region of Eastern Tyrol initiates, accompanies and evaluates a participatory process with the aim of defining and establishing criteria for regional food. The research questions are as follows; the focus for this paper is on the first question:
- Which criteria must be met by regional food from the point of view of regional actors in the food system in Eastern Tyrol?
- Which are the motives and aims of regional actors to produce, manufacture, market and provide regional food?
- Which problems do regional producers, manufacturers and providers have in meeting the criteria of regional food and which solutions do they suggest to overcome the defined problems?

Methods - The process of developing criteria for regional food

Scientific literature and regional marketing initiatives (RMIs) in Austria, Germany and Switzerland were analysed with regard to the criteria used to define regional products to prepare the basis for discussion in the workshops (Figure 1). Furthermore, short interviews with consumers were conducted to also include the Eastern Tyrolean consumers' connotations of regional food products in the discussions.

The participatory establishment of criteria for regional food was accomplished through two consecutive workshops with regional actors in Eastern Tyrol.

Explorative interviews with farmers, manufacterers and providers followed up the workshops (Figure 1) aiming to discuss and comment on the results of the workshops in specific cases and for specific products.

Subsequently, the applied methods are described in detail.

---

[13] The study is part of a pilot project realized by the Federal Government of Tyrol in the district of Lienz (Eastern Tyrol) within the Interreg IIIb-project 'Pusemor' (Public Services in Sparsely Populated Mountainous Areas). The pilot project team consists of members from the Federal Government of Tyrol, the Eastern Tyrol Competence Network for Health, the University of Natural Resources and Applied Life Sciences in Vienna, the regional management and a public relations office and works in close cooperation with local experts and interest groups.

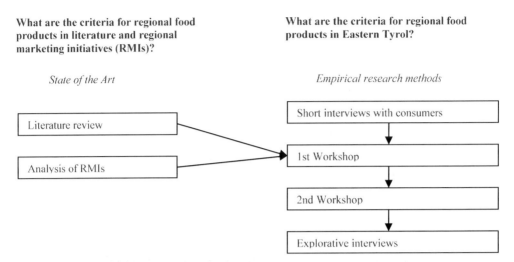

What are the criteria for regional food products in literature and regional marketing initiatives (RMIs)?

What are the criteria for regional food products in Eastern Tyrol?

State of the Art

Empirical research methods

Literature review

Analysis of RMIs

Short interviews with consumers

1st Workshop

2nd Workshop

Explorative interviews

Figure 1. Methodological process

Literature review and Analysis of RMIs

The homepages of 22 regional initiatives from Austria, Germany and Switzerland were analysed with regard to the criteria they established for their regional products. When necessary, additional or more detailed information was requested per mail. The criteria were collected and clustered as they related to (1) the provenance of ingredients, (2) process quality and (3) product quality.

Short interviews with consumers

Thirteen randomly chosen consumers were interviewed following the cultural domain analysis (Weller and Romney, 1988). First, the consumers were asked to list the regional food products they know ('Which regional food products are there in Eastern Tyrol?'). Second, they were asked to name the characteristics of regional food products ('What makes a food product regional?'). The short interviews were conducted on the weekly city market in Lienz and in front of several grocery stores in the study region.

Workshops

The participants of the workshops were farmers, manufacturers, providers (shopkeepers, gastronomes and hoteliers), representatives of the farmers' association and stakeholders of the Eastern Tyrolean food system. The regional project partners made the contact and set up the invitation list. Both workshops took place in the local agricultural school on a weekday evening and lasted about three hours each.

The first workshop with 5 participants was aimed at introducing the project and the topic of the workshops to the regional actors. The list of criteria used by the analyzed RMIs was presented, discussed and the criteria were ranked by the participants with regard to the relevance for Eastern Tyrol.

At the second workshop, 16 participants established criteria for regional food products that meet the special needs and conditions of Eastern Tyrolean based on the ranking of criteria of the first workshop.

In small groups the criteria of the analysed regional marketing initiatives were discussed, modified or rejected and the results were presented to the whole group.

Explorative, semi-structured interviews

Three regional product chains (lamb meat, bread and fruit products) were selected as case studies and depicted by interviewing 10 regional producers, manufacturers and providers. The respondents were deliberately chosen and are all well-known, committed actors in the regional food system in Eastern Tyrol. The interviews aimed at commenting, adjusting and finalizing the criteria for regional food against the background of specific products. Further topics dealt within the interviews were aims and motives of the respondents to engage in the regional food system; the weaknesses and disadvantages of regional food systems in contrary to conventional ones; and their suggestions to overcome and solve these problems.

Results and discussion

The analysis of 22 regional marketing initiatives gives a holistic impression of the spectrum of criteria that can be used to characterize a 'regional product' in general. Although the degree of accuracy and the level of detail vary significantly between the analysed initiatives, all include criteria for *regional provenance*. Many initiatives claim that not only the processing (e.g. baking the bread, producing the sausage) takes place in the region, but the raw material (e.g. the flour, the meat) is produced there, too. Often there are exceptions for ingredients that can not be produced in the region due to climate factors.

The results from the workshops in Eastern Tyrol show that unprocessed food (e.g. apples) must be entirely produced in the region, whereas the participants can not agree on the exact share of ingredients in processed products which have to be of regional provenance. They name several regional products where sufficient quantities of raw materials are not available in Eastern Tyrol, e.g. bacon (pork) or bread (cereal). Some participants regard the location of the processing of a product as more important than the location of the agricultural production, because the value added is foremost generated in processing. Furthermore, considering the conditions of agricultural production in the mountainous region of Eastern Tyrol, arable production is in many cases difficult and rare.

One of the most frequently mentioned problems for manufacturers and providers is the regional availability of some raw materials. However, the reduction of transport, which is also an important motive for several respondents to engage in the regional food system, can no longer be guaranteed when raw products are imported into the region.

Many initiatives have detailed criteria concerning process quality, mostly related to ecological and animal-friendly agricultural production. Social aspects of process quality are rarely tackled, neither in the RMIs nor in the discussion in Eastern Tyrol. In Eastern Tyrol, participants hesitate to establish concrete process quality criteria for regional products. They refer to the strict laws in Austria for agricultural production and animal welfare and to the perceived extensive and nature-oriented agriculture in the region. The current status is supposed to guarantee ecologically sound food production. The most challenging criterion uttered in this context is the abandonment of genetically modified animal feed.

While it is evident that the agricultural system in the study region is in general ecologically sound, high process quality can only be assured with criteria allowing evaluation and control.

Premium product quality is guaranteed in many of the analysed regional marketing initiatives by regulating the use of additives, the sensorial quality (taste, smell, and appearance) or aspects of food chemistry.

The Eastern Tyrolean list of criteria includes the abandonment- of chemical additives.

The discussion in the workshops clearly shows that the importance of product quality is recognised by the participants, especially as sales argument. Several interviewed processors and marketers remark a lack of professional product quality of some regional foods. Nevertheless, the results from the workshops do not include elaborated criteria to define the product quality of regional products. The reason for this might be ascribed to the fact that attributes like 'excellent taste', 'fresh' or 'attractive appearance' can be neither easily defined nor evaluated by set criteria.

Conclusion

Two short workshops turned out to be insufficient time to meet the requirements of stakeholder participation and deal with such a complex theme as regional food systems on the desired concrete level. There are also many highly emotional issues confronting the actors of regional food systems and so it was quite difficult to step into a very detailed discussion of criteria. Moreover, the participatory process was started with a priori set aims, methods and schedule due to the organisational frame of the project.

Consequently, the aim of establishing criteria for regional food products was not supported by some participants. Their interest was primarily to discuss the chances of marketing regional products or to express their despondency concerning the overwhelming superiority of supermarkets. Furthermore, some interviewed providers as well as some participants of the workshops do not regard the establishment of criteria for regional food as important. From their point of view, personal relations and trust between producers and consumers, 'pictures and stories' behind the product and simply a good taste are more powerful sales arguments than a checkable list of criteria.

Initiating the with 'example criteria' from existing regional marketing initiatives turned out to be very helpful for reaching relevant results even in short time.

The participation of experts from various fields of food production will be needed to finbalize a more detailed list of criteria. Furthermore, splitting up in different workgroups focussing on specific product groups can be advised.

The project has already generated positive effects in the region (articles, cooperations etc.). The workshops were very important starting points in connecting regional actors, building networks, starting discussions and spreading the information in the region. The process shows how much interest in criteria for regional food exists from the producers as well as regional development organisations.

Acknowledgements

The authors thank the Federal Government of Tyrol for funding and all regional partners for good cooperation and assistance.

References

Dorandt, S. (2005). Analyse des Konsumenten- und Anbieterverhaltens am Beispiel von regionalen Lebensmitteln. Empirische Studie zur Förderung des Konsumenten-Anbieter-Dialogs. Dissertation Universität Gießen. Verlag Dr. Kovac, Hamburg, Germany, 380p.

Ermann, U. (2002). Regional Essen? Wert und Authentizität der Regionalität von Nahrungsmitteln. In: Gedrich, K and U. Oltersdorf (eds.) Ernährung und Raum: Regionale und ethnische Ernährungsweisen in Deutschland. Berichte der Bundesforschungsanstalt für Ernährung, Karlsruhe, Germany, pp. 121-140.

Ermann, U. (2005). Regionalprodukte - Vernetzung und Grenzziehung bei Regionalisierung von Nahrungsmitteln. Sozialgeographische Bibliothek, Bd. 13, Stuttgart, Germany, 320p.

Schade, G. and Reuter, K. (2001). Regional - immer öfter erste Wahl? Argumente für die Verknüpfung von Regionalität und Bio aus Marketingsicht. Bioland 6, 21. http://orgprints.org/1969/01/schade-reuter-2002-regional-bio.pdf (14.01.2007)

Weller, S.C. and Romney, A.K. (1988). Systematic data collection. Qualitative research methods, 10. Sage Publications, Newbury Park, USA, 96p.

Part 12 - GMO: GM plants

Contested GMO's: how questions of global justice and basic structures matter in the debate on GM plants

Kristian Høyer Toft
Priority area BioCampus, Department of Media, Cognition and Communication, Philosophy Section, University of Copenhagen, Njalsgade 80, 2300 Copenhagen S, Denmark, khtoft@hum.ku.dk

Abstract

Since GMOs contribute to the distribution of benefits and burdens globally, they give rise to questions of justice, not only in regard to individuals, i.e. consumers and farmers, but also in regard to nation-states and regions. Hence GMOs are an example of global interdependency - environmentally and economically. For instance, it has been claimed that the EU should not prohibit plant GMOs, since this would prevent poor developing countries from exporting seed to the EU. However, counterarguments say that plant GMOs will lead poor countries' farmers to be subjugated to monopolistic seed companies. The basic disagreement is therefore not only about (scientific) facts of potential risks but also about values. Within recent theory of international justice, cosmopolitans have challenged nationalists as well as Rawlsians adhering to a 'law of peoples' conception. The disagreement concerns whether social justice has a purview beyond the nation-state. According to Rawlsian ideal-theory, the purview of social justice is the basic structure of a nation-state, i.e. the constitutional essentials and the regulation of markets. However, Rawls also admits that international society may show some resemblance to a basic structure. This article will investigate how GMOs give rise to international institutions on a par with an international basic structure, i.e. trade agreements, bio-ethical declarations, standards of approvals etc., and how this basic structure may be the subject of an assessment by the standards of justice. The point is that GMOs by themselves cannot be said to be just or unjust, only the institutional set-up regulating GMOs may be so.

Keywords: international justice, GMO, WTO, human rights, bioethics

Introduction

Within academic research on GMOs (genetically modified organisms, hereafter mainly synonymous with GM plants but including all other sorts of GMOs) several experts from the social sciences and philosophy have pointed out that the issue of risk cannot be fully accounted for by natural scientific means only; issues about ethics and values need to be taken into consideration as well (Bernauer, 2003; Gaskell and Bauer, 2006; Lassen *et al.*, 2002; Jensen, 2006; Brom, 2004). In this paper the suggestion to look at the value aspect of GMOs is expanded into a discussion about international justice and the regulatory institutions set up to control their proliferation. The primary reason for looking at GMOs from the perspective of international justice is that globally they are now a significant part of agriculture and that GM seed is traded across national borders. GMOs are no different from other biotechnological inventions in the sense that they are very much part of the broad tendency towards Globalization. Another reason for applying the perspective of international justice is normative: Given that it is considered desirable to try to regulate the development, which standards should be the guiding ones? The normative issue also has relevance to the pervasive scepticism found in the public opinion. The prevalent notion that only monopolistic seed companies may benefit from the introduction of GMOs may explain part of the resistance as well as indicate that the issue of justice is a significant concern that needs to be addressed.

The research into and development of GMOs has predominantly been carried out by multinational seed companies. During the 1990s small and medium-sized plant breeding companies were bought up by large multinationals such as Monsanto. As it became clear that future profits on pesticides were expected to decrease due to GMOs competing out the traditional crops, the chemical industry strategically chose to expand into plant breeding to get hold of the best plant varieties. The most well-known case of a multinational strategically securing its own profits is probably Monsanto's development of crops resistant to Glyphosate (Round-up). The strategy seems logical: take the best conventionally bred crops and insert a gene for resistance to Glyphosate resulting in a superior variety compared to non-GM plants. Thereby Monsanto had safeguarded continued revenues on Glyphosate as well as having created a basis for new revenues from the patented GM crops. However, it is now evident that this strategy to maximize profits has backfired; Monsanto has become the very symbol of capitalist arrogance towards all alleged victims of GMOs: the poor farmers in the developing world because capitalist exploitation has exacerbated, the environment due to cross-pollination and release into the wild (the Monarch butterfly case), the consumer who is exposed to products disputably injurious to his/her health (cases like Pusztai's disputed studies showing that GM potatoes have detrimental effects on the immune system of rats. Though the reliability of this study is widely disputed the studies have ignited public scepticism fuelled by the media. See Dixon, 2003).

Often it is said that due to its ignorance of corporate social responsibilities (CSR) Monsanto is to blame for the global controversy over GMOs. Certainly, Monsanto has faced up to taking responsibility in the strategy of turning the former chemistry company into a 'green' company with an environmental friendly image (Finegold *et al.*, 2005). However, the damage has been done and so far it seems that the public image of GMOs is irretrievably beyond repair. At least if one looks to the Eurobarometer (64.3) of 2006 where a majority of EU citizens evaluate GM foods as altogether 'morally unacceptable', 'not useful', 'risky' and something that 'should not be encouraged' as opposed to the generally positive evaluations of nanotechnology, pharmacogenetics and gene therapy.

The case of Monsanto's introduction of GMOs on the world market in a sense shows how the issue of Globalization and justice has been on the agenda since the very first cargoes of GMOs were shipped into the ports of the EU in the mid-1990s. Certainly, the dispute has been mostly about the scientifically measurable risks to human health and the environment and not much about issues of justice. However, the point to press here is not that the entire debate started off on the wrong foot in its myopic focus on scientific risk. Safety, surely, should have (the) high(est) priority, but it is becoming clear that issues of values, politics and justice are also part of the GMO agenda, as they should be.

Globalization and GMOs

There are at least three issues of global inter-dependency on the GMO agenda. The three issues are: science and scientists disperse knowledge across borders, the flows of capital and trade are global (the WTO dispute between the US and the EU exemplifies this), and the environment is globally affected by the proliferation of GMOs (issues of traceability, risk to human health and biodiversity are examples).

Only through an effective international cooperative approach can citizens within nation-states have trust in the safety of GM foods (Bernauer, 2003). However, it can be questioned whether it is realistic to expect international cooperation in the case of GMOs. The recent case of a US complaint against the moratorium of EU approval of GMOs filed at the WTO in 2003 shows how difficult it might be to resolve conflicts over levels of reasonable risk. And it shows how faint the hope for global regulation may be since the world is divided between two major regimes of regulation. The US regime is based on a minimal concept of 'scientific evidence' and 'sound science' (Bernauer, 2003: 55), whereas the EU regime is based on the 'precautionary principle' which allows for a much wider notion of risk and hence can be interpreted as a much more restrictive approval procedure compared to the US regulatory approach. The rest of the world is divided between the two dominant economies of the US and the EU. Countries like Argentina and Canada cluster around the US regime and countries like Norway, Switzerland and many Central and Eastern European countries cluster around the EU regime. The big Asian countries, China, India and Japan are to be found in the middle having moved towards stricter approval procedures, e.g. mandatory labelling. So, the global picture represents 'regulatory polarization' (Bernauer, 2003: 8), not to mention the differences found within the EU. Hence, it is still too early to tell which type of regulatory regime will prevail in the longer run.

On top of the regulatory conflict within the WTO which concerns rules of trade policy one finds the more general and abstract rules of international law, the many bio-ethical declarations that, inspired by the UN human rights regime, intend to set the baseline for acceptability at the national and regional levels. Many of these declarations, such as UNESCO's 'Universal Declaration on Bioethics and Human Rights' of 2005, are not targeted at GMOs in particular since the issues addressed are primarily human health and environmental sustainability, i.e. standards for the assessment of biotechnological products such as GMOs. However, within the range of documents inspired by human rights one finds a 'thin' and abstract framework for global regulation of GMOs and, as many of these documents are ratified by a large number of countries, they give credence to a 'thin basic structure' of international society based on the notion of a 'common humanity' (see Franconi, 2006).

Only against the backdrop of international justice theory can the status of international regulation of GMOs be assessed. This does not mean that there are no other approaches to assessing GMOs, such as bio-ethics, economics and molecular biology, but the point here is that the impact of GMOs on global agriculture should also be assessed by the standards of international justice theory. And the fact is that this task has not been taken up by many experts though there are exceptions (see Pinstrup-Andersen and Schiøler, 2001). However, most such assessments are found within legal theory and development economics, hence the normative philosophical perspective of international justice theory has not been paid much attention. One reason for this neglect is probably to be found in the factor of what political theorists call an 'essentially contested' concept. GMOs are essentially contested since there is no consensus on the proper definition of GMOs in the context of political and ethical discussion. Moreover the concept calls for seemingly irreconcilable views. The point is that the property relevant for evaluation is contested and that it is difficult to find agreement on the standards of risk-assessment. This should not be confused with the claim that nobody really agrees on what the scientific procedure of gene modification is. It merely signals that the GMO concept has become essentially contested since it is vested with deep political and economic interests. To use Sheila Jasanoff's term: GMOs are 'co-produced' (Jasanoff, 2004) by scientific as well as political actors and therefore an objective description of GMOs is contested. This point is acceptable as long as it is not taken to imply rampant constructivism and relativism since there is, in fact, some basic agreement about what GMOs are and what the features relevant for assessment are. Jasanoff's STS (science and technology studies) statement merely helps to clarify that scientific objects are politically infused. And this may also explain why the theme of international justice is perhaps best seen mainly as an NGO and civil society issue and not exclusively as an issue for scientific approaches. This is also in tune with the democratic view that, ultimately, only citizens can decide whether GMOs should be accepted or not (see Brom, 2004 for a 'public reason' argument).

Global justice theory and legal dualism

What are the outlines of international justice theory?

Two competing views are to be found on international obligations; the cosmopolitans argue that obligations are owed to individuals irrespective of national belonging (Pogge, 2002), whereas nationalists argue that obligations are foremost owed to fellow compatriots (Miller, 1995). A middle-position which admits that international society rests upon common ideals of international law is also found. However, from this position, nation-states are considered the primary unit of concern and this refers issues of justice to start at the domestic level. Such theories of international justice are often referred to as two-tier theories on a par with so-called legal dualism, since they emphasize the difference between the domestic and the international level. John Rawls' *The Law of Peoples* (1999) expresses this dualistic view.

One major reason why obligations towards foreigners are taken to exist at all is the fact of international inter-dependency. Due to international trade and international law, individuals and nations stand in some cooperative reciprocal relation with each other - such institutions of reciprocity are often referred to as the 'basic structure' of a society (Rawls, 1999). It is assumed that reciprocal relations may give rise to claims of justice: how should the benefits and burdens resulting from cooperation be distributed? As a central dispute in international justice theory it is discussed what sort of obligations are owed to foreigners. Among cosmopolitans one finds the claim that there are global obligations to discharge basic needs, i.e. to provide foreigners with social rights and thus admit the legitimacy of global social justice. Conversely, nationalists will constrain foreigner's rights to negative rights addressing primarily the issue of liberty. Rawls would opt for the nationalist view in admitting only a conditional duty of assistance in regard to helping the very needy abroad, though he also agrees with the cosmopolitans that we should comply by the rules of a well-ordered international society.

So, the major divides in international justice theory are about what sort of obligations are owed and to whom and for what reasons. All agree that the liberties needed in support of democracy are universal. However, there is disagreement as to what is positively owed to foreigners; should foreigners be entitled to a distribution of wealth from the rich countries or not? The cosmopolitans think they should. This type of disagreement gives a good framework for understanding our international obligations in regard to GMOs. It is not so much the liberties that are seen to be violated by GMOs as much as it is the issue of a just distribution of the benefits and burdens. Is it fair, for instance, that the Argentinean, poor farmers should pay royalties for seed to Monsanto in order to supply cheap soy, maize and cotton for the rich Western countries? (Joensen and Semino, 2004). From the viewpoint of a cosmopolitan this is unfair. However, if one believes that each country should take care of its own citizens and that states should not interfere with each other's internal affairs, unless their own vital interests are at stake, this is, by the standards of justice, at most a case between the state of Argentina and Monsanto. One may be in favour of several human rights declarations and yet claim that the liberties of farmers in Argentina are not violated, since it can be disputed whether the 'oppressive' structure of GMO-agriculture is to be considered under the label of agricultural development and innovation, with inevitable detrimental side-effects for traditional ways of agricultural production, or whether it should be admitted to being in fact oppressive and forced upon the poor farmers. Moreover, it can be disputed whether private companies are in fact responsible for the exploitative and morally deficient structures of the liberalized economy supported by most countries today. The case of patented AIDS-medicine and the complaints against the global pharmacy industry for not supplying poor Africans with affordable medicine shows that perhaps responsibility is to be placed elsewhere, viz. with international society (Buchanan and Decamp, 2006).

Besides the issue of what sort of obligations are owed to foreigners, international justice theory also contains a major discussion on what triggers claims to justice. All agree that issues of justice are to some extent confined to social institutions or what Rawls calls the basic structure of society. Only if citizens create basic institutions that set the framework for the distribution of their cooperative venture the assessment of justice is appropriate. In this Rawlsian vein the two realms of private life and international society fall outside the domain of justice. As long as it follows the just rules of a well-ordered society the case of private life is simply not relevant to life in society at large, hence the liberal Western legal tradition appropriate plenty of room for free contracting in the private sphere. Likewise in the international sphere. Here the reason for not seeing justice as appropriate is the reality of weak and not transparent international institutions as well as the pervasive diversity among the different cultures of nations (the 'fact of pluralism'). International society is therefore perceived as an anarchical 'state of nature' in the Hobbesian fashion. So, Rawlsians follow Kant in not thinking that a world-government is a feasible or desirable option. This is, however, not the same as there not being any obligations owed to foreigners due to our common humanity, as expressed in the human rights documents. But these obligations are not necessarily measurable by the standard of justice, according to Rawlsians. And if they are it will only be a restricted list of obligations on a par with the liberties in combination with respect for state sovereignty. In fact the national legislatures today predominantly adhere to this type of dualist conception of the relation between national and international law, giving priority to national law in cases of conflict with international law. Denmark, for instance, has incorporated and ratified many international declarations, especially EU law, but still the legal system works on the assumption of legal dualism as opposed to legal monism. Hence, most legal regimes already, and perhaps necessarily, work on the assumption of dualism.

This, of course, raises doubts about the feasible workings of international justice in the GMO case, since if cosmopolitan monism is infeasible, due to the predominant standards of legal dualism, the cosmopolitan's aspirations to global justice are far too idealistic. The typical cosmopolitan reply to this is that, historically, the world has exhibited a willingness to make transitions towards a global regulatory regime in the fashion of non-binding though effective 'soft law' on a par with the UN human rights regime and that we need to actively push for more global regulation in order to control the adversities of Globalization.

Conclusion

GMOs are globally distributed and affect global institutions, as seen in the WTO dispute between the US and the EU, as well as human rights inspired declarations. Since such institutions constitute global interdependency they should be measured by the standards of justice. This is an argument in favour of cosmopolitanism in the case of assessing GMOs. However, since the global institutions set up to regulate GMOs are showing signs of weakness, it can be disputed whether in fact they can be entitled 'global'. The WTO dispute shows 'regulatory polarization', and non-binding declarations such as UNESCO's 'Universal Declaration on Bioethics and Human Rights' suffer the legal dualism that pervades the realm of international law. Hence, a cosmopolitan view on the global regulation of GMOs seems desirable but unfeasible, whereas the nationalist view seems fully feasible, but not at all desirable. It is still too early to tell what the stable outcome of regulatory institutions of GMO will be. A stance somewhere in between cosmopolitanism and nationalism is probably what is realistically to be hoped for and in that respect, institutions of GMOs are no different from any other case of global interdependency.

References

Bernauer, T. (2003). Genes, trade and regulation - The seeds of conflict in food biotechnology. Princeton University Press, Princeton, 229p.

Brom, F.W.A. (2004). WTO, public reason and food - public reasoning in the 'trade conflict' on GM-food. Ethical Theory and Moral Practice 7: 417-431.

Buchanan, A. and Decamp, M. (2006). Responsibility for global health. Theoretical Medicine and Bioethics 27: 95-114.

Dixon, B. (2003). Genes in food - why the furore? Biochemical Society Transactions, vol. 31, part 2: 299-305.

Finegold, D.L., Bensimon, C., Daar, A.S. and Eaton, M. (2005). Monsanto company: Bio-agricultural pioneer. In: Bioindustry ethics. Elsevier, Amsterdam, pp. 267-300.

Franconi, F. (2006). Genetic resources, biotechnology and human rights: The international legal framework. EUI working paper LAW No. 2006/17.

Gaskell, G. and Bauer, M.W. (2006). Genomics and Society - Legal, ethical and social dimensions. Earthscan, UK, 261p.

Gaskell, G., Allansdottir, A., Allum, N., Corchero, C., Fischler, C., Hampel, J., Jackson, J., Kronberger, N., Mejlgaard, N., Revuelta, G., Schreiner, C., Stares, S., Torgersen, H. and Wagner, W. (2006). Europeans and Biotechnology in 2005: Patterns and Trends - Eurobarometer 64.3.

Jasanoff, S. (2004). States of Knowledge - The co-production of science and social order. Routledge, Abingdon, 317p.

Jensen, K.L. (2006). Conflict over risks in food production: A challenge for democracy. Journal of Agricultural and Environmental Ethics 19: pp. 269-283.

Joensen, L. and Semino, S. (2004). Argentina's torrid love affair with the soybean. Seedling, 5-10.

Miller, D. (1995). On nationality. Clarendon Press, Oxford. 210p.

Lassen, J., Madsen, K.H. and Sandøe, P. (2002). Ethics and genetic engineering - lessons to be learned from GM foods. Bioprocess and Biosystems Engineering 24: 263-271.

Pinstrup-Andersen, P. and Schiøler, E. (2001). Seeds of contention: World hunger and the global controversy over GM crops. The John Hopkins University Press, Baltimore, 157p.

Pogge, T. (2002). World poverty and human rights. Polity Press, Cambridge, 284p.

Rawls, J. (1999). The Law of Peoples. Harvard University Press, Cambridge Mass. 181p.

UNESCO (2005). Universal Declaration on Bioethics and Human Rights, Paris.

Benefits at risk: does the motivation of farmers counteract potential benefits of GM plants?

Jesper Lassen[1] and Peter Sandøe[2]
[1]*Danish Centre for Bioethics and Risk Assessment, Department of Human Nutrition, Faculty of Life Sciences, University of Copenhagen, Rolighedsvej 30, 1958 Frederiksberg C, Denmark, jlas@life.ku.dk*
[2]*Danish Centre for Bioethics and Risk Assessment, Department of Food and Resource Economics, Faculty of Life Sciences, University of Copenhagen, Rolighedsvej 25, 1958 Frederiksberg C, Denmark, pes@life.ku.dk*

Abstract

Herbicide resistant GM plants have been promoted as a tool in the development of more environment-friendly agriculture. The environmental benefits here, however, depend not only on farmers' acceptance of GM crops as such, but also on their willingness to use herbicides in accordance with altered spraying plans. In this paper, we will argue that factors driving the spraying practices of Danish farmers may hamper efforts to secure the environmental benefits of the new crops.

Keywords: agricultural biotechnology, farmer attitudes, Europe, GM crops, focus group interviews

Introduction

It is often said that herbicide resistant GM crops will permit more environment-friendly agriculture. Environmental benefits may be secured for at least two reasons. First, relatively harmful herbicides could be replaced with less harmful herbicide, such as RoundUp. Second, the new technology may make delayed herbicide treatment possible. This will allow weeds to grow to certain size, to the benefit of the local fauna. Thus there will allegedly be a positive effect on biodiversity in the agricultural landscape.

These benefits depend, however, on farmers doing as they are expected. First, farmers growing GM crops must actually replace the traditional herbicides and restrict themselves to recommended doses of the new herbicide. Second, GM farmers must carefully follow the suggested spraying plans that ensure delayed treatment. And of course, third, farmers at large must be willing to adopt the new GM crops in the first place.

So far, in Europe, GM food issues have mainly been framed as a conflict between a sceptical public, on the one side, and a more enthusiastic food sector, on the other. As a result the vast majority of social science studies have investigated the nature of European public scepticism, using quantitative as well as qualitative studies of public perceptions. However, the primary producers - the farmers who are to grow the GM crops - are rarely subject of such studies. As argued above, this may be a practical problem, because the alleged advantages rest, at least in part, on farmers' attitudes and practices. The lack of attention to European farmers may also raise democratic issues, since it may reflect the fact that farmers have been largely ignored in public discussion of GM crops.

A review of the scientific literature on farmers' perceptions of GM crops found that, while some articles address farmers in developing countries (e.g. Bellon and Berthaud, 2006;Chong, 2005;Yang *et al.*, 2005) and others address farmers in the developed world beyond Europe, particularly in Australia, New Zealand and the US (e.g. Fairweather and Campbell, 2003; Grice *et al.*, 2003; Kondoh and Jussaume, 2006; Norton and Lawrence, 1995; Wilson *et al.*, 2005), none examine the perceptions of European farmers.

The objective of the research presented in this paper is to develop a better understanding of Danish farmers' perceptions of GM agriculture, and to examine their willingness to adopt the farming practices that are needed to secure the promised environmental benefits of GM crops.

Design of the study

Six focus-group interviews, involving a total of 36 farmers from various regions of Denmark, were carried out during the summer 2005. In order to focus the discussions on the key issues, the participating farmers were grouped according to which crop was their main interest. Thus the six groups included two groups with corn farmers, two groups with rapeseed farmers and two groups with farmers growing sugar beet. The division of farmers into these groups, each representing variations in farming practice as well as in the challenges posed to the cultivation of GM crops, enabled the interviews to generate a picture of the different arguments for and against GM crops. The interviews were exploratory. They were carried out in accordance with a funnel-shaped guide. This ensured that the interviews moved on from discussions of the participants' farm practices, through discussions of GM crops in general, to debate about alternative spraying plans and the treatment of conventional as well as GM crops - as suggested by the moderator.

All interviews were tape-recorded and transcribed verbatim. In the subsequent analysis the transcribed interviews were coded into themes; and different arguments within each theme were identified using Toulmin's argument analysis (Toulmin, 2003). It is the result of this analysis that is presented in the following.

Agriculture is business!

One key result of the focus groups is the dominance of an economic discourse. This is, of course, unsurprising: farming is a business, and the survival of the individual farmer, like the survival of any commercial enterprise, depends on an ability to minimize costs and maximize profits. Across the interviews this could be observed in the many cases in which participants explicitly referred to economic consequences for marginal returns as the most important (or only) criterion for assessing changes in their farm practices. Awareness of the importance of business economics was expressed, not only in general terms, but also in specific discussions of the pros and cons of GM herbicide resistant crops.

Often the economic discourse was not, however, explicitly formulated, but instead remained tacit. For among farmers the importance of marginal costs is mutually understood: it need not be expressed all the time. Hence arguments which appeared, at first glance, to address other, non-economic issues often had implicit economic elements.

One example of this is the importance of what can be conceptualised as 'harmony'. A recurrent theme, when new farming practices in general as well as GM herbicide resistant crops in particular were considered, was the impact any changes might have on the way work was organised at the farm. Internal harmony, where all processes of production are joined together in a coherent unity, seemed to be a goal in itself. In the discussions the farmers used words like 'rational', 'easy' or 'practical', when describing their need for harmony, arguing that new activities that disturb harmony are not attractive. But the, often tacit, driver of this attitude is that technical and organisational harmony also reflect the economic harmony (and profitability) of a well run enterprise.

The aesthetic value of a field relatively free from weeds is a strong symbolic marker of the desired harmony. To the farmer, clean fields serve as indicators of a well run farm; conversely, the presence of weeds indicates lost control and - eventually - economic failure.

Concerns beyond the economic bottom-line

Characteristically, arguments in the discussions of the usefulness of different spraying plans for herbicide resistant corn, rapeseed and sugar beets were dominated by the question of control. In keeping with the last section, the main concern was whether, and in what way, proper weed management could be maintained within the suggested plans. At this level, where the discussion tends to be centred on farming practice, almost no other types of argument were raised. The exception was rapeseed. Here some concerns were raised about the health impact of eating GM rapeseed. There were also anxieties about possibility that the economic benefits would bypass the farmers and accrue to the seed producers and the agrochemical industry.

By contrast, at the start of the interviews, when GM crops in general were being discussed, a somewhat wider agenda emerged, although most of the arguments put by the farmers, as already mentioned, eventually zeroed in on the economic issues. Thus worries concerning the possible development of herbicide resistance and the development of new (GM) weeds were raised. In addition, economic concerns of larger scale were also flagged. Here the fear, raised by some participants, was that the introduction of herbicide resistant crops may result in the development of monopolies in the seed and agrochemical sector, and that these monopolies could be used by companies to control agricultural development. Another, similarly structural concern was whether competition will in the end force Danish farmers to adopt GM crops. Here the argument was that if GM crops are economically competitive at farm level, fierce international competition will leave the farmer no choice: if he refuses to take up the new crops, he will be priced out.

Interestingly, in the interviews the prevailing economic focus contrasted with an almost total absence of mention of the societal usefulness of GM crops, or of other moral concerns. The farmer's concerns reflect their context, i.e. the farm as a business, and are very different from wider public concerns, which typically include questions of risk, usefulness and various moral matters (Grove-White *et al.*, 1997; Lassen and Jamison, 2006).

Perceptions of nature

As was mentioned in the introduction, one of the central arguments raised by some proponents of herbicide resistant GM crops refers to the positive environmental impact on biodiversity in the agricultural landscape. This argument is most often presented as a matter of caring for nature, with increased biodiversity being treated as something of value in itself. The thought is that herbicide resistant crops are a good idea because they increase diversity among wild plants, insects and larger animals in the field and in the neighbouring spaces.

In general, concerns about nature and environmental issues were not conspicuous in the participating farmers' discussions of farming and spraying practices. When nature and environmental issues were mentioned, it was most often with reference to quantitative changes in known pollutants and the effect these may have on the conditions of agricultural production, or with reference to concerns over loss of control over nature. Hence the farmers are preoccupied with nature as a condition for production: they tend to treat environmental problems as problems either because they distort nature as a resource; or because they lead to restrictive environmental regulation with an impact on agriculture; or because they lead to public claims for environment-friendly farming practices. Within this rather anthropocentric understanding there is little or no room left for the idea that increased biodiversity is of value in itself. Consequently, when the participants did occasionally refer to biodiversity it was with the agricultural importance of that diversity in mind: a diverse flora in neighbouring areas, for example, could be important as a habitat for beneficial insects predating on agriculturally harmful insects.

Conclusion

The study raises several important questions about the alleged environmental benefits of herbicide resistant GM crops.

First, the herbicide resistant crops will have to prove their economic viability. Judging by the interviews, it is clear that a positive effect on marginal costs is a precondition of success. If the crops have a negative economic impact - or even if they are economically neutral - it is not likely that they will be adopted by the farmers.

Second, even if (everything being equal) the herbicide resistant crops have a positive economic impact, they can only be expected to be implemented to the extent that they can be harmoniously combined with other activities at the farm. As the interviews also showed, they could be implemented in a way that allows them to be adjusted to other activities in order to obtain the sought-after harmony. Here the core problem is that this adjustment may be in conflict with environment-friendly spraying plans. That would result in a situation in which GM crops are introduced, but in which the positive environmental effects are lost because other activities oblige or encourage the farmer to deviate from the appropriate spraying plans.

Third, aim of maintaining fields that are free of weeds may be at odds with the environmentally beneficial spraying plans. The problem is that the spraying plans considered most beneficial allow, at the same time, for the growth of considerable weed flora in the field. The adoption of such spraying plans will, for some farmers, be in conflict with a powerful conviction that the field must be kept weed-free. This conviction, which reflects a symbol of the well run and harmonious farm, is itself (if you will excuse the pun) deeply rooted.

Fourth, the interviews showed that farmers are not susceptible to an environmental line of argument that has been used elsewhere to advance the cause of herbicide resistant GM crops, because that argument relies on view of nature in which biodiversity is valued in itself. The dominant perception of the farmers was anthropocentric in character: the farmers valued nature and biodiversity only to the extent that they generated demonstrable benefits in the efficient and economic running of the farm.

So even if European farmers adopt herbicide resistant crops, the expected environmental benefits will not be secured automatically. Rather, it seems that only if some of the attitudes and perceptions found in farmers are changed will the benefits be obtained.

References

Bellon, M.R. and Berthaud, J. (2006). Traditional Mexican agricultural systems and the potential impacts of transgenic varieties on maize diversity. Agriculture and Human Values 23: 3-14.

Chong, M. (2005). Perception of the risks and benefits of Bt eggplant by Indian farmers. Journal of Risk Research 8: 617-634.

Fairweather, J.R. and Campbell, H.R. (2003). Environmental beliefs and farm practices of New Zealand farmers: Contrasting pathways to sustainability. Agriculture and Human Values 20: 287-300.

Grice, J., Wegener, M.K., Romanach, L.M., Paton, S., Bonaventura, P., and Garrad, S. (2003). Genetically modified sugarcane: a case for alternate products. Agbioforum. 6: 162-168.

Grove-White, R., Magnaghten, P., Mayer, S., and Wynne, B. (1997). Uncertain world. Genetically modified organisms, food and public attitudes in Britain. Lancaster, The Centre for the Study of Environmental Change.

Kondoh, K. and Jussaume, R.A. (2006). Contextualizing farmers' attitudes towards genetically modified crops. Agriculture and Human Values 23: 341-352.

Lassen, J. and Jamison, A. (2006). Genetic technologies meet the public: the discourses of concern. Science Technology and Human Values 31: 8-28.

Norton, J. and Lawrence, G. (1995). Farmers and scientists: views on agrobiotechnologies. Agricultural Science 8: 39-42.

Toulmin, S. (2003). The uses of argument. Cambridge University Press.

Wilson, T.A., Rice, M.E., Tollefson, J.J., and Pilcher, C.D. (2005). Transgenic corn for control of the European corn borer and corn rootworms: a survey of Midwestern farmers' practices and perceptions. Journal of Economic Entomology 98: 237-247.

Yang, P.Y., Iles, M., Yan, S., and Jolliffe, F. (2005). Farmers' knowledge, perceptions and practices in transgenic Bt cotton in small producer systems in Northern China. Crop Protection 24: 229-239.

Typology of ethical judgements on transgenic plants

Catherine Baudoin
INRA / ACTA, Unité Transformations Sociales et Politiques liées au Vivant (TSV), 65, boulevard de Brandebourg, 94205 Ivry-sur-Seine Cedex, France, baudoin@ivry.inra.fr

Abstract

We propose a typology of the principle ethical judgements on transgenic plants by examining three questions. Is it legitimate to tamper with the genome? What is the situation regarding patented biotechnological innovations? And finally, should one be opposed to all transgenic plants? All the moral advantages that should be promoted are considered, that is to say, the balance between costs and benefits, justice, freedom and biological diversity. This leads to a systemic evaluation of the consequences of using transgenic plants. Sustainable development aims at reconciling three requirements: economic development, social justice and the protection of nature. The evaluation we submit can serve as a preliminary to examining whether transgenic plants can contribute to sustainable development or not.

Keywords: transgenic plants, typology, ethical judgment

Introduction

The first step is to establish a framework for interpreting the principle ethical judgements on transgenic plants. The aim is to bring to light the main ethical questions raised by transgenic plants and to classify the most frequently advanced arguments into a typology depending on whether they fall within the scope of deontological ethics or consequentialist ethics. It should be noted that deontological arguments are those based on principle, such as the Kantian argument: one acts out of a sense of duty and because it is the rule. Something is considered to be good in its own right and should therefore be done for its own sake, independently of its effects. On the other hand, according to consequentialist arguments, the quality of an action should be judged by its consequences. These are teleological arguments since it is the purpose that is considered.

The ethical problems caused by transgenic plants can be approached through the following three questions. First of all, is it legitimate to tamper with the genome? Secondly, what happens when transgenesis, a laboratory technique, becomes an innovation protected by a patent? Finally, over and above the issue of the patentability of living organisms, should one be opposed to all transgenic plants?

Is it legitimate to tamper with the genome?

The question refers exclusively to the use of transgenic techniques, independently of the goals and conditions of this use: are there objections on principle to transgenesis?

There are objections on principle to transgenesis

According to the argument based on religion, nature is sacred. By modifying it, man plays at being a demiurge and displays *hubris* (Straughan, 1995). Furthermore, common morality is shocked by such operations since people were led to believe that everything is in the genome (the secret of life, physiological and psychological characteristics, deviances). Consequently, if everything is in the gene, it should not be touched, especially if the boundaries of species are infringed and 'frankenfoods' are produced, that is to say, half-animal and half-plant monsters.

There are no objections on principle to transgenesis

The religious ban on tampering with deoxyribonucleic acid (DNA) is contested by certain theologians. God granted humanity a dominating and controlling position over nature. Biotechnologies therefore provide an opportunity for humans to work with God as co-creators (Straughan, 1995).

Several counter-arguments go against the intuitions of common morality. The concept of the species is too fixist because transfers of genes do, in fact, occur in nature. For example, zoonoses and the phenomenon of tumorigenesis in plants through *Agrobacterium tumefaciens* bacteria call into question the notion of the boundaries of species. Biotechnologies consist in 'piloting natural processes' (Larrère, 2000). It is the art of 'having things done' by others, that is to say, micro-organisms, plants or animals. In consequence, it is not really an artificial process. Biotechnolgies are a continuation of the classic process of plant selection. Furthermore, if 'everything is not in the gene' (Atlan, 1999), the secret of life is not penetrated any further by manipulating the genome than by modifying the cellular metabolism (Larrère, 2004).

Finally, whether there are objections on principle or not to transgenesis, it is futile to hinder its development because the science will be inevitably achieved. This is the 'law' of Gabor: anything that can be technically achieved will be achieved, whatever its moral cost. This argument naturalises the development of techniques.

Transgenesis is a legitimate operation but only under certain conditions

This is a matter of respecting the intrinsic value of plants because each living organism possesses its own worth. There are two variations to this deontological and biocentric argument: the dignity of creation (Swiss Ethics Committee on Non-human Gene Technology, 2003) and the integrity of plants (Danish Ministry of Trade and Industry, 1999).

According to Balzer *et al.* (2000), the dignity of plants, unlike human dignity, is not absolute because it does not forbid improving plants to make them more useful to humans. The effect on their dignity is determined in relation to their phenotype. For example, as long as the resistance of a plant to herbicides does not impede either its growth or its reproduction, its dignity has not been undermined. On the other hand, if it loses its leaves at an early stage, it is indeed affected.

Respecting the integrity of plants implies making a distinction between destructive interventions and creative actions that respect nature (Danish Ministry of Trade and Industry, 1999). In that case, a threshold needs to be applied to determine if what we do is too risky or if nature is so subjugated that it can no longer react.

If it is accepted that there are no objections on principle to transgenesis, a number of questions arise.

From transgenesis to transgenic plants: the question of whether a living organism is patentable

The issue here is to evaluate the conditions that resulted in transgenesis leaving the laboratories and becoming an innovation protected by a patent. Are transformations of the world due to the innovation of genetically modified plants morally acceptable? An innovation is patentable if it is new, if it is the result of an inventive activity with potential industrial applications. Under these circumstances, a patent grants inventors a temporary exclusive right over their invention. In exchange, inventors must provide an accurate description of their invention.

Deontological arguments

In the view of some people, patents for transgenic plants are legitimate: DNA is a chemical molecule, and the transgene is obtained through a technical operation known as genetic 'construction'.

In the opinion of others, extending patents to cover genetically modified plants is not legitimate. The gene contains a set of information. Highlighting one of its functions (and deriving therapeutic or industrial applications from it) implies deciphering this information. This is a discovery and there is, consequently, nothing to patent. Furthermore, the expression of the genetic sequence depends on interactions between genes (epistatic) and interactions with the cytoplasm (epigenetic). This is the argument centred on genetic stock. It explains why scientists come up against the difficulty of attributing a function to a segment of DNA based solely on the criteria of sequence data.

Consequentialist arguments

The patent is an economically efficient strategic tool (Canadian Biotechnology Advisory Committee, 2002). It offers an alternative to the commercial secret and makes it possible to reconcile a return on investments (or the interests of the inventor) with publicity for the results (dissemination of information). It allows for exceptions to be made for protection, either in favour of scientific research or in the form of exemptions for breeders and, in France, privileges for farmers. Finally, advantageous permits can be granted to countries in the South.

But there is also a view that patents can lead to numerous problems. They can hinder research because the research tools themselves are patented. Owing to the difficulty of satisfying the criteria of industrial applicability, patents with claims that are too broad have been granted. This leads to a dependence on patents and the consequent spate of royalties. The economic effectiveness of a patent is therefore called into question. A series of mergers and acquisitions of companies have taken place in order to face the numerous disputes between the holders of rights. In view of the costs involved, the patenting system only benefits multinational firms and makes public research more expensive without ensuring its protection. The concentration of firms in the agro-chemical and seed sector increases the reliance of farmers on their suppliers. Transfers of licenses that are either free or at an advantageous price for countries of the South (the origin of most of the genetic plant resources) are not systematic. In addition, the financial gains of European, American and Japanese patent offices are directly linked to the number of patents granted. Resorting to these offices continues to be difficult. Finally, a compromise between the dissemination of knowledge and the interests of the inventor is more easily achieved through plant variety protection rights and maintenance as part of the common heritage.

Patents on living organisms seemed to be an interesting option in view of a narrow determinism that believes a gene corresponds to a protein, which itself carries out a function. The science of genomics has challenged this theory. Because of epigenetic and epistatic interactions, the same genetic sequence can carry out several functions. However, patent rights have frozen matters. Thus, for reasons that are both scientific and consequentialist, there is a tendency to avoid legal constraints, either through the merger of firms or through the mutualisation of panels of patents between research institutions (whether associated or not to companies).

If patents applied to living organisms give rise to problems, what about the actual transgenic plants themselves? Should one be opposed to all transgenic plants in general? Beyond the issue of patents, it is useful to examine the consequences of using such plants.

Transgenic plants: systemic evaluation of the consequences

A consequentialist evaluation of transgenic plants can be carried out in accordance with the moral good one seeks to maximise, in other words, the balance between costs and benefits, justice, freedom or biological diversity.

Balance between costs and benefits

Some people believe there are no health risks, transgenic products being the equivalent in substance to conventional products. Environmental risks, for instance, the resistance of insects to insecticides, and the crossing of transgenic plants with similar plants, are slight and under control. Transgenic plants obtained through genetic pollution (a very rare event) do not, in effect, provide a competitive advantage in a natural environment. Finally, agronomic risks are controllable through the diversification of ways to prevent them and the spatial organisation of parcels of land to limit the risks of genetic pollution. Advantages for consumers in terms of quality products are expected. Environmental benefits are also anticipated, for instance, a reduction in the use of pesticides and herbicides, as well as biological methods for clearing up pollution.

However, there is a controversy over all these points (COMEPRA, 2004; Commission de l'Ethique de la Science et de la Technologie, 2003). While certain people proclaim a technological revolution, others denounce the promises and the 'sophisticated makeshift approach' (Larrère and Larrère, 2000) since the effective expression of transgenesis is not controlled.

The debate on the risks and advantages of biotechnologies is nevertheless restricted to economic calculations, which are not sufficient for a moral evaluation of technology. Other moral advantages need to be considered.

Justice

The aim is to examine the injustices or inequalities that the spread of transgenic plants produces or, on the contrary, diminishes.

According to certain views, transgenic plants will make it possible to reduce hunger in the world (Nuffield Council on Bioethics, 2003). Others hold that this argument is merely rhetorical since the countries of the South are not solvent. In developing countries, political instability, problems of distributing resources, and poverty are responsible for inequalities (Sen, 1981). Consequently, even if transgenic plants are adapted to local conditions, this would not be sufficient to solve the problem of world hunger. Furthermore, biotechnologies increase inequalities among multinationals and seed companies. Public research, which finances the high cost of genetic engineering and the production of transgenic plants, finds itself at the service of a handful of multinationals with the means to buy licenses and protect their patents. The search for alternative solutions is hampered by the fact that research is at the service of private interests. Integrative approaches resulting from the standard selection of plants that take local conditions into account should, in fact, be developed (Swiss Ethics Committee on Non-human Gene Technology, 2004). This gives rise to an effect of economic domination over world agriculture.

Freedom

Transgenic plants are examined to see whether they alter the freedom of the different players. Some people believe that farmers retain their freedom to decide whether to resort to transgenic varieties or not. For others, the dependence of farmers increases as a result of mergers in the agro-chemical and seed sector. Contracts tying farmers to agro-chemical firms contain clauses that bind them to refrain from re-using seeds from one year to another. In addition, there are also problems of co-existence between transgenic and non-transgenic cultivation (especially in the case of biological crops). And finally, the lack of product labelling is a denial of the freedom of choice of consumers.

Biological diversity

Biological diversity can constitute a norm for the consequentialist evaluation of using transgenic plants. In this case, it is a question of determining the consequences of growing transgenic crops on the biological diversity of plants, whether cultivated or not.

Some people believe that transgenic plants encourage biological diversity. The higher output of transgenic plants that are resistant to insects should make it possible to increase world production and free surfaces so that they can be devoted to protecting or creating habitats for the purpose of preserving the diversity of species. But others believe that if transgenic varieties are massively distributed and if they eliminate competition from traditional varieties, there will be a decrease in the genetic diversity of cultivated plants. Erosion of the genetic diversity of cultivated plants is not conducive to adapting crops to time and space variations in production conditions. As Straughan (1995) wrote, 'a reduced number of 'supercrops' might prove to be less resilient and so more vulnerable to various forms of attack in the future'. The aim, therefore, is to improve plants, eventually through transgenesis, but without compromising their evolutionary capacities, as this would satisfy the argument of integrity.

Conclusion

Our proposed typology provides a framework for interpreting the principal ethical judgements on transgenic plants. If the absence of objections on principle to transgenesis is accepted, the evaluation of the consequences of using transgenic plants is not limited to calculating the balance between costs and benefits. It is also necessary to consider other moral advantages that one wishes to promote, such as justice, freedom and biological diversity. This would then lead to a systemic evaluation of the consequences of using transgenic plants. The effects of their use are assessed as a global project.

This kind of evaluation is vital, particularly from the viewpoint of sustainable development. In reality, the idea of sustainable development, as defined at the International Rio Conference in 1992, is to reconcile three needs: economic development, social justice and the protection of nature. One of the difficulties in implementing sustainable development is to determine precise objectives and concrete criteria for evaluating innovations. Considering the specific case of transgenic plants, the typology we propose could be a preliminary to examining whether these plants can contribute to sustainable development or not.

References

Atlan, H. (1999). La fin du tout génétique? Nouveaux paradigmes en biologie. INRA Editions Sciences en questions, Paris, France, 91p.
Balzer, P., Rippe, K.P. and Schaber, P. (2000). Two concepts of dignity for humans and non-human organisms in the context of genetic engineering. Journal of Agricultural and Environmental Ethics 13: 7-27.
Canadian Biotechnology Advisory Committee (2002). Patenting of higher life forms and related issues. Canadian biotechnology advisory committee, Ottawa, Canada, 60p.
COMEPRA (2004). Avis sur les OGM végétaux. INRA, Paris, France, 12p.
Commission de l'Ethique de la Science et de la Technologie (2003). Pour une gestion éthique des OGM, Résumé, recommandations et mises en garde. Gouvernement du Québec, Québec, Canada, 23p.

Danish Ministry of Trade and Industry (1999). An ethical foundation for genetic engineering choices. Danish Ministry of Trade and Industry, Copenhagen, Denmark, 60p.

Larrère, C. and Larrère, R. (2000). Les OGM entre hostilité de principe et principe de précaution. Le Courrier de l'environnement de l'INRA 43 : 15-23.

Larrère, R. (2000). Faut-il avoir peur du génie génétique? Cahiers Philosophiques de Strasbourg 10 : 11-48.

Larrère, R. (2004). Organismes génétiquement modifiés. In: M. Canto-Sperber (ed.) Dictionnaire d'éthique et de philosophie morale. PUF, Paris, France, pp. 1378-1381.

Nuffield Council on Bioethics (2003). The use of genetically modified crops in developing countries. Nuffield council on bioethics, London, UK, 122p.

Sen, A. (1981). Poverty and famines: an essay on entitlement and deprivation. Oxford University Press, New York, US, 257p.

Straughan, R.R. (1995). Ethical aspects of crop biotechnology. In: T.B. Mepham, G.A. Tucker and J. Wiseman (eds.) Issues in agricultural bioethics. Nottingham University Press, Nottingham, UK, pp. 163-176.

Swiss Ethics Committee on Non-Human Gene Technology (2003). Gene Technology for Food. ECNH, Bern, Switzerland, 20p.

Swiss Ethics Committee on Non-Human Gene Technology (2004). Gene Technology and Developing Countries. ECNH, Bern, Switzerland, 32p.

Marketing GM Roundup Ready rapeseed in Norway? Report from a value workshop

Ellen-Marie Forsberg
National Committees for Research Ethics, P.O. Box 522 Sentrum, 0105 Oslo, Norway,
ellen-marie.forsberg@etikkom.no

Abstract

During the spring of 2007 the secretariat of the National Committee for Research Ethics in Science and Technology (NENT) organised a so-called 'value workshop' discussing the ethical acceptability of marketing GM Roundup Ready rapeseed. As this rapeseed has been approved for marketing in the EU the Norwegian authorities must now decide whether to allow it in Norway. The Norwegian Gene Technology Act demands that 'the production and use of genetically modified organisms [...] take place in an ethically justifiable and socially acceptable manner'. The value workshop is a methodology that is thought to be suited for the task of assessing ethical justification. It involves using an ethical matrix method to structure ethical dialogue with affected parties to an issue. Relevant affected parties are e.g. Canadian farmers, Norwegian farmers, consumers, industry, the animals targeted for this feed, as well as the environment. Individuals representing these different parties and interests (of obviously different character), as well as other interested individuals and experts, were invited to discuss ethical principles, factual consequences and value trade-offs involved in accepting this product on the market. A conclusion on ethical acceptability was sought, but it turned out that all the available time was consumed in identifying the values and discussing the relevant facts. A final balancing and conclusion on the ethical justifiability of marketing GM Roundup Ready rapeseed in Norway was therefore not reached. However, the workshop did result in important input for further decision making.

Keywords: GMO, ethical tools, participatory process, ethical matrix

Background and organising of the workshop

This value workshop was organised by the secretariat of the National Committees for Research Ethics in Science and Technology, with financing from the Norwegian Research Council. The practical background was the fact that as Roundup Ready rapeseed has been accepted for marketing (for feed purposes) in the EU it is now being considered for Norway, and the Norwegian Gene Technology Act requires that ethical assessments are performed for all use of GMOs. The theoretical background was a doctorate project on methods for doing ethical assessments of GM food. This doctorate work discusses how using an ethical matrix method in participatory processes may lead to well-justified ethical advice (cf. Forsberg, 2007a, b, c).[14] The workshop was a way to test the method as it was further developed in this thesis and to gain experiences with regard to how the method works in practice for issues related to GM food.

The organisers focused initially on securing the attendance of representatives of affected parties. In the ethical matrix method it is considered important that the views and values of the affected parties are considered in the ethical assessments. It proved difficult to find someone to represent the views of the biotech seed industry, but an expert on genetic engineering agreed to present his perceptions of the values and perspectives of this industry. The organisers wanted to include perspectives also related to the production of the GM rapeseed, as the ethical justifiability of *marketing* a product is influenced

---

[14] The ethical matrix method was originally developed by Ben Mepham. Confer for instance Mepham (1996, 2004).

by the ethical justifiability of the *production* of the product. The Roundup Ready rapeseed is grown in several countries and Canada was chosen to exemplify the issues related to this rapeseed production as Canada is a large producer. A former farmer and a representative of Canadian First Nations were invited to join the discussions. Representatives of the agriculture and aquaculture industry buying the oil or meal from this rapeseed were also invited, as well as representatives for ultimate consumers and the environmental movement. In addition to this, invitations were sent to various other experts and interested individuals or organisations.

The workshop lasted two days and included four sessions:
- Session 1: establishing a value matrix.
- Session 2: weighting of most important values.
- Session 3: constructing a consequence matrix.
- Session 4: doing a final evaluation.

The work consisted of group work and plenary discussions. There were 14 participants, including the facilitators. Although some participants were invited as representatives of affected parties, and asked to present their perspective and values related to this GM rapeseed, it was stressed that all the participants spoke only on behalf of themselves in the subsequent discussions. Moreover, they were asked to take the perspective of citizens trying to determine the best solution in an overall societal perspective. The workshop was intended as consensus *seeking*, particularly for the establishment of the ethical platform (i.e. the value matrix). However, it was made clear that one would not be consensus *forcing*, and that dissent would be respected.

Results

The session establishing the value matrix had sufficient time, and this result is a good foundation for further analysis of concrete applications (see Figure 1). The participants had been given a proposal for a value matrix as a starting point for discussion and felt free to change and adjust the content and the framework of this input. The work session where weighting was performed suffered from a lack of time and some participants indicated afterwards a doubt that these beliefs were indeed prioritised by all. In Figure 1 the values that were considered to be of special importance are marked grey.

The first half of the second day was designated to noting the relevant facts. These facts were structured in a so-called consequence matrix. This matrix consists of the frames of the value matrix and all consequences relevant for the values in the value matrix are noted. It is also noted whether the consequences are in accordance with the underlying values (using + as symbol), are in conflict with the underlying values (using -) or are uncertain (using ?). The symbols were put in brackets if there was uncertainty about whether the description really applied. It turned out that this activity took the whole day, at the expense of actually using the tools (the value matrix and the consequence matrix) that were established in the course of the work. Moreover, as can be seen by noting the blank cells in the consequence matrix, the discussions had to be closed before the assessment of consequences had come to a conclusion. Still, the consequence matrix actually produced contains important information (see Figure 2).

Discussion

There may be different reasons why the discussion of the consequences took such long time. The intention of the organisers was to discuss the Roundup Ready rapeseed in particular, not genetic engineering in food production as such. The intention was that even if one did not agree on all the consequences one could note this disagreement and move on to the next consequence and ultimately to a judgement on the whole picture of this particular issue. However, it became evident that it was

	Well-being	Dignity	Equality
Farmers in GM producing areas (primary)	Safe workplace Safe income Predictable social situation.	Right to control of their work situation, right to make own decisions on choice of seed. (and respect for their choices in the chain)	Equal opportunities and requirements (rammebetingelser) for GM and non-GM farmers Coexistence
Seed industry (suppliers) (Monsanto, etc)	Adequate profit and income Predictability, consistency, simplicity, real possibility for compliance	Acknowledgement for their part of the value chain, being heard in negotiations Protection of private initiatives Freedom of innovation to improve the human condition Respect for scientific achievements	Equal terms for this industry as for the other food industries
Other users of the land	Protect the quality of current use	Respect for their needs and their use of the land Local acceptance of use Respect for traditional knowledge	Fair access to the resources
National interests (in cultivating society)	Safe and profitable use of resources No health risks or added anxieties Economic growth	Freedom to manage resources and technology for the best of the society as a whole No dependencies Respect for social acceptance of technology/practice	Equitable living conditions for urban and rural societies, and for rich and poor
Government/ regulatory authorities/national interests (in importing society)	Food security Adequate institutions and competence Complexity of society	Freedom to manage resources and technology for the best of the society as a whole No dependencies Innovation for the improvement of the human condition	Equal opportunities in decision making processes
Other affected countries	Not being imposed new burdens or liabilities	Respect for their views and rights	Harmonisation of trade rules (and other rules) internationally Capacity building
Feed industry/ processors, retailers	Predictability, consistency, simplicity, real possibility for compliance	Occasion to choose and influence the production of food products Real choice of seed (ensuing chains are split) Respect for differences in industries	Equitable trade conditions

	Well-being	Dignity	Equality
Consumers	Guaranties for healthy food in adequate amounts Adequate nutrition No health risks Food safety Not laying too much responsibility on the consumer Consumer concerns about unintended spread to other products	Occasion for the consumer to chose and influence the production of food products Adequate and understandable information Labelling and real traceability (also in processed food) Market freedom Respect for social acceptability of technology/practice	Food products of good quality available for different consumer groups
Future generations	No activities that threaten their health or living conditions Enable adequate productivity - access to gene material for local adaptation of species Precaution Food safety	Providing them with ample range of choice	The protection of the environment and resources so that future generations will have equal opportunities as we do (biodiversity)
Animals	Animal welfare Animal safety	Suitable living conditions Respect for species uniqueness/integrity	
The biosphere - in producing and consuming areas	Health of ecosystems Environmental safety	Harm and abuse of nature as limited as possible Respect for natural properties	The diffusion to a viable level of environmental burdens over a manifold of ecosystems (biodiversity)

Figure 1. Value matrix for GM crop issues.

	Well-being	Dignity	Equality
Farmers in GM producing areas (primary)	Secure workplace: GM farmers + More benign chemicals: GM farmers + Safe income: GM farmers + Non GM farmers − Predictable social situation: in the short term + in the long term −	Effects on negotiative power of non GM farmers − Of GM farmers + (Allowing GM acknowledge their professional dignity)	Coexistence even harder? ? Still, more diversity in agricultural practice today
Seed industry (suppliers) (Monsanto, etc)	Adequate profit: GM supplier + Predictability, consistency, simplicity, real possibility for compliance +	Acknowledgement for GM industry part of the value chain, being heard in negotiations +	Equal terms for this industry as for the other food industries + Small seed industry hurt −
Other users of the land	Protect the quality of current use (?) Part of a culture that reduces the quality of indigenous homeland? (?)	Disrespects indigenous needs or voices − Disregards indigenous knowledge and its role in the origin of the plants − Less negotiative power −	
National interests (in cultivating society)	Increased export revenue in short term (+) Long term consequence? ? Possible health benefit from more benign herbicide (+)	Lack of transparency and democracy in the origin of the crops −	
Government/ regulatory authorities/ national interests (in importing society)	Food security: short term + long term ? Complexity of society (expensive to handle segratation issues) (−)	Export food products fed on GMOS: possible trade disadvantage?	
Other affected countries	Better trade possibilities for GM using countries + Worse trade possibilities for non GM using countries −	GM using countries will have stronger negotiating position + Non GM using weaker −	
Feed industry/ processors, retailers			Hard and expensive to find non GM seeds +
Consumers	No evidence that there are changes to consumer safety - Relevant studies and monitoring must be done! 0? DISSENT Do consumers feel safe about it? ?	More complex for consumers (labelling, information) − Freedom to chose +	

	Well-being	Dignity	Equality
Future generations	Food safety and security may be affected by absorption of GT73 genes? ?	Will restrict future opportunity to make choice of non GM (DISSENT)	Imported seeds may be spilled - may transfer genes to wild relatives - Strengthens monoculture, decreases biodiversity? ? May increase biodiversity because better weed control? ?
Animals	No evidence yet that there are changes to animal feed safety Relevant studies and monitoring must be done! 0? DISSENT		
The biosphere - in producing and consuming areas	Less tillage in the short term + Unintended effects on insects, etc.? Weedy relatives may have fitness advantages: therefore need for stronger herbicide in the future? ?		Strengthens monoculture, decreases biodiversity? (?) Genetic erosion might increase? (?)

*Figure 2. Consequence matrix depicting the consequences of marketing GM Roundup Ready rapeseed in Norway.*A participant later said he dissented to including this consequence.*

difficult to treat this case as a separate case. As it is potentially the first GM product to enter the food chain in Norway the treatment of this case might set general standards. The discussion therefore often ended up at a principled level where it was difficult to agree to a description of the situation and then to move on. The challenge for the facilitators was therefore to balance the need for this discussion with the agenda for the meeting. It seemed in the end not possible to close the discussion on the consequences at a time making it possible to discuss a final judgement.

Although a judgement was not made by the participants in this process, two tools for drawing conclusions were indeed produced: the value matrix and the consequence matrix. The value matrix provides a map of the most important values and an ethical platform for finding facts for a number of crop issues. The consequence matrix may serve as a factual foundation for the participants to draw their own individual conclusions or for other committees or government agencies to make judgements on this particular case. As such, a thorough analysis of the ethics of marketing GM Roundup Ready rapeseed is available for decision makers or others for drawing conclusions. Taking the general level of conflict among the affected parties during the value workshop into account, one may consider inviting a committee of lay people to draw conclusions on the basis of the material produced in the value workshop. This will benefit from the input from the affected parties and the experts, but may have the advantage of avoiding the issue to turn too political. This may be a way to redesign the method in the future.

As one can note from the matrices reproduced above the ethical picture became very complex. Indeed, when compared with what was designed in advance as a proposal for a value matrix (cf. Forsberg, 2004), an already complex picture became even more complex. This of course makes judgements difficult. However, one may legitimately ask whether this is a fault of the method or if it simply realistically portrays an extremely complex value situation, - and indeed that a simpler tool would reduce this complexity in an unfortunate way leading to less holistic and robust judgements.

Tentative conclusions on the Roundup Ready rapeseed

Can we draw any tentative conclusions from the value matrix, weightings and the consequence matrix that were produced? Of course, any conclusions drawn by this author alone as a single expert will not have the same justifiability as if they were drawn by the participants themselves. Still, the basis for an expert judgement is at this point more comprehensive and more thoroughly discussed than it would have been without the value workshop.

We should first note that there are indeed benefits for other stakeholders than simply (parts of) the seed industry. This holds especially for the Norwegian feed industry, as well as aqua- and agriculture, who find it difficult to find non GM seeds at reasonable prices. With regard to farmers, the cultivating and importing societies, other affected countries and consumers there seems to be both benefits and burdens. Although other users of the land, in this case represented by Canadian indigenous people, seem to suffer from the use of this land for production of this rapeseed, it is hard to distinguish the burdens coming from this particular GM crop from the burdens of industrial farming as such. All in all, it is not an entirely clear picture appearing from the consequence matrix.[15]

Still, I will tentatively draw the conclusion, - which was indeed drawn by some of the participants in a very brief final round - that there are too many uncertainties (marked by question marks in the consequence matrix) to allow the Roundup Ready rapeseed to be marketed in Norway at the present point of time. Although the official risk assessments of effects on animal and human health show that this rapeseed is safe, a number of the participants did not feel that sufficient amount of research had been carried out. It was pointed out that most of the feeding trials had been carried out on animals that would not be consuming this rapeseed (e.g. on rats), and not on the animals most likely to be fed rapeseed oil/meal on a large scale (like salmon). Other weaknesses of the design of the available studies were also pointed out.

It was also noted that there would be no way to avoid some seed spill - with the accompanying gene transfer to the environment. Although the importance of this effect was discussed, it was acknowledged that there would be effects on the environment. As animal, human and environmental safety issues were marked as values of particular importance, it seems that question marks on these consequences should be taken seriously. As there were several other uncertainties as well (as noted in the consequence matrix), the reasonable conclusion, to my mind, is to await more information before issuing a marketing allowance.

---

[15] More could of course be said about this picture, but there is no space here to go in more depth.

Conclusion

The value workshop on Roundup Ready GM rapeseed suffered from insufficient time but two important outcomes were still produced: the value matrix and the consequence matrix. The value matrix does map the most important ethical concerns to take into account when assessing crop biotechnology. The consequence matrix notes important consequences of marketing this GM rapeseed in Norway related to the values identified. Sufficient time should be accorded to do ethical evaluations of these issues, but also to perform adequate research that fill in the caveats and uncertainties still characterising some parts of the scientific evidence relevant for making a decision on marketing allowance.

References

Forsberg, E.-M. (2004). Ethical assessment of marketing GM Roundup Ready Rapeseed GT73. In: J.D. Tavernier and S. Aerts (eds.) Science, Ethics and Society. Proceedings to the 5th European Congress on Agricultural and Food Ethics.

Forsberg, E.-M. (2007a). Value Pluralism and Coherentist Justification of Ethical Advice. Journal of Agricultural and Environmental Ethics 20: 81-97.

Forsberg, E.-M. (2007b). *A Deliberative Ethical Matrix Method - Justification of Moral Advice on Genetic Engineering in Food Production*. Dr. Art. Dissertation. Oslo, Unipub.

Forsberg, E.-M. (2007c). Pluralism, the Ethical Matrix and Coming to Conclusions. Journal of Agricultural and Environmental Ethics (in press).

Mepham, T.B. (1996). Ethical analysis of food biotechnologies: an evaluative framework. In: T.B. Mepham (ed.) Food Ethics. London, Routledge, 101-119.

Mepham, T.B. (2004). A decade of the ethical matrix: A response to criticisms. In: J.D. Tavernier and S. Aerts (eds.) Science, Ethics and Society. Proceedings to the 5th Congress of the European Society for Agricultural and Food Ethics.

GM-food: how to communicate with the public?

Ursula Hunger, Brigitte Gschmeidler and Elisabeth Waigmann
dialog<>gentechnik, Dr. Bohrgasse 9/P, 1030 Vienna, Austria, hunger@dialog-gentechnik.at

abstract

Benefits and risks of gene technology especially of genetically modified organisms are controversial. However, society daily encounters applications of these research fields and rightly expects reliable and well-founded information. On this account, dialog<>gentechnik assumes the role of a platform for the exchange between science and public.

About dialog<>gentechnik

dialog<>gentechnik is an independent non-profit organization, whose activities are exclusively financed by public funding. In order to stay independent, the organization does not accept financial funds from the gene technology industry.

dialog<>gentechnik was dedicated to provide competent and reliable scientific information and functions as an interface between science and the public. It facilitates and supports dialogue between various parties about gene technology and related topics. All activities are supported by more than one hundred experts and their know-how and by the members of the organization, eight Austrian scientific societies.

The projects of dialog<>gentechnik not only provide knowledge on science and technology behind life science issues but also enhance the understanding of the controversial debate around GMOs including issues such as ethics, legislations, safety, and risk assessment.

dialog<>gentechnik:
- provides comprehensible, competent and balanced information;
- answers questions to gene technology and related topics;
- connects to experts who are willing to answer specialized questions;
- organizes events and projects (exhibitions, discussion groups, school activities etc.);
- supports scientists in their PR work.

Projects

dialog<>gentechnik sets different activities to stimulate the dialogue on life science issues. Several projects focussing on genetically modified organisms are presented here:

Discussion at the gambling table - a game on green gene technology

This card game invites to discuss the topic of gene technology in agriculture and food. It gives an overview of the variety of positions on genetic engineering in the agriculture, offers the participants the possibility to learn more about genetically modified crops and to discuss the complex political and scientific aspects of this topic. Playing the game helps the participants to express their own opinions on green gene technology and to identify similarities and consensus in their ideas. This card game is played at events organised by dialog<>gentechnik (Figure 1).

Figure 1. Card game on green gene technology.

Films about Gene Technology - the EU-Projects 'MREFS and EMRS'

In order to make the European research about biotechnological applications in food and health more well-known and to facilitate a purposeful discussion on ethical, socio-economic and social aspects of scientific achievements in the life science range, the European commission finances two projects to impart science to laymen.

In these projects, with the acronyms EMRS and MREFS, films about European Union-promoted research projects are produced to highlight innovative aspects, possible applications, risks and advantages and also ethical and social components of the research work in a descriptive and understandable way.

dialog<>gentechnik contributed to the films on gene technology and GM-Food, which are offered at the EU-homepage www.eusem.com. The topics of the films are major research projects within the Sixth EU Framework Programme about food-quality and safety.

School Kit about Green Gene Technology

With contribution of dialog<>gentechnik the UNESCO developed a teaching toolkit (manual, DVD and posters) on the issues of GMOs with the working title 'Learning and teaching about new technology', targeted to secondary schools teachers and educators.

This kit supports teachers with background information to get informed on new technologies and supplies new ideas and hands-on experiments for the classroom as well as working sheets for the pupils.

A traveling exhibition - the EU project 'DNA-Test'

The objective of the project 'DNA-TEST' is to set up a travelling exhibition on selected life science topics combined with a short live theatre performance. Inside and outside the trailer information on gene technology is presented on various multimedia means and posters.

The application of gene technology needs to be in agreement with the basic values of society. People have to take decisions on such technologies and their applications in their capacity as consumers and citizens. Therefore a new activity will facilitate the examination of those technologies. The Flanders Interuniversity Institute for Biotechnology (Belgium), Pandemonia (Netherlands) and dialog<>gentechnik will set up a pilot show case in three different countries in Europe. The show case offers the possibility of hands-on exploration of gene technology related to GM-food.

A kit for the show case will be distributed throughout Europe and beyond the partners of the project.

Vienna Open Lab

Visitors explore the world of science and research in the 'Vienna Open Lab', which is a joint initiative by dialog<>gentechnik and IMBA. The Vienna Open Lab is the first hands-on laboratory in Austria. It offers pupils and interested people of all ages the possibility to perform molecular biological experiments on their own. Together with young scientists the visitors experience science through hands-on experiments and get exciting insights into the day-to-day experimental work in a life science laboratory.

Experiments and techniques offered in the various courses include:
- DNA extraction from fruits, vegetables, plants, food products and saline mouthwash.
- Analysis of DNA with restriction endonucleases and gel electrophoresis.
- Transformation of bacterial cells with plasmid DNA.
- Predict bitter tasting ability by PCR.
- Detection of genetically modified food by PCR.

The activities are specifically adapted to the different age groups. Training programs last from two hours up to six hours. Special programs (e.g. teachers` workshops) take one day and longer.

The Vienna Open Lab should be a place where scientists and the public meet and enter into a dialogue to discuss various aspects of research in biosciences. It should aim to substantially expand young peoples´ interest in biology and stimulate them in taking up careers in the life sciences.

Part 13 - GMO: transgenic animals

Ethical concerns related to cloning of animals for agricultural purposes

Mickey Gjerris and Peter Sandøe
Danish Centre for Bioethics and Risk Assessment, Faculty of Life, University of Copenhagen, Rolighedsvej 25, DK - 1958 Frederiksberg C, Denmark, mgj@life.ku.dk, pes@life.ku.dk

Abstract

Cloning of animals for agricultural purposes has moved one step up the ladder of probability since the release of the US Food and Drug Administration's risk analysis of meat and milk from cloned animals and their progeny. The usefulness of the technology in an agricultural setting is debated, but it seems that as a supplement to existing reproduction technologies, cloning might have a role to play in food production. However, the introduction of cloned farm animals will probably give rise to strong reactions from the public. Identifying the ethical issues at stake is therefore important. From the point of view of the US authorities the debate should be about risks to humans and animal welfare. Within this framework, there is a very strong case for allowing cloning of animals for agricultural purposes. All available research to date shows that there are no substantial differences between products from conventional animals and from cloned animals or their progeny. When it comes to animal welfare there are indeed problems. However, there is a striking discrepancy between the concerns about these welfare problems compared to the concerns about welfare problems caused by many conventional agricultural systems. This paper examines the values at play in the discussion and asks whether the concerns for animal welfare are really just concerns about animal integrity 'in disguise'. Our conclusion will be that there is a need to acknowledge values such as integrity, but also that the critique of cloning must be seen in the context of a wider critique of modern animal production.

Keywords: agriculture, cloning, ethics, integrity, welfare

Introduction

Since the presentation of the cloned Dorset ewe Dolly in 1997 by Ian Wilmut and his colleagues (Wilmut *et al.,* 1997) the notion of cloning has captured the public imagination and provoked discussions about the ethical implications of the technology. Initially it seemed that the only concern was that the technology might be applied to humans as well. Later there was a growing focus on the possible applications of the technology to animals and the ethical concerns to which this gave rise.

Initially there were also very high hopes regarding the potential applications of cloning technology. Now, 10 years after Dolly, those first hopes have been replaced by a more nuanced understanding of the difficulties in both mastering the technology and evaluating its possible use in relation to different applications (Gamborg *et al.,* 2006). Today a picture is emerging where cloning plays an important role in basic research, where it can serve as a source of information about foetal development, cell biology and epigenetics, for example. Furthermore, in biomedical research cloning can serve to produce disease models to further our understanding of human diseases, such as pigs that are genetically modified to resemble humans with Alzheimer's disease. Animals can also be reared that produce valuable medical substances, such as human proteins, in their milk, blood or eggs; so-called bioreactors (Vajta and Gjerris, 2006).

A series of more exotic applications have been suggested too, for instance using cloning to save endangered species, such as giant pandas and tigers, or to recreate extinct species such as the gaur or the mammoth.

None of the attempts made in these areas have shown any promising results (Gamborg *et al.*, 2006), and even if the technology becomes so developed as to be used in these contexts, it will only be in very specific and limited areas. The idea of cloning pets has also been explored, but the economic feasibility of this has not shown itself, and the only commercial company in the field, Genetics Savings and Clones, closed down in 2007 (see http://www.savingsandclone.com/).

In between these applications lies the possible use of cloning technology in the agricultural sector. At first cloning technology was welcomed as a new and exiting tool to use in animal breeding (Di Berardino, 2001), but as the limitations of the technology have been revealed the debate about the usefulness of cloning in agriculture has grown more heated. It thus seems logical to begin our discussion of the ethical implications of animal cloning for agricultural purposes by considering the potential benefits to be gained.

Cloning for agricultural purposes

As mentioned above, early hopes for the usefulness of cloning in agriculture were high, and some recent literature on technology development continues to be optimistic (Niemann *et al.*, 2003). However, even within the scientific community there is disagreement about the extent to which cloning will be a useful tool (Meyer, 2005).

Four main factors lie behind the different interpretations. The first is different understandings of the likelihood that the technical obstacles so far met can be overcome. So far the technology has had very low success, as measured by the number of viable offspring and the health problems faced by cloned animals, at least early in life (Vajta and Gjerris, 2006). Whether this is just a question of perfecting the technology or whether there are inherent limits to the effectiveness of the technology, due to epigenetic factors for example, is very uncertain.

The second factor is the challenge in producing genetically identical animals by cloning. If it is to be used on a large scale in agricultural breeding it will be necessary to solve questions about the mitochondrial DNA that the cloned animal inherits from the egg used to produce it, and the question of how much effect the various epigenetic factors have on the resulting genotype and phenotype. So far these questions are surrounded by scientific uncertainty (Vajta and Gjerris, 2006).

The third factor is the willingness of the agricultural sector to adopt the technology, if it can be made economically and technologically efficient. There can be no doubt that producers in the agricultural sector, and perhaps even more in the food-production sector, will hesitate before adopting a technology as controversial as cloning. Public scepticism to the use of cloning technology, both in Europe and in the USA, is widespread (Lassen, 2005) and remembering what happened in the case of genetically modified plants for food consumption, it could be risky to market meat or milk from cloned animals or their progeny (Gamborg *et al.*, 2006).

The fourth factor is the obvious fact that people expressing their expectations about this issue usually have vested interests. This can be either a direct economic interest in seeing the technology brought into large-scale use, or more indirectly, might involve research and funding interests in the area or certain values that either puts emnphasis on animal welfare questions or stresses the importance of free research.

There are basically two areas where cloning could potentially be of interest in the agricultural sector. The first is as an assisting technology to propagate animals with desired genetic traits, possibly induced through genetic modification; the second is to use the technology to produce clones of animals with high breeding value. In a recently conducted European research project (Cloning in Public), stakeholders such as breeding organizations, scientists and industrial interests were asked to identify the most likely areas of application and the extent to which the technology could be useful. Judging from their answers it seems fair to conclude that at the present state of the technology, and with the existing public attitude towards the use of biotechnology in the food sector, cloning will only play a very limited role for at least the next 7-10 years. Only a few elite animals will be cloned and the question is whether these will ever be allowed to enter commercial breeding programmes so that products from them or, more likely, their progeny, will appear in the food chain (Gamborg *et al.*, 2006).

The US Food and Drug Administration (FDA) risk assessment

Regardless of whether cloned animals and their progeny will enter the food chain and the extent to which they will do this, a number of risk analyses examining the possible health consequences for humans have been conducted in recent years. By far the most extensive was published in December 2006 by the FDA (Food and Drug Administration, 2006). The conclusion of this risk assessment was that there is no evidence that products from cloned animals and their progeny differ from products from conventionally bred animals in any way that will cause risks to human health through consumption of such products.

These findings correlate with the findings of other research projects (Takahashi and Yoshihio, 2004, Tomé *et al.*, 2004). The scientific uncertainty of the FDA draft risk assessment is, however, brought up by some (e.g. see Center for Food Safety, 2007a) and since the hearing period for the FDA risk-assessment report only ended on 3 May 2007 it would be premature to say what the final outcome will be. For the purpose of this article we have, however, assumed that no substantial changes will appear regarding the conclusion that products from cloned animals or their progeny pose no measurable risks to humans.

The FDA report also discusses at length the consequences of cloning for animal welfare. The ineffectiveness of the technology is revealed, among other situations, in serious welfare problems for many cloned animals. Many of these welfare problems have been gathered under the heading of large offspring syndrome (LOS) and include placental abnormalities, foetal overgrowth, prolonged gestation, stillbirth, hypoxia, respiratory failure and circulatory problems, lack of post-natal vigour, increased body temperature at birth, malformations in the urogenital tract (hydronephrosis, testicular hypoplasia), malformations in liver and brain, immune dysfunction, lymphoid hypoplasia, anaemia, thymic atrophy and bacterial and viral infections (Vajta and Gjerris, 2006).

The FDA acknowledges the animal-welfare problems that the technology creates. However, at the same time it states that these problems do not differ qualitatively from welfare problems experienced with normal reproduction and with other artificial reproduction technologies, although they appear in higher rates in the context of cloning. Therefore, it concludes that there is no need to single out cloning as a subject for special regulation (Food and Drug Administration, 2006).

Ethical issues related to cloning

The FDA draft risk assessment is, by means of the statutes of the FDA, limited to take 'science-based facts' (where 'science' means 'natural science') into consideration when assessing risks of a certain product (Food and Drug Administration, 2006). It is therefore no surprise that the assessment focuses on risks to human health and animal-welfare problems and does not go into the other ethical concerns that typically arise when discussing animal cloning. These issues include the socio-economic consequences for different sectors of introducing new technologies and concerns about the naturalness of the technology and the violation of the integrity of the cloned animals (Gjerris, 2006).

One of the concerns not dealt with by the FDA is animal integrity. Elsewhere we have discussed how the concept of integrity can be understood in relation to animal cloning (Gjerris and Sandøe, 2006). Here it suffices to say that ethical concerns about the violation of animal integrity can be interpreted as a way of stating that the commodification or culturalization of animals that cloning is an expression of is problematic. It can also be seen as a critique of the reduction of the phenotypic wholeness and independence of an animal to a genetic tool for use by humans. Integrity does not hinge on a 'science-based fact'. The concept of integrity is an attempt to frame the strangeness, the alien quality, the independence and the surplus of the animal that we experience when we relate to it without seeking ways to exploit it to our own benefit (Gjerris, 2006; Gjerris and Sandøe, 2006).

What is most interesting in this connection is to see that although the ethical concerns that are related to the question of integrity (broadly understood) play a significant role in the formation of the public sceptical attitude towards animal cloning (Lassen, 2005), they are rarely mentioned in the public discussions about cloning. This pattern can, for instance, be seen in many of the comments sent by non-governmental organizations and others to the FDA in connection with the draft risk-assessment report (Center for Food Safety, 2007b). A possible explanation for this will be discussed in the next section.

From integrity to welfare

As stated by the FDA, cloning gives rise to severe welfare problems. Not surprisingly, this is a major ethical concern in connection with the technology. It does, however, strike us as curious that the welfare problems of cloning are so much in focus when one takes into consideration that these problems are not specific to cloning, but can be found in other types of artificial reproductive technology as well. It thus seems that welfare problems that are to some extent accepted in some areas of agricultural production are not accepted when these problems are caused by cloning. This can easily lead to claims that opponents of the technology have double standards. However, instead, we would like to suggest two aspects of this that could not only make more sense of the scepticism than just claiming it is based on double standards, but could also explain the growing interest in broader notions of animal welfare than narrow scientific notions focusing exclusively on the mental experiences of animals (Gjerris *et al.*, 2006).

First of all, the tendency to frame the public debate as being about risks to humans and animal welfare can be seen as a way of engaging with public reports such as the draft risk assessment from the FDA. When reports such at this and others by governmental organizations (e.g. see National Research Council, 2002, 2004) focus exclusively on so-called 'science-based' concerns, it is an obvious strategy (conscious or unconscious) to frame concerns about cloning in this language. Ethical concerns about the violation of animal integrity are thus transported into the area of animal welfare. Here they encounter the ongoing discussion between different perspectives on animal welfare, where the concept of naturalness plays a significant role. To some the concept is part and parcel of the concept of animal welfare whereas others see it as at best irrelevant, since the animal does not necessarily experience a loss of welfare from not expressing its natural behaviour or living in a natural environment (Gjerris *et al.*, 2006). The concept of integrity is therefore at risk of being translated into broader notions of animal welfare, thus losing its original point and becoming perhaps more a hindrance to the debate than a concept that actually clarifies the values at stake.

At the same time the transportation of concerns from the distinctly philosophical concept of integrity into the more scientific concept of animal welfare makes it clear why the debate in the animal-welfare community between the narrower and the broader interpretations of the concept cannot come to an end. If concerns about integrity have to be framed by welfare to be legitimate in the public debate and at the same time are to reflect the basic experiences that initially gave rise to the concerns, then the concept of welfare has to be broadened, or the concept of integrity be accepted as useful and legitimate, to bring it into the public debate on par with concerns about human health and animal welfare.

The second point worth mentioning is that the scepticism about animal cloning must be seen in context with the general attitude towards agricultural production systems. It makes no sense from a philosophical point of view to not use the same standards for the rest of agriculture as one is ready to use for cloning. Thus if the animal-welfare problems experienced in cloning are unacceptable, this should, if no relevant differences can be shown between cloning and other more established artificial reproductive technologies, also be reflected in the evaluation of these technologies. Similarly, if the ethical concerns are about the degree to which cloning turns the animal into a biological factory, where only the genetics of animal play a role and all relations to the phenotypical wholeness of the animal is lost, these concerns are also relevant in many other sectors of modern agricultural production.

This brings us back to the claim about double standards. It should therefore be noted that there exists a widespread critique of other and more established artificial reproduction technologies in the agricultural sector (Det Dyreetiske Råd, 1998; Olsson *et al.*, 2006). So, one cannot say that opponents of cloning necessarily have double standards because they accept everything else that is going on within agricultural breeding (Kappel, 2002). From the point of view of the welfare problems that these technologies give rise to, in relation to the question of the naturalness of the technologies and - perhaps most importantly - as a critique of the way these technologies enhance production pressure on animals, these more established technologies are also the subject of criticism. What one can say is that these technologies are rarely as well known as cloning, but this does not mean that the reaction to them would not be the same, were they so. Our guess is that new biotechnologies, such as cloning, will function to raise awareness and make the public more sceptical towards modern agricultural methods in general.

References

Center for Food Safety (2007a): Not ready for prime time. FDA´s flawed approach to assesing the safety of food from animal clones. Center for Food Safety.

Center for Food Safety (2007b): Consumers flood FDA with over 130,000 comments opposing food from cloned animals. Center for Food Safety.

Det Dyreetiske Råd (1998): Anvendelse af 'Ovum-Pick-Up'-teknik til opsamling af oocyter fra tamkvæg. Det Dyreetiske Råd, Justitsministeriet, Copenhagen.

Di Berardino, M.A. (2001): Animal cloning-the route to new genomics in agriculture and medicine. Differentiation 68: 67-83.

Food and Drug Administration (2006): Animal Cloning: A Draft Risk Assessment. Center for Veterinary Medicine, U. S. Food and Drug Administration, Department of Health and Human Services.

Gamborg, C., Gjerris, M., Gunning, J., Hartlev, M., Meyer, G., Sandøe, P. and Tveit, G. (2006): Regulating Farm Animal Cloning. Recommendations from the project Cloning in Public. Project Report 15. Danish Centre for Bioethics and Risk Assessment.

Gjerris, M. (2006): Ethics and Farm Animal Cloning. Risks, Values and Conflicts. Danish Centre for Bioethics and Risk Assessment

Gjerris, M., Olsson, A. and Sandøe, P. (2006): Animal biotechnology and animal welfare. Ethical eye - Animal welfare. Council of Europe Publishing

Gjerris, M. and Sandøe, P. (2006): Farm animal cloning: The role of the concept of animal integrity in debating and regulating the technology, in Kaiser, M. and Lien, M.: Ethics and the politics of food. Preprints of the 6th Congress of the European Society for Agricultural and Food Ethics. Wageningen Academic Publishers.

Kappel, K. (2002): Bioteknologi og skepsis, Kritik nr. 155-156, p.127-131

Lassen, J. (2005): Public perceptions of farm animal cloning in Europe. Danish Centre for Bioethics and Risk Assessment

Meyer, G. (2005): Why clone farm animals? Goals, motives, assumptions, values and concerns among European scientists working with cloning of farm animals. Danish Centre for Bioethics and Risk Assessment.

National Research Council of the National Academies (2002): Animal Biotechnology. Science-Based Concerns. National Academy of Science.

National Research Council and Institute of Medicine of the National Academies (2004): Safety of genetically engineered foods. Approaches to assessing unintended health effects. National Academy of Science.

Niemann, H., Rath, D. and Wrenzycki, C. (2003). Advances in biotechnology: new tools in future pig production for agriculture and biomedicine. Reprod. Dom. Anim. 38: 82-89.

Olsson, IAS., Gamborg, C. and Sandøe, P. (2006). Taking ethics into account in farm animal breeding: what can the breeding companies do? Journal of Agricultural and Environmental Ethics 19:37-46.

Takahashi, S. and Yoshihio, I. (2004). Evaluation of meat products from cloned cattle: biological and biochemical properties. Cloning and Stem Cells 6: 165-171.

Tomé, D., Dubarry, M. and Fromentin, G. (2004). Nutritional value of milk and meat products derived from cloning. Cloning and Stem Cells 6: 172-177.

Vajta, G. and Gjerris, M. (2006). Science and technology of farm animal cloning: State of the art. Animal Reproduction Science 92: 211-230.

Wilmut, I., Schnieke, A.E., McWhir, J., Kind, A.J and Campbell, K.H., (1997). Viable offspring derived from fetal and adult mammalian cells., Nature 385: 810-813.

Animal welfare aspects of creating transgenic farm animals

Kristin Hagen
Europäische Akademie zur Erforschung von Folgen wissenschaftlich-technischer Entwicklungen
Bad Neuenahr-Ahrweiler GmbH, Wilhelmstr. 56, 53474 Bad Neuenahr-Ahrweiler, Germany,
Kristin.Hagen@ea-aw.de

Abstract

It is becoming increasingly practical to create transgenic animals. With regard to livestock, the technology can be applied to farm animals in food production, or for using farm animals to produce functional foods, pharmaceuticals, industrial compounds or organs. This paper addresses animal welfare aspects of creating transgenic animals, primarily with regard to the research and development stages of a transgenic livestock project. Firstly, a transgenic livestock project may have a welfare-compromising goal (e.g. production increase in a breed which already suffers from producing too intensively). Further, the goal of the project can have harmful side-effects; e.g. harm caused by the bioactivity of foreign proteins which may enter the body's circulation. Transgenic animals can be created with various methods of transgene insertion and reproduction, some of which (notably, cloning) currently lead to high abortion rates and health problems in the offspring. Genetic or epigenetic disruption and unexpected aberrant gene expression can result either from the reproductive method or the genetic modification and can affect animal welfare. Theoretically, the likelihood for unanticipated effects is very high because of the novelty and variability of the procedures. Of particular concern is the possibility of subtle dysfunctions that may at first remain undetected. Therefore, welfare evaluation schemes should be part of transgene evaluation processes, and long-term evaluation schemes including measures of animal welfare should be adopted once transgenic strains enter the production phase.

Keywords: transgenic, livestock, animal welfare

Introduction

Gene technology is rapidly developing also with regard to the traditional farm animals: genomic information is becoming available, and techniques to remove, modify, replace or add individual genes are refined. The classes of animals involved in transgenic livestock production include those that are used when a transgenic strain is created, and those belonging to a successfully created strain used in production. Currently, a lot of transgenic farm animals are created in laboratories, although few have so far entered production. Creation of transgenic animals throws up a number of ethical questions: product safety, environmental safety, preservation of genetic resources, impacts on agricultural systems, animal welfare, and animal/natural dignity and integrity. The focus will here be on animal welfare with regard to transgenic farm animals, especially in the research and development phases.

The goals of transgenic livestock projects and transgene expression

Non-food applications of this technology on farm level include the production of nutraceuticals, industrial compounds and biopharmaceuticals in transgenic animal 'bioreactors'. For example, the first recombinant biopharmaceutical produced in a transgenic animal received market approval in 2006: Atryn, a product to treat antithrombin deficiency, is produced by GTC Biotherapeutics in the milk of transgenic goats (Walsh, 2006). Typical (envisaged) applications of transgenic technology in animal food production include changes in the animals' digestive physiology, increased growth rates and fertility, altered carcass and milk compositions, increased lactational performance, enhanced disease resistance, decreased susceptibility to stress and pollution, and reduced environmental impact (Niemann *et al.*, 2005).

There can be negative animal welfare effects of transgenesis that are related to its objectives. In biomedical contexts, these are overt (e.g. in the case of organ donation for transplantation to humans or in the case of animal models of human disease) and will form part of cost-benefit analyses. However, in food production, careful analysis of the desirability of a given trait should also be carried out. For example, a 'positive' trait like increased growth rate would be negative from a welfare point of view in breeds that already suffer from side effects of too high growth rates. Such effects are not caused by the technology as such, but rather made possible by it, and should be considered at the outset of projects.

The animal welfare consequences of transgene expression strongly depend on the nature of transgenesis. For instance, in the case of biopharmaceuticals production in transgenic animals, relevant factors include the expression site (e.g. milk, urine), bioactivity, concentrations, tissue-specificity and temporal expression patterns of the pharmaceutical proteins to be produced. Harm to the animals can be caused by the bioactivity of a foreign protein either at the intended expression site, because it enters the body's circulation, or because it gets expressed at unintended sites.

One example of insufficient tissue-specificity was reported by Massoud *et al.* (1996) who found that rabbits engineered to express human erythropoietin (EPO) in their mammary glands also expressed EPO at low levels in other organs, resulting in elevated numbers of red blood cells, infertility and premature death. Insufficient tissue-specificity can be caused by lack of elements in the regulatory parts of DNA inserted with the transgene. In addition, especially if the transgene is placed randomly in the genome, there may be aberrant expression patterns due to position effects: regulatory parts of host genes nearby the transgene can influence its expression with regard to concentrations, temporal patterns and sites.

The placement of the transgene in the genome also involves the possibility of insertional mutations which disrupt endogenous gene expression. Depending on which host gene has been disturbed, there can be harmful consequences for the animals. A related problem lies in the general possibility of unintended interaction effects that could lead, for example, to subtle dysfunctions or changes in temperament

Animal welfare problems during the creation of a transgenic livestock strain

During the experimental phase, animals used or generated may include egg cell donors, foster mothers, transgenic founder animals, further generations of transgenic animals for evaluation, and excess animals (e.g. those that are poorly expressing the transgene or are of the wrong gender). Various techniques are being developed for the creation of transgenic animals. Here, the focus is on pronuclear DNA microinjection and cell nuclear transfer (cloning) as the currently most common procedures. In pronuclear DNA microinjection, a recombinant piece of DNA is injected directly into fertilised egg cells, which either are directly transferred into foster mothers or cultivated *in vitro* first. Cell nuclear transfer involves the replacement of an entire nucleus in an egg with another, into which a recombinant piece of DNA has been integrated in advance.

Donor animals, recipient animals and excess animals

In all procedures to create transgenic animals, egg cells are needed. Unless they are obtained from ovaries that are by-products in slaughterhouses, live donor animals are subjected to the effects of either the explantation of ovaries or the collection of in vivo fertilised eggs. Superovulation, which is necessary for the collection of eggs, is known to lead to some physical discomfort and abdominal pain in humans

(Boivin and Takefman, 1996). Ultrasound-guided transvaginal oocyte recovery is a mildly aversive invasive procedure.

For the recipient animals (foster mothers), hormonal priming or induction of pseudopregnancy are probably mildly aversive procedures. Transfer of embryos into the oviducts is more or less invasive depending on the species; non-surgical transfer is possible in large animals. Recipients are negatively affected in the case of abnormal foetal development and the necessity of caesarean sections for delivery, which are particularly common with cell nuclear transfer.

With pronuclear DNA microinjection, because analysis of the transgene's expression cannot, as with cell nuclear transfer, be carried out at the cell stage, large numbers of animals who do not carry the gene or do not express it are produced: the percentage of offspring carrying the transgene is 1-5%. The number of non-expressing (or poorly expressing) animals born can be reduced with various methods, but there is often a trade-off. It is possible to subject foster mothers to amniocentesis for genotyping and abort nontransgenic foetuses. *In vitro* methods that produce more transgenic embryos tend to reduce their viability, pre-disposing for developmental problems. The use of cloning rather than DNA microinjection, allowing a considerable reduction of non-transgenic numbers, leads to high incidents of foetal and perinatal diseases, abnormalities, and death (see details below).

Developmental problems

Reproductive technologies and gene insertions, replacements or knock-outs can disrupt the complex interaction between genetic and epigenetic effects controlling animal cellular development. The disruptions can lead to abnormalities in the developing embryos, which in turn can cause suffering: to the foster mothers, to foetuses if they are sentient, to the new born animals, and possibly also long-term to the offspring. It is not yet known in detail how the various reproductive and gene technological procedures contribute to abnormal development (Young and Fairburn, 2000), but abnormal foetal development and health problems around birth appear to be particularly prevalent with abundant *in vitro* manipulations. Calves from *in vitro* culture with microinjection showed increased incidence of high birth weights, malformations and perinatal mortality (Van Reenen *et al.*, 2001). However, the amount of developmental problems is substantially lower with pronuclear DNA microinjection than with cell nuclear transfer.

With cloning (cell nuclear transfer), *in vitro* production is prolonged, and the reconstructed oocytes experience genetic reprogramming and exposure to fusion facilitating and activating stimuli in addition to standard *in vitro* procedures. The effects are massive disturbances in the regulation of gene expression in early embryogenesis and notorious inefficiency in producing live offspring. Death of cloned embryos and foetuses occurs throughout pregnancy, and a high proportion of those that survive to term die soon after birth. In cattle, the highest levels of losses (>65%) occur before implantation into foster mothers (Edwards *et al.*, 2003). The same authors report that about 50% of implanted bovine cloned embryos survive the first 30 days of pregnancy, of which 50-100% then die in the next 30 days. These losses, as well as those that occur in the 2nd and 3rd trimester, are commonly characterised by abnormal placental developments. Placental dysfunction and abortions compromise foster mother welfare. Foetuses that die in the 3rd trimester have symptoms such as amniotic squames and meconium in the lungs that would be indicative of suffering, provided they have the ability to suffer. If foetuses develop to term, there is often impaired hormonal signalling in preparation for birth and offspring are often unusually large. This can lead to birth complications and often makes caesarean delivery necessary.

A typical cluster of offspring symptoms including abnormally large birth weight is called 'large offspring syndrome'. It is thought that coculture or serum in the culture medium has been the reason (Young *et al.,* 1998). Postnatal complications reported in cloned cattle also include lung dysmaturity, pulmonary hypertension, respiratory distress/failure, decreased oxygen supply in body tissues, decreased body temperature, hypoglycemia, metabolic acidosis, enlarged umbilical vessels, development of sepsis in umbilical structures or lungs, increased birth weight, asynchronously large organs, muscoskeletal abnormalities (especially contracted flexor tendons), abnormal immune function, anaemia, brain lesions, depression and prolonged recumbency (Edwards *et al.,* 2003, Chavatte-Palmer *et al.,* 2004, Li *et al.,* 2005). In some cases, the severity of such complications is not apparent directly after birth. On the other hand, not all cell nuclear transfer derived offspring are ill, and it also appears that physiological differences are less in older offspring (Chavatte-Palmer *et al.,* 2004).

Transgene analysis

At present, transgene analysis normally means analysis of transgene mRNA expression and veterinary examination of the transgenic animals to identify obvious health problems. Transgene analysis involves a number of laboratory routines such as blood sampling and tissue biopsies, which are aversive to the animals. How aversive such routines are also depends on the handling, as the restraint and the related fear can cause stress reactions in the animals. Depending on the aim of the transgenic livestock project, there can also be special procedures involved, e.g. the hormonal induction of lactation (in females not yet sexually mature and in males) in order to analyse components of the milk without too long time delays.

Animal welfare research

Management of potential welfare problems arising from generation of transgenic animals involves the development and inclusion of comprehensive (but feasible) welfare protocols into such examinations. Being a leading-edge technology, genetic engineering involves a high degree of novelty and in many cases, very little is known about the long-term impact of the suggested procedures. There is also a high degree of variability, for example with random gene insertion, where each transgenic founder animal is different. Another important source of uncertainty are expression levels and expression sites. Therefore, elaborate testing of the animals with regard to their geno- and phenotypes are part of the research routine (transgene analysis). To some extent such testing also involves animal welfare parameters (i.e. at present, mainly health parameters). Thorough welfare evaluation schemes could be developed for farm animal species as part of the general transgene evaluation (genotyping and phenotyping), in order to ensure early detection of welfare-compromising direct effects of transgenesis. Such schemes have already been developed for transgenic mouse strains, and van Reenen *et al.* (2001) have worked out in detail how such schemes could be integrated into the transgene evaluation processes common in the laboratories.

The analysis of phenotypes that is carried out at the experimental stage needs to be repeated at herd level later to detect possible long-term effects. Long-term evaluation schemes should therefore be adopted once transgenic strains enter the production phase, to allow for assessment of effects that may develop with age and in future generations. To date, there are few studies monitoring the long-term effects of *in vitro* reproductive techniques and transgenesis. With regard to cell nuclear transfer, the few long-term studies indicate that cloned animals can have normal zootechnical characteristics and healthy offspring (Enright *et al.,* 2002, Chavatte-Palmer *et al.,* 2004). However, because the long-term consequences are so poorly known, and because they may differ for the different types of transgenesis, adverse animal health and welfare effects can only be excluded by long-term monitoring.

Conclusions

The animal procedures involved in transgenic livestock production combine well-known (and not transgenesis-specific) management-, housing- and production-related welfare problems with factors that involve medium (e.g. *in vitro* reproduction techniques) or large (e.g. transgene expression) degrees of uncertainty with regard to welfare implications. Because the harm involved can be large, and because the very goal of the project may involve harm to the animals, the need for transgenic livestock projects should be evaluated in advance, not just, as currently common, in the context of animal experimentation, but also with regard to production aims. Welfare consequences genetic manipulations should be monitored carefully in the short and long term (several generations). There is not yet a lot of experience, and because each transgenic strain is different, they require case-by-case evaluation.

References

Boivin, J. and Takefman, J.E. (1996). Impact of in-vitro fertilization process on emotional, physical and relational variables. Human Reproduction 11: 903-907.

Chavatte-Palmer, P., Remy, D., Cordonnier, N., Richard, C., Issenmann, H., Laigre, P., Heyman, Y. and Mialot, J.-P. (2004). Health status of cattle at different ages. Cloning and Stem Cells 6: 94-100.

Edwards, J.L., Schrick, F.N., McCracken, M.D., van Amstel, S.R., Hopkins, F.M., Welborn, M.G. and Davies, C.J. (2003). Cloning adult farm animals: a review of the possibilities and problems associated with somatic cell nuclear transfer. American Journal of Reproductive Immunology 50: 113-123.

Enright, B.P., Taneja, M., Schreiber, D., Riesen, J., Tian, X.C., Fortune, J.E., Yang, X. (2002). Reproductive characteristics of cloned heifers derived from adult somatic cells. Biology of Reproduction 66: 291-296.

Li, S., Li, Y., Du, W., Zhang, L., Yu, S., Dai, Y., Zhao, C., Li, N. (2005). Aberrant gene expression in organs of bovine clones that die within two days after birth. Biology of Reproduction 72: 258-265.

Massoud, M., Attal, J., Thépot, D., Pointu, H., Stinnakre, M.G., Théron, M.C., Lopez, C. and Houdebine, L.M. (1996). The deleterious effects of human erythropoietin gene driven by the rabbit whey acidic protein gene promoter in transgenic rabbits. Reproduction Nutrition Development 36: 555-563.

Niemann, H., Kues, W. and Carnwath, J.W. (2005). Transgenic farm animals: present and future. Scientific and Technical Review of the Office International des Epizooties 24: 285-298.

Van Reenen, C.G., Meuwissen, T.H.E., Hopster, H., Oldenbrock, K., Kruip, Th.A.M. and Blokhuis, H.J. (2001). Transgenesis may affect animal welfare: a case for systematic risk assessment. Journal of Animal Science 79: 1763-1779.

Walsh, G. (2006). Biopharmaceutical benchmarks 2006. Nature Biotechnology 24: 769-776.

Young, L.E., Sinclair, K.D. and Wilmut, I. (1998). Large offspring syndrome in cattle and sheep. Reviews of Reproduction 3: 155-163.

Young, L.E. and Fairburn, H.R. (2000). Improving the safety of embryo technologies: possible role of genomic imprinting. Theriogenology 53: 627-648.

Cloning for meat or medicine?

C. Gamborg¹, M. Gjerris¹, J. Gunning², M. Hartlev³ and P. Sandøe¹
¹*Danish Centre for Bioethics and Risk Assessment, Faculty of Life Sciencec, University of Copenhagen, Rolighedsvej 25, DK-1958-Frederiksberg C, Denmark, chg@life.ku.dk*
²*Faculty of Law, University of Copenhagen*
³*Centre for Ethics, Law and Society, Cardiff Law School, Cardiff University*

Abstract

This paper is about the regulation of animal cloning. At present animal cloning is governed by an indirect regulatory framework at EU level. However, recent developments in farm animal cloning in the US make it necessary to consider whether the EU regulatory framework is adequate and, if not, how farm animal cloning might otherwise be regulated in the EU. Already, offspring from an animal cloned in the USA have been imported as embryos to Europe (www.telegraph.co.uk/news/main. jhtml?xml=/news/2007/01/11/nembryo11.xml). In the paper, conclusions from a two-year EU specific support action, CLONING IN PUBLIC, designed to develop recommendations for regulation and stimulate public debate are presented. Discussions at all the project workshops have clearly demonstrated an absence of consensus among participants as to what animal cloning applications are technically possible, economically sound and ethically acceptable. In the paper, various possible developments of applications and ensuing regulation options (using existing regulation, new EU regulation and national regulation) are explored, together with a discussion of the underlying value issues. It is argued (a) that there is a difference in the ethical acceptability of cloning for biomedical purposes and cloning for food production; and (b) that there are also ethically relevant differences between the existing regulatory frameworks for these two kinds of application. New biomedical applications will mainly fall under the regulation of animals for research, where there is already a clear focus on the welfare of the involved animals. New applications for food production will fall under legislation that primarily protects consumers and the environment.

Keywords: animals, biotechnology, ethics, food safety, regulation

Introduction

To date, no specific binding legal instrument explicitly concerned with animal cloning has been passed in the EU; so, unlike the genetic modification of animals, cloning is not directly regulated at EU level. However, EU regulatory instruments make up a rather complex *indirect* regulatory framework. Associated with the Treaty Establishing the European Community, there is food safety and consumer protection legislation, animal welfare and zootechnical legislation, and legislation relevant to GMOs and intellectual property rights. All of this legislation affects the permissibility of cloning, and, no less, the potential arrival of cloned-animal products on the market. Twenty-four of the twenty-five member states have no specific legislation on the cloning of farm animals. Like their European counterparts, major research-heavy countries, such as the US, Canada, Australia, China and Japan, have no specific legislation on animal cloning.

In December 2006, the US Food and Drug Administration (FDA) finally, after several years of delay, released its 678-page draft risk assessment, in which it was claimed that '[e]xtensive evaluation of the available data has not identified any food consumption risks or subtle hazards in healthy clones of cattle, swine, or goats. Thus, edible products from healthy clones that meet existing requirements for meat and milk in commerce pose no increased food consumption risk(s) relative to comparable products from sexually-derived animals ... Edible products derived from the progeny of clones pose no additional food consumption risk(s) relative to corresponding products from other animals.' (CVM, 2006: 15-16). With these recent developments in farm animal cloning in the US, we need to consider whether farm animal cloning can be adequately regulated by the present EU legislation, and if not, how farm animal cloning might otherwise be regulated. Two of the key questions are: At what level, and how, should the technology be regulated? On what kinds of consideration should regulation be based?

Last year CLONING IN PUBLIC (CiP), a specific support action within the sixth framework programme, Priority 5, Food quality and safety' (Contract no. 514059), was completed. The overall aims of CiP were to develop recommendations on European regulation of, and guidelines covering, research on farm animal cloning and its subsequent applications (e.g. in genetically modified animals for bio-reactors); and to stimulate informed public debate across Europe. These aims were of equal importance. Clearly, they are also interrelated, because if regulations and guidelines are to serve their purpose they must take public concerns into account.

Discussions at each of the CiP project workshops clearly demonstrated the absence of consensus among participants as to what applications are technically possible, economically sound and ethically acceptable. The main uncertainty from a technological point of view is whether the technology will, on any considerable scale, be of use in the production of meat and other animal products - or whether it will mainly be of use for basic research and for biomedical purposes. Given the divide between the public perception of biomedical applications of biotechnology and applications relating to agriculture and food production, it may make a significant difference which scenario turns out to be the real one. Thus the public debate about biotechnology in Europe has been characterised by a divide between, on the one hand, biomedical applications of biotechnology and, on the other hand, applications relating to agriculture and food production. Whereas the former commands a certain degree of acceptance the latter is widely rejected (Lassen, 2005).

In what remains of the present paper, various possible developments of applications and ensuing regulation options (using existing regulation, new EU regulation and national regulation) are explored, together with a discussion of the underlying value issues. It is argued (a) that there is a difference in the ethical acceptability of cloning for biomedical purposes and cloning for food production; and (b) that there are also ethically relevant differences between the existing regulatory frameworks for these two kinds of application.

Following a terminological stipulation used in the project, the term 'cloning' here refers to *asexual reproduction* - or, more precisely, to the production of individuals with virtually identical genetic material by asexual reproduction. In recent debates, interest has centred on cloning by somatic cell nuclear transfer (SCNT). The term 'farm animal' refers to farm animal species such as ruminants (e.g. cows, sheep), pigs and poultry (chicken, turkey). However, the term does *not* imply that an animal is kept or used in an agricultural setting or for agricultural purposes. Thus potential applications of 'farm animal cloning' include those in medicine.

Possible applications and regulatory options

What direction will farm animal cloning now move in? Discussions between scientific, legal and ethical experts at the workshops held as part of the project, and a series of interviews with farm animal cloning scientists (Meyer, 2005), suggest that we should consider three possible developments:
1. Basic research and biomedicine applications *only*, both in and outside the EU.
2. Basic research and biomedicine applications both in and outside the EU, and agricultural applications *only* outside the EU.
3. Basic research, biomedicine and agricultural applications both in and outside the EU.

Basic research may be aimed at understanding embryonic development, or gaining knowledge about epigenetic processes, or laying the base for developing disease models. Concerning biomedicine, reproductive cloning can be used as a tool for the efficient production of transgenic animals that will mostly serve as disease models enhancing our understanding of human diseases. The technology can also be used to produce animals that in turn produce pharmaceuticals. With regard to agriculture, reproductive cloning may be used for propagating a desirable genotype of the kind that can be used to generate individual animals with high genetic merit.

Given these potential applications of farm animal cloning, the question is how to regulate the technology. One approach would be to regulate farm animal cloning through existing mechanisms at the EU level, with perhaps some small adjustments. A second option would be to introduce new regulation at the EU level specifically on farm animal cloning. Finally, the possibility that individual member states might want to introduce specific national regulation which is not an implementation of higher-order EU regulation has to be considered.

Potentially, the three developments in application and the three regulatory responses yield nine regulatory scenarios (if we see the options as separate ones). At present, however, the third development in the way cloning might be applied seems less likely than the first two (Gamborg *et al.*, 2006).

1. So long as farm animal cloning within the EU and globally comes to be used primarily, or exclusively, for basic research purposes (e.g. understanding basic reproduction and cell formation processes or developing disease models) and/or for biomedical purposes (farm animals species as disease models or bioreactors), existing legislation will cater for an wide array of ethical concerns. Thus concerns centring on risks to humans and the environment, and on the obvious welfare problems produced by research, could be met in this way. Other ethical concerns, however, such as those relating to the integrity or naturalness of the animals, would be overlooked.

2. If farm animal cloning is used for basic research and biomedicine applications in and outside the EU, and for agricultural applications only *outside* the EU, and if the ethical concerns we are seeking to meet relate to risks other than those threatening human health or the environment, the most pressing question will be whether EU countries or regions will be able to restrict imports of cloned animals, or products derived from them or their progeny, or insist on a certain kind of labelling. Offspring from cloned animals have already been imported from the US into Europe and their products are likely to enter the market. An example of this is the recent British born calf, named Dundee Paradise, created as an embryo from a normal bull and the clone of a prize-winning dairy cow in the US and then imported as an embryo and implanted into a cow on a farm in the UK (http://news.bbc.co.uk/1/hi/sci/tech/6249613.stm). A conflict rather similar to that over genetically modified crops in the 1990s - when industry lobbied strongly against various legislative reforms, being particularly unhappy about labelling regardless of detectability - may well arise again.

3. The third development, including animal cloning for general agricultural purposes within the EU, as was mentioned above, does not seem very likely to occur in the near or distant future. But if it did occur, existing regulation would place little control on animal cloning as such. Introducing new EU legislation is another option. In practice, however, as the project's workshops testified, consensus on new EU controls on animal cloning could be difficult to achieve. With regard to the third regulatory initiative, national legislation could be enacted, as has already happened in Denmark (and in Norway, a non EU member state).

With the first two regulatory options (i.e. reliance on the present, indirect EU mechanisms and the introduction of new, direct EU regulation), the risk aspect would feature prominently. Possible risks to human health and the environment are obviously central concerns which would be addressed with regard to animal cloning for agricultural purposes. In addition, the uncertainty that products from cloned animals will be publicly acceptable ought to be considered. The acceptability here may be linked to an interest in promoting the basic values underpinning the European Community (e.g. sustainability, biodiversity and the precautionary principle) and, more specifically, to a number of other important concerns, such as animal welfare, animal integrity, and consumer rights.

From a consumer rights perspective, the worry might be that a legal gap opens if foodstuff from cloned animals is covered neither by the regulation of novel food nor by the regulation of genetically modified food; hence it might be suggested that cloning requires us to introduce more specific labelling requirements. General food law (Regulation (EC) No 178/2002) explicitly mentions the consumer's right to make informed choices, but it is questionable whether this provides the consumer with a right to receive specific cloning-related information about food products derived from cloned animals. In the area of risk assessment more comprehensive results are still pending. It is difficult to identify genuine gaps in the present legal framework until further information is available from risk assessments, and until subsequent discussion by the relevant experts to the Commission has taken place. If it is discovered that non-transgenic cloned animals and their products are substantially equivalent to animals already in commercial production, such animals and their products may fall under much of the existing food safety-related legislation (Gunning *et al.*, 2006).

Ethical implications

In assessing these application possibilities and options for regulation, it is important to obtain an understanding of what the regulation is intended to achieve and what the underlying values are. No matter how the EU chooses to act it will be necessary to explicate the values underlying the choice, and also to see the extent to which these values relate to concerns that are prominent among the public in Europe.

Although our knowledge of European perceptions of farm animal cloning is somewhat limited, owing to a lack of qualitative and quantitative studies focusing on this specific issue, we can draw some conclusions from the Eurobarometer studies that have been conducted since 1991 on European public attitudes to animal biotechnology. As Lassen (2005) notes, cloning always exists in a context. Only at the basic research level, might it be considered a stand-alone technique. Otherwise, animal cloning operates within the context of its application. Moreover, when it comes to public perception, cloning is often placed within the context of other (animal) biotechnology.

There seem to be three main groups of concern relevant to animal biotechnology: concern for animals (e.g. with regard to animal welfare or animal integrity), concern for humans (e.g. biosafety or slippery slope concerns) and concern for society, understood, for example, as a concern for the increased industrialisation of agriculture. In our view, it is possible to see a difference between medical and agricultural applications, with medical application being more positively received. As Lassen (2005) points out, public perception is difficult to gauge since it depends to a large degree on the existence of alternatives, and on perceived usefulness. Lassen (2005:17) took the responses to point to the existence of two scales: the organism involved and the type of application. 'On the first of these scales, cloning sits towards the controversial end, since its object is animals. On the second scale the position depends on the purpose and application of the cloning being considered. Taking both scales into consideration, one would expect to find farm animal cloning in food production to be controversial in all respects.'

If this is correct, there will be a difference in the ethical acceptability of cloning for biomedical purposes and cloning for food production. However, there are also ethically relevant differences between the existing regulatory frameworks for farm animal cloning for agricultural applications in the food chain. Biomedical applications will mainly fall under regulation of animals for research, where there is already a clear focus on the welfare of the involved animals. Applications for food production will fall under legislation that is mainly concerned with the protection of consumers and the environment.

In principle, EU legislation leaves room for the application of EU rules within the framework of national ethical principles. However, it remains uncertain whether this in practice provides an opportunity to call upon moral principles to intervene in trade (Gunning *et al.*, 2005). National restrictions have to comply with the specific provisions in the Treaty of the European Community - especially those covering quantitative restrictions between Member States (Articles 28-31), because the free movement of goods, services, persons and capital is an essential element of the internal market. Individual countries may, as a result of public consultation, decide that animal cloning is a moral issue and prohibit such activity or restrict it to certain applications, as is has been done in Denmark. Similar action has been taken by Norway, a member of the European Economic Area (EEA). Danish law restricts animal cloning to research purposes, and a licence will be required for the import of cloned animals, which, again, can only be used for research. The law is silent on imports of products from cloned animals, but a newly proposed legal amendment, if approved by the EU, should place a ban on products.

Consensus across the EU on aspects other than risks to human health and the environment does not seem likely. The powerful pharmaceutical sector is likely to support cloning of animals for the potential production of therapeutic substances, and countries with an important livestock industry will presumably see advantages in the latest technological advances in livestock improvement. Currently, there are no products on the market. However, it is expected that the pharmaceutical company, Pharma - following the approval of the first product based on the milk of a genetically modified goat (ATryn®, or recombinant human antithrombin), which can be used in the treatment of Hereditary Antithrombin Deficiency (HD), and which could be produced using cloning to create additional offspring that carry the favorable modified genome - will soon put its product on the market.

Conclusions

The overall conclusion of the CiP project was that 'the EU will be left with difficult processes of decision-making if the cloning of farm animals is put to use in agricultural production in countries outside (likely) or within (only likely at a low level) the EU. Whether it is decided to rely on existing regulation or to introduce new and specific legislation, concerns for free trade and concerns for social acceptability in a European context will have to be negotiated.' (Gamborg *et al.*, 2006: 7).

Depending on which of the possible developments occur, the concerns addressed, and the value issues raised, will differ. Biomedical applications of farm animal cloning will mainly fall under the regulation of animals for research, where the focus is on the welfare on the animals. Applications for food production will fall under legislation that is mainly concerned with the protection of consumers and the environment. Thus, the choice of what regulatory approach to take depends on ethically significant differences. These differences need to be included in the overall process of deciding upon regulatory responses.

References

CVM (Center for Veterinary Medicine) 2006. Animal Cloning: A Draft Risk Assessment. Center for Veterinary Medicine, U. S. Food and Drug Administration, Department of Health and Human Services, Rockville. Available online at http://www.fda.gov/cvm/ CloneRiskAssessment.htm

Sustainable food production and ethics

Gamborg, C., Gjerris, M., Gunning, J., Hartlev, M., Meyer, G., Sandøe, P. and Tveit, G. (2006). Regulating farm animal cloning. Recommendations from the project Cloning in Public. Project Report 15. Danish Centre for Bioethics and Risk Assessment, Copenhagen, 18 pp. Available online on http://www.sl.life.ku.dk/cloninginpublic/

Gunning, J., Hartlev, M. and Gamborg, C. (2006). Challenges in regulating farm animal cloning: An assessment of regulatory approaches and the legal framework within the EU. Cloning in Public project report no. 6. Danish Centre for Bioethics and Risk Assessment, Copenhagen, 52 pp. Available online on http://www.sl.life.ku.dk/cloninginpublic/

Lassen, J. (2005). Public perceptions of farm animal cloning in Europe. Project Report 9. Danish Centre for Bioethics and Risk Assessment, Copenhagen. Available online on http://www.sl.life.ku.dk/cloninginpublic/.

Meyer, G. (2005). Why clone farm animals? Goals, motives, assumptions, values and concerns among European scientists working with cloning of farm animals. Project Report 8. Danish Centre for Bioethics and Risk Assessment, Copenhagen. 27 pp. Available online on http://www.sl.life.ku.dk/cloninginpublic/

Assessing people's attitudes towards animal use and genetic modification using a web-based interactive survey

Catherine A. Schuppli and Daniel M. Weary
Animal Welfare Program, Faculty of Land and Food Systems, University of British Columbia, 2357 Main Mall, Vancouver, BC, V6T 1Z4, Canada

Abstract

The objectives of this study were to compare people's attitudes towards different uses of research animals, the use of animal models with and without genetic modification, genetically modified (GM) food versus non-food products, and the role of information in decision-making. Using an interactive, online survey, we probed participant views on 2 examples of research on domestic pigs (1) to reduce agricultural pollution and (2) to improve organ transplant success in humans. Participants were asked if they would support each use of pigs using a 6-point Likert scale. We surveyed 252 animal technicians, veterinarians, animal researchers, animal advocates, university students and others. Overall, 65% of respondents indicated that they supported (i.e. strong yes and weak yes) the research on pigs to reduce phosphorous pollution but this support declined to 49% (P=0.001) when pigs were to be fed genetically modified corn, and declined further to only 20% (P=0.0) when the study involved the creation of a new GM line of pigs. A similar pattern was found for the use of pigs in biomedical research; 49% of respondents indicated that they would support the use of pigs to improve organ transplant, but this support declined to just 29% (P=0.0) when the research required the creation of a new line of GM pigs and 31% (P=0) for the use of already genetically modified pigs. Thus, in both scenarios the level of support declined when genetic modification was proposed. Moreover, the level of support for using pigs in research was higher for the environmental scenario that did not involve GM (65% versus 49%, P=0.0), but slightly lower for research involving GM (29% versus 20%, P=0.045).

Keywords: animal welfare, research animals, pigs, animal ethics

Introduction

Pigs are an interesting model for examining people's attitudes, beliefs and values about animal use, welfare and genetic modification (GM). Research in the area of public perceptions of biotechnology in Europe has revealed high levels of negativity and worry (Towsend and Campbell, 2004), especially related to genetic modification of food (Gaskell *et al.,* 2000). Research also indicates a hierarchy in the public acceptance of GM food products with GM animals viewed as the most ethically unacceptable form of GM (Royal Society of Canada 2001, Epstein, 2002, Mepham). Adding to the complexity is a concern for the welfare of farm animals, as evidenced by the recent explosion of policies and programs - governmental, intergovernmental and corporate - pertaining to the humane rearing and handling of farm animals (Fraser, 2006). Unfortunately there has been little study to date on public attitudes regarding farm animal welfare and how these interact with concerns about genetic modification.

The multiple uses of pigs provide an opportunity to compare people's attitudes towards different uses of animals (while keeping species constant), to compare people's attitudes towards the use of animal models with and without genetic modification, to compare attitudes about genetic modification of food versus non-food products and explore the role of information in decision-making.

Methods

Using an innovative, interactive, online survey, designed to model social norms and to investigate public perceptions of new biotechnologies (NERD, 2006), we explored two examples of research on domestic pigs. These examples addressed an overlapping set of issues, but with a few key differences. The scenarios probed participant views on research to 1) reduce agricultural pollution and 2) improve organ transplant success in humans. Both scenarios concluded with the genetic modification of pigs as a potential component in addressing the research questions. Participants were asked if they would support each use of pigs using a 6-point Likert scale (strong yes, weak yes, neutral, weak no, strong no, undecided). Throughout the survey participants could access information about animal welfare, animal ethics, science and public policy.

Results

Preliminary data are from a sample of 252 animal technicians, veterinarians, animal researchers, animal advocates, and university students. The sample includes participants from Canada (41%), the US (25%) and the UK (19%) and other countries. A larger and more diverse sample is currently being recruited.

Overall, 65% of respondents indicated that they supported (i.e. strong yes and weak yes) the research on pigs to reduce phosphorous pollution (Figure 1), but this support declined to 49% (P=0.001) when pigs were to be fed genetically modified corn, and declined further to only 20% (P=0.0) when the study involved the creation of a new GM line of pigs. A similar pattern was found for the use of pigs in biomedical research; 49% of respondents indicated that they would support the use of pigs to improve organ transplant, but this support declined to just 29% (P=0.0) when the research required the creation of a new line of GM pigs and 31% (P=0) for the use of already genetically modified pigs. Thus, in both scenarios the level of support declined when genetic modification was proposed. Moreover, the level of support for using pigs in research was higher for the environmental scenario that did not involve GM (65% versus 49%, P=0.0), but slightly lower for research involving GM (29% versus 20%, P=0.045).

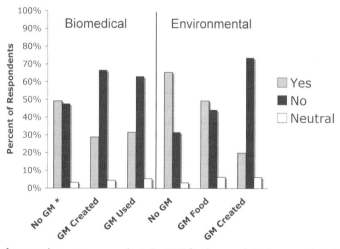

Figure 1. The level of support by survey respondents (n=252) for the use of pigs in research to study: 1) immune mechanisms of organ rejection (biomedical case) and 2) ways to reduce phosphorous pollution from pig manure (environmental case). Note: Likert scale categories strong yes and weak yes, and strong no and weak no were combined. * GM represents genetic modification.

Discussion and conclusions

These findings agree with previous studies reporting public concern regarding the genetic modification of plants and animals with greater concern about GM animals. Future analyses will explore how providing respondents information on animal welfare and ethics influences their responses to these issues. In addition, various socio-demographic were collected to help characterize respondents (for example, age, gender, nationality, education, familiarity with the WEB, familiarity with animal research, animal welfare and ethics and degree of concern for animals and the environment, etc.). These results will help to increase our understanding of the views of Canadians and others regarding the genetic modification of animals and plants, an already contentious area, which will help to inform policy decisions with more dependable information.

Acknowledgements

The research was generously funded by Genome Canada and Genome BC and by the UBC Animal Welfare Program, which is supported by the Natural Sciences and Engineering Research Council of Canada, the BCSPCA, the BC Veterinary Medical Association, and other sponsors listed on the programme website at ww.landfood.ubc.ca/animalwelfare/.

References

Epstein, R. (2002). Redesigning the world: ethical questions about genetic engineering. In: R. Sherlock and J.D. Morrey (eds.) Ethical issues in biotechnology. Rowman and Littlefield, Lanham, Boulder, New York, p. 47.70.

Fraser, D. (2006). Animal welfare assurance programs in food production: a framework for assessing the options. Animal Welfare 15: 93-104.

Gaskell, G., Allum, N., Bauer, M., Durant, J., Allansdottir, A., Bonfadelli, H., Boy, D., de Cheveigné, S., Fjaestad, B., Gutteling, J.M., Hampel, J., Jelsøe, E., Correia Jesuino, J., Kohring, M., Kronberger, N., Midden, C., Nielsen, T.H., Przestalski, A., Rusanen, T., Sakellaris, G., Torgersen, H., Twardowski, T., and Wagner, W. (2000). Biotechnology and the European public. Nature Biotechnology 18: 935-938.

Mepham, B.T. Date Unknown. Comments on Matthew Freeman' paper: Genetic modification of animals: agriculture and environment biotechnology commission. http://www.aebc.gov.uk/aebc/pdf/comments.pdf. Accessed on May 14, 2006.

NERD (2006). NERD - Norms evolving to respond to ethical dilemmas. http://robo.ethics.ubc.ca/~pad/NERD/. Accessed on October, 16, 2006.

Royal Society of Canada (2001). Elements of Precaution: Recommendations for the Regulation of Food Biotechnology in Canada. http://www.rsc.ca//files/publications/expert_panels/foodbiotechnology/GMreportEN.pdf. Accessed on October 10, 2006.

Townsend, E. and Campbell, S. (2004). Psychological determinants of willingness to taste and purchase genetically modified food. Risk Analysis 24: 1385-1393.

Part 14 - Nature conservation and ethics

Environmental values and their impact on sustainability and nature conservation

Diana Ehrenwerth and Alexander G. Haslberger
Center for Ecology, University of Vienna, Althanstraße 14, 1090 Vienna, Austria

Abstract

Globalized production and trade have exhausted the carrying capacity of the earth, and social systems may not be able to cope fast enough. In the past, different sets of environmental values have supported different approaches to nature, although it remains unclear how they corresponded to factual behaviour towards nature - as shown in a case study on late imperial China. Inexplicitly however, underlying values and conceptions influence political and economic decision making. Discursive reassessment and rebalancing might enhance its outcomes for sustainability.

Keywords: values, nature conservation, sustainability

Introduction

Up to the 19th century environmental constraints were - if at all - only a local problem. There mostly remained huge untouched landscapes and apparently inexhaustible reserves of energy and raw materials. This apparent inexhaustibility came to an end as human population exploded and economic processes expanded. Some of these processes have reached global dimensions and now make up significant parts of global material flows. The UN-Millenium Ecosystem Assessment (2005) describes the degradation and unsustainable usage of ecosystem services, the increasing likelihood of nonlinear changes in ecosystems, including accelerating, abrupt and irreversible changes, which have important consequences for human well-being and the ensuing growing inequities across groups of people causing poverty and social conflict. Present experiences suggest that established social systems may not be able to cope with these problems fast enough. Amongst others, their objectives for sustainability and nature protection are influenced by underlying values. Environmental values and their consequences for conceptions of and behaviour towards nature were analyzed by comparing cases from cultural history and present day political strategy.

Environmental (nature) values

Environmental values can be divided in instrumental and intrinsic. Callicott (2005) specifies four categories for instrumental values of biodiversity, namely goods, services, information and psycho-spiritual. Discussion of intrinsic value focuses on two issues: What sorts of things may possess it? Does it exist objectively? Intrinsic value is generally attributed to human life. More recently this attribution has been expanded to robustly conscious animals, species, biotic communities, ecosystems etc.

In environmental ethics we can differentiate between several stances with human responsibility broadening from step to step: The narrowest, anthropocentrism, attributes intrinsic value to human beings only, but instrumental value to nature. Man is considered the master of nature. The Judeo-Christian concept of dominion, represented in Genesis 1:28 ranges here. Pathocentrism extends intrinsic value to our 'fellow sufferers', higher animals. Biocentrism focuses on life and attributes intrinsic value to all individual organisms. Man is considered one among equals. Ecocentrism goes still further, conceiving man as a mere member of a greater totality (Weish, 2006).

Environmental ethics of different world cultures can be traced in their religions, as perfunctorily shown here for Islam, Hinduism and Confucianism:

Similar to the basic Judeo-Christian view, Islam teaches that human beings have a privileged place in nature. There has been a tendency among Muslims to take an instrumental approach to the human-nature relationship. Shariah, the laws of Islam, which regulate the religious as well as the secular realm, contains elements of conservation, such as the protection of animals from cruelty, forest protection, or the limit of the growth of cities. New conservation regulations in Islamic states must be grounded in the Qur'an.

The central theme of Hinduism is the search for liberation from suffering, and ultimately for release from the cycle of birth and death. All living beings are parts of Brahman, plurality and difference is illusory. Thus, to harm other beings is to harm oneself. Some animals are treated as divine and many Hindus are vegetarian because of their belief in the sanctity of life. True happiness does not come from outer possessions, but from within. The exploitation of the outer world distracts from the central purpose of life, which is to discover the spiritual nature.

Sunderlal Bahuguna, leading eco-activist of the Chipko movement, states: *'The solution of present-day problems lie in the re-establishment of a harmonious relationship between man and nature. To keep this relationship permanent we will have to digest the definition of real development: development is synonymous with culture. When we sublimate nature in a way that we achieve peace, happiness, prosperity and, ultimately, fulfilment along with satisfying our basic needs, we march towards culture.'*

Confucianism was the leading Chinese doctrine and state religion, since Han (206 BC-220) till Qing (1644-1911) times. Its founder, Kongzi (551-479 BC) established a range of values with great potential for continuing to shape East Asian societies in their quest for sustainable development as well as for environmental thought in general. Amongst them are harmony amidst change; the importance of the family including past, present, and future generations; the significance of education, and history as an element of civilizational continuity and moral rectification (Tucker, 1998).

Impact on sustainability

Values influence political and economic decisions, which often interfere with objectives of nature protection and sustainability. The underlying values are rarely made explicit, whereas a reassessment and rebalancing might lead to different outcomes. Three examples show this for food production policy and food consummation:

According to FAO (2004) food production will need to be further intensified in order to meet a growing world population's demands. Yet intensification in itself carries the risk of further degrading natural resources, ultimately leading to decreased food security. It can also have important socioeconomic consequences for the unjust distribution of burdens and benefits. In making policy decisions with the aim of achieving sustainable agricultural intensification, planners must therefore identify and evaluate alternative strategies, in terms both of their immediate and longer-term impacts and their implications for all social groups concerned. There is a clear ethical dimension to such policy formulation.

The problem of specification of nature values and damages turned out to be a major difficulty in negotiations of the parties about criteria for liability, redress and reparation for potential consequences of genetically modified organisms under the UN-Cartagena Protocol on Biosafety (Third meeting of the ad hoc open-ended working group of legal and technical experts on liability and redress in the context of protocol).

An in depth analysis of causes for resistances against implementation of genetically modified organisms of consumers in some areas such as Europe or parts of Asia indicated different understandings of concepts of nature and nature protection as a central source of opposition (Gaskell *et al.*, 2000).

Within conservation biology itself, where an important trend at the moment is the understanding and protection of 'ecosystem services', a shift towards 'returning to the protection of nature for nature's sake', in contrast to the present typically utilitarian approaches, is being discussed (McCauley, 2006).

The correspondence between values and conceptions on one hand and actual behaviour towards nature on the other hand, is hard to determine as the following case illustrates.

Personal perceptions, official dogma and the usage of nature in late imperial China

'If any one should wish to get the kingdom for himself, and to effect this by what he does, I see that he will not succeed. The kingdom is a spirit-like thing, and cannot be got by active doing. He who would so win it destroys it; he who would hold it in his grasp loses it. The course and nature of things is such that what was in front is now behind; what warmed anon we freezing find. Strength is of weakness oft the spoil; the store in ruins mocks our toil. Hence the sage puts away excessive effort, extravagance, and easy indulgence', translates James Legge chapter 29 of the Daodejing in 1891. Richard Wilhelm puts in 'world' instead of 'kingdom' in his first rendering in 1911, the year the Chinese empire collapsed. We cannot be sure, what exactly the Laozi meant when it was inscribed on bamboo sticks during the epoch of Warring States (403-221 BC), before the empire was first established. Through the centuries hundreds of learned commentaries have been contributed and during late empire dozens of different translations into foreign languages have been added. Nonetheless nature is clearly considered as part of the supreme power at the root of all things (Dao), and the wise man is required to adapt to its rhythms instead of trying to act 'against its current' and reshape it. This contemplative and soft attitude found lasting repercussion in the four arts of the Chinese literati, that is calligraphy, painting, poetry and seal carving, as well as in architecture, medicine etc. It has been a strong undercurrent of the Chinese view of the world and human life throughout the empire, and probably still lies somewhere at the bottom. Momentarily overpowered by an unprecedented economical and technical effort to propel China to the leading ranks or, seen from a more individual perspective, to gain material comfort, riches and thus social acceptance and respect, this undercurrent might resurface again more broadly, as is already manifest in the aspiration of Chinese environmentalists to ground the global discourse of sustainable development in eastern philosophical traditions (Yang, 2004).

In an early Western publication on Chinese painting Otto Fischer discerns *'a general aesthetic attitude of the whole people towards nature, for which Chinese feel [...] love, human devotion and a deep sense of connectedness, as hardly any other culture on Earth does. [...] No wonder, that [...] painting expresses since such a long time and with such animation the devotion to the landscape.'* (Fischer, 1923: 159) There are thousands and thousands of Qing poems in which nature is similarly contemplated, revered, marvelled at in its very smallest sprouts and insects or in its vastest mountain ranges and river plains. They also display a broad range of aspects and sometimes contradictory attitudes concerning the actual 'management' of nature (Elvin, 2004: 437-453)

Complementarily, in official dogma as manifest in Qing documents we find a resurging imperial cosmology, notably promoted by the early emperors Kangxi (1661-1722), Yongzheng (1722-1735) and Qianlong (1735-1796). People, most of all the Son of Heaven himself, were considered responsible for their weather, in the wording of a decree issued by the Yongzheng emperor in 1725: *'Now ... disasters arising from floods, droughts, or locusts are due either to a defect in the government at Court,... or to the chief officials in the area being unable to serve the state in a fair and correct fashion,... or else again the low moral quality of customs in some particular prefecture or county, where people's heart-minds are false and treacherous, thereby causing the Dark Force and Bright Force to be thrown into confusion, so that disasters are numerous and repeated...'* (Da Qing Shichao shengxun 1616-1874, quoted after Elvin, 2004: 417). This can well be interpreted as political opportunism. The emperor could highlight his own ethical qualities through either auspicious signs or else through devote self-reproach. He could also blame regional officials at will. But 'moral meteorology' goes way beyond political argumentation. Even as such, it could only be effective because it was deeply rooted in popular beliefs.

While Qing painting and poetry show devotion and dedication to nature, and imperial dogma upholds the immediate correspondence between human virtue and natural processes, the exceptionally strong usage of the land springs to the eye of early Western visitors, thus captured by a Jesuit missionary in the 18th century: *'Even if the land is exhausted by thirty-five centuries of harvests, it has to provide a new one every year to supply the urgent needs of a countless population. This excess of population ... here increases the need for farming to the point of forcing the Chinese to do without the help of cattle and herds, because the land that would feed these latter is needed for feeding people. This is a great inconvenience as it deprives them of fertilizer for the soil, meat for the table, horses, and almost all the advantages that can be gained from herds. Were it not for the mountains and the wetlands China would be absolutely without the benefit of woods, and without venison and game. Let us add that it is the strength and application of human beings that meets all the costs of farming [here]. More labor and more people are needed to obtain the same quantity of grain than are required elsewhere. The total quantity surpasses the imagination, but even so it is no more than sufficient....'* (Missionaires de Pékin (1776-1786): Mémoires concernant l'histoire, les sciences, les arts, les moers, les usages, etc. des Chinois. Quoted after Elvin (2004: 463). Similar and sometimes way more drastic pictures arise from Chinese texts. The land was in fact tamed, transformed and exploited to a degree that had few parallels in other parts of the world at the time.

As has been stated by various authors (Roetz, 1984, Elvin, 2004)) representation and reality appear at least superficially contradictory. The question arises, whether the actual behaviour of a collective is necessarily in accord with ideas and feelings expressed in the sources. Do these sources rather reflect the dominant tendencies of an age or are they reactions *against* these dominant tendencies? We might take one step further and ask, as Mark Elvin does at the very end of his monumental environmental history: *'Was China, among the great developed premodern civilizations of the world, unique in this? And if not, what does this imply for the realism, or otherwise, of the hope that we can escape from our present environmental difficulties by means of transformation of consciousness?'* (Elvin, 2004: 471).

Emerging global values

'The market does not automatically benefit society. We have to form it, establish rules and most of all assert values.' (Muhammad Yunus, Spiegel-interview 8.5.2007) But which values? Values are generally in flux in globalizing and globalized systems. Media and travel opportunities bring benefits within reach, which were out of sight for most people in former times. Lifestyles converge, and as industrialization and consumerism spread, the pressure on the environment increases further.

Today's environmental problems, unprecedented in quantity and quality, are not taken into account by traditional value systems. They require new moral answers and thus engender a broad discourse. This discourse embodies the search of common ground, such as in the pioneering work of the Alliance of Religions and Conservation (ARC), which links secular conservation and ecology with the faith worlds of the major religions. On the other hand it also displays the need to emphasize specific cultural values, as done for instance by the Asian Bioethics Conference. The discourse concerns global fields of work and science, as shown in the two following examples about agriculture and food biotechnology:

The relationship between agriculture and nature is a central issue in the current agricultural debate. Consideration for nature is part of the guiding principles of organic farming and many farmers are committed to protecting natural qualities. However, nature is an ambiguous concept that involves multiple interests and actors reaching far beyond farmers. An analysis showed that different understandings of nature exist within farmers, scientists and NGO's and that there is disagreement as to whether emphasis should be given to biological qualities, production values, or experiential and aesthetic perspectives. This complexity provides a challenge. The example illustrates an underlying battle for the right to define nature and nature quality and essentially decide what farmers should work towards (Hansen *et al.*, 2006).

We find various designs for ethical codes for science and technology, such as for instance a 'global code of ethics for modern food biotechnology': Food has profound cultural, social, moral, and historical meanings, which are crucial to individual and social identity and well-being. It is essential for health and connects humans with each other. It bonds humans to nature. Food is an expression of nature and embeds itself in nature. Food for humans can be a natural resource, a product, a commodity, or an income. A strong and transparent ethical framework can guide the interactions and activities of stakeholders, especially in situations, where past experiences, current norms and legislation are inadequate, where controversy and opposing views are common, and where a large power or capacity differential between stakeholders exists. Four ethical principles were proposed for the code: the principle of beneficence, the principle of non-maleficence, the principle of justice and fairness, and the principle of choice and selfdetermination (Gesche *et al.*, 2006).

On a global political level, there is a tendency of divide of interests between poorer and more affluent countries. The first strive for economical development, the second are more aware of global confines and growing environmental risks, triggered mostly by their own doings. In spite of obvious resource shortages, a Charta for environmental rights and duties remains in discussion. Possibly, values will converge further through trying to tackle the global environmental challenge together.

Conclusion

Values can be considered orientation marks for behaviour adequate for certain social constellations in space and time. Thus they are necessarily oriented backwards. In times of change, values are gradually adapted to the new situation and new values originate.

The environmental challenge we face today is unprecedented and thus requires new, now global values to be formed and old ones to be discarded. This process is painful and takes time. It might be smoothed at points if argumentation and education can build on supportive aspects of still effective traditional value systems and by making underlying values explicit and discussing them openly.

The analysis of environmental values and the measurement of their correlation with actual behaviour pose a hereto unresolved challenge. Research needs to zoom into a sizable historical and local constellation in order to assess the numerous factors at work. The comparison of different constellations might then uncover experience and produce knowledge helpful for cooperative action.

References

Callicott, B. (2005). Conservation values and ethics. In: M.J. Groom, G.K. Meffe and C.R. Carroll (eds.) Principles of conservation biology. Sinauer, Sunderland, USA, pp 111-136.
Elvin, M. (2004). The retreat of the elephants: an environmental history of China. Yale University Press, New Haven, USA.

Fischer, O. (1923). Chinesische Landschaftsmalerei. Kurt Wolff Verlag, München, Germany.

FAO (2004). The ethics of sustainable agricultural intensification. Rome, Italy.

Gaskell, G., Allum, N., Bauer, M., Durant, J., Allansdottir, A., Bonfadelli, H., Boy, D., de Cheveigné, S., Fjaestad, B., Gutteling, J.M., Hampel, J., Jelsøe, E., Jesuino, J.C., Kohring, M., Kronberger, N., Midden, C., Nielsen, T.H., Przestalski, A., Rusanen, T., Sakellaris, G., Torgersen, H., Twardowski, T. and Wagner, W. (2000). Biotechnology and the European Public. Nature Biotechnology 18: 935-938.

Gesche, A., Entsua-Mensah, M. and Haslberger, A.G. (2004). A global code of ethics for modern food and agricultural biotechnology. 5th Congress of the European Society for Agricultural and Food Ethics, Catholic University of Leuven, Belgium.

Hansen, L., Noe, E. and Horing, K. (2006). Nature and nature values in organic agriculture. An analysis of contested concepts and values among different actors in organic farming. Journal of Agricultural and Environmental Ethics 19: 147-168.

Legge, J. (1891). The Texts of Taoism. The Sacred Books of the East Volumes 39 and 40. Oxford University Press, Oxford, England.

McCauley, D.J. (2006). Selling out on nature. Nature 443: 27-28.

Millennium Ecosystem Assessment (2005). Ecosystems and Human Well-being: Synthesis. Island Press, Washington, DC, USA.

Roetz, H. (1984). Mensch und Natur im alten China: Zum Subjekt-Objekt-Gegensatz in der klassischen chinesischen Philosophie. Lang, Frankfurt am Main, Germany.

Tucker, M.E. (1998). Confucianism and ecology: potential and limits. Earth Ethics 10, no.1.

Weish, P. (2006). Zentrale Themen der Biologie: Grundfragen der Humanökologie und Umweltethik. Script. University of Vienna, Vienna, Austria.

Yang, G. (2004). Global environmentalism hits China: international and domestic groups join forces to combat environmental woes. YaleGlobal, 4 February 2004.

Aspects of ethics and animal protection in animal husbandry under semi-natural conditions, in animals returned to the wild, and in wild animals

K.M. Scheibe
Leibniz Institute for Zoo and Wildlife Research Berlin (IZW), PO.BOX 601103, 10252 Berlin, Germany

Abstract

The keeping of domestic animals under semi-natural conditions has increasingly gained in importance as an alternative, sustainable landscape utilisation concept. Besides, both animals which were returned to the wild and wild animals are more and more appreciated in representing key elements for nature conservation in many European countries. While natural conditions are often used as a reference for the evaluation of intensive animal production processes, they are not stress-free. Semi-naturally kept domestic animals as well as feral and wild animals are submitted to human influences, thus their living conditions have to be regarded under welfare aspects. However, the term 'welfare' is not appropriate for such cases and the term 'animal protection' may rather be referred to in this context. Traditionally, animal welfare is applied to domestic and laboratory animals, whereas wild animals are subject to hunting as well as nature and species conservation. These two different approaches give rise to several questions. The first question is whether domestic animals which are kept under semi-natural conditions or returned to the wild as well as wild animals are exposed to situations that are relevant in respect of animal protection and/or welfare aspects. This is followed by the question whether different living conditions may result in basic differences of the biological structure and whether such differences might change human responsibility. A conflict between the respective aims of animal welfare and nature conservation has been declared for the considered animals. In the following, the two different approaches of animal protection and nature conservation are discussed and joined to a general, concluding ethical concept.

Keywords: animal welfare, wild animals, nature conservation, husbandry conditions

Relationship between humans and animals

There is a wide spectrum of different relationships between animals and humans. Domestic animals have been selected, bred, and are mainly kept for production. Some other species are bred for sports or as pets. Accordingly, the breeding aims and husbandry conditions of domestic animals were mainly designed in relation to utilisation and production aspects, with moral attitudes being developed to guarantee a certain level of animal welfare under such conditions. The ethical principles that are applied to these animals biologically derive from human emotionality and rationality (Scheibe 1998a). Representing alternative concepts to intensive farming, nature-orientated husbandry systems have been rediscovered, reactivated, or recently developed, mainly to reduce labour and expenses. In some of these concepts, aspects of sustainability and landscape conservation are also taken into account. Given that all domestic species are able to develop behavioural patterns similar to those of wild animals when kept under semi-natural conditions (e.g. Jensen 1988, 1991), such conditions are considered as a reference to define the 'normal' i.e. natural behaviour of domestic animals. However, it is often ignored that these conditions are also not stress-free and do not per se guarantee the animals' well being. The relationship between humans and animals which are kept in semi-natural conditions is not as close as that in conventional husbandry, a fact that often results in humans neglecting their responsibility for the welfare of the respective animals. Free-living, large animals are usually considered as natural elements that are not subjected to human influence. Nevertheless, they are mainly regarded as game and thus exposed to hunting in many areas.

Furthermore, they are indirectly but substantially affected by critical, anthropogenic impacts such as climatic change or habitat restriction.

For conservationists, larger vertebrates have become increasingly important. With some relevant European wild species such as wild horses and aurochs being extinct, their domestic forms are considered for conservation concepts and kept in large enclosures to be returned to the wild. European bison (*Bison bonasus*) for example have been bred and kept in zoos for long periods and subsequently been reintroduced to natural or semi-natural conditions. Such animals are often regarded exclusively in the context of ecological aspects, but the respective concepts differ from each other as they follow different aims, including different utilisation strategies. Humans usually perceive these animals either as wild animals or as domestic animals kept under unusual conditions.

Problems in nature conservation and recently developed concepts

More and more animal and plant species become extinct, not only in the tropics but also in Central Europe. Especially the species living in wet areas, dry biotopes, alpine areas, and semi-open landscapes are threatened (Barth, 1995). This general tendency is connected to fundamental changes in the landscape structure. The extinction of a species does not only represent a singular, local effect but may also lead to, for example, a disruption of nutritional chains and, at the worst, the destabilisation of the whole ecosystem. Since many threatened species are located in open habitats, the conservation and protection of these habitats is particularly important. Hence a diverse landscape including both grassland and woodland biotopes is aimed for in nature conservation and for a sustainable utilisation. This aim corresponds to the historical landscape in 1750-1800, where the highest species diversity in Europe has been observed, and also to a potential, post-glacial landscape development in the absence of human influence. However, such a diverse landscape cannot be achieved by means of the traditional instruments in nature conservation. Therefore, new concepts have been discussed and developed not only for restricted, protected areas but also for large game reserves (Riecken *et al.*, 2001).

These concepts include:
- biotope conservation as a part of the extensive agriculture;
- semi-open pasture landscapes;
- 'New Wilderness (1)' areas that are moderately influenced by recent wild herbivores;
- 'New Wilderness (2)' areas that are influenced by reconstructed natural species communities of large mammals;
- management by natural disasters such as fire, flooding or storm.

While these concepts differ substantially in detail, most of them include either domestic or wild herbivores as a natural element of the ecosystem. The concept 'New Wilderness (2)' is based on the assumption that the present species community of wild animals in Europe does not correspond to the original i.e. natural species composition (Gerken, 1996; Drüke and Vierhaus, 1996). The natural behaviour of these species that includes grazing, browsing, and laying down as well as the treading and uprooting of plants along with their preferences for different habitat structures and the impact of associated predators are all appreciated as key factors in a semi-natural landscape. The concept 'New Wilderness (2)'excludes any significant human influence and corresponds to the idea of a national park. Nevertheless, it can only be realised in restricted areas. This is also true for the other above-mentioned concepts.

The basic aim of all these concepts is the conservation of complex ecological relationships as well as the promotion of natural evolution in more than a single species. When consequently followed, this aim includes a natural population regulation and in some cases may also a certain degree of utilisation by humans. Such an utilisation might be a purely non-consumptive one like tourism, but can also involve the restricted exploitation of an increasing animal population.

There are many different opinions concerning the value and application of animal welfare aspects in the 'New Wilderness' areas (Adrian and Orban, 2005).

Animal welfare and nature conservation - conflicts and common interests

The animal welfare concept is based on the biological attribute that animals can suffer, with its aim being the prevention or reduction of such suffering in animals. This attitude is mainly applied to domestic or laboratory animals (Sambraus, 1997; Dawkins, 1982), where humans are assumed to cause the suffering. The animal welfare concept is predominantly aimed on individuals or groups, yet breeds and species may also be concerned if they suffer as a result of breeding.

It has been discussed whether the welfare concept is actually relevant for wild animals (Kirkwood and Sainsbury, 1996; Cooper, 1998), which instead are subject to conservation biology and nature conservation respectively. The explicit aim of conservation is to maintain biological diversity, including reserve and landscape design, as well as sustainable utilisation and also the recovery of endangered species by the protection of ecosystems, captive breeding programs, and reintroduction (Beissinger, 1997). The reason for the need of conservation is the fact that approx. 87% of the already endangered bird and mammal species are still threatened by human influence (Rahmann and Kohler, 1991). However, a conflict has been declared between the aims of species conservation and animal welfare (Wünschmann, 1991), as wild animals are only perceived as ecological factors but not as individuals in the context of conservation (Precht, 1997). Even in a comprehensive review about the importance of ethological science for conservation biology, animal welfare science was only mentioned as an approach that sometimes complicates conservation efforts (Scheibe, 1998b).

Nevertheless, if the two different approaches of species conservation and animal welfare are compared, common aspects can easily be identified and the conflict between the different aims can be solved. Both the ability to suffer and the brain structures that produce emotional reactions are the results of evolution and basically identical for all higher vertebrates (Llinas and Steriade, 2006). In this context, there are no differences between domestic and wild animals in regard to their ability to suffer.

The animal welfare and nature conservation concepts both have in common that they are directed against human disturbance and/or damage of natural elements, including individuals as well as groups and species. Individuals are always concerned in the processes of extinction, as they may suffer from stressful conditions caused by humans or even die from starvation or thirst as a result of human influence on their habitat and/or increased interspecies competition. Accordingly, individuals are the basic elements on which protective measures should be addressed. However, natural processes of evolutionary importance may also cause harm and suffering. Under seasonally unfavourable conditions such as cold or shortage of food, free ranging animals have to develop adaptive, physiological mechanisms and also suffer natural death. Therefore, animals living under natural or semi-natural conditions cannot be the subject of the 'animal welfare concept' as defined above, as it is not appropriate in this context. The less common term 'animal protection' may rather be applied, since the respective animals can only be protected against disturbing influences by humans. Natural processes like predation, seasonal variation of essential parameters or interspecies competition have to be accepted as far as they are not modified by human influence. The natural processes even have to be protected since they are key elements of natural evolution.

In natural or semi-natural environments the concepts of animal protection have to be restricted to human influences. On the other hand, individuals require more attention in the context of nature and species conservation. Since many people are mainly interested in the fate of single individuals, the protection of individuals may also be an important motivation for species conservation.

Examples

There are two major problems with the wilderness concepts, feeding and population regulation. Under natural conditions, wild animals are exposed to seasonal variations in food availability and food quality. To cope with this variation animals have developed different adaptive mechanisms; in particular common are seasonal rhythms. These include rhythms of feed intake and body weight as well as food selection and digestion. In Przewalski horses (*Equus ferus przewalskii*) which were kept in a semi-reserve without any feeding, an annual rhythm of activity, feed intake and body weight has been observed (Berger *et al.,* 1999; Scheibe and Streich, 2003). For former zoo animals newly introduced to this reserve, it took one to two years until this rhythm was completely developed (Scheibe *et al.,* 2000). Red deer (*Cervus elaphus*) change their nutritional strategy from summer to winter, more following the strategy of a roughage feeder in winter. This could be identified by the seasonally different ultradian rhythms of feeding (Berger *et al.,* 2002). Since supplementary feeding would disturb these natural adaptive processes, the feeding of wild animals during winter is strictly prohibited by the hunting laws of Brandenburg and other German federal states, with the exception of declared cases of an emergency only. Accordingly, in all projects following the wilderness concept, these regulations should be applied and even when contrasting the arguments of animal welfare.

Population regulation by hunting has been accepted as rather unavoidable and is even required by the German hunting law. Since all species have a reproductive capacity that exceeds the carrying capacity of a certain area, populations are naturally regulated by population cycles based on starvation, diseases, and predation (Illius *et al.,* 1995; Holtmeier, 2002). These are also influenced by long-lasting cycles of vegetation development. In the models of wilderness, such undisturbed processes are highly important. Today, even large reserves are restricted areas, with humans being responsible for the effects of this restriction. It has to be considered separately for each area, under which circumstances and to what extent a population regulation by humans may be needed. Hunting is a human impact and not compatible with the idea of undisturbed nature (e.g. in the national park concept). Hunting also alters the behaviour of wild animals towards humans (Büker *et al.,* 1999) as well as the utilisation of vegetation structures. The decisive factor for wild animals in learning to avoid humans is the lethal danger of hunting (Petrak, 1996). An extension of the closed period may already alter this process substantially (Petrak, 1985; David, 1995). When a human population regulation is required, alternative methods to remove animals have to be found. The utilisation of stationary traps may be an adequate method (Stubbe *et al.,* 1995).

Conclusions

Our ethical relation to animals depends on the degree of human influence to which they are exposed to, not on their status (domestic or wild). If the different living conditions of animals are compared, there is a stepwise transition from the conventional keeping conditions of domestic animals to semi-natural conditions and finally the wilderness concepts (Table 1).

The degree of utilisation by humans is reduced over these steps. Nevertheless, even the wilderness concepts may include a certain degree of consumptive utilisation to enable population regulation. They may even be regarded as models for alternative forms of husbandry to produce animal products in the future. Furthermore, they offer possibilities for non-consumptive utilisation like tourism. Accordingly, these concepts also require ethical principles, including even more indirect influences. Following these

Table 1. Comparison of different living conditions for animals with respect to human aims and responsibility.

	conventional animal production	domestic animals on pasture, herded	all-year around pasture, wild animal farming	New Wilderness (1)	New Wilderness (2)	reintroduction
aim	production	production with reduced food costs	production with reduced labour and reduced food costs	ecological effects	nature, evolution	species conservation, reconstruction of natural ecological conditions
human responsibility	health and welfare of individuals	health and welfare of individuals	health and welfare of individuals	for direct and indirect influence	no direct influence, only for indirect influence	for preparation of individuals
population regulation	following agricultural aims	following agricultural aims	following agricultural aims	following ecological aims, drawback less	none, only as compensation for restricted space	none
feeding	food distribution	pasture management	pasture management, supplemental feeding	no feeding, in exceptional situations only	no feeding	no feeding
curative treatment	complete	complete	complete	none	none	none
prophylactic treatment	complete	complete	complete	none	none	as preparation only

principles, decisions have to be made concerning all activities concerning these animals. This includes processes such as the reintroduction of zoo animals or artificially bred individuals into the wild. It seems to be essential to reach a consensus between the aims and methods of animal protection and species as well as nature conservation. For the new semi-natural living conditions, appropriate laws and regulations must be developed that take the special ecological aims into account and at the same time protect the animals against inadequate human influence and disturbance.

References

Adrian, U. and Orban, S. (2005). Extensive Tierhaltung zur Landschaftspflege. TVT-Nachrichten 1/2005: 17-21.

Barth, W.-E. (1995). Naturschutz: Das Machbare. Hamburg, Germany, 467p.

Beissinger, S.R. (1997). Integrating behavior into conservation biology: Potentials and limitations. In: Clemmons, J.R., Buchholz, R. (eds.) Behavioral approaches to conservation in the wild. Cambridge Univ. Press, Cambridge, GB, pp. 23 - 47.

Berger, A., Scheibe, K.M., Eichhorn, K., Scheibe, A., Streich, J. (1999) Diurnal and ultradian rhythms of behaviour in a group of Przewalski horse (Equus ferus przewalskii), measured through one year under semi-reserve conditions. Applied Animal Behaviour Science 64: 1-17.

Berger, A., Scheibe, K.-M., Brelurut, A., Schober, F., Streich, W.J. (2002): Diurnal and ultradian rhythms in red deer behaviour - longtime and continuous measurements under quasi-natural conditions. Biological Rhythm Research 33: 237-253.

Büker, A., Scheibe, K. M., Streich, J., Eichhorn, K., Scheibe, A. (1999). Reaktion von freilebenden Rehen (Capreolus capreolus) auf anthropogene Aktivitäten in Abhängigkeit von der Landschaftsstruktur. Natur- und Kulturlandschaft 3: 298-309.

Cooper, J.E. (1998). Minimal invasive health monitoring of wildlife. Animal Welfare 7: 35-44.

David, A. (1995). Rotwild als Pflegefaktor. Wild und Hund 18, H. 95: 26-27.

Dawkins, M. S. (1982). Leiden und Wohlbefinden bei Tieren. Ulmer, Stuttgart , Germany 129p.

Drüke, J., Vierhaus, H. (1996). Welche Beziehungen bestehen zwischen Naturschutzprojekten im Kreis Soest und den verschwundenen Großtieren? Natur- und Kulturlandschaft 1: 153-158.

Gerken, B. (1996). Einige Fragen und mögliche Antworten zur Geschichte der mitteleuropäischen Fauna und ihrer Einbindung in ein Biozönosespektrum. Natur- und Kulturlandschaft 1: 7-15.

Holtmeier, F.-K. (2002). Tiere in der Landschaft. Ulmer, Stuttgart, Germany, 367p.

Illius, A. W., Albon, S.D., Pemberton, J.M., Gordon, I.J., Clutton-Brock, T.H. (1995). Selection for forage efficiency during a population crash in Soay sheep. Journal of Animal Ecology 64: 481-492.

Jensen, P. (1988). Maternal behaviour and mother-young interactions during lactation in free-ranging domestic pigs. Applied Animal Behavior Science 20: 297-308.

Jensen, P. (1991). Back to nature: the use of studying the ethology of free-ranging domestic animals. Proceedings of the International Congress of the Society for Veterinary Ethology, Edinburgh, pp.62-64.

Kirkwood, K.J., Sainsbury, A.W. (1996). Ethics of interventions for the welfare of free-living wild animals. Animal Welfare 5, 235-243.

Llinas, R. R., Steriade, M. (2006). Bursting of thalamic neurons and states of vigilance. Journal of Neurophysiology 95: 3297-3308.

Petrak, M. (1985). Wild im Erholungswald (2). Niedersächsischer Jäger, 1/85: 37-42.

Petrak, M. (1996). Der Mensch als Störgröße in der Umwelt des Rothirsches. Zeitschrift für Jagdwissenschaft 42: 180-194.

Precht, R.D. (1997). Noahs Erbe: Vom Recht der Tiere und den Grenzen des Menschen. Rotbuch Verlag, Hamburg, Germany, 405p.

Rahmann, H., Kohler, A. (1991). Tier und Artenschutz. Hohenheimer Umwelttagung 23. Weikersheim, Germany, 211p.

Riecken, U., Finck,P., Schröder, E. (2001). Tagungsbericht zum Workshop 'Großflächige halboffene Weidesysteme als Alternative zu traditionellen Formen der Landschaftspflege'. Natur und Landschaft 76: 125-130.

Sambraus, H.H. Ed. (1997). Das Buch vom Tierschutz. Enke Verl., Stuttgart, Germany.

Scheibe, K.M. (1998a). The basis of animal welfare - contributions of behavioural science. Animal Research and Development 47: 53-70.

Scheibe, K. M. (1998b). Review: 'Behavioral approaches to conservation in the wild'. Animal Welfare 7: 335-337.

Scheibe, K.M.; Berger, A.; Budras, K.; Bull, J.; Eichhorn, K.; Dehnhard, M.; Kalz, A.; Sieling, C.; Weber, S.; Zimmermann, W. (2000): Research on Przewalski horses in the semireserve Schorfheide. In: Havrylenko, V.S.: New biosphere reserve 'Askania Nova': protection and preservation of rare species. Ukrains´ka Mis´ka Drukarnja, Cherson, Ukraina, pp. 31-39.

Scheibe, K. M., Streich, W. J. (2003). Annual rhythm of body weight in Przewalski horses (Equus ferus przewalskii). Biological Rhythm Research 34: 383-395.

Stubbe, Ch., Ahrens, M., Stubbe, M., Goretzki, J. (1995). Lebendfang von Wildtieren : Fangtechniken - Methoden - Erfahrungen. : Dt. Landwirtschaftsverl., Berlin, Germany Wünschmann, A. (1991). Tier- und Artenschutz aus der Sicht internationaler Naturschutzorganisationen (WWF). In: Rahmann and H.; Kohler, A. (Eds.): Tier- und Artenschutz. Hohenheimer Umwelttagung 23, Weikersheim, Germany: 47-52.

Sustainable food production and ethics

The ethics of catch and release in angling: human recreation, resource conservation and animal welfare

Cecilie M. Mejdell and Vonne Lund
National Veterinary Institute, P.O. Box 8156 Dep., N-0033 Oslo, Norway, cecilie.mejdell@vetinst.no

Abstract

A form of angling where fish is first caught and then released back into the water, is known as 'catch and release'. While hobby fishing in the sea is rarely regarded as a threat to fishing stocks, recreational fishing in rivers and lakes may involve considerable drain on some fish populations. In several places in the world catch and release fishing has gradually become the normal procedure. This allows fishing to continue as a sport while ensuring that a vulnerable fishing stock is not overtaxed. Thus, this is perceived as a compromise that takes into consideration both the interests of the anglers and the environmental interests to preserve a threatened resource. And the fish likely prefers life to death and is believed to be 'happy' when released. So what is the problem? The main problem is that in order to release the fish, it must first be caught: the good deed (release) depends on a preceding bad action (catch). The fighting against hook and line results in an immense stress reaction in the fish. The fish may become totally exhausted and the post release mortality may be considerable. Probably fish has the capacity to experience pain and fear and thus suffer. The interests of the parties involved: human (joy of anglers and income for property owners); animal (life and welfare); environment (safe-guarding of biological and genetic resources, and biodiversity); as well as societal values and attitudes (respect for life and resources, educational aspects, and tradition) are discussed.

Introduction

Angling is a popular leisure activity enjoyed by many people, both children and adults. There are also commercial interests associated with recreational fishing, primarily salmon fishing, since landowners with fishing rights can make a good income from the sale of fishing licences. However, hobby fishing in rivers and lakes may involve a considerable drain on some fish populations and may represent a threat to the fish stock. In several places in the world, among them North America, New Zealand, Great Britain and Ireland, it has gradually become the normal procedure for anglers to release fish after catching them ('catch and release'). This is usually perceived as a sensible compromise that takes into consideration both the interests of the anglers and a weak resource base. Thus, hobby fishing can be allowed to continue as a sport while ensuring that the fish population is not overtaxed. In these countries, both anglers and public generally accept 'catch and release' as a model for resource management. Actually, people may feel that releasing the fish that gave them such a thrilling angling experience is a morally good act (Muniz, 1997). And surely, the fish would prefer life to death if given the choice. But is it this simple?

The ethical dilemma

The main ethical dilemma is that in order to release the fish, it must first be caught, i.e. hooked and landed. Thus, the good deed (the release) depends on a preceding bad action (the catch). As previously mentioned, there are also other interests involved that may have moral relevance.

Interests of anglers and landowners

Angling is a recreational sport. People enjoy being outdoors in a peaceful and beautiful nature, and at the same time they experience the anticipation and excitement while waiting for something to happen.

The bite and the following fight with the fish on the hook represent a special kind of kick. Anglers, like hunters, often say that the fishing/hunting situation gives them a feeling of being real, in contact with nature and life and death. The original impetus for fishing, to procure food, is no longer so important for people in our modern society. Even though it is a pleasure to serve self caught fish for dinner, many people say that the angling experience is just as or even more important to them than the catch itself. Thus, catch and release is a solution allowing recreational angling to continue even if the resource basis is too low to allow extensive harvesting. Also, landowners with fishing rights can continue to get an income from angling related tourism.

In countries where catch and release has become generally accepted, people may feel that they show respect to the fish by releasing it. However, in Norway, where hunting and fishing traditions are strong and catch and release fishing still is uncommon only 38% are in favour of and 55% against catch and release fishing (Gallup, 2006). In comparison, more than 70% of the population support game hunting (Gallup, 2006).

Natural resource management

The main aim of most national regulations of both game hunting and fishing has been to make sure that nature's surplus resources are harvested in a sustainable manner, securing nature's productivity. Legislation and restrictions have been used to regulate hunting and fishing in order to ensure sustainable exploitation of natural resources and prevent local overexploitation. For this reason fishing, particularly fresh water fishing in rivers and lakes, has traditionally been subjected to various restrictions. Examples from Norway are the ban on trout fishing during the spawning period in the autumn, a minimum or maximum size for certain species, and the ban on using certain types of fishing gear, e.g. otter boards and gill nets, in some lakes. Catch and release is different from these types of regulations, because statutory catch and release eliminates harvesting of food as a motivation for fishing. Fishing takes place, even when it is clear that no self-caught fish will become dinner of the day.

From the view of natural resource management, e.g. protecting a fish stock from being overtaxed, catch and release is an option. However, the precondition is of course that released fish actually survive and have reproductive success.

The interests of the fish

The interest of fish does not mainly concern species conservation, but rather individual animal welfare. Fish has the anatomy and physiology necessary for perceiving noxious stimuli and they demonstrate behaviours which in other species are characteristic of painful experiences. An increasing number of scientists hold it for likely that fish experiences pain and fear and thus are capable of suffering (e.g. Chandroo et al., 2004, Huntingford et al., 2006). And, even if fish sentience may not be scientifically proved, a reasonable risk management strategy would be to implement animal welfare considerations in angling, because the consequences in terms of suffering, if ignoring this possibility, is great (Lund et al., 2007).

Thus, the interests of the fish will not only be to survive but to avoid suffering. Fighting against hook and line results in an immense stress reaction in the fish. The fight may last for minutes up until an hour, and the fish may become totally exhausted with depleted glycogen reserves in the muscle. Numerous studies have documented consequences of catch and release on the fish. These include mortality, reproductive fitness, physiological stress reactions, physical injury, and alterations in behaviour (Cooke and Sneddon, 2006, Meka and McCormick, 2005, Muniz, 1997, Thorstad et al., 2003). Mortality rate after release may be as high as 90%, but in other cases close to zero. Mortality depends on several factors, e.g. fish species, presence of predatory species, water temperature and salinity, and also human behaviour dependent factors as choice of fishing gear, and care at handling. Handling out of water is negative, as is keeping caught fish live in a net with the aim to release it at the end of the day. Fish hooked in vital organs such as gills or stomach have very high mortality. Natural bait increases the risk for the hook to be swallowed, causing serious injury. Fish which is actively feed seeking is more prone to swallow the bait compared to a mature salmon with very low food intake. Special hooks have been developed that reduce the likelihood of the fish being fatally injured. The hooks may lack barbs, enabling them to be easily removed, or they may be made of a material that rusts within weeks (Tsuboi *et al.*, 2006) and which are not removed. Recommendations for catch and release fishing are given to reduce strain on the fish and the risk of mortality (e.g. Cooke and Sneddon, 2006; Reiss *et al.*, 2003).

The same fish risks being caught more than once. In a river at Yellowstone National Park where 'catch and release' is practised and where there are many anglers, a fish is caught on average 9.7 times during one season (referred by Muniz 1997).

Discussion

If fish mortality after release is considerable, catch and release is obviously neither good natural resource management nor an acceptable practice from an animal welfare point of view. However, for the discussion, let us assume that mortality is close to zero, so that the catch and release does not represent a threat to the fish population. The case is then a question about human interests and fish welfare.

The first point is about angling as a recreational activity. The 'catch and release' concept turns fishing into a pure sport or entertainment, disengaged from its original purpose, that of providing food. Recreation for humans becomes the central aspect. Also, catch and release involves something fundamentally different in relation to the catch regulation measures that have been applied by the Norwegian authorities to date. Suspending fishing for certain periods, protecting individual species or applying rules concerning mesh size are all regulations that apply to fishing for food, whereas 'catch and release' signalizes a change in the main purpose of fishing to being purely a matter of recreation. The use of animals for human entertainment, in a manner that may compromise animal welfare, e.g. bull fighting, circuses, rodeo shows and dog racing, is increasingly criticised from an animal welfare point of view. The argument is that it is not ethically acceptable to torment an animal just for human entertainment. Considering the increasing evidence that fish are sentient beings, the same argument can be applied to catch and release fishing. In addition, unlike the use of animals in circuses or in bull fighting, this practice cannot be defended referring to tradition. Rather, catch and release fishing is a new phenomenon spreading at the rate by which fish stocks are depleted.

The second point is about safeguarding natural resources. It is important to support and develop public attitudes that safeguard natural resources and encourage sustainable management. However, there is little reason to believe that the 'catch and release' concept promotes such attitudes. On the contrary, the danger is that catch and release will conceal the necessity of respecting nature's own limits, as long as the objective is to allow unlimited fishing to take place.

The central question is whether it is more ethical to kill the fish than to release it. If the fish could choose for itself, it would undoubtedly choose to live. However, this is not the point. The good deed, i.e. releasing the fish, namely depends on first inflicting suffering on the fish, in the form of stress, pain and fear. Just as it is not acceptable to trap birds or mammals purely for the fun of it, fish should not be subjected to angling for no other reason than to satisfy people's need for excitement. If the fish stock is too small to be harvested, the alternative is to stop fishing.

In our view, neither regard for anglers pleasure nor for landowners' income can defend a management system where no importance is attached to the suffering inflicted on the fish.

References

Chandroo, K.P., Yue, S. and Moccia, M.D. (2004). An evaluation of current perspectives on consciousness and pain in fish. Fish and Fisheries 5: 281-295.

Cooke, S.J. and Sneddon, L. (2006). Animal welfare perspectives on recreational angling. Applied Animal Behaviour Science 104: 176-198.

Gallup TNS (2006). Natur- og miljøbarometer 2006. http://stillhet.stoyforeningen.no/docs/barometer.ppt.

Huntingford, F.A., Adams, C., Braithwaite, V.A., Kadri, S., Pottinger, T.G., Sandøe, P. and Turnbull, J.F. (2006). Current issues in fish welfare. Review paper. Journal of Fish Biology 68: 332-372.

Lund, V., Mejdell, C.M., Röcklinsberg, H., Anthony, R. and Håstein, T. (2007). Expanding the moral circle: farmed fish as objects of moral concern. Diseases of Aquatic Organisms 75: 109-118. http://www.int-res.com/abstracts/dao/v75/n2/.

Meka, J.M. and McCormick, S.D. (2005). Physiological response of wild rainbow trout to angling: impact of angling duration, fish size, body condition, and temperature. Fisheries Research 72: 311-322.

Muniz, I.P. (1997). Forvaltningstiltak ved rekreativt fiske på anadrom laksefisk. En litteraturgjennomgang over 'fang og slipp'. (Management measures related to recreational fishing for anadromous salmonids. A literature review on the 'Catch and release' concept) NINA Report 482: 1-28.

Reiss, P., Reiss, M. and Reiss, J. (2003). Catch and Release Fishing Effectiveness and Mortality. (http://www.acuteangling.com/Reference/CandRMortality.html).

Thorstad, E.B., Naesje, T.F., Fiske, P. and Finstad, B. (2003). Effects of hook and release on Atlantic salmon in the River Alta, northern Norway. Fisheries Research 60:293-307.

Tsuboi, J., Morita, K. and Ikeda, H. (2006.) Fate of deep-hooked white-spotted charr after cutting the line in a catch-and-release fishery. Fisheries Research 79: 226-230.

Part 15 - Other contributions

Why farmers need professional autonomy

F.R. Stafleu and F.L.B. Meijboom
Ethics Institute, Utrecht University, Heidelberglaan 2, 3584 CS Utrecht, The Netherlands,
F.R.Stafleu@uu.nl

Abstract

After the World War II the scale of agricultural production increased enormously but also caused problems for the environment and animal welfare. In reaction the concept of sustainability was introduced. To reach a sustainable agriculture all stakeholders have their own role in the discussion. The concept of 'professional autonomy' for farmers can be a tool for farmers to play this role. In this paper we redefine 'professional autonomy' as a political and ethical concept, aimed at giving farmers the room they need to play their role in the moral debate in society. There suggests to be good reasons why farmers should claim professional autonomy. The core reason is that farmers are members of a profession. Members of a profession have to play a shaping role in the ethical discussions of society. Furthermore, farmers seems to have their own 'farmers' morality' that may have an important input in the public discussion. To play their role in the discussion, farmers, as members of a profession, cannot be bound by too strict ethical protocols as this would eliminate their ability to create new insights, which are needed as input to the debate. This illustrates the importance of professional autonomy. Whether farmers are allotted autonomy by society depends on future developments like making a farmer's morality explicit and the development of a code of conduct. If it can be justified, the allotment of professional autonomy will enhance and shape the 'voice' of the farmers in the public discussion and increase the quality of the debate.

Keywords: farmer's morality, professional autonomy, professionalism, public debate

Introduction

After World War II the agricultural production in the Netherlands had to be enhanced to help to rebuild the country and enable the primary production and the ability to compete on European and world markets. The government started programs to 'rationalise' the agricultural production. These were successful and resulted in efficient, intensive farming methods. The scale of production increased enormously, resulting in a clear cost reduction. However this rationalisation and the increase of productivity also resulted in severe problems with respect to the environment, animal welfare, and the rural area (cf. Brom, 2002; Lang 1999; RLG, 1998).

The problems caused by the strict focus on rationalisation and the increase of productivity showed that the emphasis on primary production and processing should not only focus on quantity but also on quality. There is a transition within the agro-food sector which has as a result that other issues than productivity-related ones are seen as essential. Money, raw materials and export are still considered as crucial for the sector, yet other themes also surface in the discussion, such as responsibility, trust, animal welfare, and the ethics of stakeholders like farmers (cf. Dagevos, 2002). In this context the concept of sustainability has been introduced.

Sustainable agriculture aims at reaching an equilibrium between 'people, profit and planet'. To reach this goal the issue of responsibility immediately surfaces: structural developments towards a more sustainable agricultural production require that all stakeholders cooperate and play their own role. The distribution of responsibility, however, is a tricky issue. In analysing debates on the distribution of responsibilities in the agro-food chain, it is striking that each part of the food chain tries to put responsibility on another part of the chain. Producers justify disputable methods of production, for instance regarding animal welfare, by pointing to economic pressure and to the consumers that still buy the products. Consumers point to the difference in prices and the responsibility of the government. Politicians emphasise the responsibility of producers, and so on, until a deadlock is reached (Meijboom *et al.,* 2006). This can also be recognised in the debate on sustainability. Consumers, for instance are criticised for the discrepancy between what they say they find important, and the behaviour they exhibit in the grocery store. They only incidentally buy environment and animal friendly products. This is problematic, since a lack of consumer support is a serious problem for farmers pioneering new sustainable agricultural practices (cf. Brom *et al.,* 2007). On the other hand, consumers criticise farmers' attitudes and practices: especially the welfare of animals and the quality of the environment is thought to be sacrificed by farmers for economic reasons (Waelbers *et al.,* 2002). The farmers' voice is seen as limited to economic considerations only while the quality of the farmers' ethics is doubted. Consequently, their voice in the public debate is limited.

This position of farmers is problematic for three reasons. First, farmers play a crucial role in the development of a sustainable agriculture and thus should fully participate in this process. Second, a closer look at the morality of farmers shows that they have an ethics, even though it is often implicit. Finally, the fact that farmers do not have an adequate moral voice in the discussion is a major problem in a democratic-pluralistic society where stakeholders are expected to defend their own position (Brom, 1997).

To address these problems we propose in this paper the concept of 'professional autonomy' for farmers as an important tool for farmers to play their part the development of a sustainable agricultural production.

Ethical competence and sustainability

Sustainable agricultural production, defined as an equilibrium between people, planet and profit, is not merely a technical issue. The social criticism on farmers already indicates that it has a moral character. The basic question of which equilibrium is the right one is a fundamentally moral question: for example, moral deliberation is needed to find a balance between human profit and animal welfare. Such a balance is not self-evident, nor the mere result of scientific research, rather is it reached in a public discussion between stakeholders. Hence it is crucial that, like all stakeholders, farmers have an independent and substantial position in this public debate on the content of sustainability.

The clear moral element of the debate requires of each stakeholder the ethical competence to play their own role in this discussion, as ethical competence is needed more than communication skills. Awareness of and reflection on one's own moral point of view is basic to having 'voice' in the public discussion. In an earlier paper we showed that there certainly is a morality that is typical for farmers, but that this morality is highly implicit. Consequently, it cannot play a role in the debate (Stafleu *et al.,* 2004). This highlights the importance of making the implicit farmers' morality more explicit, and so to help to empower farmers in the debate (De Greef *et al.,* 2006). Even though it is still subject of our current research, for this paper we presume that farmers have an ethical view that is a useful contribution to the ethical public debate and that they can improve the quality of the debate on sustainable production.

To enable farmers to play this role we analyse whether the allotment of autonomy by society is useful, since professional autonomy is often considered as 'an essential attribute of a discipline striving for full professional status.' (Wade, 1999: 310) We define 'professional autonomy' as a certain room to move for the members of a profession in order to play their role in the moral debate in society.

Professional autonomy and farmers as members of a profession

Professional autonomy is a concept that is well-known in the medical and health sector. Physicians are allotted with autonomy by society. This provides them the freedom to exercise their professional judgement in the care and treatment of their patients.

When professionals claim this autonomy, they claim within their practice a certain freedom of action, a certain freedom from interference from the rest of society, so they may work in their own way guided by their own moral and technical norms and values, derived from unique experience and specialised education (RvVZ, 2000). Professionals may claim this responsibility, but to be effective it has to be allotted by society, i.e., they have to be trusted by society. Certainly in times were social responsibility is the norm, society does not give a 'carte blanche'! Professional autonomy and responsibility are inextricably related (cf. GR, 2006, 9).

The examples from the medical and health sector illustrates that attributing professional autonomy to certain groups in society is subject to conditions. Therefore, the question is whether there are good reasons why farmers should claim professional autonomy? To answer this question, we first have to distinguish between a 'professionalised occupation' and a 'profession'. When an occupation gets professionalised this may mean a number of things: the basic meaning is that the members get paid for their work. But to professionalise also means that internal quality norms are formulated, professional organisations are founded, and education for the occupation is organised in official schools and training. From a sociological point of view the occupation gets (or defends) its own independent status and niche in society (Roodbol, 2005). However, not every professionalised occupation is a 'profession'. A profession distinguishes itself from professionalised occupations by dealing with fundamental human needs and touching upon ethical issues that are subject to debate in society. For example, doctors and lawyers are traditionally seen as members of a profession. Members of a profession play a vital role in shaping the discussion and influencing societal scenarios of 'the good life'. This means that members of a profession are more than technical experts. They also need an ethical expertise and motivation aimed at playing this role (Carr, 1999). To play this role, it has been argued that members of a profession cannot be bound by too strict ethical protocols and cannot be subjected to overly rigid ethical rules from society, as this would eliminate their freedom and ability to create new insights and experiences which are needed as input to the public debate.

Thus, professional autonomy is only ascribed to members of a profession. Consequently, to know whether farmers can be allotted with professional autonomy we have to answer the question whether farmers are members of a profession. On the one hand, farmers certainly do professionalize. They organize themselves and there are special agricultural schools, which provide specialized theoretical and practical training. But, as mentioned above, this is not enough to say farming is a profession. To be members of a profession farmers must deal with fundamental human needs and touch upon basic ethical issues that play in society. In our view they do so: when we focus on livestock farming, farmers have extensive experiences with fundamental issues like animal welfare, which these days play an important role in the public debate on human-animal relations. Moreover, from the results of an explorative study (Stafleu *et al.*, 2004a) we may expect that farmers have a typical 'farmers' morality' touching upon human-animal relations. For this reason the role of farmers in the public debate is not only needed because they are the primary producers of food and caretakers of the animals, but also because they may have a unique input in the public moral discussion.

In conclusion, farmers are members of a profession and thus fulfil the formal condition for the ascription of professional autonomy. As professionals they may have an important role to play in the discussion and may reasonably claim professional autonomy.

The allotment of professional autonomy to farmers

Claiming professional autonomy is one thing, being allotted such autonomy is quite another. Under what circumstances may a claim have success?

First of all the hypothesis that farmers have one or more special and distinguishing moralities must be affirmed. After all, such a morality may give a special input in the public debate, and the fact that such a morality must be evaluated and elaborated in practice is an important reason for claiming professional autonomy.

Second, as mentioned above carte blanche will not be given: society certainly wants to know how far this autonomy will reach. In general, professions will have a code of conduct in which the boundary conditions are described (Carr, 1999). Therefore, our research project is not only aimed at researching and analysing farmers' morality, but also wants to develop a code for good farming based on this morality. This code of conduct, developed in cooperation with the farmers, will describe the boundary conditions to which the farmers will commit themselves. If it is not possible to develop such a code of conduct, or if the farming community does not want to commit themselves to it, then an autonomy claim will be problematic.

Last but not least, farmers must develop an ethical competence so they may get a bigger and more effective 'voice' than they have now. This is necessary because the justification of the claim to professional autonomy is rooted in the assumption that farmers not only should play, but also do play an important role in the public debate. Such a competence contains more than knowledge of their own morality and code of conduct. It also contains the psychological and communicational skills needed to participate in an ethical discussion. These skills must be acquired by training and education.

Conclusion

There are good reasons to expect that farmers have a justified claim to professional autonomy. Whether this expectation will become actuality is partly dependent on future developments like making a special farmers morality explicit, the development of a code of conduct and the enhanced participation of farmers in the public debate. The project 'A new ethics for livestock farming: towards value based autonomy in livestock farming?' may contribute to these developments. The allotment of professional autonomy will enhance and shape the 'voice' of the farmers in the public debate. There will also be a practical consequence: the allotment may stop the avalanche of rules that are these days foisted upon the farmers, with all the resulting negative bureaucratic and enforcement problems. Detailed rules are not longer necessary when farmers are entrusted with a certain amount of 'ethical self governance'.

Acknowledgement

This paper is part of the project 'A new ethics for livestock farming: towards value based autonomy in livestock farming?', funded by the Dutch Organization for Scientific Research (NWO). Thanks are also due to Karel de Greef, Carolien de Lauwere, Sabine de Rooij, and Jan Douwe van der Ploeg who participate in this NWO project.

References

Brom, F.W.A. (1997). Onherstelbaar verbeterd. Biotechnologie bij dieren als moreel probleem. Van Gorcum, Assen, The Netherlands.

Brom, F.W.A., T. Visak and F. Meijboom. (2007). Food, citizens and market: the quest for responsible consuming. In: L. Frewer and H. van Trijp (Eds.). Understanding consumers of food products. Woodhead Publishing, Cambridge, UK, pp. 610-623.

Crr, D. (1999). Professional Education and Professional Ethics. Journal of Applied Philosophy. 16/1: 247-260.

De Greef, K.H., de Lauwere, C.C., Stafleu, F.R., Meijboom, F., de Rooij, S., Brom, F.W.A. and van der Ploeg, J.D.. (2006). Towards value based autonomy in livestock farming. In M. Kaiser and M. Lien (eds.). Ethics and the politics of food. Wageningen University Press, Wageningen, The Netherlands, pp. 61-65.

Dagevos, J.C. (2002). Panorama Voedingsland: Traditie en transitie in discussies over voedsel. Working document 88. Rathenau Instituut, Den Haag, The Netherlands.

GR: Gezondheidsraad/Raad voor de Volksgezondheid and Zorg. (2006). Vertrouwen in verantwoorde zorg? Effecten van en morele vragen bij het gebruik van prestatie-indicatoren. Signalering ethiek en gezondheid 2006/1, Centrum voor ethiek en gezondheid, Den Haag, The Netherlands.

Lang, T. (1999). 'The complexities of globalization: The UK as a case study of tensions within the food system and the challenge to food policy. Agriculture and Human Values, 16, pp. 169-185.

Meijboom F.L.B., Visak, T., Brom, F.W.A. (2006). From trust to trustworthiness: why information is not enough in the food sector. Journal of Agricultural and Environmental Ethics. 19/5: 427-442.

RLG: Raad voor het Landelijk Gebied. (1998). Care and Trust, the basis for food production in the twenty-first century. Publication 98/7, www.rlg.nl.

Roodbol, P.F.R. (2005). Dwaalichten, struikeltochten, tolwegen en zangsporen. Onderzoek naar de taak herschikking tussen verpleging en artsen. Thesis Rijksuniversiteit Groningen, Rijksuniversiteit Groningen, Groningen, The Netherlands.

RvVZ: Raad voor de Volksgezondheid and Zorg. (2000). Professionals in de gezondheidszorg. Raad voor de Volksgezondheid en Zorg, Zoetermeer. The Netherland, pp. 95-107.

Stafleu, F.R., de Lauwere, C.C. and de Greef, K.H. (2004). Respect for functional determinism A farmers' interpretation of respect for animals. In: J. de Tavernier and S. Aerts (eds). Science, Ethics and Society. 5th Congress of the European Society for Agricultural and Food Ethics. CABME, Leuven, Belgium 355p.

Stafleu, F.R., de Lauwere, C.C., de Greef, K.H., Sollie, P., Dudink, S.. (2004a). Boerenethiek, eigen waarden als basis voor een 'nieuwe ethiek'. Een inventarisatie. Verkennende studie, NWO, 's Gravenhagen, The Netherlands.

Wade, G.H. (1999). Professional nurse autonomy: concept analysis and application to nursing education. Journal of Advanced Nursing. 30/2: 310-318.

Waelbers, K, Stafleu, F.R., Brom, F.W.A. 2002. Het ene dier is het andere niet! Verschillen in (morele) grondslagen in veterinairbeleid voor Landbouwhuisdieren en dieren die in natuurgebieden leven. CBG reeks 14. CBG Universiteit Utrecht, Utrecht, The Netherlands.

Reflections from cultural history: the story of the Bohemian landscape in Romania - a sustainable past?

Pavel Klvac and Zbynek Ulcak
Department of Enviromental Stufies, School of Social Studies, Masaryk University, Gorkeho 7, 602 00 Brno, Czech Republic, klvac@fss.muni.cz

Abstract

In the 1820s, during the colonisation of the wild borderland of the Austrian empire, several thousand Czechs moved to the Carpathian Mountains region near the Danube River. They built six villages, each quite distant from the other. Today these villages contain a population of about 2000. Strict ethnic endogamy helped to conserve their cultural distinction. The Czech minority still possesses its unique culture, including language, religion, traditions, crafts, farming, and food production. This paper is based on approximately 20 visits to the region from 2000-2007, during which techniques of participant observation and qualitative interviews were used.

Due to the geographic distance between the villages high above the Danube in the hills of the southernmost Carpathians, and thanks to their unique cultural character, traditional agriculture and food production have been preserved in a form which in many aspects remains the system practiced at the beginning of 20th century.

Proponents of the present concept of organic farming, often, with romantic nostalgia, admire such traditional agriculture as a model of nature- and landscape-friendly practices, closely connected to sustainability principles. But to what extent is this attitude just an uncritically accepted 'myth of the perfect steward' - a variation on the eternal theme of a 'disappeared golden age'? Study of the continuing Czech agricultural system in Romania provides an opportunity to examine this theme. This paper demonstrates the connections between the lifestyle of the Czech inhabitants, their landscape and principles of sustainability.

Keywords: Carpathians, Banat region, farm animals, fossil fuels, landscape, organic farming

Introduction

The Banat region is situated in the south-west part of Romania. In the first third of the 19th century when Czech colonists were coming to Banat, the territory had belonged to the Austrian Military Border with military garrisons as administrative centres established to protect it against Turkish invasions. (Honzak *et al.*, 1997). The low density of population of its south part (only six inhabitants per square kilometre) was the reason for authorities' interest in the settlement of this territory. The arrival of the Czech colonists to this southernmost area at the Carpathian foothills took place in two waves. The first wave of migration (1820-1824) was connected to an appeal from the local timber merchant Magyarly. He sent agents to Bohemia to find skilled woodcutters and wood workers. Their task was to lumber wood from the impenetrable beech forest, to transport it and then to burn charcoal. Around 150-200 families of woodcutters and craftsmen from the Plzen, Domazlice, Klatovy, Kladno and Caslav regions received promises of some concessions and privileges. Dreaming of a better life, they set out for Banat.

The first two Czech villages were named Svata Elizabeta and Svata Helena, named allegedly after Magyarly's daughters. In 1826 Magyarly's lease ran out, and subsequently he cancelled the supplies to villages, took the tools away and disappeared. The Czech emigrants fell into deep despair and saw no other choice but to ask the military officers for admission to the frontier village alliance. Their request was granted.

The second wave of colonisation was organised by the command of the Austrian Military Border itself, which needed to increase the number of its frontier guards. The emigrants were encouraged by the offer of reimbursement of travel costs, exemption from the compulsory military service and taxes, a financial contribution up to the first harvest, the provision of land for pastures and fields, and by the possibility to obtain land and wood to build a house. From 1826 to 1830 the second wave of colonists went by wagon from Bohemia to Vienna and from there they sailed across the Danube for Banat. New Czech villages were founded - Bigr, Eibental, Rovensko, Gernik, Sumice, a Frauvizn. It is estimated that about 4000 people moved to these villages (Jech *et al.*, 1992; Secka, 1996).

The first years in the mountains of Banat were very hard for the settlers. Frantisek Karas, an historian, describes this period in his text, published in 1937, in this way: 'And what was their later fate like? For one hundred years from the morning to the night our man had lumbered woods, resisted the hard and severe climate and drudged on stony hillsides, still afraid of the rain carrying away the thin layer of the soil, promising a weak harvest' (Karas, 1937:17).

It is certain that at the time of the colonists' arrival in the wild Carpathian area, there was only a thin system of mountain paths connecting individual Romanian sheepfolds. The area was covered by deciduous, mostly beech forests. The memories of the difficult struggle in the uncultivated lonely backwoods are preserved in legends, which in the 1960's were recorded by Czech ethnographers: 'There was nothing but forest. And everybody going to the forest for the first time said goodbye to his wife and children. To overturn meant the death of the horses and of everything' (Jech and Svoboda, 1996:21). It is still common to see in those mountains animals that have been exterminated in other European countries. For example, in the winter, hungry wolves sometimes stray very close to villages, and from time to time bears come for their quarry. Exponents of the contemporary concept of organic agriculture tend to admire traditional agriculture, because the nature- and landscape-friendly lifestyle associated with such practices is often seen as a romantic example of rural idyll (Berry, 1997, 2002; Henderson and Van En, 1999; Charles, 2003). To what extent are these ideas just an uncritically accepted 'myth of the perfect farmer' - a kind of variation on the never-ending search for a 'bygone Golden Age?' The Romanian Socialist Republic's attempt to establish an agricultural cooperative in one of the Czech villages - Svata Helena - went broke at the end of the 1950's.

Data and methods

This paper is based on approximately 20 visits to the region from 2000-2007, during which techniques of participant observation and qualitative interviews with local inhabitants were used.

The local food system was observed. Its sustainability was firstly studied as a property of the system, balancing the flow of material resources, assimilative capacities and life-supporting services according to Reid (1995). Social and ethical aspects were assessed by using the Principles for organic agriculture (IFOAM, 2005).

'When the devil sends his regards.'

The region is formed by broken hilly terrain reaching the height of 400-900 metres above sea level. The hills rise steeply from the level of the Danube River which flows in the region only several dozen metres above sea level and broadens up to an admirable width of two kilometres. The dramatically varied limestone relief consists of deep valleys, rock formations, abysses and gorges. The surroundings of Svata Helena, Gernik and Rovensko is rich in specific karst formations such as caves, waterfalls, subterranean rivers, small karst lakes or sink holes. In the karst area, especially after wood harvesting, there was a shortage of drinking water, which caused the disappearance of two Czech villages, Svatá Elizabeta and Frauvizn.

The remoteness of the Czech villages in the hard to access mountain terrain (approximately 5-20 km from the Danube) was an important factor of their isolation. People living there say: 'The devil here sends his regards.' Mobile phone coverage reached this area much earlier than the system of telephone lines. Due to the isolation and separation from the surrounding world, the character of agricultural farming has been preserved much in the same form as it was in Bohemia before the period of communist-era collectivisation. The yields of the fields and farms of the families are predominantly used for their own consumption. At times of bountiful harvest, the small surplus is sold in order to buy necessities needed in harder times. The important institutions are markets, where, as the locals say, you can buy 'everything from all quarters of the world'. On every last Sunday of the month, these markets are held in Romanian towns and villages in the valley, and also on the village square in Gernik. The Czech women go there to sell mainly milk, butter, cream, cottage cheese and cheese.

Most of the villages were electrified about a half century ago. Until recently in the area of Rovensko and Bigr, electricity was generated by a diesel engine which could be operated only for several hours a day. Nowadays every household owns a television, and viewing serials and soap-operas had become a trend there as it has also in the Czech Republic. Occasionally a satellite dish can be spotted. The most modernised area is Svata Helena, which is close to the petrol station. According to the statistics of Josef Pek, the chairman of the local Democratic Union of Slovaks and Czechs in Romania, there are 144 households in Santa Helena which in total own 51 cars, 14 tractors and three combine harvesters. Nevertheless, the work in the fields remains very strenuous. Many families live at a distance of over two hours' walk from their fields. Every year they worry about the crop yield, which depends on the weather - 'As much The Lord gives, as much we harvest.'

Earth - the provider

The basis of successful farming and peasant identity is the farmland. The land from the virgin forests first had to be conquered by native colonists and then protected from successive pressures from nature. Due to the rugged terrain, the fields and greenlands of the settlements are scattered at a distance of several hours' walk away from the villages (and also from each other). The fields of Bigr were so far away from the villages that during the agricultural season, the peasants left their homes for several days.

In the karst areas there is a shallow layer of arable land. The depth of the ploughing is very often 10-20 cm. The lower layers consist mainly of stone or 'dead land.' Despite the saying that this is the worst quality soil - 5th grade (*categoria a cincea*) according to the Romanian scale - the villagers do not complain. Within its bounds they use their own classification of soil types: 'smolnice' - when there is a drought, the soil is hard, and when it is rainy, 'it is sticking to my shoes.' 'Garinska' - black, light, fertile soil - exists in only a few places, the most fertile of which can be found at the Garina locality. 'Kremel' is poor, sandy soil and needs to be manured, in contrast to the fertile 'belizna.' All these types of farmland are exploited for growing various kinds of field crops - mainly corn, wheat and potatoes. The system of annual crop rotation - for example corn-wheat-oats, clover or Lucerne - is individual for the fields and for the farmers. Sometimes the peasants also use a fallow to interrupt the rotation. Farmland is abundantly manured - 'if you do not provide manure, nothing will grow' - mainly with natural manure, but sometimes artificial fertilisers are added.

In comparison with previous periods of drought, the year 2002 produced a highly abundant harvest, especially of corn - with 'corn-cobs like flails.' The season was also beneficial for pests - 'mice were fat like warm slippers.' In Bigr, herds of wild boars devastated the woods and at night the harvest had to be protected from them. In spite of all their dependence on the ups and downs of the nature, uneasy living conditions and the results of the farming, in talking about the residents of these villages we would rather discuss relative rather than absolute poverty. Memories of a period of struggle for existence are still vivid in the minds of the local eyewitnesses. Because of an earlier lack of farmland, the individual fields had become smaller and smaller, as they were divided among a growing population, thereby causing part of the population to migrate. Some migrated to neighbouring Romanian villages and towns (such as Berzasca, Liubcova, Zlatita), and others migrated abroad - to Serbia, Bulgaria and America (Secka, 1996). Today there is an abundance of farmland, mainly because many inhabitants departed to the Czech Republic after 1989 - 'if they only want work', say the old to the young. This migration is the reason for the increase in uncultivated land. This process has increased dramatically since 2007, when Romania became an EU member and the immigration of local inhabitants to the Czech Republic has become less influenced by formal restrictions. Some fields have been bought by or are rented to Romanians from the neighbouring village Coronini. Many fields are being turned into pastures.

Animal husbandry

Gone are the times when 'herds of sheep and goats were streaming the beaten track' and 'children were weighed down with knapsacks and walking after them with sticks in their hands...' This is how Mr. Frantisek Hruza, a native of Svata Helena, remembers his childhood in the book *From Svata Helena to Czech* (Hruza, 1995:16). Only several herds - three herds of cows (each of almost thirty head), two flocks of goats (around 200 head) and two flocks of sheep (altogether about twenty) - were grazing on the common pastures ('hutvajda') around the village in 2002. When people began to leave the villages, the number of animals started to decrease. By comparison, in the first half of the 20th century, 'there was not a house that did not have sheep - everybody over fifty!' Almost every homestead also owned a sheepfold. Pigs were bred in the Romanian way: in the morning they were put out to the pastures and in the evening they returned by themselves. Up to now, this custom was kept only in Rovensko. For this reason the pigs there have in their snouts several iron rings so they will graze rather than dig to the ground. Slaughtering of pigs is conducted in the Czech spirit (Ulcak, 2006). Pigs from the other villages live in small, one, two or three-sectional (sometimes timbered) pigsties built from split and hewn boards and beams. Sometimes villages have a small free-range space. In addition, various kinds of poultry such as hens, geese, ducks and turkey are still bred. Horses and cows are used for both pulling wagons and for ploughing. Horse-drawn vehicles are still the most useful means of transport. In 1991 there were about 370 horses in Gernik (Secká, 1992).

(Un)sustainable agriculture?

How are things with the local cultural landscape? This is by no means the key question about any landscape ecological idyll. The lay tourist is astounded by the picturesque mosaic of small plots which for him resembles very well known pictures and images of landscapes of the past (Stibral, 2005).

A landscape ecologist assessing sustainability as a property of the system, where material resources, assimilative capacities and life-supporting services are in balance (Reid, 1995) could not be satisfied with this point of view, however. The layer of fertile topsoil has become thinner due to erosion caused by a lack of soil conservation elements in the landscape. Some of the pastures are in a similar situation. Water disappears together with forest. Some of the springs have decreased in water output, and others disappeared from the landscape forever. The villages have suffered from a shortage of water - only some houses have their own wells, and service water is collected in barrels.

On the other hand, the level of fossil energy use in agricultural inputs, represented both by direct fossil fuels use and fertiliser consumption, remains surprisingly low. The level of mechanisation is still very low. The cereal crop is sown and mainly harvested by hand, and similarly potatoes are planted by hand. The fields are furrowed by a single-share plough with draught horses and cows. Motor threshing machines and winnowing machines have already become commonplace. Hand power saws have also started to appear.

A truly sustainable technology is represented by the continuing use of small water-powered grain mills ('vodenice'). This sophisticated mechanism uses water from a local creek and it is unique from technical, historical and cultural points of view.

We can also look at 'sustainability' in a broader sense (compare to the approach of D. Reid) as not just a materially but also a socially and ethically sound system. Could we in this case speak about 'happy' farm animals? Farm animals are stalled in small and dark cowsheds. Frequent pasturage and individual care by farmers goes without saying. Cows and sows know what bulls and boars look like. Farm animals are fed only with natural fodder. But surely the standards of modern organic agriculture would not be met in terms of some animal welfare requirements, such as stable space for individual animals, free access to open space, etc.

The food for human consumption is produced and processed in natural ways, using human energy and animal power with little mechanisation. The raw material for production of food is obtained from local sources. The quality of home products is exactly what the inhabitants of many developed countries, fed up with industrial food production, are looking for. The cycle of production and consumption in the local community is closed and could be the object of envy of the 'Western environmentally responsible consumer.' The isolation of this region has contributed to relatively little import of foods. But this situation has been changing. For example, until the middle of 1990's bread was still a genuine home product - made from grain sown by hand to a bread loaf home-baked and consumed. This tradition (with surely positive sustainable consequences) has in the last few years been rapidly neglected. The reason is simple - bread making is a time- and work-intensive process, as expressed by local inhabitants - 'why should we kill ourselves over this?'

Conclusions

To think too much about 'objective measures of sustainability' of the local, micro-level world would not make much sense. The migration of the villagers to the Czech Republic provides a sufficient answer. With the end of socialism, the Czechs in Romania could return to the Czech Republic, and it became easier to get a permanent residency there. Many of the Czechs, especially the younger generation, took advantage of the opportunity. During the last 17 years, thousands of compatriots arrived in the Czech Republic and most of them have settled permanently. That is why the size of the population in the Banat villages is still falling. The demographic structure of the population is changing dramatically - in 1990 about a thousand inhabitants lived in Svata Helena; today less then half of them remain. The only means of earning money remains hard labour on one's own farm: 'In Bohemia you work 8 hours a day, we work 24 hours', said the locals. With the departure of a large part of the population, the share of the cultivated landscape has decreased, and the land has returned to its original stage of wilderness.

Acknowledgements

This work was supported by the National Research Programme of the Czech Republic - NPVII - project 2B06126. Editing assistance was provided by Benjamin Vail.

References

Berry, W. (1997). The agricultural crisis: a crisis of culture. The Myrin Institute, New York, U.S.A., 41 p.

Berry, W. (2002). The art of the common place. Counterpoint, Washington, D.C., U.S.A., 330 p.

Charles, The Prince of Wales (2003). In the name of progress. Resurgence 220: 26-27.

Henderson, E. and Van En, R. (1999). Sharing the harvest. Green Publishing Company, White River Junction, U.S.A., 254 p.

Honzak, F., Pecenka, M., Stellner, F. and Vlckova, J. (1997). Evropa v promenach staleti. (Europe in the centuries). Libri, Praha, Czech Republic, 767 p.

Hruza, F. (1995). Ze Svate Heleny do Cech. (From Svata Helena to Bohemia). A-ALEF, Ostrava, Czech Republic, 16 p.

IFOAM. (2005): Principles for Organic Agriculture. Available at: http://www.ifoam.org/about_ifoam/principles/index.html

Jech, J., Secka, M., Scheufler, V. and Skalnikova, O. (1992). České vesnice v rumunskem Banate. (Czech villages in Romanian Banat). Ustav pro etnografii a folkloristiku AV CR. Praha, Czech Republic, 202 p.

Jech, J. and Svoboda, J. (1996). Byl tady samej les. (There was just forest here). Vydavatelstvo Ivana Krasku, Nadlak, Slovak Republic, 78 p.

Karas, F. (1937): Ceskoslovenská vetev zapomenuta nebem i zemi. Cechove v Rumunsku. (Czechoslovakian branch forgotten by the heaven and country. Czechs in Romania). Spolek Komensky, Praha, Czech Republic, 56 p.

Reid, D.(1995) Sustainable Development. An Introductory Guide. Earthscan, London, U.K., 261 p.

Secka, M. (1996). Cesi v Rumunsku. (The Czechs in Romania) In: Cesi v cizine. (The Czechs abroad). Ustav pro etnografii a folkloristiku AV CR, Praha, Czech Republic, pp. 96-104.

Stibral, K. (2005) Proc je priroda krasna? Esteticke vnimani prirody v novoveku (Why is nature beautiful? Aesthetic perception of nature in modern period). Dokoran, Prague, 202p.

Svoboda, J. (1999): Ceska mensina v Rumunsku. Sdruzení Banat. Praha, Czech Republic, 21 p.

Ulcak, Z. (2006). Lookin' for some down-home cookin'? A case study of household pork production in the Czech Republic. In: M. Kaiser and M. Lien (eds.) Ethics and politics of food. Wageningen Academic Publisher, Wageningen, The Netherlands, pp. 297-300

Landscape, land use and soul: ecopsychology: mending a troubled relationship

Diana Voigt¹, Thomas Lindenthal² and Andreas Spornberger³
¹NATUR and SEELE - Ökopsychologie - Tiefenökologie - 'The Work that Reconnects', Ernst Wolf Gasse 3, A-3011 Neu Purkersdorf, Austria, diana.voigt@chello.at
²BOKU - University of Natural Resources and Applied Life Sciences, Vienna, Department of Sustainable Agricultural Systems, Division of Organic Farming, Gregor-Mendel-Strasse 33, A-1180 Vienna, Austria, thomas.lindenthal@boku.ac.at
³BOKU - University of Natural Resources and Applied Life Sciences, Vienna, Department of Applied Plant Sciences and Plant Biotechnology, Institute of Horticulture and Viticulture, Gregor-Mendel-Strasse 33, A-1180 Vienna, Austria, andreas.spornberger@boku.ac.at

Abstract

The essay provides an overview of the human relationship to land and of humans' resulting actions in influencing, shaping and manipulating landscape in agricultural and industrial societies. A growing disconnection between human nature and surrounding nature has its roots in psychological components, emotions and mindsets. The consequence, in its external dimension, is an estranged relationship between science (esp. agricultural science), economy and nature. In the internal human dimension it is a disconnection between rationality and mind on one hand and emotions, bodily senses and intuition on the other. These divisions are seen by the authors as characteristic for Western industrial societies and as the root of the growing ecological crisis. Theories of ecopsychology are provided to raise the awareness of the issue and explain it further. First steps for overcoming this estrangement in the fields of agricultural research and teaching are also presented. They include the application of ecopsychological methods such as attention- and experience-oriented activities outdoors, increasing self-awareness in direct encounters with nature, integrating ecopsychology and an ethic of responsibility (land ethic) in (agricultural) research and teaching, motivation-oriented learning and research, participatory research work involving emotional and creative techniques in educational research projects.

Keywords: ecopsychology, depth psychology, agriculture, organic farming

Problems and objectives

The disconnection between human and non-human nature - and consequently a split within human nature between rationality and mind on one side and emotions, bodily senses and intuition on the other - results in land use that is estranged and exploitative, an economy that is based on the illusion of continuous growth and a science that is rationally dominated and disconnected. The result is broadly seen in impoverished landscapes, degradation, loss of diversity and system breakdown (Abt, 1992, Egger, 2003, McNeill, 2003).

The objectives of this contribution are:
- to highlight the split between land use, science (esp. agricultural science) and economy on one hand and nature on the other;
- to raise awareness of the unconscious, archetypal dimension of this split;
- to describe the role of ecopsychology in bridging this gap and offering a different worldview and options for a sustainable lifestyle;
- to show first steps out of this disconnection in academic research and teaching.

Methods

The methods presented in this essay are derived from literature review, experience in working with clients in ecopsychology seminars, ecopsychological field work and teaching experiences in agricultural sciences.

Literature review
The search for relevant literature encompassed scientific and popular publications in the fields of ecopsychology, deep ecology, depth psychology and environmental psychology as well as journals of agricultural sciences.

Seminar and field work
The authors of this essay have either practical experience with the use of ecopsychology methods in seminars and outdoor self-awareness courses, or with conducting scientific and technical seminars for organic farmers and teaching students in agricultural science courses. Although not quantitatively evaluated, these experiences represent a core aspect of the paper.

Results

Critical review of the human relationship to landscape and its consequences for land use, agricultural science and economy from the perspective of ecopsychology

Human relationships, attitudes and actions towards non-human nature have always been influenced by the psychological components of the individual, or in their collective form by those of human society. They have been influenced and formed by emotions, experiences, opinions and mindsets, i.e. by worldviews.

With the Neolithic Revolution, the relationship between humans and the land changes, becoming ambivalent and burdened. New psychological patterns and attitudes arise. The relative abundance and diversity that hunter-gatherer societies have experienced in their migrational wanderings disappears with the rise of the first agricultural societies in the desert areas of Mesopotamia, Palestine and Egypt. The former perception of plenty gives way to the experience of scarcity. Nature is no more the provider of abundance and variety, but gives little, and that has to be struggled for. Humans start to manipulate and control nature. They select and domesticate plants and animals, with only a few species remaining for the use and fulfilment of human needs. All other nature is useless, a disturbance that has to be repressed. The land does not belong to everybody anymore, it has become private property. For the first time there is a division between civilised, controlled nature and the 'wilderness', i.e. potentially dangerous nature. The human being becomes a producer and defines his personality through ownership. Land becomes the instrument of production. The world is no longer given, it is made; the land is not a 'you' but an 'it' (Shepard, 1998). The basis for an estranged, utilitarian and unsustainable use of land has been established.

Here, for the first time, the thought pattern that is regarded as the main reason for today's environmental crisis by many ecopsychologists becomes visible - the division or split of human nature from the surrounding nature (Conn, 1995; Metzner, 1995; Roszak, 1995; Fisher, 2002; Macy, 2003). This theme will find its most pointed expression in the capitalist industrial society of our beginning 21st century.

Another factor in the history of human estrangement from nature derives from the influence of patriarchal religions. A mighty, engineer-like god, far above the earth, is believed to have created humans, and only humans, in his image, entrusting them with the role of the ruler or at least the administrator of the earth (Meyer-Abich, 1997; Shepard, 1998).

This worldview would be reinterpreted by the mechanistic philosophers of the 17th century: Man is the god-like head of Creation, the only being bestowed with intelligence and reason. He is surrounded by dead material and becomes an engineer manipulating the huge resource-machine earth. With his reason, as expressed in science, he can approach divine knowledge. This ideology builds the basis for the modern industrial society (Meyer-Abich, 1997; Roszak, 1995). It dovetailed with early and present-day capitalism and, with the help of fossil energies and ongoing technological developments, it has resulted in unprecedented possibilities for interfering in and manipulating nature. The 20th century was unique in the intensity, dimension and speed of human influence on and change of natural systems. Hand in hand with it came the big and well-known man-made environmental problems in agriculture and the standardisation of agricultural landscapes to the 'total landscape' of the industrial growth society (McNeill, 2000; Sieferle, 1997).

The mindset that has caused this development continues to be one of alienation of the human mind and emotions from nature. But meanwhile this split has been consolidated, becoming an established mode of perceiving the world. It runs through all areas of human society and shapes the sciences, the economy, the educational systems and the religious institutions. An unquestioned anthropocentrism serving only the interests and wellbeing of the human species has degraded all other species, as well as the ecosystems and the entire biosphere, to a stockpile for limitless economical growth. The environmental crisis presents itself as a crisis of consciousness. 'We regard it (the earth) as a thing, a big thing, an object to be owned, mined, fenced, guarded, stripped, built upon, dammed, ploughed, burned, blasted, bulldozed, and melted to serve the material needs and desires of the human species at the expense, if necessary, of all other species, which we feel at liberty to kill, paralyse, or domesticate for our own use. Among the many forms of egoism ... this form of species arrogance has received little scrutiny.' (Mack, 1995).

Ecopsychology challenges and criticises this mindset in its theoretical approach. In its consciousness-raising, practical work with clients it tries to facilitate another worldview and its lived application. Sustainable behaviour becomes a question of consciousness.

The central thesis of ecopychology can be summarised in the old alchemist phrase: inside like outside - outside like inside. This means there is no separation between humans and nature; man is no special creation outside of evolution. There is only one nature and man is part of it. This approach is interdisciplinarily supported by the sciences of ecology, systems theory and (micro)biology (Lovelock, 1979; Sahtouris, 1999; Harding, 2006). The damaged and endangered status of planet earth is therefore a sign and result of the split discussed above, and of the estranged and damaged psychological status of human inner nature.

Reciprocally, the crisis of the outer world, the environment, has deep psychological impact on humans and can result in emotional disorders, anxieties, depressions, denial and destructive behaviour. In resolving these fields of crisis ecopsychology is faced with two major tasks, a critical-political task and a healing therapeutic task.
- The critical task involves analysing the causes behind the historical, political and social systems that are damaging and threatening human and non-human nature. Means of consciousness-raising and self-reflection are required here.
- The healing therapeutic task goes to the roots of our human existence and of our being as part of the natural world. It supports reconnection to everything wholesome and the healing of estrangement and division (Fisher, 2002).

Unconscious archetypal motives behind ecological problems

In Jungian psychology, there are three unconscious motives or archetypes that cause the problematic attitude of our civilisation towards nature: the belief that 'everything is possible' (omnipotence), the wish to rule nature (dominance) and the fascination with technology (control).

Behind two of these, the fascination with technology and the belief that 'everything is possible', are two deeper archetypes (Jung, 1954) constellated in the collective unconscious. These two archetypes, which - due to the lack of conscious awareness - have destructive consequences, are (Abt, 1988; Egger, 2003):
- the archetype of the (divine) creator;
- the archetype of the hero or the ruler of nature.

The modern way of rational thinking is deeply influenced by some of the positive aspects of (technological) progress, but its destructive consequences are very often ignored or denied on an emotional and even on a rational level (Abt, 1988).

The unconscious fascination with technology became apparent in our practical work as agricultural scientists during six seminars with organic farmers in the context of soil fertility in the years 2000 and 2001. Based on semi-quantitative group interviews and descriptive statistical methods (Atteslander, 2000), the majority of the interviewees (N=135) pointed out how important the improvement of soil fertility - as a major principle of organic farming (IFOAM, 2005) - was in their practical work. However, this majority also admitted that powerful tractors had a big fascination. They knew that the extreme weight of these powerful machines is damaging to the soil, and nonetheless nearly all of the farmers interviewed had bought such machines. In these seminars several farmers became aware of the contradiction between their conscious attitudes (improvement of soil fertility and therefore the need to avoid soil compaction) and their actual behaviour (use of huge tractors with the danger of compacting the soil). This contradictory behaviour is formed by the unconscious influence of the specific fascination with technology.

Ecopsychology and its role in reconnection with nature

Like all psychologies, ecopsychology deals with the question how estrangement and separation can be altered and overcome. It is a psychology of relationship. Healing is considered the ability to live in relation, to be connected with all there is and to unfold or let unfold the whole potential of human and non-human nature (Cohen, 1997; Clinebell, 1996). Ecopsychology is therefore based in nature ethics. It rejects anthropocentrism and highlights the importance of biocentric behaviour towards nature, which acknowledges the inherent value of all beings in the biosphere and their right to their own good and wellbeing (Taylor, 1997; Weihs, 2003).

The practise of ecopsychology suggests a holistic approach to the sensuous and perceptual abilities of human nature. Through exercises and therapeutic interventions, it works toward the revival of all bodily senses and toward developing emotional sensitivity, helping to refine these inborn qualities so as to create wakefulness, attentiveness, compassion, caring and respect for both inner nature and surrounding nature.

C. G. Jung's model of the Four Domains of Human Nature (Jung, 1955) is also a useful tool for this kind of work: the collective human abilities to think (intellectual capacity), sense (sensual capacity), feel (emotional capacity), and intuit (spiritual capacity) are used as equally important instruments in working toward a sustainable world. Fritjof Capra (www.ecoliteracy.org) puts it more simply, calling these features 'head - hand - heart - spirit'. Both concepts help to avoid an unbalanced, ratio-dominated perception of the world and to support sustainable attitudes through activating our mind-body-emotion-intuition potential in its entirety.

The following ecopsychological methods can play an important role in practical work with nature and in the work of agricultural scientists and teachers: the central importance of contact and reconnection with our own nature is emphasised in exercises that train our sensory perception, including our body and breath awareness, our contact to the ground with our feet and our perception of our environment with all senses. Of further importance is the revival of a broad emotional sensitivity and its consideration as a guiding, decision-making quality - for example, concern, shock and pain when faced with the destruction of diverse agricultural landscapes or compassion towards farm animals. The application of such emotional-sensual awareness to the entirety of surrounding nature - in moments of slowing down, sensing and intuiting - is crucial for a new understanding of nature and the human role in it (Roszak, 1995).

Implementing a reconnection with nature in agricultural research and teaching

The literature review revealed no scientific studies quantifying the influence of ecopsychological methods on the behaviour of agricultural sciences students or farmers toward nature.

Nonetheless, a methodology for overcoming the disconnection between humans and nature in the field of applied agricultural research and teaching is of crucial importance. Its principles could include:

- Experience-oriented learning outdoors and in direct encounters with nature: Techniques of working with the whole of our senses in a living environment are important tools which are accepted and appreciated by students. They encompass both rational reflection on experiences and the development of an awareness for unconscious impulses.
- Integration of ecopsychology and an ethic of responsibility in (agricultural) research and teaching via pilot projects and a discourse on environmental ethics: These methods have not yet been applied, but are required in order to find solutions toward sustainable agriculture.
- Motivation-oriented learning and research imply more creative answers to ecological and social problems in agriculture (Bernard, 2002): Valuing of and focussing on joy and enthusiasm are often neglected in the daily work of agricultural scientists. A negative consequence of this deficit is the low proportion of research projects with new (ground-breaking) and system-oriented methods and objectives. In the field of teaching, project-oriented learning is an important method for increasing motivation and joy. Another advantage of this form of learning is the higher capacity for reflection within the project groups.
- Connection of theory and practice by integrating people who are affected by the current debate and decisions into research work, e.g. participatory research projects with farmers (Baars, 2007): The increasing development towards interdisciplinary research in agriculture (DFG, 2005) shows the necessity and the success of intensive cooperation between farmers and scientists in conducting research for sustainable agriculture, e.g. solutions involving welfare-oriented animal husbandry or preventive plant protection in organic farming.
- Creative techniques in teaching and conducting research projects (films, expositions, demonstrations of solutions) in cooperation with artists and creative, active farmers (Baars, 2007): These techniques increase awareness of and respect for environmental problems in agriculture while facilitating an understanding of difficult interdisciplinary approaches and helping to adapt scientific solutions to the specific conditions of soil and climate in different regions.

References

Abt, T. (1988). Fortschritt ohne Seelenverlust. 2. Auflage. Verlag Hallwag Bern.

Abt, T. (1992). Auf der Suche nach einem Dialog mit der Natur. Leitbilder aus der Innenwelt zum Übergang in eine nachhaltige Gesellschaft. Gaia 6, 318-332.

Atteslander, P. (2000). Methoden der empirischen Sozialforschung. 9. Aufl., Walter de Gruyter, Berlin New York.

Bernard, R. (2002). Research Methods in Anthropology - Qualitative and Quantitative Approaches, Altamira Press, Walnut Creek.

Baars, T. (2007). How biographical experiences affect a research and training programme in biodynamic agriculture at Kassel University. In:B. Havertkort and C. Reijntjes (eds.) Moving worldviews - Reshaping sciences, policies and practices for endogenous sustainable development. Compas, Leusden, Netherlands, pp. 364-379.

Clinebell, H. (1996). Ecotherapy. Healing Ourselves, Healing the Earth. Haworth Press, Binghamton, USA.

Cohen, M.J. (1997). Reconnecting with Nature. Finding Wellness through Restoring your Bond with the Earth. Ecopress, Lakeville, USA.

Conn, S.A. (1995). When the Earth Hurts, Who Responds? In: T. Roszak, M.E. Gomes and A.D. Kanner (eds.) Ecopsychology. Restoring the Earth, Healing the Mind. Sierra Club Books, San Francisco, USA.

DFG (German Research Foundation) (2005). Future Perspectives of Agricultural Science and Research, White paper. Wiley-VCH, Weinheim, Germany.

Egger, B. (2003). Reading Collective Events. Ecological Issue of Energy and Globalisation of the Market. Proceedings of the 15. Internat. Congress for Analytical Psychology, Cambridge, 2001, Daimon, CH-8840 Einsiedeln, Switzerland, pp. 669-680.

Fisher, A. (2002). Radical Ecopsychology. Psychology in the Service of Life. State University of New York Press, Albany, USA.

IFOAM (2005). Principles of organic agriculture. Press-Release. Bonn. www.ifoam.org.

Jung, C.G. (1954). Archetypes of the Collective Unconscious. In: C.G. Jung (1968). The Archetypes and the Collective Unconscious. Collect. Works 9/1. Princeton Univ. Press, 13.

Jung, C.G. (1955). Mysterium Coniunctionis. Untersuchung über die Trennung und Zusammensetzung der seelischen Gegensätze in der Alchemie. Rascher, Zürich, Switzerland.

Harding, S. (2006). Animate Earth. Science, Intuition and Gaia. Green Books, Devon, UK.

Lovelock, J. (2000). Gaia. A New Look at Life on Earth. Oxford Univ. Press, Oxford, UK.

Mack, J. E. (1995). The Politics of Species Arrogance. In: T. Roszak M.E. Gomes and A.D. Kanner (eds.) Ecopsychology. Restoring the Earth, Healing the Mind. Sierra Club Books, San Francisco, USA.

Macy, J. (2003). Die Rückkehr ins lebendige Leben. Strategien zum Aufbau einer zukunftsfähigen Welt. Junfermann, Paderborn, Germany.

McNeill, J.R. (2003). Blue Planet. Eine Umweltgeschichte des 20. Jahrhunderts. Campus Verlag, Frankfurt/Main, Germany.

Metzner, R. (1995). The Psychopathology of the Human-Nature Relationship. In: T. Roszak, M.E. Gomes and A.D. Kanner (eds.) Ecopsychology. Restoring the Earth, Healing the Mind. Sierra Club Books, San Francisco, USA.

Meyer-Abich, M. (1997). Praktische Naturphilosophie. C. H. Beck, München, Germany.

Roszak, T. (1995). Ökopsychologie. Der entwurzelte Mensch und der Ruf der Erde. Kreuz Verlag, Stuttgart, Germany.

Sahtouris, E. (1999). Earth Dance. Living Systems in Evolution. Self published manuscript, www.ratical.org

Shepard, P. (1998). Nature and Madness. The University of Georgia Press, Athens, USA.

Sieferle, R. P.(1997). Rückblick auf die Natur. Luchterhand Verlag, München, Germany.

Taylor, P. W. (1997). Die Ethik der Achtung gegenüber der Natur. In. Krebs, A. (ed.). Naturethik. Grundtexte der gegenwärtigen tier- und ökoethischen Diskussion. Suhrkamp TB Verlag, Frankfurt/Main, Germany.

Weihs, P. (2003). Umweltethik, Skript, BOKU, Sommersemester 2003, Vienna, Austria.

Organic farmers' experiments and innovations: a debate

Susanne Kummer, Racheli Ninio, Friedrich Leitgeb and Christian R. Vogl
Working Group for Knowledge Systems and Innovations, Institute of Organic Farming, Department for Sustainable Agricultural Systems, Univ. of Natural Resources and Applied Life Sciences, Vienna, Gregor-Mendel Straße 33, A-1180 Wien, Austria

Abstract

Farming activities demand continuous modifications and adaptations of practices to fit changing agro-ecological and socioeconomic conditions. Experimentation has always been part of farming. Farmers hold detailed and valuable knowledge about their environments, they experiment actively and have their own research tradition. As the organic farming system has been developed mainly by farmers themselves, there is a considerable experimental potential within the organic movement leading to significant innovations. To understand the motives, topics, methods and results of organic farmers' experiments and to identify the factors influencing the experimental process is an important contribution for a better understanding and support of the organic farming movement.

Keywords: farmers' experiments, organic farming, farmers' innovations, local knowledge

Introduction

Experiments are central features of farming activities (Chambers, 1999, Van Veldhuizen *et al.*, 1997). History of farming shows how farmers constantly developed and adapted their farming systems to changing agro-ecological and socioeconomic conditions. Farmers hold detailed and valuable knowledge about their environments, actively do experiments, and have their own research tradition (Sumberg and Okali, 1997).

The attention that farmers' experimentation obtained during the last century was minor. However, since the 1980s an increasing number of policy initiatives oriented towards improving the sustainability of agriculture. This process developed due to the recognition of the threats that exist to future welfare and the environment; the growing dissatisfaction with the performance of many agricultural projects in these years; and the rising interest in participatory research. Due to these developments, the preoccupation with farmers' local knowledge, empowerment of farmers, and sustainable farming systems increased (Sumberg and Okali, 1997). For the development of a sustainable agriculture (Pretty, 1995), but also for organic farming, it is critical that local knowledge and skills in experimentation are brought to bear on the processes of research.

Organic farming

The transformation to sustainable agriculture requires a fundamental change in learning processes, which are different from the processes of adoption of add-on innovations in the conventional farm management (Röling and Wagemakers, 2000). Prior to the late 1980s, grassroots organisations, farmers and traders drove the constant development of organic farming. Organic farming research first developed through pioneer farmers and scientists in the 1920s. Formal scientific research activities began in the 1970s through a few private research institutes. Organic farming chairs at universities and organic farming projects at state research institutes were established only later, and attract little funding (Niggli and Willer, 2000). It took a long time before the standards established by the organic agriculture community were echoed by national and supranational legislation and control systems (Scialabba and Hattam, 2002). Organic agriculture improved and developed despite inexistent national policies, and these were the farmers that have defined organic agriculture long before the formal research systems did (Bull, 2000). Hence, organic farmers have always experimented on topics pertinent to their production systems and have done most of the advances and innovations in organic farming by themselves.

The example of organic farmers´ experiments in Austria

Pioneer farmers in Austria started practicing bio-dynamic farming already in the 20s of the last century (Pirklhuber and Gründlinger, 1993). After the Second World War, some farmers were concerned about the negative effects that synthetic pesticides and fertilizers, as well as intensive crop and livestock production had on their soils, crops, farm animals, or even on human health. They followed the principles of organic farming, and through their experiments with farming practices that were new for them, they were actively contributing to the development of organic farming in Austria. In these early times of the organic movement, farmers conducted experiments on the production and elaboration of compost and various manure treatment, in order to enhance health and productivity of soils. These experimentation processes on compost and manure treatment were often initiated by the founders of organic farming, especially within the bio-dynamic movement, and were further developed by the farmers. Other important topics of experiments were the breeding of locally adapted crop varieties that fitted the requirements of organic farming. Organic farmers themselves started producing their own seeds, for example of spelt, vegetables or seed mixtures for different types of grassland and pastures.

Current example for experiments and initiatives within the organic movement is the establishment of marketing initiatives like regional cooperatives, as well as tourism on organic farms. Other important topics for current experimentation processes of organic farmers are the production and use of renewable energy resources, such as self-made organic fuel and biomass (Grasser-Elias *et al.*, 2005).

Some of the experiments of organic farmers failed, as many organic farmers explain in personal communication talking about their farming activities. But others as for example those described above yielded important innovations for organic farming.

An innovation is an idea, practice, or object that is perceived as new by an individual or a group. It matters little, so far as human behaviour is concerned, whether or not an idea is objectively new as measured by the lapse of time since its first use or discovery. The perceived newness of the idea for the individual determines his or her reaction to it. If the idea seems new to the individual, it is an innovation. Diffusion is the process by which an innovation is communicated through certain channels over time among the members of a social system. Diffusion can be planned or spontaneous spread of new ideas, which will consequently lead to a change (Rogers, 1995).

Technical and non-technical information, as well as new and modified agricultural practices clearly move within and between communities (Sumberg and Okali, 1997). E.g. Bajwa *et al.* (1997) noticed in their interaction with farmers in Punjab, that when an innovation is worthwhile, it seems to be adopted by farmers whether it is formally recommended or not.

Innovations might spread very rapidly through innumerable personal trials, sometimes without any intervention of the official research and extension systems. On the other hand, some innovations might not succeed to spread out of the farm borders. Many factors have an influence on the rate of adoption of an innovation within a social system: the system's social structure, the role of opinion leaders and change agents, the types of innovation-decisions as well as the consequences of the innovation (Rogers, 1995).

Summary

The above described processes of experimentation and innovation have not yet been assessed scientifically in organic farming. More research has to be done to generate empirical knowledge on the processes by which organic farmers experiment, and how diffusion of innovations in organic farming is triggered. To understand the motives, topics, methods and results of organic farmers' experiments and to identify the factors influencing the experimental process is an important contribution for a better understanding and support of the organic farming movement. This is significant for an improved collaboration between the formal research institutions and the local farmers' research, as well as for the further development of organic farming.

Acknowledgements

Research about organic farmers' experiments in Austria, Cuba and Israel is funded by FWF (Austrian Science Foundation) between January 2007 and December 2008.

References

Bajwa, H.S., Gill, G.S. and Malhotra, O.P. (1997). Innovative farmers in the Punjab. In: L. Van Veldhuizen, A. Waters-Bayer, R. Ramirez, D.A. Johnson and J. Thompson (eds) Farmers' Research in Practice. Lessons from the Field. Intemediate Technology Publications, London, 67-79.

Bull, C.T. (2000). Research models for maximizing the impact of organic research conducted with limited resources. In: T. Alföldi (ed.) Future of organic farming. Proceedings 13th IFOAM Scientific Conference in Basel, 29.-31.8.2000. FiBL, Zürich.

Chambers, R. (1999). Rural development. Putting the last first. Essex, UK, Pearson Education Longman Ltd.

Grasser-Elias, C., Plakolm, G. and Weiss, H. (2005). Der Bauer als Forscher. Bio-Austria. Fachzeitschrift für Landwirtschaft und Ökologie 2/05: 40-42.

Niggli, U. and Willer, H. (2000). Organic agricultural research in Europe - present state and future prospects. In: T. Alföldi (ed.) Future of organic farming. Proceedings 13th IFOAM Scientific Conference in Basel, 29.-31.8.2000. FiBL, Zürich.

OED (1992). Oxford English Dictionary. Oxford University Press, Oxford.

Pirklhuber, W. and Gründlinger, K. (1993). Der biologische Landbau in Österreich. Ein Beitrag zur umweltverträglichen Landbewirtschaftung. BMLFUW, Monographien Band 35, 03/1993. Umweltbundesamt (Hrsg), Wien.

Pretty, J.N. (1995). Sustainable Agriculture. London, UK, Earthscan Publications Ltd.

Rogers, E.M. (1995). Diffusion of Innovations. Fourth Edition. The Free Press, New York.

Röling, N.G. and Wagemakers, M.A.E. (2000). Introduction. In: N.G. Röling and M.A.E. Wagemakers (eds.) Facilitating sustainable agriculture. Cambridge University Press, Cambridge, 1-22.

Scialabba, N. and Hattam, C. (eds.) (2002). Organic agriculture, environment and food security. Environment and Natural Resources Series No. 4, FAO, Rome.

Sumberg, J and Okali, C. (1997). Farmers' Experiments. Creating local knowledge. London, Lynne Rienner Publishers, Inc.

Van Veldhuizen, L., Waters-Bayer, A., Ramirez, R., Johnson, D.A. and Thompson, J. (1997). Introduction. In: L. Van Veldhuizen, A. Waters-Bayer, R. Ramirez, D.A. Johnson and J. Thompson (eds.) Farmers' Research in Practice. Lessons from the Field. Intermediate Technology Publications, London, 13-27.

Detecting pathogens using real-time PCR: a contribution to monitoring animal health and food safety

Sophie Kronsteiner[1], Konrad J. Domig[1], Silvia Pfalz[1], Philipp Nagel[2], Werner Zollitsch[2] and Wolfgang Kneifel[1]
[1]*Department of Food Science and Technology, Division of Food Microbiology and Hygiene, BOKU - University of Natural Resources and Applied Life Sciences, Gregor Mendel Str. 33, A-1180 Vienna, Austria*
[2]*Department of Sustainable Agricultural Systems, Division of Livestock Sciences, BOKU - University of Natural Resources and Applied Life Sciences, Gregor Mendel Str. 33, A-1180 Vienna, Austria*

Abstract

It is well-known that gut microbial changes in pigs clearly influence health and welfare of these animals, since it sensitively reacts to environmental and stress factors. Due to the fact that a balanced gut microbiota is of crucial importance for the safety and quality of animal food, a study was undertaken to clarify whether different diets (containing probiotics, maize silage and grass silage, respectively) may influence the development as well as the persistence of pathogens in different segments of the intestine (stomach, duodenum, ileum, caecum, colon). Thirty-six healthy piglets were assigned to four different dietary treatments. The basal feed mixture was modified by the addition of either probiotics, whole plant maize silage or grass silage. No special supplements were added to the feed of the control group. Real-time PCR based methods were used to screen for the presence of different important pathogens such as *Brachyspira hyodysenteriae*, *Lawsonia intracellularis*, and *Salmonella* spp. The developed test system can be considered as a useful monitoring tool for the assessment of pathogen presence which constitutes an important factor of food safety in pork production.

Keywords: animal health, food safety, pig, pathogen, real-time PCR

Introduction

Food safety and animal health are essential issues in organic food production. Balanced nutrition of pigs is not only of economic importance, but also needs to consider welfare and environmental traits. Therefore the aim of a sound feeding strategy is to assure all goals of high quality, safe product, eco- and bio-sustainability, animal welfare and profit (Zollitsch *et al.*, 2003; Yang, 2007).

A balanced gut microbiota provides the basis for the safety and quality of animal food. The microbiota benefits host health and performance because they influence many physiological, developmental, nutritional and immunological processes of the animal and participate in organ, tissue and immune development. Moreover, the microbiota aids to protect the host from colonisation by pathogens and thus the risk of shedding of human pathogen microorganisms and the contamination of the carcass could be reduced (Richards, 2005).

Swine dysentery is a severe disease, causing diarrhoea, decreased rate of growth, poor feed utilisation and costs for antimicrobial therapy. This disease is caused by the anaerobic spirochaete *Brachyspira hyodysenteriae* (Calderaro *et al.*, 2001). Proliferative enteritis is an enteric disease that occurs in a variety of animals. The obligate intracellular bacterium *Lawsonia intracellularis* has been identified as the causative agent in swine (Cooper *et al.*, 1997). Dysentery, as well as porcine salmonellosis caused by *Salmonella* spp. is acknowledged as hazards to animal health and welfare.

It is the aim of this study to point out, if there is an influence on the development and the persistence of pathogens in different segments of the intestine, when different feedstuffs (whole plant maize silage, grass silage, probiotics) are present in the diet of organic growing-fattening pigs. Furthermore, this study should add some information on the debate whether the methods given below can be used to analyse the presence of pathogens in the gut as a potential animal health and food safety aspect.

Material and methods

The presence of the pathogens *Brachyspira hyodysenteriae*, *Lawsonia intracellularis* and *Salmonella* spp. was studied in thirty-six healthy piglets by testing faeces as well as intestinal samples (from stomach, duodenum, ileum, caecum and colon). The tested piglets were assigned to four different dietary treatments. The basal feed mixture was modified by the addition of either probiotics (*Bifidobacterium animalis*), corn silage or grass silage. While probiotics are frequently used to maintain microbial balance in the gut in conventional swine production, the provision of forage (such as whole plant maize and grass silage) is strongly advocated in oder to increase animal welfare in organic pig husbandry (Council of the European Union, 1999). No special supplements were added to the diet of the control group.

To analyse the possible appearance of the pathogens *Brachyspira hyodysenteriae*, *Lawsonia intracellularis* and *Salmonella* spp., DNA was extracted from all faeces and intestinal samples using the QIAamp DNA Stool Mini Kit (QIAGEN, Hilden, Germany) (La *et al.*, 2005; Rasbäck *et al.*, 2005). This ready-to-use kit provides a fast purification of total DNA from stool samples. Moreover it contains InhibitEX tablets, which remove substances that could degrade DNA or inhibit enzymatic reactions.

Real-Time PCR based methods were used for the detection of these pathogens. The identification of the tested pathogens was realised by using species-specific primer pairs.

Oligonucleotide primers LIR1 5'-GCA GCA CTT GCA AAC AAT AAA CT-3' and LIR2 5'-TTC TCC TTT CTC ATG TCC CAT AA-3' described by Suh and Song (2005) were applied to amplify a 210 bp fragment from *Lawsonia intracellularis*. The amplification of a 298 bp fragment from *Salmonella* spp. was performed by using oligonucleotide primers SAF 5'-TTG GTG TTT ATG GGG TCG TT-3' and SAR 5'-GGG CAT ACC ATC CAG AGA AA-3' (Suh and Song, 2005). To amplify a 435 bp fragment of *Brachyspira hyodysenteriae* the oligonucleotide primers Herbst UP1 5'-GCT AGT CCT GAA AGT TTG AGA GG-3' and Herbst LP2 5'-AGC TTC ATC AGT GAT TTC TTT ATC A-3' were applied (Herbst *et al.*, 2004).

Real-Time PCR is a very reliable method that is able to detect even very low concentrations. To test the detection limit, *Salmonella* spp. was cultivated on nutrient agar, followed by setting up the optical density and dilution series. After mixing aliquots of these dilutions to faeces samples a Real-Time PCR was done, which revealed a minimum concentration of 2.5×10^2 CFU/PCR reaction. This showed that the sensitivity of the PCR method is sufficient for the monitoring of intestinal pathogens (Rasbäck *et al.*, 2006; Suh and Song, 2005).

Results

All the samples tested showed negative results regarding *Salmonella* spp. and *Brachyspira hyodysenteriae*. However, in the samples from three pigs of different feeding groups *Lawsonia intracellularis* were detected. While the faeces samples of these animals were negative, *Lawsonia intracellularis* was detected in the intestinal samples from ileum, caecum and colon.

Conclusion

In conclusion, Real-Time PCR analysis turned out to be a very useful method for fast monitoring of pathogens in particular regarding the very fastidious species *Brachyspira hyodysenteriae* and *Lawsonia intracellularis*.

Acknowledgement

The authors gratefully acknowledge funding from the European Community financial participation under the Sixth Framework Programme for Research, Technological Development and Demonstration Activities, for the Integrated Project QUALITYLOWINPUTFOOD, FP6-FOOD-CT-2003-506358.

References

Calderaro, A., Bommezzadri, S., Piccolo.G., Zuelli, C., Dettori, G. and Chezzi, C. (2001). Rapid isolation of *Brachyspira hyodysenteriae* and *Brachyspira pilosicoli* from pigs. Veterinary Microbiology 105: 229-234.

Cooper, D.M., Swanson, D.L., Barns, S.M. and Gebhart, C.J. (1997). Comparison of the 16S ribosomal DNA sequences from the intracellular agents of proliferative enteritis in a hamster, deer, and ostrich with the sequence of a porcine isolate of *Lawsonia intracellularis*. International Journal of Systematic Bacteriology 47: 635-639.

Council of the European Union (1999). Council Regulation (EC) No 1804/1999 of 19 July 1999 supplementing Regulation (EEC) No 2092/91 on organic production of agricultural products and indications referring thereto on agricultural products and foodstuffs to include livestock production. Official Journal L 222, 24/08/1999. p. 0001-0028.

Herbst, W., Willems, H. and Baljer G. (2004). Verbreitung von *Brachyspira hyodysenteriae* und *Lawsonia intracellularis* bei gesunden und durchfallerkrankten Schweinen. Berliner und Münchener tierärztliche Wochenschrift 117: 493-498.

La, T., Colins, A.M., Phillips, N.D., Oksa, A. and Hampson, D.J. (2006). Development of a multiplex-PCR for rapid detection of the enteric pathogens *Lawsonia intracellularis*, *Brachyspira hyodysenteriae* and *Brachyspira pilosicoli* in porcine faeces. Letters in Applied Microbiology 42 284-288.

Rasbäck, T., Fellström, C., Gunnarsson, A. and Aspán, A. (2006). Comparison of culture and biochemical tests with PCR for detection of *Brachyspira hyodysenteriae* and *Brachyspira pilosicoli*. Journal of Microbiological Methods 66: 347-353.

Richards, J., Gong, J. and de Lang, C.F.M. (2005). The gastrointestinal microbiota and its role in monogastric nutrition and health with an emphasis on pigs: current understanding, possible modulations, and new technologies for ecological studies. Canadian Journal of Animal Science 85: 421-435.

Suh, D.K. and Song, J.C. (2005). Simultaneous detection of *Lawsonia intracellularis*, *Brachyspira hyodysenteriae* and *Salmonella* spp. in swine intestinal specimens by multiplex polymerase chain reaction. Journal of Veterinary Science 6: 231-237.

Yang, T. (2007). Environmental sustainability and social desirability issues in pig feeding. Asian-Australasian Journal of Animal Sciences, 20: 605 - 614.

Zollitsch, W., Kristensen, T., Krutzinna, C., MacNaeihde, F. and Younie, D. (2003): Feeding for Health and Welfare: the Challenge of Formulating Well-balanced Rations in Organic Livestock Production. In: M. Vaarst, S. Roderick, V. Lund and W. Lockeretz (eds.) Animal Health and Welfare in Organic Agriculture, Wallingford, UK, pp. 329 - 356.

Analysis of diet-induced changes in intestinal microbiota: microbial balance as indicator for gut health

Silvia Pfalz[1], Konrad J. Domig[1], Sophie Kronsteiner[1], Philipp Nagel[2], Werner Zollitsch[2] and Wolfgang Kneifel[1]
[1]*Department of Food Science and Technology, Division of Food Microbiology and Hygiene, BOKU - University of Natural Resources and Applied Life Sciences, Gregor Mendel Str. 33, A-1180 Vienna, Austria*
[2]*Department of Sustainable Agricultural Systems, Division of Livestock Sciences, BOKU - University of Natural Resources and Applied Life Sciences, Gregor Mendel Str. 33, A-1180 Vienna, Austria*

Abstract

Pigs are very sensitive to changes in feeding and other stress factors and therefore are affected frequently by diarrhoea. A stable gut microbiota is considered as a prerequisite for healthy animals and a high level of animal welfare. Furthermore, animal health affects the economic efficiency as well as the safety and quality of food from animal origin. Consumers are willing to pay extra for pork with certain assurances, including a high welfare status of pigs. In this project, the influence of probiotics, whole plant maize silage and grass silage on the composition of the intestinal microbiota in different parts of the intestine (stomach, duodenum, ileum, caecum, colon) and the faecal microbiota of organic growing-finishing pigs was investigated. Thirty-six healthy piglets were assigned to four different dietary treatments. The basal feed mixture (control group) was modified by adding either probiotics, whole plant maize silage or grass silage. The microbial composition of the samples was examined by cultural methods in an anaerobic workstation. The following microbial parameters were examined: Total aerobic and anaerobic counts, lactobacilli, bifidobacteria, clostridia, *E. coli*, enterococci and enterobacteria.

Keywords: pig, feed, intestinal microbiota, cultural methods, organic farming

Introduction

Food safety, sensory quality, animal and environmentally friendly production are very important to pork consumers. Therefore the quality and welfare of growing-fattening pigs has become a major point (Millet *et al.*, 2005). The consumers are also interested in the ethics of treating animals and the effect on the environment and themselves (Yang, 2007). Not least, the prohibition of growth promoters in conventional pig farming has led to an intensified search for alternatives (e.g. probiotics, prebiotics) which support the microbiota of the gastrointestinal tract in their approach to control pathogenic bacteria. In this context different nutritional measures and feed additives were tested regarding their effect on the host and on the gut microbiota (Metzler *et al.*, 2005).

The intestinal microbiota plays an important role in animal health. The modification of the gut microbiota could have positive effects for the animal, which also find their expression in improved performance (Simon *et al.*, 2004). The composition of the feed directly or indirectly influences the composition of the microbes. However, many potential stressors exist that may interfere with the microbial balance between harmless and pathogenic bacteria. Among others, stress factors may include parasites, moulds and mycotoxins, low feed quality and social stress (Metzler *et al.*, 2005). Especially the time around weaning is a critical phase in the life of a pig. Separation from the sow, abrupt change from milk to a diet based on cereals and fighting until a social rank is established are the main social stress factors for a piglet (Melin *et al.*, 2004). Coliforms are of special interest because they are the first group of bacteria which colonise the gut after birth and because they play an important role in establishing a climax community of the normal microbiota in the gut (Katouli *et al.*, 1999).

Material and methods

To examine the effects of different feeding strategies on the composition of microbiota in the gastrointestinal tract the different groups of bacteria were characterised by cultural methods in this study. While most previous studies focussed only on one or a very small number of bacterial groups, it was attempted to better characterise the complex composition of gastrointestinal microbiota by analysing the content of the most important groups of gut microbes.

Experimental design

A total of 36 growing-finishing pigs with an initial bodyweight of approximately 38 kg were analysed. The animals were kept in groups of 4 or 5 pigs per pen on straw bedding and were assigned to a control group receiving standard organic growing-fattening diets and three treatments: (1) probiotic supplement (*Bifidobacterium animalis*), (2) whole plant maize silage and (3) grass silage.

Monthly (three times) a sample of faeces was collected from each animal. At an average body weight of 119 kg the animals were slaughtered, and samples from five parts of the gastrointestinal tract (stomach, duodenum, ileum, caecum, colon) were taken. The samples of faeces and digesta from the intestinal tract were stored under anaerobic conditions at -80 °C.

Microbiological examination

1 g of the sample (faeces, digesta samples) was homogenised with 9 ml of Wilkins-Chalgreen Bouillon (Oxoid, United Kingdom) in an anaerobic workstation. Then, 10 fold dilutions were prepared and 0.1 ml was plated on selective and non-selective media. Total anaerobic bacteria were enumerated by culturing on Wilkins-Chalgreen agar (Oxoid). Lactobacilli were enumerated on de Man, Rogosa and Sharp agar (Merck, Germany). Bifidobacteria were enumerated on Wilkins-Chalgreen agar (Oxoid) supplemented with 100 mg/l Mupirocin (LGC Promochem, Germany) and 5 ml glacial acetic acid (Applichem, Germany). Clostridia were enumerated on Wilkins-Chalgreen agar (Oxoid) supplemented with Colistin, Novobiocin and Polymyxin E (Sigma, Germany). All anaerobic bacteria were incubated in an anaerobic workstation (Don Whitley Scientific, United Kingdom) for 3 days at 37 °C.

Total aerobes were enumerated on Plate Count agar (Merck) following aerobic incubation at 30 °C for 3 days. *E. coli* was enumerated on Chromocult Coliformen agar ES (Merck), enterobacteria on McConkey agar (Merck) and enterococci on D-Coccosel agar (Biomerieux, France) following aerobic conditions at 37 °C for 1 day.

Results

Almost no significant differences regarding the viable cell counts were detected between the different treatments. The variation within treatment was higher than between treatments. The viable cell counts of the samples collected from faeces were very homogenous. However, those from the different parts of the gastrointestinal tract underwent distinct fluctuations.

Based on enhanced anaerobic cultivation techniques, the culturable part of the intestinal microbiota was examined with a broad set of microbial parameters. The results show that the gut microbiota of pigs kept under the conditions as described herein is very stable. Therefore the normal microbiota should provide resistance to colonisation by pathogenic and other non-indigenous microbes (Richards *et al.*, 2005). Neither *E. coli* nor clostridia that may impair intestinal health could be found in unusually high concentrations. In the study of Katouli *et al.* (1999) it is shown that the colonisation of pigs by the coliform bacteria is strain and host specific. Therefore it can be concluded that the gut microbiota was normal and that the different dietary treatments did not have a significant effect on the gastrointestinal microbiota of the pigs. Furthermore, the rather complex method used in this study proved to be a suitable approach to the characterisation of gastrointestinal microbiota and to estimate whether the microbial composition can be characterised as 'balanced'. The latter could be an indicator for a high level of gut health, positively contributing to a high animal health and welfare status and to food safety.

Acknowledgement

The authors gratefully acknowledge funding from the European Community financial participation under the Sixth Framework Programme for Research, Technological Development and Demonstration Activities, for the Integrated Project QUALITYLOWINPUTFOOD, FP6-FOOD-CT-2003-506358.

References

Katouli, M., Melin, L., Jensen-Waern, M., Wallgren and P., Möllby, R. (1999). The effect of zinc oxide supplementation on the stability of the intestinal flora with special reference to composition of coliforms in weaned pigs. Journal of Applied Microbiology 87: 564-573.

Melin, L., Mattsson, S., Katouli, M. and Wallgren, P. (2004). Development of post-weaning diarrhoea in piglets. Relation to presence of *Escherichia coli* strains and Rotavirus. Journal of Veterinary Medicine, B, Infectious Diseases and Veterinary Public Health 51: 12-22.

Metzler, B., Bauer, E. and Mosenthin, R. (2005). Microflora management in the gastrointestinal tract of piglets. Asian-Australasian Journal of Animal Sciences 18: 1353-1362.

Millet, S., PH Moons, CH., Van Oeckel, M. J. and Janssens, G. PJ. (2005). Welfare, performance and meat quality of fattening pigs in alternative housing and management systems: a review. Journal of the Science of Food and Agriculture 85: 709-719.

Richards, J., Gong, J. and de Lang, C.F.M. (2005). The gastrointestinal microbiota and its role in monogastric nutrition and health with an emphasis on pigs: current understanding, possible modulations, and new technologies for ecological studies. Canadian Journal of Animal Science 85: 421-435.

Simon, O., Vahjen, W. and Taras, D. (2004). Ernährung und intestinale Mikrobiota bei Schwein und Geflügel. 20. Hülsenberger Gespräche der Schaumann Stiftung, 112-120.

Yang, T. (2007). Environmental sustainability and social desirability issues in pig feeding. Asian-Australasian Journal of Animal Sciences 20: 605-614.

Analysis of food associated bacterial diversity with molecular methods comparing lettuce from organic and conventional agriculture and cheese from pasteurized *versus* unpasteurized milk

Jutta Zwielehner[1], Michael Handschur[1], Norbert Zeichen[1], Selen Irez and Alexander G. Haslberger[1]
[1]University of Vienna; Austria; Department for Nutritional Sciences; Althanstrasse 14; 2D541; 1090 Wien, Austria

Abstract

Microbial diversity was assessed in lettuce from organic and conventional agriculture and in cheese from pasteurized and unpasteurized milk on the basis of 16S rDNA comparison, cloning and q-PCR quantification of lactic acid bacteria (LAB). A considerable diversity of Lactobacillales could be identified, mainly from organic lettuce phyllosphere samples and in all cheese samples from unpasteurized milk. Only one third of randomly picked clones retrieved could be identified as previously described cultivable species. Cheese from unpasteurized milk could be shown to harbour a greater diversity of microorganisms than did cheese from pasteurized milk. Lactic acid bacteria previously described in the intestinal tracts of humans were also found in lettuce and cheese. These results suggest that different agricultural practices and production techniques do result in different health-relevant characteristics of food. Therefore different agricultural methods and a political environment enabling different agricultural practices need to be supported.

Keywords: lettuce, cheese, microflora, organic agriculture

Introduction

Consequences of different agricultural practices continue to be discussed for food safety and food quality aspects controversially. Modern molecular methods enable a better characterisation of food characteristics such as safety and health relevant microbial diversity in foods. Scientific evidence needs to be collected on promoting further understanding of the complex microbial systems of alimentary products. It is crucial not only to aim at reducing hygiene risks but also to aim at potentially health-promoting bacterial communities that even might act as a pathogen defence in the sense of competitive exclusion.

To date little is known about food-associated microorganisms and their fate in the human GI tract. Lactic acid bacteria (LAB) described in food have also been identified from faeces (Heilig *et al.*, 2002), e.g. *Lactobacillus rhamnosus, Lactobacillus reuteri, Leuconostoc mesenteroides, Streptococcus salivarius* subsp. *hermophilus, Bifidobacterium spp.* as well as some *Enterococcus species* (Karimi und Peata, 2003). Experiences with probiotic strains suggest that newly introduced strains cannot permanently establish in the human gut system without continuous delivery. Despite this limited colonization ability it was demonstrated that probiotic bacteria exert immune-modulatory effects and may improve the immune functions damaged by immunosuppressive agents (Bujalance *et al.*, 2007).

Several bacteria, including lactic acid bacteria (LAB), make a positive contribution to health promoting qualities of cheeses or fermented milk, while pathogens like *Listeria ssp.* constitute a health risk. Ripened cheeses are characterized by a succession of largely undefined microbial communities on their surface (Brennan *et al.*, 2002, Ercolini *et al.*, 2003). Culture independent methods offer new opportunities to obtain a more reliable image of microbial communities (Ercolini *et al.*, 2003). *Brevibacterium linens* has previously been considered to be the major species on the cheese surface (Mounier *et al.*, 2005). Valde´s-

Stauber *et al.* (1997) found that *Arthrobacter nicotianae*, *B. linens*, *Corynebacterium ammoniagenes*, *Corynebacterium variabile*, and *Rhodococcus fascians* were the dominant organisms in 21 brick cheeses from six German dairies. Phillips and Harrison (2005) compared conventionally and organically grown salad vegetables using culturing techniques without finding significant differences. They described considerable counts of Lactobacilli at a scale of 10^5 cfu/g in spring mix. Counts of LAB on several minimally processed, frozen and prepacked vegetables have been found to range from 2.9 to 5.6 cfu/g (Manani *et al.*, 2006). The Austrian food report 2006 indicated that 6.7 kg of lettuce is consumed per person per year, which indicates a significant delivery of beneficial bacteria to consumers. In Austria the organically managed acreage has reached 13.5% of the total area under cultivation. A regular consumption of cheese and leafy vegetables like lettuce is typical in many areas worldwide. The bacterial flora of cheese from pasteurized and unpasteurized milk and lettuce (*Latuca sativa*) grown either by organic or conventional cultivation practices were therefore analyzed using molecular methods focusing on potentially health promoting species.

Material and methods

Sampling

Lettuce originating from these fields was sampled on fields and in supermarkets in Vienna and soil samples were taken at the sampling sites. 35 leaves of lettuce from organic and conventional agriculture respectively were analyzed from every sampling site. Cheeses from organic and conventional production facilities were sampled in order to compare cheeses of the same variety from pasteurized and un-pasteurized milk. For core- and rind samples two comparable cheeses were taken. The rind was abraded with a sterile knife, whereas the core samples were directly used for DNA-extraction. All samples were transported in a cooled box and processed immediately.

Extraction of bacterial DNA

For DNA extraction the FastDNA-Spin-Kit for Soil (MP-Biomedicals) was used following a slightly modified protocol of the manufacturer.

Qualitative PCR Amplification

Fragments of the small subunit ribosomal RNA gene were amplified using a ready-to-use PCR Mastermix (Promega).

DGGE Analysis

8% (Vol./Vol., in 0,5 x TAE buffer) polyacrylamide gels of 10 x 20 cm size, 1mm thin were prepared for DGGE with increasing content of denaturing agents.

Quantitative real-time PCR (rt-PCR)

DNA was subjected to rt-PCR with a *Taq*Man system. Lactobacilli specific primers and probe were used as described elsewhere (Haarman and Knol, 2006). Standard-dilutions of a suspension of *Lactobacillus casei* DSM 20011T in saline solution starting from a concentration of 10^6cfu/mL were used for amount determination.

Clone libraries

PCR products were inserted into a p-GEM Easy Vector (Promega) following the instructions of the manufacturer.

rRNA Gene Analysis

Nucleotide sequences were analyzed with the FASTA (www.ebi.ac.uk/fasta/) search and compared to previously published sequences in the PHYLIP function of RDP 8.1 (Rdp8.cme.msu.edu/) (Cole *et al.*, 2003). Sequences were corrected using the function 'trim vector' in CodonCode Aligner (CodonCode Corporation).

Results and discussion

Bacterial communities in lettuce were analyzed taking into account consequences from the food chain from the field to the market as well as different cultivation and production methods. Cheese samples were taken from organic and conventional markets, so that the same varieties of cheese from pasteurized and unpasteurized milk could be compared. Lettuce samples were taken from two neighbouring fields, one under conventional, the other under organic management.

We used the method of 16S rDNA amplification in PCR plus nested PCR, DGGE analysis and cloning to study microbial populations in phyllosphere of lettuce and cheeses from organic and conventional agriculture. Quantitative PCR was performed with primers amplifying the intergenic spacer of 16S-23S rRNA gene because this region is less conserved than the 16S rRNA gene sequence (Haarman and Knol, 2006). rt-PCR showed that Lactobacilli were present in phyllosphere of lettuce at a range of $2,7*10^2$ to $1,7*10^4$ copies and in core samples of cheese from unpasteurized milk up to $1,4*10^4$ copies/500mg sample. In cheese from pasteurized milk no lactobacilli could be detected in rt-PCR. The abundance of eubacterial species could be estimated from the abundance of sequences in 50 randomly picked clones. The majority of all sequences identified had similarities below 95% to previously described organisms, indicating that a vast diversity of food-associated microorganisms is yet unknown.

Much speculation has been done concerning the microbial quality of food from organic agriculture. In this work some facultative pathogenic organisms could be identified but no pathogenic organisms could be detected. Supporters of organic agriculture claimed that low input farming methods resulted in elevated bacterial diversity. In organically grown lettuce we could identify bacteria from 11 different bacterial genus, from conventionally grown lettuce phyllosphere samples 8 different bacterial genus could be identified. In general more *Lactobacillales* could be identified from organic lettuce. Further analysis need to confirm the reproducibility and consequence of this observation. No apparent changes in the bacterial populations in phyllosphere of lettuce from the field to the retailer could be observed. In green mold cheese from unpasteurized milk an elevated microbial diversity in the rind compared to core could be observed whereas in green mold cheese from pasteurized milk the composition of the bacterial communities of rind and core samples were observed to be identical in DGGE analysis. This elevated microbial diversity in the core of green mold cheese was -on the basis of DGGE analysis-observed to be attributable to the emergence of fungal species and lactic acid bacteria. Fungi and yeasts like *Kluyveromyces lactis, Saccharomyces cerevisiae, Yarrowia lipolytica* and *Debaryomyces hansenii* have previously been described in cheeses throughout the entire maturation process (Gardini *et al.*, 2006). As a result of the sequencing of the V3 region fragments, the unpasteurised cheese flora was found to consist of closest relatives of *Agrococcus casei, Brachybacterium alimentarium, Microbacterium* sp., *Arthrobacter protophormiae, Corynebacterium variabile, Brevibacterium aurantiacum* and *Brevibacterium lines*. In the pasteurised sample all bands observed in DGGE analysis were identified as *Staphylococcus equorum*.

The most abundant species in organic cheese made from unpasteurised milk were *Brevibacterium aurantiacum* (16/40 randomly picked clones) and *Arthrobacter protophormiae* (8/40).

Some LAB identified from phyllosphere of lettuce enter the food chain *via* insects, others such as *Leuconostoc* spp. have previously been described as epiphytic organisms as confirmed in this work. Sequences attributable to *Lactobacillus apis* (AY667701.1) and *Lactobacillus alvei* (AY667698.1) have been identified in phyllosphere of lettuce. These species have previously been isolated and described from larvae, faeces and guts of healthy honeybees. The health relevance of these lactobacilli originating from animal sources prevalent in phyllosphere of lettuce needs to be elucidated. *Leuconostoc mesenteroides, Leuconostoc gelidum* and *Leuconostoc citreum* could also be isolated from lettuce. These LAB commonly found on fruits and vegetables are also responsible for the fermentation of cabbage and cheese. Leuconostoc species are used as starter cultures in the dairy- and bakery industry. The consumption of fermented milk products has been observed to be beneficial to human beings long before the term 'probiotic' emerged. Probiotic properties of *Leuconostoc* species have been investigated and an increase in β-galactosidase activity and enhanced pathogen resistance of the host have been observed in animal trials (Singh and Kansal, 2003). *Leuconostoc mesenteroides* and *Leuconostoc citreum* have not only been identified in lettuce but have also been isolated from human GI tracts (Singh and Kansal, 2003, Heilig *et al.*, 2002) indicating that these strains are capable of surviving the passage through the alimentary tract. One sequence identified was attributable to *Lactobacillus reuteri*. This species has previously been isolated and described from several food sources like sourdough and the human microbiota. Its probiotic properties have been tested and proven in various publications (Peran *et al.*, 2007, Simpson *et al.*, 2000). The development of the immune system after birth is highly dependent on the composition of the microbiota. Therefore it is hardly surprising that practically all probiotic strains tested so far displayed immunmodulatory properties. *L.reuteri* has been described in faeces of infants as a scale of 1,3 to 6,4% of total lactobacilli (Haarman and Knol, 2006).

Arguments for supporting diversity of production methods

Beside the broadly discussed potential hygiene risks, there are various arguments supporting diversity of food production methods such as the practice of cheese production from unpasteurized milk. Comparing the bandpattern of different cheeses it becomes obviously that their bacterial ecosystem differs significantly. Microorganisms are responsible for a broad diversity of tastes, aromas and textures, but could also contribute to a health promoting effect of the product. During cheese production starter cultures are added and other microorganisms are adventitiously present in the environment and selected during the maturation process. During pasteurization not only pathogenic bacteria but also the presumptive health relevant species and bacteria responsible for ripening and taste are eliminated. Production without pasteurization is shown to support the presence of beneficial lactobacilli in cheese. Similarly a greater diversity of lactic acid bacteria has been found in organically grown lettuce, where the reproducibility of these results in other settings needs to be determined.

The overall quantification of lactobacilli in phyllosphere of lettuce with rt-PCR resulted in cell counts in the range of $2,7*10^2$ to $1,7*10^4$ copies for lettuce and up to $1,4*10^4$ copies/500mg green mold cheese from unpasteurized milk. These cell counts resulted from lettuce phyllosphere that had been rinsed just as a consumer would prior to consumption. Intensive processing of lettuce like exposure to marinade further decreases bacterial load of lettuce (Nacimiento *et al.*, 2004). The probiotic *Lactobacillus reuteri* has been identified in phyllosphere of lettuce, the health relevance of other organisms specified remains to be determined.

Food production is a multi-dimensional system with economic, ecologic, social and health relevant implications. Sustainability is imperative in a holistic perspective where food production in a globalised economy relies on economic niches, ecosystems face depletion and an increasingly sterile environment and nutrition lead to an increase in atopic diseases. Present results suggest that diverse production methods favour a greater diversity of microorganisms in food items and should therefore be supported.

References

Brennan, N.M., Ward, A.C., Beresford, T.P., Fox, P.F., Goodfellow, M. and Cogan, T.M. (2002). Biodiversity of the bacterial flora on the surface of a smear cheese. Appl. Environ. Microbiol. 68: 820-830.

Bujalance, C., Moreno, E., Jimenez-Valera, M. and Ruiz-Bravo, A. (2007). A probiotic strain of Lactobacillus plantarum stimulates lymphocyte responses in immunologically intact and immunocompromised mice. Int J Food Microbiol. 113: 28-34.

Cole, J.R., Chai, B., Marsh, T.L., Farris, R.J., Wang, Q., Kulam, S.A., Chandra, S., McGarrell, D.M., Schmidt, T.M., Garrity, G.M. and Tiedja, J.M. (2003). The Ribosomal Database Project (RDP-II): previewing a new autoaligner that allows regular updates and the new prokaryotic taxonomy. Nucleic Acids Res 31: 442-443.

Ercolini, D., Hill, P.J. and Dodd, C.E.R. (2003). Bacterial Community Structure and Location in Stilton Cheese. Appl. Environ. Microbiol. 69: 3540-3548.

Gardini, F., Tofalo, R., Belletti, N., Iucci, S., Suzzi, G., Torriani, S., Guerzoni, M.E. and Lanciotti, R. (2006). Characterization of yeasts involved in the ripening of Pecorino Crotonese cheese. Food Microbiol. 23: 641-648.

Haarman, K. and Knol, J. (2006). Quantitative Real-Time PCR Analysis of Fecal Lactobacillus Species in Infants Receiving a Prebiotic Infant Formula. Appl Environm Microbiol 72 : 2359-2365.

Heilig, H.G.H.J., Zoetendal, E.G., Vaughan, E.E., Marteau, P., Akkermans, A.D.L. and DeVos, W.M. (2002). Molecular Diversity of Lactobacillus spp. and Other Lactic Acid Bacteria in the Human Intestine as Determined by Specific Amplification of 16S Ribosomal DNA. Appl. Environm. Microbiol. 68: 114-123.

Karimi, O. and Peata, A.S. (2003). Probiotics: Isolated Bacteria Strain or Mixtures of Different Strains? Drugs of Today 39: 565-597.

Manani, T.A., Collison, E.K. and Mpuchane, S. (2006). Microflora of minimally processed frozen vegetables sold in Gaborone, Botswana. J food Prot. 69: 2581-2586.

Mounier, J., Gelsomino, R. and Goerges, S. (2005). Surface Microflora of Four Smear-Ripened Cheeses. Appl. Environ. Microbiol 71 : 6489-6500.

Nacimiento, M.S., Silva, N., Catanozi, M.P. and Silva, K.C. (2004). Effects of different disinfection treatments on the natural microbiota of lettuce. J Food Prot. 66:1697-700.

Peran, L., Sierra, S., Comalada, M. and Lara-Villoslada, F. (2007). A comparative study of the preventative effects exerted by two probiotics, Lactobacillus reuteri and Lactobacillus fermentum, in the trinitrobenzenesulfonic acid model of rat colitis. Br J Nutr. 97: 96-103.

Phillips, C.A. and Harrison, M.A. (2005). Comparison of the Microflora on Organically and Conventionally Grown Spring Mix from a California Processor. J Food Protection 68: 1143-1146.

Simpson, J.M., McCracken, V.J., Gaskins, H.R. and Mackie, R.I. (2000). Denaturing Gradient Gel Electrophoresis Analysis of 16S Ribosomal DNA Amplicons To Monitor Changes in Fecal Bacterial Populations of Weaning Pigs after Introduction of Lactobacillus reuteri Strain MM53. Appl Environm Microbiol 66: 4705-4714.

Singh, R. and Kansal, V.K. (2003). Augmentation of immune response in mice fed with dhi: a fermented milk containing Leuconostoc citrovorum and Lactococcus lactis. Milchwirtschaft 58: 480-482.

Valdes-Stauber, N., Scherer, S. and Seiler, H. (1997). Identification of yeasts and coryneform bacteria from the surface microflora of brick cheeses. Int. J. Food Microbiol. 34: 115-129.

Non-physical interaction between humans and plants and its impact on plant development

F. Leitgeb, C. Arvay, K. Dolschak, A. Spornberger and K. Jezik
Department of Applied Plant Sciences and Plant Biotechnology, Institute of Horticulture and Viticulture, BOKU - University of Natural Resources and Applied Life Sciences, Gregor-Mendel-Strasse 33, A-1180 Vienna, Austria, friedrich.leitgeb@boku.ac.at, clemens.arvay@gmx.at, tschauko@gmx.at, andreas.spornberger@boku.ac.at, karoline.jezik@boku.ac.at

Abstract

In this project the positive and negative verbal impact of humans on flowering plants was investigated. The aim was to find out, if there is a measurable relationship between people and growth, vitality and behaviour of blooming of *Calendula officinalis*. Therefore an experiment with four treatments and six repetitions (in total 36 plants per treatment) was established in the greenhouse of the Institute of Horticulture and Viticulture (BOKU). In each unit six plants were arranged in pots, with a distance of 60 cm between single units. One group of plants was treated by the project team with positive, another group with negative words and phrases 36 times within the planting period. A third group was treated by a human healer, who applied a remote treatment. One group represented the control and was not attended with any special treatment. On the tested plants the following characters were evaluated: The amount of flowers, pest and disease infestations and weight. The data collected was analysed statistically using SPSS. Results did not show significant differences between control group, positively treated group and healer group. But it was manifest that the negatively treated group showed higher pest and disease infestation as well as fewer flowers; however no obvious difference could be found between treatments according to dry weight. These results demonstrate that further experiments are required to study this topic.

Keywords: biocommunication, plant perception, *Calendula officinalis*

Introduction

The question if plants have emotions and are able to interact in a non-physical way with human beings cannot be answered satisfactorily with the aid of scientific literature. But the fact that plants interact with their physical environment is widely recognized within the scientific community. However, common belief often neglects that plants could have feelings and the possibility of mental and verbal human-plant interaction. Results from several experiments, conducted by academics and non-academics, indicated that plants were able to perceive auditory and emotional stimuli and were able to distinguish between harmony and dissonance (Tompkins and Bird, 1977, Kerner and Kerner, 1992) One of the most famous researchers concerning bio-communication is Cleve Backster, who conducted experiments with polygraphs and measured physiological reactions of plants when treated with external stimuli. He found out that the lighting of a match caused similar reactions in *Dracaena dermensis* as can be observed when humans are scared (Backster, 1968). This phenomenon is known as the Backster effect. Although there is no scientific prove and the findings of Backster as well as other scientists or self proclaimed scientists have been criticized because of deficient scientific methodology (Kmetz, 1977; Carroll, 2003). The issue of verbal and mental human-plant interaction is related to plant intelligence which is based on communication between the various parts of the plants and cell-to-cell communication. The information flow within and between plants involves a wide range of bio-chemical volatiles (Trewavas, 2003).

Although scientifically not proven, farmers all over the world experiment with sounds to stimulate plant growth and health. One example can be mentioned from Spain where an organic farmer runs the radio the whole day for the purpose to create a favourable atmosphere for his plants in the greenhouses (www.aiguaclara.org). Another example is the Maharashi Vedic Organic Agriculture Institute which applies vedic sounds to improve the development of food crops (www.mvoai.org). A literature review of scientific and non-scientific sources shows that there could be a reaction of plants to sounds and/or non-verbal human influence. That was the initial point for conducting this research. The main focus was laid on the question if there is a measurable verbal, non-physical interaction between humans and plants, which are the impacts on the development of the plants and which differences can be observed between different treatments. The results of this experiment should contribute to a positive human-plant-relationship which in turn could lead to a reduction of chemical inputs and thereby conduce to sustainable food production. The study aims to draw attention to reconsider conventional agricultural practice and common methods in agricultural research, especially in pest and disease control.

Methods and material

The experiment was established in the greenhouse of the Institute of Horticulture and Viticulture at the University of Natural Resources and Applied Life Sciences (BOKU). Besides the regular project assistants, a person who works as a spiritual healer contributed to the research due to special capacities in alternative treatments. For the transaction of the experiment four comparative groups were set up (Figure 1):
A = Control group (without linguistic and emotional treatment)
B = Positive group (treated with positive words and thoughts)
C = Negative group (treated with negative words and thoughts)
D = Healer group (treated by healer)

A total amount of 144 plants (*Calendula Officinalis*) were arranged in this experiment. To counterbalance different site conditions treatments were repeated six times. Each plant unit was set up in a triangle and consisted of six plants which were separated by a minimum distance of 60 cm, i.e. each treatment consisted of 36 plants. The plant units on one table were set up in a way that the repetitions altered in their sequence. Every plant of a unit was numbered from one to six to guarantee an unbiased final evaluation of the variables. During the experiment no fertilizers or pesticides were used and the plants were irrigated by the projects assistants according to the needs of the plants.

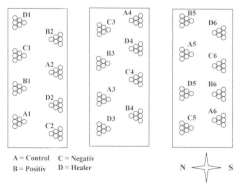

Figure 1. Arrangement of the experiment.

The regular project assistants treated the positive and the negative plant groups 36 times within the total growth duration of 51 days beginning with the 7th of April 2005. Each treatment for a whole plant group took at least 15 minutes. Between the positive and the negative treatments the project assistants took a break to avoid the transmission of emotions. This time was usually used for irrigation. Each plant unit was treated from a distance of approximately 30 cm which permitted a focussed concentration on the plant unit. During the treatments of the positive units project assistants tried to create a friendly, harmonious and facilitating atmosphere with phrases containing words like love, life, air, friendship, health, power, beauty, resistance, richness in flowers etc., whereas terms of reproach were used to create an unfavorable atmosphere for the negative group. Phrases contained words like hate, disease, weakness, pest infestation, rottenness, ugliness etc. The healer treated the plants only three times personally but applied remote treatments using a detailed map and photographs. The healers' statement to the question how he treated the plant group was: 'As a healer I am capable of focusing the cosmic energy which Chinese people call Chi, Japanese people call Ki, and we call it vitality in a way that the quantum-physical structure in an organism is set in order again. In this way a dynamic equilibrium is created which enables the energy flow'.

The course of the experiment was documented in a research diary. Date and time of the treatments and irrigation were recorded. During the experiment duration two intermediate evaluations were realized, just counting the flowers considering their development. For the final evaluation of the experiment on 28th of May 2005, all plants were numbered according to a code scheme which was set up by one project assistant without knowledge of the other participants which guaranteed an unbiased analysis. During the final evaluation the following characteristics were analysed for each plant: number of flowers, fresh weight (analytical balance), dry weight (analytical balance), aphid and fungal infestation. Measuring the development of the flowers as well as aphid and fungal infestation was realized on the basis of pre-defined classes.

Analysis of variance followed by a Student-Newman-Keuls-Test (S-N-K) at a level of significance of α =0.05, was realized in SPSS for Windows 12.0.

Results

The highest amount of flowers was found in the control group (A), whereas the lowest amount of flowers was observed in the negatively treated group (C). There was a significant difference between the control group (A) and the negative group (C) at a level of $\alpha = 0.05$. The positive group (B) and the group treated by the healer (D) did not show significant differences (Figure 2). The results below show the summarized amount of flowers.

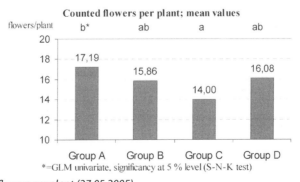

Figure 2. Counted flowers per plant (27.05.2005).

The highest aphid infestation was observed in the negative group (C) and the lowest in the control group (A). The positive group (B) and the healer group (D) showed the same percentage. However, on a level of α =0.05 there were no significant differences between the groups (Figure 3).

The negative group (C) showed a significantly higher fungal infestation than all the other groups. The lowest degree of fungal infestation was observed in the control group (A) (Figure 4).

The control group (A) reached the highest mean value in fresh weight, whereas in the negative group (C) the lowest fresh weight was documented. At a level of alpha = 0.05 there was a significant difference between the control group (A) and the negative group (C; Figure 5).

The highest mean value of dry weight was recorded for the positive group (B), the lowest for the healer group (D). But on a level of α = 0.05 there were no significant differences between the groups (Figure 6).

The highest percentage of dry matter content was calculated for the negative group (C). The lowest percentage was found in the control group (A). On the level of α = 0.05 there was a significant difference between group A and group C (Figure 7).

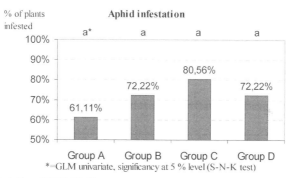

Figure 3. Aphid infestation (27.05.2005).

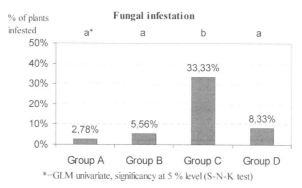

Figure 4. Fungal infestation (27.05.2005).

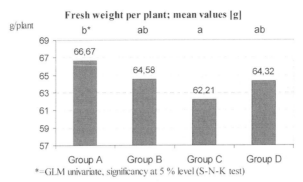

Figure 5. Fresh weight per plant (27.05.2005).

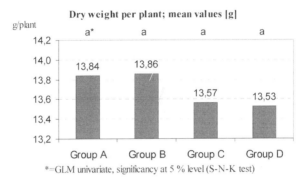

Figure 6.Dry weight per plant (28.05.2005).

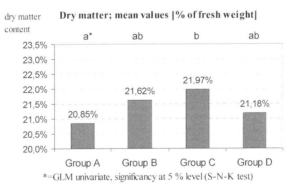

Figure 7. Dry matter content (28.05.2005).

Conclusion and discussion

The control group (A) and the negative group (C) showed significant differences in four of the investigated parameters. Therefore a significant influence of verbal and non-verbal interaction on flowering, vitality and plant growth of *Calendula officinalis* can be stated. The results can be interpreted in the way that

Sustainable food production and ethics

it was not possible for the project assistants and for the healer to influence the plants in a positive way, but plants could have been influenced in a negative way. The results reflect the assumption, that it was easier to transmit negative emotions than positive ones. The control group showed the highest amount of flowers and was less affected by pest and fungal infestation. This leads to the conclusion that a positive treatment did not have a positive impact on plant vitality. Further, the fungal infestation of the negatively treated group (C) has to be emphasized which showed a significant difference to the control group. As a matter of fact this leads to the assumption that the fungal infestation was responsible for the lower fresh and dry weights of the negatively treated group.

As the experiment was realized in a greenhouse, spatial separation of the treated groups was not possible. Although the distance between the trial groups was measured with a pendulum to assure the focus of the treatments on one plant unit only, impact of the treatments on neighbouring groups cannot be fully neglected. For a repetition of the experiment it has to be considered to spatially separate the different treatments and to cultivate the plants under controlled atmosphere and artificial light.

References

Backster, C. (1968). Evidence of a Primary Perception in Plant Life. International Journal of Parapsychology 10: 329-348.

Carroll, R.T. (2003). The skeptic's dictionary. Wiley, UK.

Kerner, D. and Kerner, I. (1992). Der Ruf der Rose. Kiepenheuer and Witsch, Köln, Germany.

Kmetz, J.M. (1977). A study of primary perception in plants and animal life. Journal of the American Society for Psychical Research 71: 157-170.

Tompkins, P. and Bird, C. (1977). Das geheime Leben der Pflanzen. Pflanzen als Lebewesen mit Charakter und Seele und ihre Reaktionen in den physischen und emotionalen Beziehungen zum Menschen. Fischer, Frankfurt, Germany.

Trewavas, A.J. (2003). Aspects of plant intelligence. Annals of Botany 92, 1-20.

A tailor-made molecular biological strategy for monitoring the gut microbiota of pigs reflecting animal health

Agnes Petersson[1], Konrad J. Domig[1], Elisabeth Moser[1], Karl Schedle[2], Wilhelm Windisch[2] and Wolfgang Kneifel[1]

[1]BOKU - University of Natural Resources and Applied Life Sciences, Department of Food Science and Technology, Division of Food Microbiology and Hygiene, Gregor Mendel Str. 33, A-1180 Vienna, Austria

[2]BOKU - University of Natural Resources and Applied Life Sciences, Department of Food Science and Technology, Division of Animal Food and Nutrition[2,] Gregor Mendel Str. 33, A-1180 Vienna, Austria

Abstract

The aim of this project was to optimise the methodology for determining the influence of different feeding strategies on the gut microbiota of pigs by using Real-time PCR-based tools. The composition and the physiology of the farm animals gut microbiota are of great interest to science because of their influence on animal health. Animal health again is very important for the economic efficiency and mainly for the safety and quality of animal food. Experience has shown that culture methods often are not capable of satisfactorily describing the real microbial situation in the intestine. Hence, although numerous studies have been introduced during the last years applying this microbial strategy, recent findings have demonstrated that the detection of microbial changes upon feeding can only be accomplished using molecular tools in connection with culture methods. In this study we have compared different methods for monitoring microbial changes in the gastrointestinal tract of pigs as influenced by different feeding strategies. Starting from faecal samples and samples from different gut segments, the total microbial DNA was isolated and different bacterial parameters were quantified by Real-time PCR. The results were compared with those obtained by culture methods and the results of the Fluorescence *in-situ* Hybridisation (FISH). It was shown that Real-time PCR can be regarded as a fast alternative to conventional techniques, also complementing results obtained from conventional culture methods.

Keywords: real-time PCR, fluorescence *in-situ* hybridisation, pig, gut, microbiota

Introduction

Intensive animal production is characterised by a high biological and economic productivity with a simultaneously low input of labour, energy and space per animal. Chances of intensive pig production are based on competitiveness, decrease of environmental pollution due to optimised productivity and eased controls of process quality. Risks can be found in animal health because of high densities of animals (Mennerich-Bunge, 2003). This indicates not only a health threat for animals but also for humans (Yang, 2007). The intensive use of in-feed subtherapeutic antibiotics by the pork industry to increase growth performance and support disease prophylaxis has led to microbial resistances to specific antibiotics. Within the EU the use of in-feed subtherapeutic antibiotics is banned since the beginning of the year 2006, because these resistances also affect the efficacy of human therapeutics (Adjiri-Awere and Van Lunen, 2005; Baker, 2006).

Feeding pigs is not only an economic function, but also needs to consider welfare and environmental traits. So the aim of a perfect feeding strategy is to assure all goals of high quality, safe product, eco- and bio-sustainability, animal welfare and profit (Yang, 2007).

Sustainable agriculture is defined as practices that meet current and future societal needs for food and fibre, for ecosystem services, and for healthy lives, and that do so by maximising the net benefit to society when all costs and benefits of the practices are considered (Tilman *et al.*, 2002).

Pigs are important growing-finishing animals because pork is an important protein source for human nutrition. Like humans they belong to the group of monogastric species and so they are excellent models to describe connections between gut microbiota and indigestions. Pigs are sensitive to changes and give useful information about the intestinal effect of the feed. The gastrointestinal tract of pigs is closely colonised with bacteria and impacts growth performance and animal health (Simpson *et al.*, 1999; Leser *et al.*, 2000).

Microbiota benefits host health and performance because they influence a lot of physiological, developmental, nutritional and immunological processes of the animal and participate in organ, tissue and immune development. Moreover the microbiota aids to protect the host from colonisation by pathogens thus the risk of excretion of human pathogen microorganisms and consequential the contamination of the carcass could be reduced (Richards *et al.*, 2005).

The gastrointestinal tract of pigs represents a dynamic ecosystem, which contains a complex community of microaerophil and anaerobe microorganisms. The gastrointestinal tract of pigs contains approximately 10^{14} microorganisms. Over 90% of these microorganisms are grampositive anaerobe bacteria and belong to the genera *Streptococcus, Lactobacillus, Eubacterium, Clostridium, Fusobacterium* and *Peptostreptococcus* whereas the gramnegative part of the microbiota is dominated by *Bacteroides* and *Prevotella*. The phylogenetic diversity of the intestinal bacterial community was examined by comparative 16S ribosomal DNA-sequence analysis and the results showed the complexity of the microbiota. It was determined that the majority of the bacteria species colonising the gastrointestinal tract of the pig are not characterised yet (Leser *et al.*, 2002; Konstantinov *et al.*, 2004; Metzler *et al.*, 2005).

The development of the intestinal microbiota is a gradual and sequential process and is influenced by non-dietary factors as well as dietary factors (Inoue *et al.*, 2005).

Before birth the gastrointestinal tract of pigs is sterile. During and immediately after birth it is colonised by bacteria from the mother, diet and environment. The colonisation of bacteria varies in different areas of the gastrointestinal tract whereas population densities show an increase from the proximal to distal gastrointestinal tract. Because of high acidity and rapid movement of contents there are only low numbers of bacteria (10^3 to 10^5 CFU g^{-1} digesta) in stomach and proximal small intestine particularly acid-tolerant groups, such as lactobacilli and streptococci. By contrast in ileum dominates a neutral pH and digesta is slower moved. A greater variety of bacterial species and higher numbers of bacteria can be found here (10^8 to 10^9 CFU g^{-1} digesta). Highest bacteria densities are found in cecum and colon (10^{10} to 10^{12} CFU g^{-1} digesta) where strict anaerobic bacteria dominate (Simon *et al.*, 2004; Richards *et al.*, 2005).

Intensification of the pig industry has led to increased risk of both clinical and subclinical enteric diseases. The occurrence of diarrhoea in weaning piglets associated with increased mortality is an important problem in pig farming (Awati *et al.*, 2005).

Weaning is a dangerous phase for a pig because of several factors like separation from the sow, feed change from milk to grain and social rank fights. At this critical phase pigs often decrease the amounts of feed and water and so their growth is also decreased. Consequences of which are structural, functional and barrier function changes in the intestines which can result in massive diarrhoea, less weight increase and increased mortality because pathogenic bacteria can easily colonise the gut. This means economic loss. A lower morbidity and mortality and so greater production can be achieved if a stable microbiota is assured (Katouli *et al.*, 1999; Collinder *et al.*, 2003; Melin *et al.*, 2004; Metzler *et al.*, 2005; Richards *et al.*, 2005; Roth and Ettle, 2005).

Identification of gut bacteria with conventional methods is labour-intensive, time consuming and expensive. Culture-based methods detect only a part of the actual gut bacteria because it is not possible to isolate all bacteria and to cultivate the bacteria outside the alimentary system (Blaut *et al.*, 2002; Khan *et al.*, 2002; Simon *et al.*, 2004).

Because of the limits of culture-based methods, such as medial selectivity for readily culturable bacteria and the presence of non-culturable bacteria, our understanding of the gut microbiota based on these methods may be inaccurate and is certainly incomplete. To overcome the limitations of culture-based techniques, molecular approaches are used increasingly to characterise the gut-microbiota (Li *et al.*, 2003; Vanhoutte *et al.*, 2006).

Rapid enumeration and identification of bacteria can be achieved by molecular techniques, such as FISH, Denaturing Gradient Gel-Electrophoresis (DGGE), competitive quantitative PCR and Real-time PCR (Vanhoutte *et al.*, 2006; Fu *et al.*, 2006). Real-time PCR is some kind of polymerase chain reaction, a method to amplify DNA, which allows quantification of DNA concentrations. Its advantages are increased specificity, sensitivity and accuracy. Furthermore risk of cross-contamination is reduced because of its closed tube system. Compared to conventional PCR, there is no need to perform a gel electrophoresis (Jensen *et al.*, 2005; He and Jiang, 2005). Real-time PCR is a suitable technique to study the composition of such complex communities as the gastrointestinal tract (Gueimonde *et al.*, 2004).

Materials and methods

In this project 45 pigs were fed common diets enriched with varying amounts of different sources of insoluble fiber (wheat bran, Chinese pine pollen). Samples from colon and ileum were collected in sterile plastic bags and stored anaerobic at -80°C. Starting from these samples DNA was isolated using the QIAamp DNA Stool Mini Kit (Qiagen, Germany) which is appropriate for stool samples (Li *et al.*, 2003). Samples were analysed with fluorescence *in-situ* hybridisation using commercial FISH-Kits (Ribotechnologies/Microscreen, The Netherlands). Bacterial counts were determined for all bacteria, lactobacilli and bifidobacteria. Real-time PCR analyses were performed using the Rotorgene 3000 cycler (Corbett Research, Australia) whereas SybrGreen was used as fluorescent marker. The real-time PCR experiments were carried out at the following conditions: 3 minutes at 95 °C, 20 seconds at 95 °C, 20 seconds at the annealing temperature, 20 seconds at 72 °C. *Lactobacillus fermentum*, *Escherichia coli*, *Bacteroides vulgatus*, *Bacteroides thetaiotaomicron*, *Streptococcus bovis*, *Bifidobacterium thermophilum* and *Clostridum difficile* were cultivated as positive controls. Serial dilutions of *E. coli* were used as standards to quantify real-time PCR products.

Results and discussion

The aim of this project is to develop a fast and reliable culture-independent method to monitor certain microorganism groups in the intestine of pigs. This is important to examine, because the intestinal microbiota influences animal health. A balanced gut microbiota helps to reduce the risk of infectious diseases and therefore contributes to animal welfare. Both, pathogens and commensal bacteria, which influence the gut health positively can be detected using the described method. Real-time PCR is used because it allows acquiring quantitative data in a short time period.

Additionally a culture-based and a FISH method were accomplished to evaluate the results achieved with real-time PCR. All three determined parameters (lactobacilli, bifiobacteria and all bacteria) differed considering FISH and culture-based results. In general the results achieved with culture-based method were higher which maybe due to the higher specificity of the FISH method.

Typical anaerobic representatives of the pigs gut microbiota were successfully cultivated and taken for the development of the real-time PCR approach. The following species were applied as reference strains: *Streptococcus bovis*, *Clostridium difficile*, *Escherichia coli*, *Bacteroides vulgatus*, *Bacteroides thetaiotaomicron*, *Lactobacillus fermentum*, *Bifidobacterium thermophilum*.

After optimisation of the PCR reaction conditions these reference strains could be detected with species specific primers and approved with a gel electrophoresis. DNA was isolated out of the colon and ileum samples and bifidobacteria, lactobacilli and further anaerobe could be identified whereas real-time PCR results correlated with gel electrophoresis results.

Recovery analysis was performed to check the reliability of the DNA isolation kit and the PCR method. For this ileum samples in which *Bacteroides vulgatus* could not be detected were added different concentrations of *Bacteroides vulgatus*. After DNA isolation and real-time PCR *Bacteroides vulgatus* could be found in the samples up to 25 CFU/PCR-approach which relates to 10^4 CFU/ml. Different concentrations of *E. coli* ought to serve as standard to quantify microorganisms.

It could be shown that the presented method is a promising tool to detect pathogens and groups of the balanced microbiota to visualise changes due to different feeding strategies. This method allows the monitoring of microbial gut health based on a fast molecular biological method and therefore contributes to animal welfare.

Acknowledgement

The H. Wilhelm Schaumann Stiftung (Hamburg, Germany) is gratefully acknowledged for its financial support regarding A. Petersson and her work presented in this paper.

References

Adjiri-Awere, A. and Van Lunen, T.A. (2005). Subtherapeutic use of antibiotics in pork production: Risks and alternatives. Canadian Journal of Animal Science 85: 117-130.

Awati, A., Konstantinov, S.R., Williams, B.A., Akkermans, A.D.L., Bosch, M.W., Smidt, H. and Verstegen, M.W.A. (2005). Effect of substrate adaptation on the microbial fermentation and microbial composition of faecal microbiota of weaning piglets studied in vitro. Journal of the Science of Food and Agriculture 85: 1765-1772.

Baker, R. (2006). Health management with reduced antibiotic use - the U.S. experience. Animal Biotechnology 17: 195-205.

Blaut, M., Collins, M.D., Welling, G.W., Doré, J., Van Loo, J. and De Vos, W. (2002). Molecular biological methods for studying the gut microbiota: the EU human gut flora project. The British Journal of Nutrition 87: 203-211.

Collinder, E., Cardona, M.E., Berge, G.N., Norin, E., Stern, S. and Midtvedt, T. (2003). Influence of Zinc Bacitracin and *Bacillus licheniformis* on microbial intestinal functions in weaned piglets. Veterinary Research Communications 27: 513-526.

Fu, C.J., Carter, J.N., Li, Y., Porter, J.H. and Kerley, M.S. (2006). Comparison of agar plate and real-time PCR on enumeration of *Lactobacillus*, *Clostridium perfringens* and total anaerobic bacteria in dog faeces. Letters in Applied Microbiology 42: 490-494.

Gueimonde, M., Tölkkö, S., Korpimäki, T. and Salminen, S. (2004). New real-time quantitative PCR procedure for quantification of bifidobacteria in human fecal samples. Applied and Environmental Microbiology 70: 4165-4169.

He, J.-W. and Jiang, S. (2005). Quantification of enterococci and human adenoviruses in environmental samples by real-time PCR. Applied and Environmental Microbiology 71: 2250-2255.

Inoue, R., Tsukahara, T., Nakanishi, N. and Ushida, K. (2005). Development of the intestinal microbiota in the piglet. The Journal of General and Applied Microbiology 51: 257-265.

Jensen, A.N., Andersen, M.T., Dalsgaard, A., Baggesen, D.L. and Nielsen, E.M. (2005). Development of real-time PCR and hybridization methods for detection and identification of thermophilic *Campylobacter* spp. in pig faecal samples. Journal of Applied Microbiology 99: 292-300.

Katouli, M., Melin, L., Jensen-Waern, M., Wallgren, P. and Möllby, R. (1999). The effect of zinc oxide supplementation on the stability of the intestinal flora with special reference to composition of coliforms in weaned pigs. Journal of Applied Microbiology 87: 564-573.

Khan, A.A., Nawaz, M.S., Robertson, L., Khan, S.A. and Cerniglia, C.E. (2001). Identification of predominant human and animal anaerobic intestinal bacterial species by terminal restriction fragment patterns (TRFPs): a rapid, PCR-based method. Molecular and Cellular Probes 15: 349-355.

Konstantinov, S.R., Favier, C.F., Zhu, W.Y., Williams, B.A., Klüß, J. Souffrant, W.-B., De Vos, W.M., Akkermans, A.D.L. and Smidt, H. (2004). Microbial diversity studies of the porcine gastrointestinal ecosystem during weaning transition. Animal Research 53: 317-324.

Leser, T.D., Lindecrona, R.H., Jensen, T.K., Jensen, B.B. and Moller, K. (2000). Changes in bacterial community structure in the colon of pigs fed different experimental diets and after infection with *Brachyspira hyodysenteriae*. Applied and Environmental Microbiology 66: 3290-3296.

Leser, T.D., Amenuvor, J.Z., Jensen, T.K., Lindecrona, R.H., Boye, M. and Moller, K. (2002). Culture-independent alysis of gut bacteria: the pig gastrointestinal tract microbiota revisited. Applied and Environmental Microbiology 68: 673-690.

Li, M., Gong, J., Cottrill, M., Yu, H., De Lange, C., Burton, J. and Topp, E. (2003). Evaluation of QIAamp® DNA Stool Mini Kit for ecological studies of gut microbiota. Journal of Microbiological Methods 54: 13-20.

Melin, L., Mattsson, S., Katouli, M. and Wallgren, P. (2004). Development of post-weaning diarrhoea in piglets. Relation to presence of *Escherichia coli* strains and Rotavirus. Journal of Veterinary Medicine, B, Infectious Diseases and Veterinary Public Health 51: 12-22.

Mennerich-Bunge, B. (2003). Intensive Schweinehaltung - Chance oder Risiko? Züchtungskunde 75: 452-458.

Metzler, B., Bauer, E. and Mosenthin, R. (2005). Microflora management in the gastrointestinal tract of piglets. Asian-Australasian Journal of Animal Sciences 18: 1353-1362.

Richards, J.D., Gong, J. and De Lange, C.F.M. (2005). The gastrointestinal microbiota and its role in monogastric nutrition and health with an emphasis on pigs: Current understanding, possible modulations, and new technologies for ecological studies. Canadian Journal of Animal Science 85: 421-435.

Roth, F. and Ettle, T. (2005). Organische Säuren: Alternative zu antibiotischen Leistungsförderern. In: Abteilung Tierische Lebensmittel, Tierernährung und Ernährungsphysiologie, BOKU Wien: 4. BOKU-Symposium Tierernährung, 27.10.2005, Universität für Bodenkultur Wien, pp. 28-33.

Simon, O., Vahjen, W. and Taras, D. (2004). Ernährung und intestinale Mikrobiota bei Schwein und Geflügel. 20. Hülsenberger Gespräche 'Mikrobiologie und Tierernährung'. Lübeck, 9.-11.6.2004. Schriftenreihe der H. Wilhelm Schaumann Stiftung: 112-124.

Simpson, J.M., McCracken, V.J., White, B.A., Gaskins, H.R. and Mackie, R.I. (1999). Application of denaturant gradient gel electrophoresis for the analysis of the porcine gastrointestinal microbiota. Journal of Microbiological Methods 36: 167-179.

Tilman, D., Cassman, K.G., Matson, P.A., Naylor, R. and Polasky, S. (2002). Agricultural sustainability and intensive production practices. Nature 418: 671-677.

Vanhoutte, T., De Preter, V., De Brandt, E., Verbeke, K., Swings, J. and Huys, G. (2006). Molecular monitoring of the fecal microbiota of healthy human subjects during administration of lactulose and *Saccharomyces boulardii*. Applied and Environmental Microbiology 72: 5990-5997.

Yang, T.S. (2007). Environmental sustainability and social desirability issues in pig feeding. Asian-Australasian Journal of Animal Sciences 20: 605-614.

Evaluating novel protein sources for organic laying hens

Heleen A. van de Weerd and Sue H. Gordon
ADAS UK Ltd. Gleadthorpe, Meden Vale, Mansfield, Nottinghamshire, NG20 9PF, United Kingdom,
heleen.vandeweerd@adas.co.uk

Abstract

In view of an increasing growing world population with increasing demands for food, sustainable sources of food production, such as organic agriculture, have to be developed. The challenge for organic egg production is meeting the hens' nutritional requirements with organically sourced ingredients. The imminent move to 100% organic feed requires an increase in the number of sources of organic proteins. The aim of this project was to examine sustainable and innovative methods for meeting the organic laying hen's protein requirements. A literature review was performed to find information on novel protein sources taking into consideration crude protein content, amino acid supply, bird health and welfare and food safety. Methods for producing, harvesting and processing the most promising novel protein sources were also investigated. The results showed that none of the novel proteins studied had optimal methionine concentrations for organic egg production, but for some their singular or combined use in simulated organic layer diets allowed target methionine concentrations to be met, and without over-supplying crude protein. There was no evidence of any adverse effects of feeding earthworm meal, house fly larvae meal or microalgae *Chlorella* meal on bird health or welfare. It is concluded that novel protein sources are a huge biomass that is under-utilised, so there is great potential to develop some as feedstuffs for organic laying hens using existing knowledge on production methods.

Keywords: organic egg production, alternative proteins

Introduction

Due to recent world population growth, urbanisation and income growth in developing countries, these is an increasing demand for animal protein (Bellaver and Bellaver, 1999; Bradford, 1999; Delgado *et al.*, 1999). With the expanding demand for meat and milk production, farm animals are increasingly the main consumers of grain formerly eaten directly by humans. At present, livestock consume approximately 40% of the global cereal grain supply (Dyson, 1999). This is having a decreasing effect on meat and grain imports into the developing countries of Africa and Asia (Van der Zijpp, 1999). These developments raise ethical concerns adding to the need to develop more sustainable means of meeting global food demand. Other reasons are that increasing global food production will have effects on climate, atmospheric composition, land use and other global change drivers (Gregory and Ingram, 2000). Global strategies for livestock output growth must consider the environmental impact and should be compatible with sustainable resource management (Bellaver and Bellaver, 1999).

Organic farming is based on the principles of sustainability. It is an approach to agriculture with the aim to create integrated, humane, environmentally and economically sustainable agricultural production systems (Organic Farm Management Handbook, 2007). Organic agriculture is not practised without difficulties and one of the most challenging aspects of organic egg production is the supply of nutrients to the hen in a balanced manner so that her requirements for maintenance, production, health and welfare are fully met. There is a limited range of organic ingredients for feeding organic laying hens and this offers less scope for balancing nutrient supply. The range of organic proteinaceous ingredients available is relatively small and all sources are deficient in certain essential amino acids, particularly methionine and lysine (Hancock *et al.*, 2003; Gordon, 2004). Imbalances in naturally occurring amino acids, derived from feeds based entirely on available plant-based proteins, can lead to the diet being deficient in certain essential amino acids. Synthetic methionine, which is relied upon in feeds for conventional laying hens is not permitted in organic feeds. Diets low in methionine reduce egg mass output, but importantly there may be adverse effects on immunocompetence and feather pecking with implications for animal welfare. Another aspect of unbalanced diets is that total protein can be oversupplied in order to meet the requirements for the most limiting amino acids and this will result in excess N excretion. This increases the risk of N emissions from the excreta to the air and water environments.

To date it has not been possible to identify complete solutions to nutrient supply difficulties through the use of typical organic crops (e.g. oilseed meals). Therefore, at present a EU derogation (Annex 1B, 4.8 of Council Regulation (EEC) No. 2092/91) allows the feeding of a small percentage of non-organic feedstuffs. However, there is an increasing consumer desire and legislative requirement to move to 100% organic feed for laying hens. In order to do this successfully and without compromising animals welfare or the environment, an increase in the number of sources of organic proteins are needed. The aim of the present project was to examine sustainable and innovative methods for meeting the organic laying hen's protein requirements.

Methods

An extensive literature review was performed to find information on novel protein sources, which are not usually fed to chickens in Europe. The ingredients being considered fell within four categories: (1) insects, earthworms and gastropods (2) algae (3) aquatic plants and (4) herbs. Aspects of novel proteins that were considered were crude protein content, protein digestibility and amino acid supply. Effects on bird health and welfare (including feeding studies), egg quality and food safety were reviewed. Methods for producing, harvesting and processing the most promising novel protein sources were also investigated.

Results and discussion

The most promising novel protein sources were house fly pupae and larvae and earthworm meal (Table 1). The crude protein content of insects is comparable to that of conventional meats and the amino acid profile of earthworm meal and housefly pupae meal is similar to that of fishmeal (Calvert *et al.*, 1969; Ravindran and Blair, 1993). Earthworms contain high levels of protein, rich in essential amino acids, but the contents of protein and amino acids are variable both between and within earthworm species (Zhenjun *et al.*, 1997). Furthermore, earthworm meal is higher in protein and has a better amino acid profile than soya bean meal (loc. cit.). An average protein digestibility of 90% may be assumed for well-prepared worm meals (Fisher, 1988).

Various insect meals and earthworm meal have been used as protein supplements in feeds for chickens and their impact on performance assessed in feeding studies. Examples of these feeding studies are

Table 1. Main aspects of nutritional value of insects as alternative protein sources for feeding organic chickens (Calvert et al., 1969, 1971; Sabine, 1983; Fisher, 1988; Ravindran and Blair, 1993; Zhenjun et al., 1997).

Composition	House fly pupae and larvae meal	Earthworm meal
Crude protein (g/kg DM)	600-630	580-750
ME value (MJ/kg DM)	10.47	10.05
Amino acid content (g/100 g protein) - Methionine	2.3 (larvae), 2.6 (pupae)	2.0
Amino acid content (g/100 g protein) - Lysine	5.9 (larvae), 5.8 (pupae)	7.2

Reinecke *et al.* (1991), who examined the protein quality of three different species of (dried) earthworms in terms of net protein utilisation and relative nutritive values in growing chickens. Fifteen test diets were prepared and each diet was fed from 11 days of age for a period of 7 days. There were no differences in growth between chicks fed the different earthworm species, it was concluded that they were equal in their ability to promote growth (loc. cit.). There are some risk associated with feeding earthworms as they have the potential to accumulate toxic residues, particularly heavy metals and agrochemicals (Sabine, 1983). Feeding earthworm meals could potentially transfer these accumulated toxins to the egg. Overall however, no ill effects of feeding earthworm meals on chicken health have been reported (Ravindran and Blair, 1993).

House fly pupae meal has been fed to White Leghorn chicks as an alternative protein source to soya bean meal (Calvert et al., 1969; Teotia and Miller, 1974). In both studies the performance of the chicks was similar to that of control chicks.

A promising alternative plant protein source was the microalgae *Chlorella* (Table 2). The nutritional value of algae seems to vary and Lipstein and Hurwitz (1983) investigated this by comparing repeated samples from different locations and during different seasons. In general, the mean nutrient content varied by less than 10% except for fibre and xanthophyll contents. The crude protein content varied only slightly.

Several experimenters have fed *Chlorella* to chickens (e.g. Lipstein *et al.*, 1980, Lipstein and Hurwitz, 1983). Despite the fact that *Chlorella* is deficient in methionine egg production could be supported by feeding diets including *Chlorella* meal and no apparent effects on health or mortality were found (Lipstein and Hurwitz, 1980). There is a potential risk of feeding algae as they can accumulate heavy metals at high concentrations (Becker, 2004). This can pose a problem in the large-scale production of algae. However, microalgae fed at moderate concentrations are unlikely to be damaging to bird health.

Chlorella and other microalgae are rich sources of xanthophyll pigments, which can be readily transferred to the egg yolk and can lead to a deep orange colour depending on the concentration fed. *Chlorella* meal may be fed at concentrations up to 100 g/kg, before yolk colour is adversely affected, and possibly at higher concentrations. When including algae in organic diets, the pigmenting potential of other ingredients needs to be considered as organic diets are likely to contain maize or maize products, and grassmeal.

Table 2. Main aspects of nutritional value of microalgae as an alternative protein source for feeding organic chickens (Lipstein and Hurwitz, 1980; Lipstein and Hurwitz, 1983; Becker, 2004).

Composition	Microalgae *Chlorella* meal
Crude protein (g/kg DM)	395.0
ME value (MJ/kg DM)	9.6-11.6
Amino acid content (g/16 g N) - Methionine	1.89-1.97
Amino acid content (g/16 g N) - Lysine	5.26-5.86

Diet simulations

Simulated diets were formulated as part of the project, using these promising novel proteins or a combination of them, and compared with typical organic diets. The results showed that ME values could be met but were often marginally below target and the dietary crude protein content was too high in some cases. Methionine requirements were often met, but tryptophan and lysine were regularly over-supplied. In some cases the total P content was above the target concentration and the electrolyte balance was often not optimal.

Production methods

The production of earthworms (vermiculture) involves converting layers of organic wastes into earthworm tissue protein. Existing systems range from small-scale windrows or boxes (low-cost), to efficient mechanised systems (high cost, less labour intensive) (e.g. Edwards, 1988). Maximum productivity can be achieved by maintaining optimal moisture and temperature, aerobic conditions and avoiding excessive amounts of ammonia and salts (e.g. Frederickson, 2004). Processing comprises harvesting, blanching, drying and grinding of the worms (Mason et al., 1992; Edwards and Bohlen, 1996). Efficient harvesting can be difficult.

The biodegradation of poultry manure by the common house fly involves the inoculation of manure with fly eggs from a disease-free stock on a regular basis (e.g. Calvert et al., 1970; Teotia and Miller, 1973; Miller et al., 1974; Muller, 1980; El Boushy et al., 1985). Processing comprises separating the pupae or larvae from the digested manure, drying and grinding. The digested manure can be dried and pelleted to produce a fertiliser or soil conditioner (with 20% less nitrogen than fresh manure). Limiting factors are harvesting and drying methods and the large number of fly eggs needed to provide enough pupae/larvae.

Chlorella can be grown in temperate climates. Production methods are thin-layered sloping ponds, flat-plate bioreactors, tubular photo-bioreactors, or open raceway ponds. Growth requirements are fresh water, irradiance, temperature (15-30 °C), stirring of the culture, regular harvesting (in batches) and re-inoculation (Grobbelaar et al., 1996; (Lívanský and Doucha, 1996; Nedbal et al., 1996). There can be problems with algal weed species. Limiting factors may be the costs associated with drying.

In conclusion, novel protein sources are a huge biomass and they are under-utilised, so there is great potential to develop some as feedstuffs for organic laying hens. There is existing (technical) knowledge that can potentially be used to produce these novel proteins. In these systems, recycling of nutrients is possible, which is in keeping with the organic ethos. However, practical research now needs to be done to provide answers to remaining issues such as whether it is economically viable to utilise these novel protein sources.

Acknowledgements

To ADAS project contributors: B. Cottrill, M. Tomiczek, T. Verhoeven, R. Weightman, R. Safford, F. Nicholson, S. Holmes, T. Turner and to Defra who funded this project (project OF0357).

References

Becker, W. (2004). Microalgae in human and animal nutrition. In: A. Richmond (ed.) Handbook of microalgal culture: biotechnology and applied phycology. Blackwell Publishing, Oxford, Uk, pp. 312-351.

Bellaver C. and Bellaver, I.H. (1999). Livestock production and quality of societies' life in transition economies. Livestock Production Science 59: 125-135.

Bradford, G.E. (1999). Contributions of animal agriculture to meeting global human food demand. Livestock Production Science 59: 95-112.

Calvert, C.C., Martin, R.D. and Morgan, N.O. (1969). House fly pupae as food for poultry. Journal of Economic Entomology 62(4): 938-939.

Calvert, C.C., Morgan, N.O. and Eby, H.J. (1971). Biodegraded hen manure and adult house flies: their nutritional value to the growing chick. Proceedings of International Symposium on Livestock Wastes. American Society of Agricultural Engineering, pp. 319-320.

Calvert, C.C., Morgan, N.O. and Martin, R.D. (1970). House fly larvae: Biodegradation of hen excreta to useful products. Poultry Science 49:588-589.

Delgado, C.L., Rosegrant, M.W., Steinfeld, H., Ehui, S.K. and Courbois, C. (1999). Livestock to 2020: the next food revolution. Washington, D.C. Rome Nairobi, Kenya: International Food Policy Research Institute (IFPRI); Food and Agriculture Organization of the United Nations (FAO); International Livestock Research Institute (ILRI).

Dyson, T. (1999). World food trends and prospects to 2025. Proceedings of the National Academy of Sciences of the USA 96: 5929-5936.

Edwards, C.A. (1988). Breakdown of animal, vegetable and industrial organic wastes by earthworms. In: C.A. Edwards and E.F. Neuhauser (eds.) Earthworms in waste and environmental management. Academic Publishing, The Hague, The Netherlands, pp. 21-32.

Edwards, C.A. and Bohlen, P.J. (1996). Biology and Ecology of Earthworms. Chapman and Hall, London, UK.

El Boushy, A.R., Klaassen, G.J. and Ketelaars, E.H. (1985). Biological conversion of poultry and animal waste to a feedstuff for poultry. World Poultry Science Journal 41 (2):133-45.

Fisher, C. (1988). The nutritional value of earthworm meal for poultry. In: C.A. Edwards and E.F. Neuhauser (eds.) Earthworms in waste and environmental management. SPB Academic Publishing, The Hague, Netherlands, pp. 181-192.

Frederickson, J. (2004). Organic Food Waste Treatment Development Project. Final Report. Open University (B/1859).

Gordon, S.H. (2004). Defra-funded project OF0327. Validation of the HEN biological model for organic laying hens and an assessment of nutritional issues in organic poultry production.

Gregory, P.J. and Ingram, J.S.I. (2000). Global change and food and forest production: future scientific challenges. Agriculture, Ecosystems and Environment 82: 3-14.

Grobbelaar, J.U., Nedbal, L. and Tichý, V. (1996). Influence of high frequency light/dark fluctuations on photosynthetic characteristics of microalgae photoacclimated to different light intensities and implications for mass algal cultivation. Journal of Applied Phycology 8: 343.

Hancock, J., Weller, R. and McCalman, H. (2003). 100% Organic livestock feeds - Preparing for 2005. A report prepared for the Organic Centre Wales by Soil Association Producer Services and assisted by IGER.

Lipstein, B. and Hurwitz, S. (1980). The nutritional value of algae for poultry. Dried Chlorella in broiler diets. British Poultry Science 21: 9-21.

Lipstein, B. and Hurwitz, S. (1983). The nutritional value of sewage grown samples of Chlorella and Micrantinium in broiler diets. Poultry Science 62: 1254-1260.

Lipstein, B., Hurwitz, S. and Bornstein, S. (1980). The nutritional value of algae for poultry. Dried Chlorella in layer diets. British Poultry Science 21: 23-27.

Lívanský, K. and Doucha, J. (1996). CO_2 and O_2 gas exchange in outdoor think-layer high density microalgal cultures. Journal of Applied Phycology 8: 353-358.

Mason, W.T., Rottmann, R.W., and Dequine, J.F. (1992). Culture of Earthworms for Bait or Fish Food. Florida Cooperative Extension Service, Institute of Food and Agricultural Sciences, University of Florida.

Miller, B.F., Teotia, J.S. and Thatcher, T.O. (1974). Digestion of poultry manure by Musca domestica. British Poultry Science 15(2):231-234.

Muller, Z.O. (1980). Insect cultures in Feed from Animal Wastes: State of Knowledge. FAO Animal Production and Health Paper 18, Chapter 4.

Nedbal, L., Tichý, V., Xiong, F. and Grobbelaar, J.U. (1996). Microscopic green algae and cyanobacteria in high-frequency intermittent light. Journal of Applied Phycology 8: 235-333.

Organic Farm Management Handbook 2007 (2006). N. Lampkin, M. Measures, S. Padel (eds.) University of Wales, Organic Advisory Service (EFRC).

Ravindran, V. and Blair, R. (1993). Animal protein sources for poultry production in Asia and the Pacific. III. Animal protein sources. World's Poultry Science Journal 49: 219-235.

Reinecke, A.J., Hayes, J.P. and Cilliers, S.C (1991). Protein quality of three different species of earthworms. South African Journal of Animal Science 21: 99-102.

Sabine, J.R. (1983). Earthworms as a source of food and drugs. In: J.E. Satchell (ed.) Earthworm Ecology - From Darwin to Vermiculture. Chapman and Hall, New York, USA, pp. 285-296.

Teotia, J.S. and Miller, B.F. (1973). Environmental conditions affecting development of house fly larvae in poultry manure. Environmental entomology 2: 329-333.

Teotia, J.S. and Miller, B.F. (1974). Nutritive content of house fly pupae and manure residue. British Poultry Science 15: 412-417.

Van der Zijpp, A.J. (1999). Animal food production: the perspective of human consumption, production, trade and disease control. Livestock Production Science 59(2-3): 199-206.

Zhenjun, S., Xianhun, L., Lihui, S. and Chunyang, S. (1997). Earthworm as a potential protein source. Ecology of Food and Nutrition 36: 221-236.

Authors index

A

Aerts, S.	109, 279, 285, 365
Algers, A.	244
Algers, B.	244
Andersen, H.	127
Anthony, R.	69
Arvay, C.	526

B

Baudoin, C.	436
Beekman, V.	95
Bizaj, M.	382
Bohländer, K.	145
Boivin, X.	233
Bokkers, E.A.M.	229
Bos, A.P.	293
Bovenkerk, B.	121
Broberg, M.	184
Brom, F.W.A.	337, 342
Brulé, A.	233

C

Cohen, N.E.	337, 342
Coleman, G.	233

D

De Bakker, E.	95
De Boer, I.J.M.	229
De Graaff, R.	95
De Greef, K.H.	293
De Lauwere, C.	198
De Rooij, S.	198
De Schrijver, A.	57
De Tavernier, J.	365
Devos, Y.	57
Dietze, K.	360
Dolschak, K.	526
Domig, K.J.	515, 518, 532
Driessen, C.	219, 249

E

Ehrenwerth, D.	477
Evers, J.	365

F

Feng, K.	139
Folker, A.P.	127

Forsberg, E.-M.	442
Franzén, U.	244
Fraser, D.	31
Fraser, E.D.G.	139
Frecheville, N.	169

G

Gamborg, C.	354, 466
Gazsó, A.	407
Gesche, A.H.	163, 376
Gjerris, M.	239, 455, 466
Gordon, S.H.	537
Grasser, S.	382
Gremmen, B.	180
Gressler, S.	163, 407
Grimm, H.	300
Gschmeidler, B.	450
Guan, D.	139
Gunnarsson, S.	157
Gunning, J.	466

H

Hagen, K.	461
Handschur, M.	521
Hartlev, M.	466
Haslberger, A.G.	145, 163, 376, 477, 521
Heutinck, L.F.M.	249
Høyer Toft, K.	425
Huber, R.	239
Hunger, U.	450

I

Irez, S.	521

J

Jensen, K.K.	316
Jezik, K.	526

K

Kaiser, M.	53
Kaliwoda, J.	416
Kjærnes, U.	43
Klvac, P.	500
Knaus, W.	412
Kneifel, W.	515, 518, 532
Kochetkova, T.	309
Korzen-Bohr, S.	371

Kronsteiner, S. 515, 518
Kulø, M. 273
Kummer, S. 512

L
L'hotellier, N. 233
Lassen, J. 371, 431
Leitgeb, F. 512, 526
Leitner, H. 416
Lindencrona, M. 244
Lindenthal, T. 506
Lips, D. 109, 285
Lund, V. 37, 489
Luttikholt, L.W.M. 49

M
MacMillan, T. 169
Maxim, L. 115
McGlone, J.J. 223
Meijboom, F.L.B. 132, 342, 495
Mejdell, C.M. 489
Menke, C. 321
Millar, K.M 328, 354
Minteer, B.A. 21
Moen, O. 244
Mollenhorst, E. 229
Moser, E. 532

N
Nagel, P. 515, 518
Niebuhr, K. 321
Ninio, R. 512

O
Ohnell, S. 244
Olsson, I.A.S. 239

P
Padel, S. 26
Pallari, M. 395
Pasquali, M. 101
Paulesich, R. 145, 163
Petersson, A. 532
Pfalz, S. 515, 518
Phocas, F. 233
Pouteau, S. 75
Proyer, M. 163

R
Robertson, I.A. 263

Roep, D. 174

S
Sagl, V. 376
Sandøe, P. 127, 239, 354, 431, 455, 466
Schedle, K. 532
Scheibe, K.M. 483
Schuppli, C.A. 472
Simons, D. 389
Sneyers, M. 57
Sodano, V. 151
Sonesson, U. 157
Spornberger, A. 506, 526
Stafleu, F.R. 495
Stassen, E.N. 337
Sun, N. 139
Sundrum, A. 257, 360
Sutherland, M. 223
Sveinson Haugen, A. 81

T
Termansen, M. 139
Terragni, L. 401
Thaler, R. 376
Thompson, P.B. 63
Tomkins, S. 328
Tsioumanis, A. 347
Tuyttens, F. 204

U
Ulcak, Z. 500

V
Van der Ploeg, J.D. 198
Van der Ploeg, S. 229
Van der Sluijs, J.P. 115
Van de Weerd, H.A. 537
Vanhonacker, F. 204
Van Poucke, E. 204
Verbeke, W. 204, 210
Visak, T. 193
Vogl, C.R. 382, 416, 512
Vogl-Lukasser, B. 382
Voigt, D. 506
Vramo, L.M. 273

W
Waiblinger, S. 233, 321
Waidringer, J. 244
Waigmann, E. 450

Weary, D.M. 472
Werner, C. 360
Wiberg, S. 244
Windisch, W. 532
Wiskerke, J.S.C. 174

Y
Yeatman, H.R. 269
Yu, Y. 139

Z
Zeichen, N. 521
Zichy, M. 89
Zollitsch, W. 515, 518
Zwielehner, J. 521

Keyword index

A

adaptation	139
aetiology	360
agricultural biotechnology	431
agriculture	412, 455, 506
alternative proteins	537
analysis	89
animal	466
– disease	342
– disease epidemics	337
– ethics	69, 193, 239, 472
– health	515
– husbandry	193
– practice	337
– production	285
– transport	244
– welfare 37, 63, 157, 193, 198, 223, 233, 239, 244, 269, 273, 279, 293, 321, 461, 472, 483	
assessment	389
assurance scheme	279
auto-organisation	75
automatic milking system	249
aviary	229

B

Banat region	500
battery cage	229
beak-trimming	321
best practice	285
biocommunication	526
biodiversity	75, 81, 180
bioethics	425
biosafety	57
biotechnology	121, 198, 466
bird flu	347

C

Calendula officinalis	526
CAP	169
Carpathians	500
cheese	521
climate change impacts	139
cloning	455
co-evolution	249
co-existence	53, 57
cognitive values	75
consensus conferences	121

consumer	204, 210
– concerns	132
– expectations	257
– role	401
consumption	43
contestable concepts	101
control policy	337
conventionalisation	273
cooperative	151
corporate social responsibility	151, 180
criteria	416
cultivated areas	316
cultural methods	518

D

dairy farming	249
debate	89
decision-support framework	95
deep litter	229
dehorning	321
depoliticization	121
depth psychology	506
developing countries	184
diagnostic	360

E

Eastern Tyrol	416
eco-positive thinking	395
ecological integrity	63
ecological literacy	81
economics	157
ecopsychology	506
ecosystem health	316
embedding	174
emergence	75
emissions	244
environment	223, 389
environmental	
– ethics	81
– health	376
– virtue theory	69
epigenetics	376
epistemology	300
ethical	
– assumptions	309
– consumption	401
– judgment	436

– matrix	95, 442
– room for manoeuvre	219
– tools	442
ethic of care	69
ethics	43, 109, 376, 455, 466
– committees	121
ethno-veterinary medicine	382
Europe	431
European Union	184
evaluation concept of regional ability to	
respond external shocks	145

F

farm animals	500
farm animal welfare	204, 210
farmer	
– attitudes	431
– experiments	512
– innovations	512
– morality	495
– values	198
feed	518
feeding	412
Flanders	204, 210
fluorescence in-situ hybridisation	532
focus group interviews	431
food	43, 412
– ethics	95
– safety	347, 466, 515
– supply chain	174
– system	151
– systems	139
fossil fuels	500
framing	101
free trade	169

G

Gaucho®	115
genetic resources	75
GM	
– crops	431
– crops ethics	53
– maize	57
GMO	425, 442
governance	174
grazing	249
gut	532

H

harm principle	342

home made remedies	382
homogenization	163
honeybee	115
human-animal relationship	233, 239, 337
human rights	425
husbandry conditions	483

I

impact	376
implicit normativity	127
import requirements	184
inconsistencies	257
integral	309
integrity	455
intellectual property rights	163, 180
intensive production	37
international justice	425
intestinal microbiota	518
intrinsic value	316

K

keepers of backyard animals	342

L

landscape	500
lawfulness	184
laying hens	229
lettuce	521
liberalisation	169
life cycle assessment	157
livestock	461
livestock production	365
local	389
– biodiversity	163
– knowledge	512

M

marketing	174
marketing communications	395
meat safety	371
media coverage	347
methodology	89
microbiota	532
microflora	521
moral ideals	132
multi-causality	115
multi-criteria mapping	95

N

nanobiosensors	365

nature conservation 477, 483
NIABY 53
NIMBY 53
nutrition 376
nutrition advice 127

O
organic 229, 273
– agriculture 309, 521
– aquaculture 328
– egg production 537
– farming 37, 382, 500, 506, 512, 518
– food 184
– livestock 257
– pig production 360
– standards 328
– values 321

P
participation 95
participatory process 442
pathogen 515
perception 204
perception studies 371
philosophy of technology 69
phytotherapy 382
pig 223, 239, 472, 515, 518, 532
pig tower 219
plant perception 526
policy 169
– agenda 269
– implementation 269
political consumerism 43
pollen flow 57
post-normal science 115
power 151
practice 43, 109
practice-oriented ethics 285
producer 204
production diseases 360
professional autonomy 495
professionalism 495
professional role 293
public debate 495
public perception 371

R
real-time PCR 515, 532
reflexive design 293
regional food system 416

regulation 466
research animals 472
resilience trade 163
resource sufficiency 63
retail 279
reverence 239
risk 342
risk information 347
ruminant 412

S
scientific
– advice 127
– communication 101
– evidence 127
– reductionism 300
segmentation 210
sheep production 273
SMEs 395
social 389
– movements 63
– sustainability 371
soil association 328
status 337
stockmanship 233
supply 389
survey 210
sustainability 198, 223, 233, 279, 309, 328, 365, 376, 412, 477
– analysis 229
– trajectory 174
sustainable
– agriculture 37
– development 169
– green marketing 395
system innovation 219

T
tail-docking 321
theory 109
transgenic 461
transgenic plants 436
trust 43, 151
trustworthiness 132
typology 337, 436

U
uncertainty 115

Keyword index

V

value-laden facts	300
values	81, 127, 477
vegan agriculture	193
vulnerability	139

W

welfare	455
wild animals	483
wild nature	316
women activism	401
WTO	425

Printed in the United States
by Baker & Taylor Publisher Services